A System of Mechanical Philosophy

VOLUME 1

JOHN ROBISON
EDITED BY DAVID BREWSTER

CAMBRIDGE
UNIVERSITY PRESS

University Printing House, Cambridge, CB2 8BS, United Kingdom

Cambridge University Press is part of the University of Cambridge.
It furthers the University's mission by disseminating knowledge in the pursuit of education, learning and research at the highest international levels of excellence.

www.cambridge.org
Information on this title: www.cambridge.org/9781108070379

This edition first published 1822
This digitally printed version 2014

ISBN 978-1-108-07037-9 Paperback

CAMBRIDGE LIBRARY COLLECTION

Books of enduring scholarly value

Technology

The focus of this series is engineering, broadly construed. It covers technological innovation from a range of periods and cultures, but centres on the technological achievements of the industrial era in the West, particularly in the nineteenth century, as understood by their contemporaries. Infrastructure is one major focus, covering the building of railways and canals, bridges and tunnels, land drainage, the laying of submarine cables, and the construction of docks and lighthouses. Other key topics include developments in industrial and manufacturing fields such as mining technology, the production of iron and steel, the use of steam power, and chemical processes such as photography and textile dyes.

A System of Mechanical Philosophy

After a brief career at sea, during which he tested Harrison's chronometer for the Board of Longitude, John Robison (1739–1805) became lecturer in chemistry at the University of Glasgow. In 1774, having spent a period teaching mathematics in Russia, he returned to Scotland as professor of natural philosophy at Edinburgh. Despite his busy schedule, he contributed major articles on the sciences to the *Encyclopaedia Britannica*, giving an overview of contemporary scientific knowledge for the educated layperson. After his death, these and other pieces of his scientific writing were edited by his former pupil David Brewster (1781–1868) and were finally published in four volumes in 1822, with a separate volume of illustrative plates. This reissue incorporates those plates in the relevant volumes of text. Volume 1 contains articles on dynamics and on the construction of roofs, arches and bridges, as well as a previously unpublished manuscript on projectile motion.

Cambridge University Press has long been a pioneer in the reissuing of out-of-print titles from its own backlist, producing digital reprints of books that are still sought after by scholars and students but could not be reprinted economically using traditional technology. The Cambridge Library Collection extends this activity to a wider range of books which are still of importance to researchers and professionals, either for the source material they contain, or as landmarks in the history of their academic discipline.

Drawing from the world-renowned collections in the Cambridge University Library and other partner libraries, and guided by the advice of experts in each subject area, Cambridge University Press is using state-of-the-art scanning machines in its own Printing House to capture the content of each book selected for inclusion. The files are processed to give a consistently clear, crisp image, and the books finished to the high quality standard for which the Press is recognised around the world. The latest print-on-demand technology ensures that the books will remain available indefinitely, and that orders for single or multiple copies can quickly be supplied.

The Cambridge Library Collection brings back to life books of enduring scholarly value (including out-of-copyright works originally issued by other publishers) across a wide range of disciplines in the humanities and social sciences and in science and technology.

A

SYSTEM

OF

MECHANICAL PHILOSOPHY.

BY JOHN ROBISON, LL. D.

LATE PROFESSOR OF NATURAL PHILOSOPHY IN THE
UNIVERSITY OF EDINBURGH.

WITH NOTES,

BY DAVID BREWSTER, LL. D.

FELLOW OF THE ROYAL SOCIETY OF LONDON, AND SECRETARY TO THE
ROYAL SOCIETY OF EDINBURGH.

IN FOUR VOLUMES,

AND A VOLUME OF PLATES.

VOL. I.

EDINBURGH:
PRINTED FOR JOHN MURRAY, LONDON.

1822.

CONTENTS

OF

VOLUME FIRST.

TO

THE REV. WILLIAM TRAIL, LL.D.

FELLOW OF THE ROYAL SOCIETY OF EDINBURGH,

MEMBER OF THE ROYAL IRISH ACADEMY, CHANCELLOR
OF ST SAVIOUR'S, CONNOR,

And the last of the Fathers of Scottish Science,

THE WORKS OF A FELLOW STUDENT,

WHOSE TALENTS AND VIRTUES HE ADMIRED,

ARE RESPECTFULLY INSCRIBED

BY HIS SINCERE FRIEND,

THE EDITOR.

PREFACE OF THE EDITOR.

AFTER the death of Dr Robison, in 1805, his friend and successor, the late Professor Playfair, undertook to draw up an account of his life and writings, and to arrange and edite the various articles which he had composed for the Encyclopædia Britannica. The public already know with how much ability Mr Playfair executed the first part of his task; and it is much to be regretted that he did not complete it, by superintending the publication of the present work. When we consider, however, his advanced age, and the numerous pursuits of his own, which he did not live to finish, we cannot be surprised at his declining to occupy his time with a species of labour by which he could neither add to his own reputation nor to that of Dr Robison.

Under these circumstances, I was requested by Dr Robison's family to superintend the publication of his scientific works, which consisted of some manuscript articles on Projectiles and Corpuscular Action,* and of the articles which he had contributed to the fourth edition of the Encyclopædia Britannica. Having enjoyed the advantage of being one of Dr Robison's pupils, I could not decline a task which naturally devolved upon me; nor should I have felt myself at liberty to do so, had I been able to foresee the difficulties which I had to encounter in its execution.

As the work could not be extended beyond four volumes, it was necessary to select the most important articles for publication; and even when this selection was made, I could not confine them within the prescribed limits, without a process of abridgment, which was both troublesome and difficult. In doing this, however, I generally confined myself to the omission of those digressions of a political and religious nature, which, however appropriate they might have been at the time, were, in every respect, unsuitable to scientific discussions; though sometimes, from a diffuseness of style, and a redundancy of illustration, allowable in an Encyclopædia, I was enabled to abridge, without omitting any essential step in the investigation. The repetitions so unavoidable in articles written and published at different times,

* The manuscript articles are printed in vol. I. from p. 159 to 368.

I have in some cases omitted; but they still exist to a considerable degree, and, I am persuaded, they will not be regarded as defects by the reader who has occasion to study separately the articles in which they occur. Had the works of Dr Robison been put into my hands in MS. to be published for the first time, I should have felt that the responsibility of the author was transferred to the editor; but, in the present case, almost all the articles had been previously before the public; and had received from the hands of the author various corrections and additions. Under these circumstances, I was freed from every editorial responsibility, excepting that of the most humble kind.

In order to render this work as much as possible a system of mechanical philosophy, I was anxious that it should contain a complete treatise on astronomy. The short articles on Astronomy, and the articles on the Tides and the Precession of the Equinoxes, which Dr Robison had written for the Encyclopædia Britannica, were unfit to supply this desideratum. I found it necessary, therefore, to delay the work till the year 1820, when the copy-right of his System of Astronomy had expired. This work has therefore been used as a substitute for the astronomical articles contained in the Encyclopædia, and will be found one of the most valuable treatises on Physical Astronomy that has for a long time been given to the public.

Being desirous of making the work as complete as possible, I had proposed to give an account of the recent discoveries in science in the form of notes. I found, however, as I proceeded, that there was not room for any additional matter, excepting a few notes; and references to more recent works; and I felt that I could make no apology to the reader for inserting compositions of my own, while I was under the necessity of abridging the original work. In the article on the Steam-Engine, however, I deviated from this rule. The great improvements which had been made upon this engine since Dr Robison's article was written, rendered it necessary that considerable additions should be made to it. I had the good fortune to prevail upon our late celebrated countryman, Mr James Watt, to undertake the revision of the article; and though he intended only to correct imperfections, and supply some of the most prominent defects, yet he was gradually led to extend his views, and to compose those valuable additions on the History, the Principles, and the Construction of the Steam-Engine which enrich that part of the work.

To those who may examine Dr Robison's dissertations with a critical eye, it may be necessary to state, that they were all composed under the influence of that cruel disease with which he was afflicted for a long period of years. The knowledge of mechanical philosophy which they everywhere display possesses the rare quality of being at once practical and profound, and they are often enriched with original views and ingenious inventions, which it required only the tranquillity of health to perfect and mature. It was his destiny, however, to enjoy but at distant intervals that calm of mind which can alone sustain the ardour of discovery. At such periods, his ambition constantly reverted to those original pursuits which he was desirous of bringing to a close; but they were no sooner begun, than they were interrupted by renewed attacks of that painful disease which ultimately deprived him of his life.

Although Dr Robison's name, therefore, cannot be associated with the great discoveries of the century which he adorned, yet the memory of his talents and his virtues will be long cherished by his country. Imbued with the genuine spirit of the philosophy which he taught, he was one of the warmest patrons of genius, wherever it was found. His mind was nobly elevated above the mean jealousies of rival ambition, and his love of science and of justice was too ardent to allow him either to depreciate the labours of others, or to transfer them to himself.

To these great qualities as a philosopher, Dr Robison added all the more estimable endowments of domestic and of social life. His friendship was at all times generous and sincere. His piety was ardent and unostentatious. His patriotism was of the most pure and exalted character; and, like the immortal Newton, whose memory he cherished with a peculiar reverence, he was pre-eminently entitled to the high distinction of a Christian patriot and philosopher.

EDINBURGH, *Dec.* 22, 1821.

DYNAMICS.

1. This name marks that department of physico-mathematical science which contains the abstract doctrine of MOVING FORCES; that is, whatever necessarily results from the relations of our ideas of motion, and of the immediate causes of its production and changes.

2. All *changes* of motion are considered by us as the indications, the characteristics, and the measures of changing causes. This is a physical law of human thought, and therefore a principle to which we may refer, and from which we must derive all our knowledge of those causes. When we appeal to our own thoughts or feelings, we do not find in ourselves any disposition to refer mere existence to any cause, although the beginning of existence certainly produces this reference in an instant. Had we always observed the universe in motion, it does not appear that we should have ascribed it to a cause, till the observation of relative rest, or something leading to it, had enabled us to separate, by abstraction, the notion of matter from that of motion. We might then perceive, that rest is not incompatible with matter; and we might

even observe, by means of relative motions, that absolute rest might be produced by the concourse of equal and opposite motions. But all this requires reflection and reasoning; whereas we are now speaking of the first suggestions of our minds.

3. We cannot have any notion of motion *in abstracto*, without considering it as a state or condition of existence, which would remain, if not changed by some cause. It is from changes alone, therefore, that we infer any agency in nature; and it is in these that we are to find all that we know of their causes.

4. When we look around us, we cannot but observe, that the motions of bodies have, in most cases, if not always, some relation to the situation, the distance, and the discriminating qualities of other bodies. The motions of the moon have a palpable relation to the earth; the motions of the tides have as evident a relation to the moon; the motions of a piece of iron have a palpable dependence on a magnet. The vicinity of the one seems to be the occasion, at least, of the motions of the other. The causes of these motions have an evident connection with or dependence on the other body. We are even disposed to imagine, that they are inherent in that body, and that it possesses certain qualities which are the causes of those modifications of motion in other bodies. These serve to distinguish some bodies from others, and may therefore be called PROPERTIES; and, since the condition of other bodies so evidently depends on them, these properties express very interesting relations of bodies, and are chiefly attended to in the enumeration of the circumstances which ascertain what we call the *nature* of any thing. We do not mean to say, that these inferences are always just; nay, we know that many of them are ill-founded: but they are real, and they serve abundantly for informing us what we may expect from any proposed situation of things. It is enough for us to know, that when a piece of iron is

so and so situated in relation to a magnet, it will move in a certain manner.

This mutual relation of bodies is differently considered, according to the interest that we chance to take in the phenomenon. The cause of the approach of the iron to a magnet is generally ascribed to the magnet, which is said to attract the iron, because we commonly employ the magnet in order that these motions may take place. The similar approach of a stone to the earth is ascribed to the stone, and we say that it tends to the earth. In all probability, the procedure of nature is the same in both ; for they are observed, in every instance, to be mutual between the related bodies. As iron approaches a magnet, so the magnet approaches the iron. The same thing is observed in the motions of electrified bodies; also in the case of the stone and the earth. Therefore the cause of the motions may be conceived as inherent in either, or in both.

The qualities thus inherent in bodies, constituting their mechanical relations, have been called the MECHANICAL AFFECTIONS OF MATTER. But they are more commonly named POWERS or FORCES ; and the event which indicates their presence, is considered as the effect and mark of their agency. The magnet is said to ACT on the iron, the earth is said to ACT on the stone ; and the iron and the stone are said to ACT on the magnet and on the earth.

All this is figurative or metaphorical language. All languages have begun with social union, and have improved along with it. The first collections of words expressed the most familiar and the most interesting notions. In the process of social improvement, the number of words did not increase in the same proportion with the notions that became interesting and familiar in their turn : for it often happened that relations of certain ideas so much resembled the relations of certain other ideas, that the word expressing one of them served very well for expressing the other ; because the dissimilar circumstances of the two cases prevented all chance of mistake. Thus we are said to *surmount* a difficulty, without attaching to the word the notion of *getting over* a steep hill. Languages are thus filled with figurative expressions.

5. Power, Force, and Action, are words which must have appeared in the language of the most simple people; because the notions of personal ability, strength, and exertion, are at once the most familiar and the most interesting that can have a place in the human mind. These terms, when used in their pure, primitive sense, express the notions of the power, force, and action of a sentient, active, being. Such a being only is an agent. The exertion of his power or force is (exclusively) action: But the relation of cause and effect so much resembles in its results the relation between this force and the work performed, that the same term may be very intelligibly employed for both. Perhaps the only case of pure unfigurative action is that of the mind on the body. But as this is always with the design of producing some change on external bodies, we think only of them; the instrument or tool is overlooked, and we say that we act on the external body. Our *real* action, therefore, is but the first movement in a long train of successive events, and is but the remote cause of the interesting event. The resemblance to such actions is very strong indeed in many cases of mechanical phenomena. A man throws a ball by the motion of his arm. A spring impels a ball in the same manner by unbending. These two events resemble each other in every circumstance but the action of the mind on the corporeal organ—the rest of it is a train of pure mechanism. In general, because the ultimate results of the mutual influence of bodies on each other greatly resemble the ultimate results of our actions on bodies, we have not invented appropriated terms, but have contented ourselves with those already employed for expressing our own actions, the exertions of our own powers or forces. The

relation of physical cause and effect is expressed metaphorically in the words which belong properly to the relation of agent and action. This has been attended by the usual consequences of poverty of language, namely, ambiguity, and sometimes mistake, both in our reflections (which are generally carried on by mental discourse), our reasonings, and our conclusions. It is necessary to be on our guard against such mistakes; for they frequently amount to the confounding of things totally different. Many philosophers of great reputation, on no better foundation than this metaphorical language, have confounded the relations of activity and of causation, and even denied that there is any difference; and they have affirmed, that there is the same invariable relation between the determinations of the will and the inducements that prompt them, as there is between any physical power and its effect. Others have maintained, that the first mover in the mechanical operations, and indeed through the whole train of any complicated event, is a percipient and intending principle in the same manner as in our actions. According to these philosophers, a particle of gravitating matter perceives its relation to every other particle in the universe, and determines its own motion according to fixed laws, in exact conformity to its situation. But the language, and even the actions of all men, shew that they have a notion of the relation of an agent to the action, easily distinguishable (because all distinguish it) from the relation between the physical cause and its effect.

6. When we speak of powers or forces as residing in a body, and the effect as produced by their exertion, the body, considered as possessing the power, is said to ACT on the other. A magnet is said to act on a piece of iron; a billiard ball in motion is said to act on one that is hit by it: but if we attempt to fix our attention on this action, as distinct both from the agent and the thing acted on, we find no object of contemplation—the exertion or procedure

of nature in producing the effect does not come under our view. When we *speak* of the action as distinct from the agent, we find that it is not the action, properly speaking, but the act, that we speak of. In like manner, the action of a mechanical power can be conceived only in the effect produced.

7. A man is not said to act unless he produces some effect. Thought is the act of the thinking principle; motion of the limb is the act of the mind on it. In mechanics, also, there is action only in so far as there is mechanical effect produced. I must act violently in order to begin motion on a slide: I must exert force, and this force exerted produces motion. I conceive the production of motion, in all cases, as the exertion of force; but it requires no exertion to continue the motion along the slide; I am conscious of none, therefore I ought to infer that no force is necessary for the continuation of any motion. The continuation of motion is not the production of any new effect, but the permanency of an effect already produced. We indeed consider motion as the effect of an action; but there would be no effect if the body were not moving. Motion is not the action, but the effect of the action.

8. Mechanical actions have been usually classed under two heads: they are either PRESSURES or IMPULSIONS. They are generally considered as of different kinds; the exertions of different powers. PRESSURE is supposed to differ essentially from IMPULSE.

Instead of attempting to define, or describe, these two kinds of forces and actions, we shall just mention some instances. This will give us all the knowledge of their distinctions that we can acquire.

When a ball lies on a table, and I press it gently on one side, it moves toward the other side of the table. If I follow it with my finger, continuing my pressure, it accelerates continually in its motion. In like manner, when I press on the handle of a common kitchen jack, the fly *begins* to move. If I continue to urge or press round the handle, the fly accelerates continually, and may be brought into a state of very rapid motion. These motions are the effects of genuine pressure. The ball would be urged along the table in the same manner, and with a motion continually accelerated, by the unbending of a spring. Also, a spring coiled up round the axis of the handle of the jack would, by uncoiling itself, urge round the fly with a motion accelerating in the same way. The more I reflect on the pressure of my finger on the ball, and compare it with the effect of the spring on it, the more clearly do I see the perfect similarity; and I call these influences, exertions, or actions, by one name, PRESSURE, taken from the most familiar instance of them.

Again, the very same motion may be produced in the ball or fly, by pulling the ball or the machine by means of a thread, to which a weight is suspended. As both are motions accelerated in the same manner, I call the influence or action of the thread on the ball or machine by the same name PRESSURE, and WEIGHT is considered as a pressing power. Indeed, I feel the same compression from the real pressure of a man on my shoulders, that I would feel from a load laid on them. But the weight in our example is acting by the intervention of the thread. By its pressure, it is pulling at that part of the thread to which it is fastened; this part is pulling at the next by means of the force of cohesion; and this pulls at a third, and so on, till the most remote pulls at the ball or the machine. Thus may elasticity, weight, cohesion, and other forces, perform the office of a genuine power; and since their result is always a motion beginning from nothing, and accelerating by perceptible degrees to any velocity, this resemblance makes us call them by one familiar name.

But farther, I see that if the thread be cut, the weight will fall with an accelerated motion, which will increase to any degree, if the length of the fall be great enough. I ascribe this also to a pressing power acting on the weight. Nay, after a very little refinement, I consider this power as the cause of the body's weight; which word is but a distinguishing name for this particular instance of pressing power. Gravitation is therefore added to the list of pressures; and, for similar reasons, the attractions and repulsions of magnets or electric bodies may be added to the list; for they produce actual compressions of bodies placed between them, and they produce motions gradually accelerated, precisely as gravitation does. Therefore all these powers may be distinguished by this descriptive name *pressures*, which, in strict language, belongs to one of them only.

Several writers, however, subdivide this great class into pressions and solicitations. Gravity is a solicitation *ab extra*, by which a body is urged downward. In like manner, the forces of magnetism and electricity, and a vast variety of other attractions and repulsions, are called *solicitations*. We see little use for this distinction, and the term is too like an affection of mind.

9. Impulsion is exhibited when a ball in motion puts another ball into motion by hitting, or (to speak metaphorically) by striking it. The appearances here are very different. The body that is struck acquires, in the instant of impulse, a sensible quantity of motion, and sometimes a very rapid motion. This motion is neither accelerated nor retarded after the stroke, unless it be affected by some other force. It is also remarked, that the rapidity of the motion depends, *inter alia*, on the previous velocity of the striking body For instance, if a clay ball, moving with any velocity, strike another equal ball which is at rest, the struck ball moves with half the velocity of the other. And it is farther remarkable, that the striking body always loses as much motion as the struck body gains. This universal and remarkable fact seems to have given rise to a confused or indistinct notion of a sort of transference of motion from one body to another. The phraseology in general

use on this subject expresses this in the most precise terms. The one ball is not said to cause or produce motion in the other, but to *communicate* motion to it; and the whole phenomenon is called the *communication of motion.* We call this an *indistinct* notion; for surely no one will say that he has a clear conception of it. We can form the most distinct notion of the communication of heat, or of the cause of heat; of the communication of saltness, sweetness, and a thousand other *things;* but we cannot conceive how part of that identical motion which was formerly in A, is now infused into B, being given up by A. It is in our attempt to form this notion that we find that motion is not a *thing*, not a substance which can exist independently, and is susceptible of actual transference. It appears in this case to be a state, or condition, or mode of existence, of which bodies are susceptible, which is producible, or (to speak without metaphor) causable, in bodies, and which is the effect and *characteristic* of *certain* natural qualities, properties, or powers. We are anxious to have our readers impressed with clear and precise notions on this subject, being confident that such, and only such, will carry them through some intricate paths of mechanical and philosophical research.

10. The remarkable circumstance in this phenomenon is, that a rapid motion, which requires for the effecting it the action of a pressing power, continued for a sensible, and frequently a long time, seems to be effected in an instant by impulsion. This has tended much to support the notion of the actual transference of something formerly possessed exclusively by the striking body, inhering in it, but separable, and now transfused, into the body stricken. And now room is found for the employment of metaphor, both in thought and language. The *striking* body affects the body which it thus impels: It therefore possesses the *power* of impulsion, that is, of *communicating* motion. It possesses it only while it is in motion. This *power*, there-

fore, is the efficient distinguishing cause of its motion, and its only office must be the continuation of this motion. It is therefore called the INHERENT FORCE, the force inherent in a moving body, VIS INSITA *corpori moto*. This force is transfused into the body impelled; and *therefore* the transference is instantaneous, and the impelled body continues its motion till it is changed by some other action. All this is at first sight very plausible; but a scrupulous attention to those feelings which have given rise to this metaphorical conception, should have produced very different notions. I am conscious of exertion in order to begin motion on a slide; but if the ice be very smooth, I am conscious of no exertion in order to slide along. My power is felt only while I am conscious of exerting it: Therefore I have no primitive feeling or notion of power while I am sliding along. I am certain that no exertion of power is necessary here. Nay, I find that I cannot think of my moving forward without effort otherwise than as a certain mode of my existence. Yet we imagine that the partisans of this opinion did really deduce it in some shape from their feelings. We must continue the *exertion* of walking in order to walk on; our power of walking must be continually exerted, otherwise we shall stop. But this is a very imperfect, incomplete, and careless observation. Walking is much more than mere continuance in progressive motion. It is a continually repeated lifting our body up a small height, and allowing it to come down again. This renewed ascent requires repeated exertion.

11. We have other observations of importance yet to make on this force of moving bodies, but this is not the most proper occasion. Meanwhile we must remark, that the instantaneous production of rapid motion by impulse has induced the first mechanicians of Europe to maintain, that the power or force of impulse is unsusceptible of any comparison with a pressing power. They have asserted, that impulse is infinitely great when compared with pres-

sure; not recollecting that they held them to be things totally disparate, that have no proportion more than weight and sweetness. But these gentlemen are perpetually enticed away from their creed by the similarity of the ultimate results of pressure and impulse. No person can find any difference between the motion of two balls moving equally swift, in the same direction, one of which is descending by gravity, and the other has derived its motion from a blow. This struggle of the mind to maintain its faith, and yet accommodate its doctrines to what we see, has occasioned some other curious forms of expression. Pressure is considered as an *effort* to produce motion. When a ball lies on a table, its weight, which they call a *power*, continually and repeatedly *endeavours* (mark the metaphorical word and thought) to move the ball downward. But these efforts are ineffectual. They say that this ineffectual power is *dead*, and call it a VIS MORTUA: but the force of impulsion is called a VIS VIVA, a living force. But this is very whimsical and very inaccurate. If the impelling ball falls perpendicularly on the other lying on the table, it will produce no motion any more than gravity will; and if the table be annihilated, gravity becomes a *vis viva*.

We must now add, that, in order to prove that impulse is infinitely greater than pressure, these mechanicians turn our attention to many familiar facts which plead strongly in their favour. A carpenter will drive a nail into a board with a very moderate blow of his hammer. This will require a pressure which seems many hundred times greater than the impelling effort of the carpenter. A very moderate blow will shiver into pieces a diamond which would carry the weight of a mountain. Seeing this prodigious superiority in the impulse, how shall they account for the production of motion by means of pressure? for this motion of the hammer might have been acquired by its falling from a height; nay, it is actually acquired by means of the continued pressure of the carpenter's arm. They consider it as the aggregate of an infinity of succeeding pressures in *every* instant of its continuance, so that the insignificant smallness of each effort is compensated by their inconceivable number.

On the whole, we do not think that there is clear evidence that there are two kinds of mechanical force essentially different in their nature. It is virtually given up by those who say that impulse is infinitely greater than pressure. Nor is there any considerable advantage to be obtained by arranging the phenomenon under those two heads. We may perhaps find some method of explaining satisfactorily the remarkable difference that is really observed in the two modes of producing motion; namely, the gradual production of motion by acknowledged pressure, and the instantaneous production of it by impulse. Indeed, we should not have taken up so much of our readers attention with this subject, had it not been for some inferences that have been made from these premises, which meet us in our very entry on the consideration of first principles, and that are of extensive influence on the whole science of mechanical philosophy, and, indeed, on the whole study of nature.

12. Mechanicians are greatly divided in their opinion about the nature of the sole moving force in nature. Those whom we are now speaking of, seem to think that all motion is produced by pressure: For when they consider impulse as equivalent to the aggregate of an infinity of repeated pressures, they undoubtedly suppose any pressure however insignificant, as a moving force. But there is a, party, both numerous and respectable, who maintain that mulipsion is the sole cause of motion. We see bodies in motion, say they, and we see them impel others; and we see that this production of motion is regulated by such laws, that there is but one absolute quantity of motion in the universe which remains unalterably the same. It must

therefore be transfused in the acts of collision. We also see, with clear evidence, in some cases, that motion can produce pressure. Euler adduces some very whimsical and complicated cases, in which an action, precisely similar to pressure, may be produced by motion. Thus, two balls connected by a thread, may be so struck that they shall move forward, and at the same time wheel round. In this case the connecting thread will be stretched between them. Now, say the philosophers, since we see motion, and see that pressure may be produced by motion, it is preposterous to imagine that it is any thing else than a result of certain motions; and it is the business of a philosopher to inquire and discover what motions produce the pressures that we observe.

They then proceed to account for those pressing powers, or solicitations to motion, which we observe in the acceleration of falling bodies, the attractions of magnetism and electricity, and many other phenomena of this kind, where bodies are put in motion by the vicinity of other bodies, or (in the popular language) by the action of other bodies at a distance. To say that a magnet can act on a piece of remote iron, is to say that it can act *where* it is not; which is as absurd as to say, that it can act *when* it is not. *Nihil movetur*, says Euler, *nisi a contiguo et moto.*

The bulk of these philosophers are not very anxious about the way in which these motions are produced, nor do they fall upon such ingenious methods of producing pressure as the one already mentioned, which was adduced by Euler. The piece of iron, say they, is put in motion when brought into the neighbourhood of a magnet, because there is a stream of fluid issuing from one pole of the magnet, which circles round the magnet, and enters at the other pole: This stream impels the iron, and arranges it in certain determined positions, just as a stream of water would arrange the flote grass. In the same manner, there is a stream of fluid continually moving towards the centre of the earth, which impels all bodies in lines perpendicular to the surface; and so on with regard to other like phenomena. These motions are thus reduced to very simple cases by impulsion.

It is unnecessary to refute this doctrine at present; it is enough that it is contrary to all the dictates of common sense, to *suppose* an agent that we do not see, and for whose existence we have not the smallest argument; with equal propriety we might suppose ministering spirits, or any thing that we please.

13. Other philosophers are so dissatisfied with this notion of the production of pressure, that they, on the other hand, affirm that pressure is the only moving force in nature; not according to the popular notion of pressure, by the mutual contact of solid bodies, but that kind of pressure which has been called *solicitation*; such as the power of gravity. They affirm, that there is no such thing as contact on instantaneous communication of motion by real collision. They say (and they prove it by very convincing facts) that the particles of solid bodies exert very strong repulsions to a small distance; and therefore, when they are brought by motion sufficiently near to another body, they repel it, and are equally repelled by it. Thus is motion produced in the other body, and their own motion is diminished. And they then shew, by a scrupulous consideration of the state of the bodies while the one is advancing and the other retiring, in what manner the two bodies attain a common velocity, so that the quantity of motion before collision remains unchanged, the one body gaining as much as the other loses. They also shew cases of such mutual action between bodies, where it is evident that they have never come into contact; and yet the result has been precisely similar to those cases where the motion appeared to be changed in an instant. Therefore they conclude, that there is no such thing as instantaneous *communication*, or transfusion of motion, by con-

tact in collision or impulse. The reason why previous motion of the impelling body is necessary, is not that it may have a *vis insita corpori moto*, a force inherent in it *by its being in motion*, but that it may continue to follow the impelled and retiring body, and exert on it a force inherent in itself, whether in motion or at rest.—According to these philosophers, therefore, all moving forces are of that kind which has been named *solicitation*; such as gravity. We shall know it afterwards by the more familiar and descriptive name of ACCELERATING or RETARDING force.

14. The exertions of mechanical forces are differently termed, according to the reference that we make to the result. If, in boxing or wrestling, I strike, or endeavour to throw my antagonist, I am said to ACT; but if I only parry his blows, or prevent him from throwing me, I am said to RESIST. This distinction is applied to the exertions of mechanical powers. When one body A changes the motion of another B, we may consider the change in the motion of B either as the indication and measure of A's power of producing motion, or as the indication and measure of A's resistance to the being brought to rest, or having its motion any how changed. The distinction is not in the thing itself, but only in the reference that we are disposed, by other considerations, to make of its effect. They may be distinguished in the following manner: If a change of motion follow when one of the powers ceases to be exerted, that power is conceived as having resisted. The whole language on this subject is metaphorical. Resistance, effort, endeavour, &c. are words which cannot be employed in mechanical discussions without figure, because they all express notions which relate to sentient beings; and the unguarded indulgence of this figurative language has so much affected the imagination of philosophers, that many have almost animated all matter. Perhaps the word REACTION, introduced (we think) by Newton, is the best term for expressing that mutual force which is perceived

in all the operations of nature that we have investigated with success. As the magnet attracts iron, and in so doing is said to *act* on it; so the iron attracts the magnet, and may be said to *react* on it.

15. With respect to the difficulty that has been objected to the opinion of those who maintain that all the mechanical phenomena are produced by the agency of attracting or repelling forces; namely, that this supposes the bodies to act on each other at a distance, however small those distances may be, which is thought to be absurd, it may be observed, that we may ascribe the mutual approaches or recesses to tendencies to or from each other. What we call *the attraction of the magnet* may be considered as a tendency of the iron to the magnet, somewhat similar to the gravitation of a stone toward the earth. We surely (at least the unlearned) can and do conceive the iron to be affected by the magnet, *without thinking* of any intermedium. The thing is not therefore inconceivable; which is all that we know about absurdity: and we do not know any thing about the nature or essence of matter which renders this tendency to the magnet impossible. That we do not see intuitively any reason why the iron should approach the magnet, must be granted; but this is not enough to entitle us to say, that such a thing is impossible or inconsistent with the nature of matter. It appears, therefore, to be very hasty and unwarrantable, to suppose the impulse of an invisible fluid, of which we know nothing, and of the existence of which we have no proof. Nay, if it be true that bodies do not come into contact, even when one ball hits another, and drives it before it, this invisible fluid will not solve the difficulty; because the same difficulty occurs in the action of any particle of the fluid on the body. We are obliged to say, that the production of motion without any *observed* contact, is a much more familiar phenomenon than the production of motion by impulsion. More motion has been produced in

this way by the gravitation of a small stream of water, running ever since the creation, than by all the impulses in the world twice told. We do not mean by this to say, that the giving to this observed mutual relation between iron and a loadstone the name *tendency* makes it less absurd, than when we say that the loadstone attracts the iron; it only makes it more conceivable: It suggests a very familiar analogy: but both are equally figurative expressions; at least as the word tendency is used at present. In the language of ancient Rome, there was no metaphor when Virgil's hero said, *Tendimus in Latium. Tendere versus solem* means, in plain Latin, to *approach the sun*. The safe way of conceiving the whole is to say, that the condition of the iron depends on the vicinity of the magnet.

16. When the exertions of a mechanical power are observed to be always directed toward a body, that body is said to attract; but when the other body always moves off from it, it is said to repel. These also are metaphorical expressions. I attract a boat when I pull it toward me by a rope; this is purely ATTRACTION; and it is pure, unfigurative REPULSION, when I push any body from me. The same words are applied to the mechanical phenomena, merely because they resemble the results of real attraction or repulsion. We must be much on our guard to avoid metaphor in our conceptions, and never allow those words to suggest to our mind any opinion about the *manner* in which the mechanical forces produce their effects. It is plain, that if the opinion of those who maintain the existence and action of the above mentioned invisible fluid be just, there is nothing like attraction or repulsion in the universe. We must always recur to the simple phenomenon, the motion to or from the attracting or repelling body; for this is all we see, and generally all that we know.

17. We conceive one man to have twice the strength of

another man, when we see that he can withstand the united effort of two others. Thus animal force is conceived as a quantity, made up of, and measured by, its own parts. But we doubt exceedingly whether this be an accurate conception. We have not a distinct notion of one strain added to another; though we have of their being joined or combined. We want words to express the difference of these two notions in our own minds; but we imagine that others perceive the same difference. We conceive clearly the addition of two lines or of two minutes; we can conceive them apart, and perceive their boundaries, common to both, where one ends and the other begins. We cannot conceive thus of two forces combined; yet we cannot say, that two equal forces are *not* double of one of them. We measure them by the effects which they are known to produce. Yet there are not wanting many cases where the action of two men, equally strong, does not produce a double motion.

In like manner, we conceive all mechanical forces as measurable by their effects; and thus they are made the subjects of mathematical discussion. We talk of the proportions of gravity, magnetism, electricity, &c.; nay, we talk of the proportion of gravity to magnetism:—Yet these, considered in themselves, are disparate, and do not admit of any proportion; but they produce effects, some of which are measurable, and whose assumed measures are susceptible of comparison, being quantities of the same kind. Thus, one of the effects of gravity is the acceleration of motion in a falling body; magnetism will also accelerate the motion of a piece of iron; these two accelerations are comparable. But we cannot compare magnetism with heat; because we do not know any measurable effects of magnetism that are of the same kind with any effects of heat.

When we say, that the gravitation of the moon is the 3600th part of the gravitation at the sea-shore, we mean that the fall of a stone in a second is 3600 times greater than the fall of the moon in the same time. But we also mean (and this expresses the proportion of the *tendency* of gravitation more purely), that if a stone, when hung on a spring steelyard, draw out the rod of the steelyard to the mark 3600, the same stone, taken up to the distance of the moon, will draw it out no further than the mark 1. We also mean, that if the stone at the sea-shore draw out the rod to any mark, it will require 3600 such stones to draw it out to that mark, when the trial is made at the distance of the moon. It is not, therefore, in consequence of any immediate perception of the proportion of the gravitation at the moon to that at the surface of the earth that we make such an assertion; but these motions, which we consider as its effects in these situations, being magnitudes of the same kind, are susceptible of comparison, and have a proportion which can be ascertained by observation. It is these proportions that we contemplate; although we speak of the proportions of the unseen causes, the forces, or endeavours to descend. It will be of material service to the reader to peruse Dr Reid's judicious and acute dissertation on *quantity* in the 45th volume of the Philosophical Transactions; where, we trust, he will see clearly how force, velocity, density, and many other magnitudes of very frequent occurrence in mechanical philosophy, may be made the subjects of mathematical discussion, by means of some of those proper quantities, measurable by their own parts, which are to be assumed as their measures. Pressures are measurable only by pressures. When we consider them as moving powers, we should be able to measure them by *any* moving powers, otherwise we cannot compare them; therefore it is not as pressures that we then measure them This observation is momentous.

One circumstance must be carefully attended to. That those assumed measures may be accurate, they must be invariably connected with the magnitudes which they are employed to measure, and so connected, that the degrees of the one must change in the same manner with the degrees of the other. This is evident, and is granted by all. But we must also *know* this of the measure we employ; we must see this constant and precise relation. How can we know this? We do not perceive force as a separate existence, so as to see its proportions, and to see that these are the same with the proportions of the measures, in the same manner that Euclid sees the proportions of triangles and those of their bases, and that these proportions are the same, when the triangles are of equal altitudes. How do we discover that to every magnitude which we call *force* is invariably attached a corresponding magnitude of acceleration or deflection? Clearly. In fact, the very existence of the force is an inference that we make from the observed accelerations; and the degree of the force is, in like manner, an inference from the observed magnitude of the acceleration. Our measures are therefore necessarily connected with the magnitudes which they measure, and their proportions are the same; because the one is always an inference from the other, both in species and in degree.

18. It is now evident, that these disquisitions are susceptible of mathematical accuracy. Having selected our measures, and observed certain mathematical relations of those measures, every inference that we can draw from the mathematical relations of the proportions of those representations, is true of the proportions of the motions, and therefore of the proportions of the forces. And thus dynamics becomes a demonstrative science, one of the *disciplinæ accuratæ.*

19. But moving forces are considered as differing also in kind; that is, in direction. We assign to the force the direction of the observed change of motion; which is not only the indication, but also the characteristic, of

the changing force. We call it an *accelerating*, *retarding*, *deflecting*, *force*, according as we observe the motion to be accelerated, retarded; or deflected.

These denominations shew us incontestably that we have no knowledge of the forces different from our knowledge of the effects. The denominations are all either descriptive of the effects, as when we call them accelerating, penetrating, protrusive, attractive, or repulsive forces; or they are names of reference to the substances in which the accelerating, protrusive, &c. forces, are supposed to be inherent, as when we call them *magnetism*, *electricity*, *corpuscular*, &c.

20. When I struggle with another, and feel, that in order to prevent being thrown, I must exert force, I learn that my antagonist is exerting force. This notion is transferred to matter; and when a moving power which is *known* to operate, produces no motion, we conceive it to be opposed by another equal force; the existence, agency, and intensity of which is detected and measured by these means. The quiescent state of the body is considered as a change on the state of things that would have been exhibited in consequence of the known action of one power, had this other power not acted; and this change is considered as the indication, characteristic, and measure, of another power, detected in this way. Thus forces are recognised not only by the changes of motion which they produce, but also by the changes of motion which they prevent. The cohesion of matter in a string is inferred not only by its giving motion to a ball which I pull toward me by its intervention, but also by its suspending that ball, and hindering it from falling I know that gravity is acting on the ball, which, however, does not fall. The solidity of a board is equally inferred from its stopping the ball which strikes it, and from the motion of the ball which it drives before it. In this way we learn that the particles of tangible matter cohere by means

of moving forces, and that they resist compression with force; and in making this inference, we find that this corpuscular force exerted between the particles is mutual, opposite, and equal; for we must apply force equally to *a* or to *b*, in order to produce a separation or a compression. We learn their equality, by observing that no motion ensues while these mutual forces are known to act on the particles; that is, each is opposed by another force, which is neither inferior nor superior to it.

OF THE LAWS OF MOTION.

SUCH, then, being our notions of mechanical forces, the causes of the sensible changes of motion, there will result certain consequences from them, which may be called axioms, or laws, of motion. Some of these may be intuitive, offering themselves to the mind as soon as the notions which they involve are presented to it. Others may be as necessary results from the relations of these notions, but may not readily offer themselves without the mediation of axioms of the first class. We shall select those which are intuitive, and may be taken for the first principles of all discussions in mechanical philosophy.

FIRST LAW OF MOTION.

Every body continues in a state of rest, or of uniform rectilineal motion, unless affected by some mechanical force.

21. This is a proposition, on the truth of which the whole science of mechanical philosophy ultimately depends. It is therefore to be established on the firmest foundation; and a solicitude on this head is the more justifiable, because the opinions of philosophers have been, and still are, extremely different, both with respect to the truth of this

law, and with respect to the foundation on which it is built. These opinions are, in general, very obscure and unsatisfactory; and, as is natural, they influence the discussions of those by whom they are held through the whole science. Although of contradictory opinions one only can be just, and it may appear sufficient that this one be established and uniformly applied; yet a short exposition, at least, of the rest is necessary, that the greatest part of the writings of the philosophers may be intelligible, and that we may avail ourselves of much valuable information contained in them, by being able to perceive the truth in the midst of their imperfect or erroneous conceptions of it.

22. It is not only the popular opinion that rest is the natural state of body, and that motion is something foreign to it, but it has been seriously maintained by the greatest part of those who are esteemed philosophers. They readily grant that matter will continue at rest, unless some moving force act upon it. Nothing seems necessary for matter's remaining where it is, but its continuing to exist. But it is far otherwise, say they, with respect to matter in motion. Here the body is continually changing its relations to other things; therefore the continual agency of a changing cause is necessary (by the fundamental principle of all philosophical discussion), for there is here the continual production of an effect. They say that this metaphysical argument receives complete confirmation (if confirmation of an intuitive truth be necessary) from the most familiar observation. We see that all motions, however violent, terminate in rest, and that the continual exertion of some force is necessary for their continuance.

23. These philosophers therefore assert, that the continual action of the moving cause is *essentially* necessary for the continuance of the motion: but they differ among themselves in their notions and opinions about this cause

Some maintain, that all the motions in the universe are produced and continued by the immediate agency of Deity; others affirm, that in every particle of matter there is inherent a sort of mind, the φυσις and αυτεψυχη of Aristotle, which they call an ELEMENTAL MIND, which is the cause of *all* its motions and changes. An overweaning reverence for Greek learning has had a great influence in reviving this doctrine of Aristotle. The Greek and Roman languages are affirmed to be more accurate expressions of human thought than the modern languages are. In those ancient languages, the verbs which express motion are employed both in the active and passive voice; whereas we have only the active verb *to move*, for expressing both the state of motion and the act of putting in motion. "The stone *moves* down the slope, and *moves* all the pebbles which lie in its way:" but in the ancient languages, the mere state of motion is always expressed by the passive or middle voice. The accurate conception of the speakers is therefore extolled. The state of motion is expressed as it ought to be, as the result of a continual action." Κινειται, *movetur*, is equivalent to "it is moved." According to these philosophers, every thing which *moves* is mind, and every thing that is *moved* is body.

The argument is futile, and it is false; for the modern languages are, in general, equally accurate in this instance: "*se mouvoir*," in French; "*ſich bewegen*," in German; "*dvigatsu*," in Slavonic; are all passive or reflected. And the ancients said, that "rain falls, water runs, "smoke rises," just as we do. The ingenious author of *Ancient Metaphysics* has taken much pains to give us, at length, the procedures of those elementary minds in producing the ostensible phenomena of local motion; but it seems to be merely an abuse of language, and a very frivolous abuse. This elemental mind is known and characterised only by the effect which we ascribe to its action; that is, by the motions or changes of motions. Uniform and unexcepted experience shews us that these are regulated by laws as precise as those of mathematical truth. We consider nothing as more fixed and determined than the common laws of mechanism. There is nothing here that indicates any thing like spontaneity, intention, purpose; none of those marks by which mind was first brought into view: but they are very like the effects which we produce by the exertions of our corporeal forces; and we have accordingly given the name *force* to the causes of motion. It is surely much more apposite than the name *mind*, and conveys with much more readiness and perspicuity the very notions that we wish to convey.

24. We now wish to know what reason we have to think that the continual action of some cause is necessary for continuing matter in motion, or for thinking that rest is its natural state. If we pretend to draw any argument from the nature of matter, that matter must be known, as far as is necessary for being the foundation of argument. Its very existence is known only from observation; all our knowledge of it must therefore be derived from the same source.

If we take this way to come at the origin of this opinion, we shall find that experience gives us no authority for saying that rest is the natural condition of matter. We cannot say that we have ever seen a body at rest; this is evident to every person who allows the validity of the Newtonian philosophy, and the truth of the Copernican system of the sun and planets; all the parts of this system are in motion. Nay, it appears from many observations, that the sun, with his attending planets, is carried in a certain direction, with a velocity which is very great. We have no unquestionable authority for saying that any one of the stars is absolutely fixed; but we are *certain* that many of them are in motion. Rest is therefore so rare a condition of body, that we cannot say, from any experience, that it is its natural state.

25. It is easy, however, to see, that it is from observation that this opinion has been derived; but the observation has been limited and careless. Our experiments in this sublunary world do indeed always require continued action of some moving force to continue the motion; and if this be not employed, we see the motions slacken every minute, and terminate in rest after no long period. Our first notions of sublunary bodies are indicated by their operation in cases where we have some interest. Perpetually seeing our own exertions necessary, we are led to consider matter as something not only naturally quiescent and inert, but sluggish, averse from motion, and prone to rest (we must be pardoned this metaphorical language, because we can find no other term). What is expressed by it, on this occasion, is precisely one of the erroneous or inadequate conceptions that are suggested to our thoughts by reason of the poverty of language. We animate matter in order to give it motion, and then we endow it with a sort of moral character in order to explain the appearance of those motions.

26. But more extended observation has made men gradually desert their first opinions, and at last allow that matter has no peculiar aptitude to rest. All the retardations that we *observe* have been discovered, one after another, to have a distinct reference to some external circumstances. The diminution of motion is always observed to be accompanied by the removal of obstacles, as when a ball moves through sand, or water, or air; or it is owing to opposite motions which are destroyed; or it is owing to roughness of the path, or to friction, &c. We find that the more we can keep those things out of the way, the less are the motions diminished. A pendulum will vibrate but a short while in water; much longer in

air; and in the exhausted receiver, it will vibrate a whole day. We know that we cannot remove *all* obstacles; but we are led by such observations to conclude that, if they *could be completely removed*, our motions would continue for ever. And this conclusion is almost demonstrated by the motions of the heavenly bodies, to which we know of no obstacles, and which we really observe to retain their motions for many thousand years without the smallest sensible diminution.

27. Another set of philosophers maintain an opinion directly opposite to that of the inactivity of matter, and assert, that it is essentially active, and continually changing its state. Faint traces of this are to be found in the writings of Plato, Aristotle, and their commentators. Mr Leibnitz is the person who has treated this question most systematically and fully. He supposes every particle of matter to have a principle of individuality, which he therefore calls a MONAD. This monad has a sort of *perception* of its situation in the universe, and of its relation to every other part of this universe. Lastly, he says that the monad acts on the material particle, much in the same way that the soul of man acts on his body. It modifies the motion of the material atom (in conformity, however, to unalterable laws), producing all those modifications of motion that we observe. Matter, therefore, or, at least, particles of matter, are continually active, and continually changing their situation.

It is quite unnecessary to enter on a formal confutation of Mr Leibnitz's system of monads, which differs very little from the system of elemental minds, and is equally whimsical and frivolous; because it only makes the unlearned reader stare, without giving him any information. Should it even be granted, it would not, any more than the action of animals, invalidate the general proposition which we are endeavouring to establish as the fundamental law of motion. Those powers of the monads, or of the elemental minds, are the causes of all the changes of motion; but the mere material particle is subject to the law, and requires the exertion of the monad in order to exhibit a change of motion.

28. A third sect of philosophers, at the head of which we may place Sir Isaac Newton, maintain the doctrine enounced in the proposition. But they differ much in respect of the foundation on which it is built.

Some assert that its truth flows from the nature of the thing. If a body be at rest, and you assert that it will not remain at rest, it must move in some one direction. If it be in motion in any direction, and with any velocity, and do not continue its equable, rectilineal, motion, it must either be accelerated or retarded; it must turn either to one side, or to some other side. The event, whatever it be, is individual and determinate; but no cause which can determine it is supposed: therefore the determination cannot take place, and no change will happen in the condition of the body with respect to motion. It will continue at rest, or persevere in its rectilineal and equable motion.

But considerable objections may be made to this argument, of *sufficient reason*, as it is called. In the immensity and perfect uniformity of space and time, there is no determining cause why the visible universe should exist in the place in which we see it rather than in another, or at this time rather than at another. Nay, the argument seems to beg the question. A cause of determination is required as essentially necessary—a determination may be without a cause, as well as a motion without a cause.

29. Other philosophers, who maintain this doctrine, consider it merely as an experimental truth; and proofs of its universality are innumerable.

When a stone is thrown from the hand, we press it forward while in the hand, and let it go when the hand has acquired the greatest rapidity of motion that we can give it. The stone continues in that state of motion which it acquired gradually along with the hand. We can throw a stone much farther by means of a sling; because, by a very moderate motion of the hand, we can whirl the stone round till it acquire a very great velocity, and then we let go one of the strings, and the stone escapes, *by continuing* its rapid motion. We see it still more distinctly in shooting an arrow from a bow. The string presses hard on the notch of the arrow, and it yields to this pressure and goes forward. The string alone would go faster forward. It therefore *continues* to press the arrow forward, and accelerates its motion. This goes on till the bow is as much unbent as the string will allow. But the string is now a straight line. It came into this position with an accelerated motion, and it therefore goes a little beyond this position, but with a retarded motion, being checked by the bow. But there is nothing to check the arrow; therefore the arrow quits the string, and flies away.

These are simple cases of perseverance in a state of motion, where the procedure of nature is so easily traced that we perceive it almost intuitively. It is no less clear in other phenomena which are more complicated; but it requires a little reflection to trace the process. We have often seen an equestrian showman ride a horse at a gallop, standing on the saddle, and stepping from it to the back of another horse that gallops alongside at the same rate; and he does this seemingly with as much ease as if the horses were standing still. The man has the same velocity with the horse that gallops under him, and keeps this velocity while he steps to the back of the other. If that other were standing still, the man would fly over his head. And if a man should step from the back of a horse that is standing still to the back of another that gallops past him, he would be left behind. In the same manner, a slack wire dancer tosses oranges from hand to hand while the wire is in full swing The orange, swinging along with the hand, retains the velocity; and when in the air follows the hand, and falls into it when it is in the opposite extremity of its swing. A ball, dropped from the mast head of a ship that is sailing briskly forward, falls at the foot of the mast. It retains the motion which it had while in the hand of the person who dropped it, and follows the mast during the whole of its fall.

We also have familiar instances of the perseverance of a body in a state of rest. When a vessel filled with water is drawn suddenly along the floor, the water dashes over the posterior side of the vessel. It is left behind. In the same manner, when a coach or boat is dragged forward, the persons in it find themselves strike against the hinder part of the carriage or boat. Properly speaking, it is the carriage that strikes on them. In like manner, if we lay a card on the tip of the finger, and a piece of money on the card, we may nick away the card, by hitting it neatly on its edge; but the piece of money will be left behind, lying on the tip of the finger. A ball will go through a wall and fly onward; but the wall is left behind. Buildings are thrown down by earthquakes; sometimes by being tossed from their foundations, but more generally by the ground on which they stand being hastily drawn sidewise from under them, &c.

30. But common experience seems insufficient for establishing this fundamental proposition of mechanical philosophy. We must, on the faith of the Copernican system, grant that we never saw a body at rest, or in uniform rectilineal motion; yet this seems absolutely necessary before we can say that we have established this proposition experimentally.

What we imagine, in our experiments, to be putting a body, formerly at rest, into motion, is, in fact, only changing a most rapid motion, not less, and probably much greater, than 90,000 feet per second. Suppose a cannon pointed east, and the bullet discharged at noon day with 60 times greater velocity than we have ever been able to give it. It would appear to set out with this unmeasurable velocity to the eastward; to be gradually retarded by the resistance of the air, and at last brought to rest by hitting the ground. But, by reason of the earth's motion round the sun, the fact is quite the reverse. Immediately before the discharge, the ball was moving to the westward with the velocity of 90,000 feet per second nearly. By the explosion of the powder, and its pressure on the ball, some of this motion is destroyed, and at the muzzle of the gun, the ball is moving slower, and the cannon is hurried away from it to the westward. The air, which is also moving to the westward 90,000 feet in a second, gradually communicates motion to the ball, in the same manner as a hurricane would do. At last (the ball dropping all the while) some part of the ground hits the ball, and carries it along with it.

Other observations must therefore be resorted to, in order to obtain an experimental proof of this proposition. And such are to be found. Although we cannot measure the absolute motions of bodies, we can observe and measure accurately their relative motions, which are the differences of their absolute motions. Now, if we can shew experimentally, that bodies shew equal tendencies to resist the augmentation and the diminution of their relative motions, they, *ipso facto*, shew equal tendencies to resist the augmentation or diminution of their absolute motions. Therefore let two bodies, A and B, be put into such a situation, that they cannot, (by reason of their impenetrability, or the actions of their mutual powers) persevere in their relative motions. The change produced on A is the effect and measure of B's tendency to persevere in its former state; and therefore the proportion of these changes will shew the proportion of their tendencies to maintain their former states. Therefore let the following experiment be made at noon.

Let A, apparently moving westward three feet per second, hit the equal body B apparently at rest. Suppose, *1st*, That A impels B forward, without any diminution of its own velocity. This result would shew that B manifests no tendency to maintain its motion unchanged, but that A retains its motion undiminished.

2dly, Suppose that A stops, and that B remains at rest. This would shew that A does not resist a diminution of motion, but that B retains its motion unaugmented.

3dly, Suppose that both move westward with the velocity of one foot per second. The change on A is a diminution of velocity, amounting to two feet per second. This is the effect and the measure of B's tendency to maintain its velocity unaugmented. The change on B is an augmentation of one foot per second made on its velocity; and this is the measure of A's tendency to maintain its velocity undiminished. This tendency is but half of the former; and this result would shew, that the resistance to a diminution of velocity is but half of the resistance to augmentation. It is perhaps but one quarter; for the change on B has produced a double change on A.

4thly, Suppose that both move westward at the rate of $1\frac{1}{2}$ feet per second. It is evident that their tendencies to maintain their states unchanged are now equal.

5thly, Suppose A = 2 B, and that both move, after the collision, two feet per second, B has received an addition of two feet per second to its former velocity. This is the effect and the measure of A's whole tendency to retain its motion undiminished. Half of this change on B measures the persevering tendency of the half of A; but A, which formerly moved with the apparent or relative velocity three, now moves (by the supposition) with the velocity two, having lost a velocity of one foot per second. Each half of A therefore has lost this velocity, and the whole loss of motion is two. Now this is the measure of B's tendency to maintain its former state unaugmented; and this is the same with the measure of A's tendency to maintain its own former state undiminished. The conclusion from such a result would therefore be, that bodies have equal tendencies to maintain their former states of motion without augmentation and without diminution.

What is supposed in the *4th* and *5th* cases is really the result of all the experiments which have been tried; and this law regulates all the changes of motion which are produced by the mutual actions of bodies in impulsions. This assertion is true without exception or qualification. Therefore it appears, that bodies have no preferable tendency to rest, and that no fact can be adduced which should make us suppose that a motion once begun should suffer any diminution without the action of a changing cause.

But we must now observe, that this way of establishing the first law of motion is very imperfect, and altogether unfit for rendering it the fundamental principle of a whole and extensive science. It is subject to all the inaccuracy that is to be found in our best experiments; and it cannot be applied to cases where scrupulous accuracy is wanted, and where no experiment can be made.

Let us therefore examine the proposition, and we shall find it to be an axiom or intuitive consequence of the relations of those ideas which we have of motion, and of the causes of its production and changes.

31. It has been fully demonstrated that the powers or forces, of which we speak so much, are never the immediate objects of our perception. Their very existence, their kind, and their degree, are instinctive inferences from the motions which we observe and class. It evidently follows from this experimental and universal truth, *1st*, That where no change of motion is observed, no

such inference is made; that is, no power is *supposed* to act. But whenever any change of motion is observed, the inference is made; that is, a power or force is supposed to have acted.

In the same form of logical conclusion, we must say that, 2*dly*, When no change of motion is *supposed* or thought of, no force is *supposed*; and that whenever we *suppose* a change of motion, we, in fact, though not in terms, suppose a changing force. And, on the other hand, whenever we suppose the action of a changing force, we suppose the change of motion; for the action of this force, and the change of motion, is one and the same thing. We cannot think of the action without thinking of the indication of that action; that is, the change of motion.—In the same manner, when we do not think of a changing force, or suppose that there is no action of a changing force, we, in fact, though not in terms, suppose that there is no indication of this changing force: that is, that there is no change.

Whenever, therefore, we suppose that no mechanical force is acting on a body, we, in fact, suppose that the body continues in its former condition with respect to motion. If we suppose that nothing accelerates, or retards, or deflects the motion, we suppose that it is not accelerated, nor retarded, nor deflected. Hence follows the proposition in express terms—*We suppose that the body continues in its former state of rest or motion, unless we suppose that it is changed by some mechanical force.*

Thus it appears, that this proposition is not a matter of experience or contingency, depending on the properties which it has pleased the Author of Nature to bestow on body: it is, to us, a necessary truth. The proposition does not so much express any thing with regard to body, as it does the operations of our mind when contemplating body. It may perhaps be essential to body to move in some particular direction. It may be essential to body to stop as soon as the moving cause has ceased to act; or it may be essential to body to diminish its motion gradually, and finally come to rest. But this will not invalidate the truth of this proposition. These circumstances in the nature of body, which render those modifications of motion essentially necessary, are the causes of those modifications; and, in our study of nature, they will be considered by us as changing forces, and will be known and called by that name. And if we should ever see a particle of matter in such a situation that it is affected by those essential properties alone, we shall, from observation of its motion, discover what those essential properties are.

This law turns out at last to be little more than a tautological proposition: But mechanical philosophy, as we have defined it, requires no other sense of it; for, even if we should suppose that body, of its own nature, is capable of changing its state, this change must be performed according to some law which characterises the nature of body; and the knowledge of the law can be had in no other way than by observing the deviations from uniform rectilineal motion. It is therefore indifferent whether those changes are derived from the nature of the thing, or from external causes; for in order to consider the various motions of bodies, we must first consider this nature of matter as a mechanical affection of matter, operating in every instance; and thus we are brought back to the law enounced in this proposition. This becomes more certain when we reflect that the external causes (such as gravity or magnetism), which are acknowledged to operate changes of motion, are equally unknown to us with this essential original property of nature, and are, like it, nothing but inferences from the phenomena.

The above very diffuse discussions may appear superfluous to many readers, and even cumbersome; but we trust that the philosophical reader will excuse our anxiety on this head, when he reflects on the complicated, indistinct, and inaccurate notions commonly had of the subject; and more especially when he observes, that of those who maintain the truth of this fundamental proposition, as we have enounced it, many (and they too of the first eminence) reject it in fact, by combining it with other opinions which are inconsistent with it, nay, which contradict it in express terms. We may even include Sir Isaac Newton in the number of those who have at least introduced modes of expression which mislead the minds of incautious persons, and suggest inadequate notions, incompatible with the pure doctrine of the proposition. Although, in words, they disclaim the doctrine that rest is the natural state of body, and that force is necessary for the continuation of its motion, yet in words they (and most of them in thought) likewise abet that doctrine; for they say, that there resides in a moving body a *power* or *force*, by which it perseveres in its motion. They call it the VIS INSITA, the INHERENT FORCE OF A MOVING BODY. This is surely giving up the question; for if the motion is supposed to be continued in consequence of a force, that force is *supposed* to be exerted; and it is supposed, that if it were not exerted, the motion would cease; and therefore the proposition must be false. Indeed it is sometimes expressed so as seemingly to ward off this objection. It is said, that the body continues in uniform rectilineal motion, unless affected by some *external cause*. But this way of speaking obliges us, at first setting out in natural philosophy, to assert that gravity, magnetism, electricity, and a thousand other mechanical powers, are external to the matter which they put in motion. This is quite improper: It is the business of philosophy to discover whether they be external or not; and if we assert that they are, we have no principles of argumentation with those who deny it. It is this one thing that has filled the study of nature with all the jargon of æthers and other invisible intangible fluids, which has disgraced philosophy, and greatly retarded its progress.

32. We must observe, that the terms *vis insita*, inherent *force*, are very improper. There is no dispute among philosophers in calling every thing a force that produces a change of motion, and in inferring the action of such a force whenever we observe a change of motion. It is surely incongruous to give the same name to what has not this quality of producing a change, or to infer (or rather to suppose) the energy of a force when no change of motion is observed. This is one among many instances of the danger of mistake when we indulge in analogical discussions. All our language, at least on this subject, is analogous. I feel, that in order to oppose animal force, I must exert force. But I must exert force in order to oppose a body in motion: Therefore I imagine that the moving body possesses force. A bent spring will drive a body forward by unbending: Therefore I say that the spring exerts force. A moving body impels the body which it hits: Therefore I say, that the impelling body possesses and exerts force. I imagine farther, that it possesses force only by being in motion, or because it is in motion; because I do not find that a quiescent body will put another into motion by touching it. But we shall soon find this to be false in many, if not in all cases, and that the communication of motion depends on the mere vicinity, and not on the motion, of the impelling body; yet we ascribe the exertion of the *vis insita* to the circumstance of the continued motion. We therefore conceive the force as arising from, or as consisting in, the impelling body's being in motion; and, with a very obscure and indistinct conception of the whole matter, we call it *the force by which the body preserves itself in motion*. Thus, taking it for granted that a force resides in the body, and being obliged to give it some office, this is the only one that we can think of.

But philosophers imagine that they perceive the necessity of the exertion of a force in order to the continuation of a motion. Motion (say they) is a continued action; the body is every instant in a new situation; there is the continual production of an effect, therefore the continued action of a cause.

33. But this is a very inaccurate way of thinking. We have a distinct conception of motion; and we conceive that there is such a thing as a moving cause, which we distinguish from all other causes by the name *force*. It produces motion. If it does this, it produces the character of motion, which is a continual change of place. Motion is not action, but the effect of an action; and this action is as complete in the instant immediately succeeding the beginning of the motion as it is a minute after. The subsequent change of place is the continuation of an effect already produced. The *immediate* effect of the moving force is a DETERMINATION, by which, if not hindered, the body would go on for ever from place to place. It is in this determination only that the state or condition of the body can differ from a state of rest; for in any instant, the body does not describe any space, but has a determination by which it will describe a certain space uniformly in a certain time. Motion is a condition, a state, or mode, of existence, and no more requires the continued agency of the moving cause than yellowness or roundness does. It requires some chemical agency to change the yellowness to greenness; and it requires a mechanical cause or a force to change this motion into rest. When we see a moving body stop short in an instant, or be gradually, but quickly, brought to rest, we never fail to speculate about a cause of this cessation or retardation. The case is no way different in itself although the retardation should be extremely slow. We should always attribute it to a cause. It requires a cause to put a body out of motion as much as to put it into motion. This

cause, if not external, must be found in the body itself; and it must have a self-determining power, and may as well be able to put itself into motion as out of it.

If this reasoning be not admitted, we do not see how any effect can be produced by any cause. Every effect supposes something *done;* and any thing done implies that the thing done may remain till it be *undone* by some other cause. Without this, it would have no existence. If a moving cause did not produce continued motion by its instantaneous action, it could not produce it by any continuance of that action; because in no instant of that action does it produce continued motion.

We must therefore give up the opinion, that there resides in a moving body a force by *which it is kept in motion;* and we must find some other way of explaining that remarkable difference between a moving body and a body at rest, by which the first causes other bodies to move by hitting them, while the other does not do this by merely touching them. We shall see, with the clearest evidence, that motion is necessary in the impelling body, in order that it may permit the forces inherent in one or both bodies to continue this pressure long enough for producing a sensible or considerable motion. But these moving forces are inherent in bodies, whether they are in motion or at rest.

34. The foregoing observations shew us the impropriety of the phrase *communication of motion.* By thus reflecting on the notions that are involved in the general conception of one body being made to move by the impulse of another, we perceive, that there is nothing individual transferred from the one body to the other. The determination to motion, indeed, existed only in the impelling body before collision; whereas, afterwards, both bodies are so conditioned or determined. But we can form no notion of the thing transferred. With the same metaphysical

impropriety, we speak of the communication of joy, of fever.

35. Kepler introduced a term INERTIA, VIS INERTIÆ, into mechanical philosophy; and it is now in constant use. But writers are very careless and vague in the notions which they affix to these terms. Kepler and Newton seem generally to employ it for expressing the fact, the perseverance of the body in its present state of motion or rest; but they also frequently express by it something like an indifference to motion or rest, *manifested by its requiring the same quantity of force to make an augmentation of its motion as to make an equal diminution of it.* The popular notion is like that which we have of actual resistance; and it always implies the notion of force *exerted* by the resisting body. We suppose this to be the exertion of the *vis insita*, or the *inherent force* of a body in motion. But we have the same notion of resistance from a body at rest which we set in motion. Now, surely it is in direct contradiction to the common use of the word *force*, when we suppose resistance from a body at rest; yet *vis inertiæ* is a very common expression. Nor is it more absurd (and it is very absurd) to say, that a body maintains its state of rest by the exertion of a *vis inertiæ*, than to say, that it maintains its state of motion by the exertion of an *inherent force.* We should avoid all such metaphorical expressions as *resistance*, *indifference*, *sluggishness*, or *proneness to rest* (which some express by *inertia*), because they seldom fail to make us indulge in metaphorical notions and thus lead us to misconceive the *modus operandi*, or procedure of nature.

There is *no resistance whatever observed* in these phenomena; for the force employed always produces its *complete* effect. When I throw down a man, and find that I have employed no more force than was sufficient to throw down a similar and equal mass of dead matter, I know by this that he *has not* resisted; but I conclude that he *has* resisted, if I have been obliged to employ much more force. There is therefore no resistance, properly so called, when the exerted force is observed to produce its full effect. To say that there *is* resistance, is therefore a real misconception of the way in which mechanical forces have operated in the collision of bodies. There is no more resistance in these cases than in any other natural changes of condition. We are guilty, however, of the same impropriety of language in other cases, where the cause of it is more evident. We say that colours in grain *resist* the action of soap and of the sun, but that the Prussian blue does not. We all perceive, that in this expression the word *resistance* is entirely figurative; and we should say that Prussian blue *resists* soap, if we are right in saying that a body resists any force employed to change its state of motion; for soap must be employed to discharge or change the colour; and *it does change it.* Force must be employed to change a motion; *and it does change it.* The impropriety, both of thought and language, is *plain* in the one case, and it is no less *real* in the other. Both of the terms, *inherent force* and *inertia*, may be used with safety for abbreviating language, if we be careful to employ them only for expressing, *either the simple fact of persevering in the former state*, or *the necessity of employing a certain determinate force, in order to change that state, and if we avoid all thought of resistance.*

36. From the whole of this discussion we learn, that the deviations from uniform motions are the indications of the existence and agency of mechanical forces, and that they are the only indications. The indication is very simple, mere change of place; it can therefore indicate nothing but what is very simple, the something competent to the production of the very motion that we observe. And when two changes of motion are precisely similar, they indicate the same thing. Suppose a mariner's compass on the table, and that by a small tap with my

finger I cause the needle to turn off from its quiescent position 10 degrees. I can do the same thing by bringing a magnet near it; or by bringing an electrified body near it; or by the unbending of a fine spring pressing it aside; or by a puff of wind; or by several other methods. In all these cases, the indication is the same; therefore the thing indicated is the same, namely, a certain intensity and direction of a moving power. How it operates, or in what manner it exists and exerts itself in these instances, outwardly so different, is not under consideration at present. Impulsiveness, intensity, and direction, are all the circumstances of resemblance by which the affections of matter are to be characterised; and it is to the discovery and determination of these alone that our attention is now to be directed. We are directed in this research by the

SECOND LAW OF MOTION.

Every change of motion is proportional to the force impressed, and is made in the direction of that force.

37. This law also may almost be considered as an identical proposition; for it is equivalent to saying, that the changing force is to be measured by the change which it produces, and that the direction of this force is the direction of the change. Of this there can be no doubt, when we consider the force in no other sense than that of the cause of motion, paying no attention to the form or manner of its exertion. Thus, when a pellet of tow is shot from a pop-gun by the expansion of the air compressed by the rammer, or where it is shot from a toy pistol by the unbending of the coiled wire, or when it is nicked away by the thumb like a marble—if, in all these cases, it moves off in the same direction, and with the same velocity, we cannot consider or think of the force, or at least of its exertion, as any how different. Nay, when it is driven forward by the instantaneous percussion of a smart

stroke, although the manner of producing this effect (if possible) is essentially different from what is conceived in the other cases, we must still think that the propelling force, considered as a propelling force, is one and the same. In short, this law of motion, as thus expressed by Sir Isaac Newton, is equivalent to saying, "That we take the changes of motion as the measures of the changing forces, and the direction of the change for the indication of the direction of the forces:" For no reflecting person can pretend to say, that it is a deduction from the acknowledged principle, that effects are proportional to their causes. We do not affirm this law, from having observed the proportion of the forces and the proportion of the changes, and that these proportions are the same; and from having observed that this has obtained through the whole extent of our study of nature. This would indeed establish it as a physical law, an universal fact; and it is in fact so established. But this does not establish it as a law of motion, according to our definition of that term; as a law of human thought, the result of the relations of our ideas, as an intuitive truth. The injudicious attempts of philosophers to prove it as a matter of observation, have occasioned the only dispute that has arisen in mechanical philosophy. It is well known, that a bullet, moving with double velocity, penetrates four times as far. Many other similar facts corroborate this: and the philosophers observe, that four times the force has been expended to generate this double velocity in the bullet; it requires four times as much powder. In all the examples of this kind, it would seem that the ratio of the forces employed has been very accurately ascertained; yet this is the invariable result. Philosophers, therefore, have concluded, that moving forces are not proportional to the velocities which they produce, but to the squares of these velocities. It is a strong confirmation, to see that the bodies in motion seem to possess forces in this very proportion, and produce effects in this proportion; penetrating four times as deep when the velocity is only twice as great, &c.

But if this be a just estimation, we cannot reconcile it to the concession of the same philosophers, who grant that the velocity is proportional to the force impressed, in the cases where we have no previous observation of the ratio of the forces, and of its equality to the ratio of the velocities. This is the case with gravity, which these philosophers always measure by its accelerating power, or the velocity which it generates in a given time. And this cannot be refused by them; for cases occur, where the force can be measured, in the most natural manner, by the actual pressure which it exerts. Gravity is thus measured by the pressure which a stone exerts on its supports. A weight which at Quito will pull out the rod of a spring steelyard to the mark 312, will pull it to 313 at Spitzbergen. And it is a fact, that a body will fall 313 inches at Spitzbergen in the same time that it falls 312 at Quito. Gravitation is the cause both of the pressure and the fall; and it is a matter of unexcepted observation, that they have always the same ratio. The philosophers who have so strenuously maintained the other measure of forces, are among the most eminent of those who have examined the motions produced by gravity, magnetism, electricity, &c.; and they never think of measuring those forces any other way than by the velocity. It is in this way that the whole of the celestial phenomena are explained in perfect uniformity with observation, and that the Newtonian philosophy is considered as a demonstrative science.

There must, therefore, be some defect in the principle on which the other measurement of forces is built, or in the method of applying it. Pressure is undoubtedly the immediate and natural measure of force; yet we know that four springs, or a bow four times as strong, give only a double velocity to an arrow.

The truth of our law rests on this only, that we assume

the changes of motion as the measure of the changing forces; or, at least, as the measures of their exertions in producing motion. In fact, they are the measures only of a certain circumstance, in which the actions of very different natural powers may resemble each other; namely, the competency to produce motion. They do not, perhaps, measure their competency to produce heat, or even to bend springs. We can surely consider this apart from all other circumstances; and it is worthy of separate consideration. Let us see what can be, and what ought to be, deduced from this way of treating the subject.

38. The motion of a body may certainly remain unchanged. If the direction and velocity remain the same, we perceive no circumstance in which its condition, with respect to motion, differs. Its change of place or situation can make no difference; for this is implied in the very circumstance of the bodies being in motion.

But if either the velocity or direction change, then surely is its mechanical condition no longer the same; a force has acted on it, either intrinsic or from without, either accelerating, or retarding, or deflecting it. Supposing the direction to remain the same, its difference of condition can consist in nothing but its difference of velocity. This is the only circumstance in which its condition can differ, as it passes through two different points of its rectilineal path. It is this determination by which the body will describe a certain determinate space uniformly in a given time, which defines its condition as a moving body: the changes of this determination are the measures of their own causes;—and to those causes we have given the name *force*. Those causes may reside in other bodies, which may have other properties, characterised and measured by other effects. Pressure may be one of those properties, and may have its own measures; these may, or may not, have the same proportion with that property which is the cause of a change of velocity: and therefore changes of velocity may not be a measure of pressure. This is a question of fact, and requires observation and experience; but, in the mean time, velocity, and the change of velocity, is the measure of moving force and of changing force. When therefore the change of velocity is the same, whatever the previous velocity may be, the changing force must be considered as the same: therefore, finally, if the previous velocity is nothing, and consequently the change on that body is the very velocity or motion that it acquires, we must say that the force which produces a certain change in the velocity of a moving body, is the same with the force which would impart to a body at rest a velocity equal to this change or difference of velocity produced on the body already in motion.

This manner of estimating force is in perfect conformity to our most familiar notions on these subjects. We conceive the weight or downward pressure of a body as the cause of its motion downwards; and we conceive it as belonging to the body at all times, and in all places, whether falling, or rising upwards, or describing a parabola, or lying on a table; and, accordingly, we observe, that in every state of motion it receives equal changes of velocity in the same, or an equal time, and all in the direction of its pressure.

All that we have now said of a change of velocity might be repeated of a change of direction. It is surely possible that the same change of direction may be made on any two motions. Let one of the motions be considered as growing continually slower, and terminating in rest. In every instant of this motion it is possible to make one and the same change on it. The same change may therefore be made at the very instant that the motion is at an end. In this case, the change is the very motion which the body acquires from the changing force. Therefore, in this case also, we must say, that a change of motion is itself a motion, and that it is the motion which the force would produce in a body that was previously at rest.

The result of these observations is evidently this, that we must ascertain, in every instance, what is the change of motion, and mark it by characters that are conspicuous and distinguishing; and this mark and measure of change must be a motion: Then we must say, that the changing force is that which would produce this motion in a body previously at rest. We must see how this is manifest, as a change of motion is the difference between the former motion and the new motion; and, on the other hand, we must see how the motion produceable in a quiescent body may be so combined with a motion already existing, as to exhibit a new motion, in which the agency of the changing force may appear.

Suppose a ship at anchor in a stream; while one man walks forward on the quarter deck at the rate of two miles per hour, another walks from stem to stern at the same rate, a third walks athwart ship, and a fourth stands still. Let the ship be supposed to cut or part her cable, and float down the stream at the rate of three miles per hour. We cannot conceive any difference in the change made on each man's motion in absolute space; but their motions are now exceedingly different from what they were: the first man, whom we may suppose to have been walking westward, is now moving eastward one mile per hour; the second is moving eastward four miles per hour; and the third is moving in an oblique direction, about three points north or south of due east. All have suffered the same change of condition with the man who had been standing still. He has now got a motion eastward three miles per hour. In this instance, we see very well the circumstance of sameness that obtains in the change of these four conditions. It is the motion of the ship, which is blended with the other motions. But this circumstance is equally present whenever the same previous motions are changed into the same new motions. We must learn to expiscate this; which we shall do, by considering the manner in which the motion of the ship is blended with each of the men's motions.

42. This kind of combination has been called the COMPOSITION OF MOTION; because, in every point of the motion really pursued, the two motions are to be found.

The fundamental theorem on this subject is this:—Two uniform motions in the sides of a parallelogram compose an uniform motion in the diagonal.

Suppose that a point A (Plate I. fig. 1.) describes AB uniformly in some given time, while the line AB is carried uniformly along AC in the same time, keeping always parallel to its first position AB. The point A, by the combination of these motions, will describe AD, the diagonal of the parallelogram ABDC, uniformly in the same time.

For it is plain, that the velocities in AB and AC are proportional to AB and AC, because they are uniformly described in the same time. When the point has got to E, the middle of AB, the line AB has got into the situation GH, half way between AB and CD, and the point E is in the place *e*, the middle of GH. Draw E *e* L parallel to AC. It is plain, that the parallelograms ABDC and AE *e* G are similar; because AE and AG are the halves of AB and AC, and the angle at A is common to both. Therefore, (Euclid, Book VI. Prop. xxvi.) they are about the same diagonal, and the point *e* is in the diagonal of AD. In like manner, it may be shewn, that when A has described AF, $\frac{3}{4}$ths of AB, the line AB will be in the situation IK, so that AI is $\frac{3}{4}$ths of AC, and the point *f*, in which A is now found, is in the diagonal AD. It will be the same in whatever point of AB the describing point A be supposed to be found. The line AB will be on a similar point of AC, and the describing point will be in the diagonal AD.

Moreover, the motion in AD is uniform; for A *e* is described in the time of describing AE; that is, in half the time of describing AB, or in half the time of describing AD. In like manner, A *f* is described in $\frac{3}{4}$ths of the time of describing AD, &c. &c.

Lastly, the velocity in the diagonal AD is to the velocity in either of the sides as AD is to that side. This is evident, because they are uniformly described in the same time.

This is justly called *a composition of the motions* AB *and* AC, as will appear by considering it in the following manner: Let the lines AB AC be conceived as two material lines like wires. Let AB move uniformly from the situation AB into the situation CD, while AC moves uniformly into the situation BD. It is plain, that their intersection will always be found on AD. The point *e*, for example, is a point common to both lines. Considered as a point of EL, it is then moving in the direction *e* H or AB; and, considered as a point of GH, it is moving in the direction *e* L. Both of these motions are therefore blended in the motion of the intersection along AD. We can conceive a small ring at *e*, embracing loosely both of the wires. This material ring will move in the diagonal, and *will really partake* of both motions.

Thus we see how the motion of the ship is actually blended with the motions of the three men; and the circumstance of sameness which is to be found in the four changes of motion is this motion of the ship, or of the man who was standing still. By composition with each of the three former motions, it produces each of the three new motions. Now, when each of two primitive motions is the same, and each of the new motions is the same, the change is surely the same. If one of the changes has been brought about by the actual composition of motions, we know precisely what that change is; and this informs us what the other is, in whatever way it was produced. Hence we infer, that

43. *When a motion is any how changed, the change is that motion which, when compounded with the former motion, will produce the new motion.* Now, because we assume the change as the measure and characteristic of the changing force, we must do so in the present instance; and we must say,

44. That *the changing force is that which will produce in a quiescent body the motion which, by composition with the former motion of a body, will produce the new motion.*

And, on the other hand,

When the motion of a body is changed by the action of any force, the new motion is that which is compounded of the former motion, and of the motion which the force would produce in a quiescent body.

When a force changes the direction of a motion, we see that its direction is transverse in some angle BAC; because a diagonal AD always supposes two sides. As we have distinguished any change of direction by the term DEFLECTION, we may call the transverse force a DEFLECTING FORCE.

In this way of estimating a change of motion, all the characters of both motions are preserved, and it expresses every circumstance of the change; the mere change of direction, or the angle BAD, is not enough, because the same force will make different angles of deflection, according to the velocity of the former motion, or according to its direction: but in this estimation, the full effect of the deflecting force is seen; it is seen *as a motion;* for when half of the time is elapsed, the body is at *e* instead of E; when three-fourths are elapsed, it is at *f* instead of F; and at the end of the time it is at D instead of B. In short, the body has moved uniformly away from the points at which it would have arrived independent of the change; and this motion has been in the same direction, and at the same rate, as if it had moved from A to C by the changing force alone. Each force has produced its full effect; for when the body is at D, it is as far from AC as if the force AC had not acted on it; and it is as far from AB as it would have been by the action of AC alone.

For all these reasons, therefore, it is evident, that if we are to abide by our measure and character of force as a mere producer of motion, we have selected the proper characteristic and measure of a changing force; and our descriptions, in conformity to this selection, must be agreeable to the phenomena of nature, and retain the accuracy of geometrical procedure; because, on the other hand, the results which we deduce from the supposed influence of those forces are formed in the same mould. It is not even requisite that the real exertions of the natural forces, such as pressure of various kinds, &c. shall follow these rules; for their deviations will be considered as new forces, although they are only indications of the differences of the real forces from our hypothesis. We have obtained the precious advantage of mathematical investigation, by which we can examine the law of exertion which characterises every force in nature.

On these principles we establish the following fundamental elementary proposition, of continual and indispensable use in all mechanical inquiries.

45. *If a body or material particle be subjected at the same time to the action of two moving forces, each of which would separately cause it to describe the side of a parallelogram uniformly in a given time, the body will describe the diagonal uniformly in the same time.*

For the body, whose motion AB was changed into AD, had gotten its motion by the action of some force. It was moving along NAB; and, when it reached the point A, the force AC acted on it. The primitive motion is the same, or the body is in the same condition in every instant of the primitive motion. It may have acquired this motion when it was in N, or when at O, or any other point of NA. In all these cases, if AC act on it when it is in A, it will always describe AD; therefore it will describe AD when it acquires the primitive motion also in A; that is, if the two forces act on it at one and the same instant. The demonstration may be neatly expressed thus: The change induced by each force on the motion produced by the other, is the motion which it would produce in the body if previously at rest. Therefore the motion resulting from their joint action is the motion which is compounded of these two motions; or it is a motion in the diagonal of the parallelogram, of which these motions are the sides.

This is called the COMPOSITION OF FORCES. The forces which produce the motions along the sides of the parallelogram are called the SIMPLE FORCES, or the CONSTITUENT FORCES; and the force which would alone produce the motion along the diagonal is called the COMPOUND FORCE, the RESULTING FORCE, the EQUIVALENT FORCE.

46. On the other hand, the force which produces a motion along any line whatever, may be conceived as resulting from the combined action of two or more forces. We may *know* or *observe* it to be so; as when we see a lighter dragged along a canal by two horses, one on each side: Each pulls the boat directly toward himself in the direction of the track-rope; the boat cannot go both ways; and its real motion, whatever it is, results from this combined action. This might be produced by a single force; for example, if the lighter be dragged along the canal by a rope from another lighter which precedes it, being dragged by one horse, aided by the helm of the foremost lighter. Here the real force is not the resulting, or the compound, but the equivalent force.

This view of a motion, mechanically produced, is called the RESOLUTION OF FORCES. The force in the diagonal is said to be *resolved* into the two forces, having the directions and velocities represented by the sides. This practice is of the most extensive and multifarious use in all mechanical disquisitions. It may frequently be exceedingly difficult to manage the complication of the many real forces which concur in producing a phenomenon; and by substituting others, whose combined effects are equivalent, our investigation may be much expedited. But more of this afterwards.

We must carefully remember, that when the motion AD is once begun, all composition is at an end, and the motion is a simple motion. The two determinations, by one of which the body would describe AB, and by the other of which it would describe AC, no longer *co-exist* in the body. This was the case only *in the instant*, in the very act of changing the motion AB into the motion BD; yet is the motion AD equivalent to a motion which is produced by the *actual composition* of two motions AB and AC; in which case the two motions co-exist in every point of AD.

47. Accordingly, this is the way in which the composition of forces is usually illustrated, and thought to be demonstrated. A man is supposed (for instance) to walk uniformly from A to C on a sheet of ice, while the ice is carried uniformly along AB by the stream. The man's real motion is undoubtedly along AD; but this is by no means a demonstration that the instantaneous or short-lived action of two forces would produce that motion; the man must continue to exert force in order to walk, and the ice is dragged along by the stream. Some indeed express this proof in another way, saying, let a body describe AB, while the space in which this motion is performed is carried along AC. The ice may be carried along, and may, by friction, or otherwise, drag the man along with it; but a space cannot be removed from one place to another, nor, if it could, would it take the man

with it. Should a ship start suddenly forward while a man is walking across the deck, he would be left behind, and fall toward the stern. We must *suppose* a transverse force, and we must *suppose* the composition of this force without proof. This is no demonstration.

We apprehend, that the demonstration given above of this fundamental proposition is unexceptionable, when the terms *force* and *deflection* are used in the abstract sense which we have affixed to them; and we hope, by these means, to maintain the rigour of mathematical discussion in all our future disquisitions on these subjects. The only circumstance in it which can be the subject of discussion is, whether we have selected the proper measure and characteristic of a change of motion.—We never met with any objection to it.

48. But some have still maintained, that it does not evidently appear, from these principles, that the motion which results from the joint action of two natural powers, whose known and measurable intensities have the same proportions with AB and BC, and which also exert themselves in those directions, will produce a motion, having the direction and proportion of AD. They will not, if the velocities produced by these forces are not in the proportion of those intensities, but in the subduplicate ratio of them. Nay, they say, that it is not so. If a body be impelled along AC by one spring, and along AB by two springs equally strong, it will not describe the diagonal of a parallelogram, of which the side AB is double the side AC. Nay, they add, that an indefinite number of examples can be given where a body *does not* describe the diagonal of the parallelogram by the joint action of two forces, which, separately, would cause it to describe the sides. And, lastly, they say, that, at any rate, it does not appear evident to the mind, that two *incitements* to motion, having the directions and the same proportion of intensity with that of the sides of a parallelogram, ac-

tually generate a third, which is the immediate cause of the motion in the diagonal. An equivalent force is not the same with a resulting force.

49. Yet we see numberless cases of the composition of incitements to motion, and they seem as determinate, and as susceptible of being combined by composition, as the things called moving forces, which are measured by the velocities: we see them actually so combined in a thousand instances, as in the example already given of a lighter dragged by two horses pulling in different directions. Nay, experiment shews, that this composition follows precisely the same rule as the composition of the forces which are measured by the velocities; for if the point A (fig. 1.) be pulled by a thread, or pressed by a spring, in the direction AB, and by another in the direction AC, and if the pressures are proportional to AB and AC, then it will be withheld from moving, if it be pulled or pressed by a third force, acting in the direction A*d*, opposite to AD, the pressure being also proportional to AD. This force, acting in the direction A*d*, would certainly withstand an equal force acting in the direction AD; therefore we must conclude, that the two pressures AB and AC really generate a force AD. This uniform agreement shews that the composition is deducible from fixed principles; but it does not appear that it can be held as demonstrated by the arguments employed in the case of motions. A demonstration of the composition of pressures is still wanted, in order to render mechanics a demonstrative science.

Accordingly, philosophers of the first eminence have turned their attention to this problem. It is by no means easy; being so nearly allied to first principles, that it must be difficult to find axioms of greater simplicity by which it may be proved.

Mechanicians generally contented themselves with the solution given by Aristotle; but this is merely a composition of motions: indeed he does not give it for any thing else, and calls it "συνθεσις των φυγαν." The first writer who appears to have considered it as different from the mere composition of motions, was the celebrated Dutch engineer Stevinus in his work *on Sluices;* but his solution is obscure. It was sufficient, however, to convince Daniel Bernoulli of the necessity and the difficulty of the problem. He has given the first complete demonstration of it in the first volume of the Commentaries of the Imperial Academy of Sciences at St Petersburgh. It is extremely ingenious; but it is tedious and intricate, requiring a series of fifteen propositions to demonstrate that two pressures, having the directions and magnitudes of the sides of any parallelogram, compose a third, which has the direction and magnitude of its diagonal. His first proposition is, that *two equal pressures, acting at right angles, compose a third, in the direction of the diagonal of a square, and having to either of the other two the proportion of the diagonal of a square to its sides.*

Mr D'Alembert has greatly simplified and improved this demonstration, by beginning with a case that is self-evident; namely, *If three equal forces are inclined to each other in equal angles of* 120 *degrees, any one of them will balance the combined action of the other two.* Surely; for neither of them can prevail. Therefore *two* equal forces, inclined in an angle of 120 degrees, produce a third, which has the direction and proportion of the diagonal of the rhombus; for this is equal and opposite to one of the three above mentioned. He then demonstrates the same thing of two equal forces inclined in *any* angle; and by a series of eight propositions more, demonstrates the general theorem. This dissertation is in the Memoirs of the Academy at Paris for 1769. He improves it still farther in a subsequent memoir.

Mr Riccati and Mr Fonsenex, in the Commentaries of the Academy of Turin, have given analytical demonstrations, which are also very ingenious and concise, but require acquaintance with the higher mathematics.—There is another very ingenious demonstration in the *Journal des Sçavans* for June 1764, but too obscure for an elementary proposition. It is somewhat simplified by Belidor in his *Ingenieur François*. Frisius, in his *Cosmographia*, has given one, which is perhaps the best of all those that are easily comprehended without an acquaintance with the higher mathematics: but we imagine that, although no one can doubt of the conclusion, it has not that intuitive evidence for every step of the process that seems necessary.

50. We here offer another, composed by blending together the methods of Bernoulli and D'Alembert; and we imagine that no objection can be made to any step of it. We limit it entirely to pressures, and do not at all consider nor employ the motions which they may be supposed to produce.

(A) If two equal and opposite pressures or incitements to motion act at once on a material particle, it suffers no change of motion; for if it yields in either direction by their joint action, one of the pressures prevails, and they are not equal.

Equal and opposite pressures are said TO BALANCE each other; and such as balance must be esteemed equal and opposite.

(B) If a and b are two magnitudes of the same kind, proportional to the intensities of two pressures which act in the same direction, then the magnitude $a + b$ will measure the intensity of the pressure, which is equivalent, and may be called equal, to the combined effort of the other two; for when we try to form a notion of pressure as a measurable magnitude, distinct from motion or any other effect of it, we find nothing that we can measure it by but another pressure. Nor have we any notion of a double or triple pressure different from a pressure that is equivalent to the joint effort of two or three equal pressures. A pressure a is accounted triple of a pressure b, if it balances three pressures, each equal to b, acting together. Therefore, in all proportions which can be expressed by numbers, we must acknowledge the legitimacy of this measurement; and it would surely be affectation to omit those which the mathematicians call *incommensurable*.

In like manner, the magnitude $a - b$ must be acknowledged to measure that pressure which arises from the joint action of two pressures a and b acting in opposite directions, of which a is the greatest.

(C) Let ABCD and Ab Cd (fig. A bottom of Plate I.) be two rhombuses, which have the common diagonal AC. Let the angles BAb, DAd, be bisected by the straight lines AE and AF.

If there be drawn from the points E and F the lines EG, EH, Fg, Fh, making equal angles on each side of EA and FA, and if Gg, Hh be drawn, cutting the diagonal AC in I and L: then AI + AL will be greater or less than AQ, the half of AC, according as the angles GEH, gFh, are greater or less than GAH, gAh.

Draw GH, gh, cutting AE AF, in O and o, and draw Oo, cutting AC in K.

Because the angles AEG and EAG are respectively equal to AEH and EAH, and AE is common to both triangles, the sides AG, GE are respectively equal to AH, HE, and GH is perpendicular to AE, and is bisected in O; for the same reason, gh is bisected in o. Therefore the lines Gg, Oo, Hh, are parallel, and IL is bisected in K. Therefore AI + AL is equal to twice AK. Moreover, if the angle GEH be greater than GAH, AO is greater than EO, and AK is greater than KQ. Therefore AI + AL is greater than AQ; and if the angle GEH be less than GAH, AI + AL is less than AQ.

(D) Two equal pressures, acting in the directions AB and AC (fig. 2.), at right angles to each other, compose a pressure in the direction AD, which bisects the right angle; and its intensity is to the intensity of each of the constituent pressures as the diagonal of a square is to one of the sides. It is evident, that the direction of the pressure, generated by their joint action, will bisect the angle formed by their directions; because no reason can be assigned for the direction inclining more to one side than to the other.

In the next place, since a force in the direction AD does, in fact, arise from the joint action of the equal pressures AB and AC, the pressure AB may be conceived as arising from the joint action of two equal forces similarly inclined and proportioned to it. Draw EAF perpendicular to AD. One of these forces must be directed along AD, and the other along AE. In like manner, the pressure AC may arise from the joint action of a pressure in the direction AD, and an equal pressure in the direction AF. It is also plain, that the pressures in the directions AE and AF, and the two pressures in the direction AD, must be all equal. And also, any one of them must have the same proportion to AB or to AC, that AB or AC has to the force in the direction AD, arising from their joint action.

Therefore, if it be said that AD does not measure the pressure arising from the joint action of AB and AC, let A *d*, greater than AD, be its just measure, and make A *d* : AB = AB : A *g*   AB : A *e*. Then A *g* and A *e* have the same inclination and proportion to AB that AB and AC have to A *d*. We determine, in like manner, two forces A *f* and A *g* as constituents of AC.

Now A *d* is equivalent to AB and AC, and AB is equivalent to A *e* and A *g*; and AC is equivalent to A *f* and A *g*. Therefore A *d* is equivalent to A *e*, A *f*, A *g* and A *g*. But (A) A *e* and A *f* balance each other, or annihilate each other's effect; and there remain only the two forces or pressures A *g*, A *g*. Therefore (B) their measure is a magnitude equal to twice A g. But if A d be greater than the diagonal AD of the square, whose sides are AB and AC; then A g must be less than AI, the side of the square whose diagonal is AB. But twice A g is less than AD, and much less than A d. Therefore the measure of the equivalent of AB and AC cannot be a line A d greater than AD. In like manner, it cannot be a line A δ that is less than AD. Therefore it must be equal to AD, and the proposition is demonstrated.

(E) *Cor.* Two equal forces AB, AC, acting at right angles, will be balanced by a force AO, equal and opposite to AD, the diagonal of the square whose sides are AB and AC; for AO would balance AD, which is the equivalent of AB and AC.

(F) Let AECF (fig. 3.) be a rhombus, the acute angle of which EAF is half of a right angle. Two equal pressures, which have the directions and measures AE, AF, compose a pressure, having the direction and measure AC, which is the diagonal of the rhombus.

It is evident, in the first place, that the compound force has the direction AC, which bisects the angle EAF. If AC be not its just measure, let it be AP less than AC. Let ABCD be a square described on the same diagonal, and make AP : AQ = AE : AO = AF : A o. Draw KOG, K o g perpendicular to AE, AF; draw GI g, OH o, EG, EK, F g, FK, PF, and PE.

The angles CAB and FAE are equal, each being half of a right angle. Also the figures AEPF and AGEK are similar, because AP : AQ = AE : AO. Therefore FA : AP = KA : AE, and EA : AP = GA : AE. Therefore, in the same manner that the forces AE, AF are affirmed to compose AP, the forces AG and AK may compose the force AE, and the forces A g and AK may compose the force AF. Therefore (B) the force AP is equivalent to the four forces AG, AK, A g, AK. But (D) AG and A g are the sides of a square, whose diagonal is equal to twice AI: and the two forces AK, AK are equal to, or are measured by, twice AK. Therefore the four forces AG, AK, Ag, AK, are equivalent to 2 AI + 2 AK, = 4 AH.

But because AP was supposed less than AC, the angle FPE is greater than FAE, and GEK is greater than GAK, AO is greater than OE, and AH is greater than HQ, and 2 AH is greater than AQ; and therefore 4 AH is greater than AC, and much greater than AP. Therefore AP is not the just measure of the force composed of AE and AF.

In like manner, it is shewn, that AE and AF do not compose a force whose measure is greater than AC. It is therefore equal to AC; and the proposition is demonstrated.

(G) By the same process it may be demonstrated, that if BAD be half a right angle, and EAF be the fourth of a right angle, two forces AE, AF will compose a force measured by AC. And the process may be repeated for a rhombus whose acute angle is $\frac{1}{8}$th, $\frac{1}{16}$th, &c. of a right angle; that is, any portion of a right angle that is produced by continual bisection. Two forces, forming the sides of such a rhombus, compose a force measured by the diagonal.

(H) Let ABCD, Abcd (fig. 4.) be two rhombuses formed by two consecutive bisections of a right angle. Let AECF be another rhombus, whose sides AE and AF bisect the angles BAb and DAd.

The two forces AE, AF, compose a force AC.

Bisect AE and AF in O and o. Draw the perpendiculars GOH, $g\,o\,h$, and the lines GIg, OKo, HLh, and the lines EG, EH, Fg, Fh.

It is evident, that AGEH and AgFh are rhombuses; because AO = OE, and Ao = oF. It is also plain, that since bAd is half of BAD, the angle GAH is half of bAd. It is therefore formed by a continual bisection of a right angle. Therefore (G) the forces AG, AH, compose a force AE; and A *g*, A *h*, compose the force AF. Therefore the forces AG, AH, A *g*, A *h*, acting together, are equivalent to the forces AE, AF acting together. But AG, A *g* compose a force = 2 AI; and the forces AH, A *h* compose a force = 2 AL. Therefore the four forces acting together are equivalent to 2 AI + 2 AL, or to 4 AK. But because AO is $\frac{1}{2}$ AE and the lines G *g*, O *o*, H *h*, are evidently parallel, 4 AK is equal to 2 AQ, or to AC; and the proposition is demonstrated.

(I) *Cor.* Let us now suppose, that by continual bisection of a right angle we have obtained a very small angle *a* of a rhombus; and let us name the rhombus by the multiple of *a* which forms its acute angle.

The proposition (G) is true of *a*, 2 *a*, 4 *a*, &c. The proposition (H) is true of 3 *a*. In like manner, because (G) is true of 4 *a* and 8 *a*, proposition (H) is true of 6 *a*; and because it is true of 4 *a*, 6 *a*, and 8 *a*, it is true of 5 *a* and 7 *a*. And so on continually till we have demonstrated it of every multiple of *a* that is less than a right angle.

(K) Let RAS (fig. 5.) be perpendicular to AC, and ABCD be a rhombus, whose acute angle BAD is some multiple of 2 *a* that is less than a right angle. Let A *b* C *d* be another rhombus, whose sides A *b*, A *d* bisect the angles RAB, SAD. Then the forces A *b*, A *d* compose a force AC.

Draw *b* R, *d* S parallel to BA, DA. It is evident, that AR *b* B and AS *d* D are rhombuses, whose acute angles are multiples of *a*, that are each less than a right angle. Therefore (I) the forces AR and AB compose the force A *b*, and AS, AD compose A *d*; but AR and AS annihilate each other's effect, and there remains only the forces AB, AD. Therefore A *b* and A *d* are equivalent to AB and AD, which compose the force AC; and the proposition is demonstrated.

(L) *Cor.* Thus is the corollary of last proposition extended to every rhombus, whose angle at A is some multiple of *a* less than two right angles. And since *a* may be taken less than any angle that can be named, the proposition may be considered as demonstrated of every rhombus: and we may say,

(M) *Two equal forces, inclined to each other in any angle, compose a force which is measured by the diagonal of the rhombus, whose sides are the measures of the constituent forces.*

(N) Two forces AB, AC (fig. 6.), having the direction and proportion of the sides of a rectangle, compose a force AD, having the direction and proportion of the diagonal.

Draw the other diagonal CB, and draw EAF parallel to it; draw BE, CF parallel to DA.

AEBG is a rhombus; and therefore the forces AE and AG compose the force AB. AFCG is also a rhombus, and the force AC is equivalent to AF and AG. Therefore the forces AB and AC, acting together, are equivalent to the forces AE, AF, AG, and AG acting together, or to AE, AF, and AD acting together: But AE and AF annihilate each other's action, being opposite and equal (for each is equal to the half of BC). Therefore AB and AC acting together, are equivalent to AD, or compose the force AD.

(O) Two forces, which have the direction and proportions of AB, AC (fig. 7.) the sides of any parallelogram, compose a force, having the direction and proportion of the diagonal AD

Draw AF perpendicular to BD, and BG and DE perpendicular to AC.

Then AFBG is a rectangle, as is also AFDE; and AG is equal to CE. Therefore (N) AB is equivalent to AF and AG. Therefore AB and AC acting together, are equivalent to AF, AG, and AC acting together;

that is, to AF and AE acting together; that is (N) to AD; or the forces AB and AC compose the force AD.

Hence arises the most general proposition,

51. *If a material particle be urged at once by two pressures or incitements to motion, whose intensities are proportioned to the sides of any parallelogram, and which act in the directions of those sides, it is affected in the same manner as if it were acted on by a single force, whose intensity is measured by the diagonal of the parallelogram, and which acts in its direction: Or, two pressures, having the direction and proportion of the sides of a parallelogram, generate a pressure, having the direction and proportion of the diagonal.*

Thus have we endeavoured to demonstrate from abstract principles the perfect similarity of the composition of pressures, and the composition of forces measured by the motions which they produce. We cannot help being of the opinion, that a separate demonstration is indispensably necessary. What may be fairly deduced from the one case, cannot always be applied to the other. No composition of pressures can explain the change produced by a deflecting force on a motion already existing; for the changing pressure is the only one that exists, and there is none to be compounded with it. And, on the other hand, our notions and observations of the composition of motions will not explain the composition of pressures, unless we take it for granted that the pressures are proportional to the velocities; but this is perhaps a gratuitous assumption. At any rate, it is not an intuitive proposition; and we have mentioned some facts where it seems that they do not follow the same proportion. The pressure of four equal springs produces only a double velocity. It would appear, therefore, that there are circumstances which oblige us to say, that the *exertion* of pressure, as a cause of motion, is not (always at least) proportional to the real measurable pressure. We are therefore anxious to discover in what the difference consists; and in the mean time must allow, that the pressure exerted on a body at rest is different from its exertion in producing motion. We cannot indeed state any immediate comparison between pressure and motion, nor have we any clear conception of the connection between them. It is only by our sensations of touch that we have any notion of pressure, and it is experience that teaches us that it always accompanies every cause of motion. We can, however, observe the proportions of pressures, and compare them with the proportions of motion. We very often observe them different; and therefore it was indispensably necessary to investigate the laws of combined pressure as we did the laws of combined motion in consequence of pressure. Yet we should err, if we hastily asserted that pressures are not proportional to the motions which they produce; all that we are entitled to call in doubt is, whether the pressures in their exertion, while they actually produce motion, or changes of motion, continue to be the same as when they do not produce motion, being withstood or balanced by opposite pressures. Considered as causes of motion, we ought to think that they do not vary while they produce motion, and that the actual pressure, while it produces a double motion, is really double, although it may be quadruple when the body exerting it is made to act on a body that it cannot move. We are confirmed in this opinion by observing, that other facts shew us, that even while producing motion, the pressure which we call quadruple, because we have measured it by four equal pressures balancing it, is really quadruple, considered as the cause of motion, and produces a quadruple motion. A bow which requires four times the force to draw it to any given extent, will communicate the same velocity to a bundle of four arrows that a bow four times easier drawn communicates to one arrow, and will therefore produce a quadruple motion. Yet it will only produce a double velocity in the arrow that acquired a simple velocity from a bow having one-fourth of the strength.

These discrepancies should excite the endeavours of mechanicians to investigate the laws observed in the action of pressures in producing motion. Had this been done with care and with candour, we should not have had the great difference of opinion, which still divides philosophers, about the measures of moving forces. But a spirit of party, which had arisen from other causes, gave importance to what was at first only a difference of expression, and made the partisans of Mr Leibnitz avail themselves of the figurative language which has dóne so much harm in all the departments of philosophy. Notwithstanding all our caution, it is hardly possible to avoid metaphorical conceptions when we employ the language of metaphor. The abettors of the Leibnitzian measure of moving forces, or perhaps, to speak more properly, the abettors of the Leibnitzian measure of that force which is supposed to preserve bodies in their condition of motion —insist, that the force which is exerted in producing any change of motion is greater in proportion as the motion changed is greater: and they give a very specious argument for their assertion. They appeal to the exertions which we ourselves make. Here we are conscious of the fact. Then they give similar examples of the action of bodies. A clay ball, moving six feet per second, will make the addition of one foot to the velocity of an equal clay ball that is already moving four feet per second in the same direction. But if this last ball be already moving ten feet per second, we must follow it with a velocity of twelve feet in order to increase its velocity one foot. But, without insisting on the numberless paralogisms and inconsistencies which this way of conceiving the matter would lead us into, it suffices to observe, that the phenomena give us abundant assurance that there has been the same exertion in both these cases. This acceleration is always accompanied by a compression of the balls, and the compression is the same in both. This compression is a very good measure of the force employed to produce it; and in the present case, we need not even trouble ourselves with any rule for its measurement: for surely when the compression is not different, but the same, the force exerted is the same. This is farther confirmed by observing, that it requires the same force to make the same pit, or to give the same motion, to a piece of clay lying on the table of a ship's cabin, whether the ship be sailing two miles or ten miles per hour.

Thus we see that there are strong reasons for believing, that the exertions of pressure in producing motion, or that the pressures *actually exerted*, are proportional to the changes of motion observed, and that they coincide in this respect with our abstract conceptions of moving forces.

But we have still better arguments. None of the Leibnitzians think of denying the equal exertions of gravity, or of any of those powers which they call *solicitations* or *accelerating forces*. They all admit, that gravity, or any constant accelerating force, produces equal increments of velocity in equal times, and that a double gravity will produce a double increment in an equal time, and an equal increment in half of the time; and that a quadruple gravity will produce a double velocity in half the time. All these things are granted by them, and their writings are full of reasonings from this principle. Now from the fact, acknowledged by the Leibnitzians, that the quadruple force of a bow gives a double velocity to the arrow, in every instant of its action, it indisputably follows, that it has acted on it only for half the time of the action of the four times weaker bow; which gives the arrow only half the velocity; and thus has the discrepancy between the effects of pressures and of our abstract moving forces entirely disappeared. For this circumstance of the difference in the time of acting will be found, on strict examination, in all the cases of the change of motion by pressures which we measure by their effects on a body at rest. When this and the appreciable changes of actual pressure, during the time of producing the motion, are taken into consideration, all difference vanishes, and the composition of pressures is in perfect harmony with the composition of motions, or of abstract moving forces. DYNAMICS is thus made a demonstrative science, and affords the opportunity of investigating, by observation and experiment, the nature of those mechanical powers which reside in bodies, and which appear to us under the form of pressure, inducing us to consider pressure as a cause of motion.

In this, however, we are rather inaccurate. Pressure is one of the sensible effects of that property which is also the cause of motion. It is not the pressure of a piece of lead, but its heaviness, which is the reason that it gives motion to a kitchen jack. Pressure is merely a generic name, borrowed from a familiar instance, and given to moving forces, which have the same nature, but different names that serve to mark their connection with certain substances, in which they may be supposed to reside. Natural philosophy is almost entirely employed in examining the nature of these various pressures or accelerative forces; and the general doctrines of dynamics, by ascertaining what is common to them all, enable us to mark with precision what is characteristic of each.

53. We have now advanced very far in this investigation; for we have obtained the criterion by which we learn the direction and the magnitude of every changing force: and, on the other hand, we see how to state what will be the effect of the exertion of any force that is known or suspected to act. All this we learn by the composition of forces; and the greatest part of mechanical disquisition consists in the application of this doctrine. For such reasons it merits minute consideration; and therefore we must

point out some general conclusions from the properties of figure, which will greatly facilitate the use of the parallelogram of forces.

54. 1*st*. The constituent and the resulting forces, or the simple and compound forces, act in the same plane; for the sides and diagonal of a parallelogram are in one plane.

55. 2*d*. The simple and the compound forces are proportional to the sides of any triangle which are parallel to their directions. For if any three lines, *ab*, *bd*, *ad*, be drawn parallel to AB, AC, and AD (fig 7, No. 2.), they will form a triangle similar to the triangle ABD. For the same reasons they are proportional to the sides of a triangle *a'b'd'*, which are respectively perpendicular to their directions.

56. 3*d*. Therefore each is proportional to the sine of the opposite angle of this triangle; for the sides of any triangle are proportional to the sines of the opposite angles.

57. 4*th*, Each is proportional to the sine of the angle contained by the directions of the other two; for AD is to AB as the sine of the angle ABD to the sine of the angle ADB. Now the sine of ABD is the same with the sine of BAC contained between the directions AB and AC, and the sine of ADB is the same with the sine of CAD; also AB is to AC, or BD, as the sine of ADB (or CAD) to the sine of BAD.

58. We now proceed to the application of this fundamental proposition. And we observe, in the first place, that since AD may be the diagonal of an indefinite number of parallelograms, the motion or the pressure AD may result from the joint action of many pairs of forces. It may be produced by forces which would separately produce the motions AF and AG. This generally gives us the means of discovering the forces which concur in its production. If one of them, AB, is known in direction

and intensity, the direction AC, parallel to BD, and the intensity, are discovered. Sometimes we know the directions of both. Then, by drawing the parallelogram or triangle, we learn their proportions. The force which deflects any motion AB into a motion AD, is had by simply drawing a line from the point B (to which the body would have moved from A in the time of really moving from A to D) to the point D. The deflecting force is such as would have caused the body to move from B to D in the same time. And, in the same manner, we get the compound motion AD, which arises from any two simple motions AB and AC, by supposing both of the motions to be accomplished in succession. The final place of the body is the same, whether it moves along AD or along AB and BD in succession.

59. This theorem is not limited to the composition of two motions or two forces only; for since the combined action of two forces puts the body into the same state as if their equivalent alone had acted on it, we may suppose this to have been the case, and then the action of a third force will produce a change on this equivalent motion. The resulting motion will be the same as if only this third force and the equivalent of the other two had acted on the body. Thus, in plate I. fig. 8. the three forces AB, AC, AE, may act at once on a particle of matter. Complete the parallelogram ABDC; the diagonal AD is the force which is generated by AB and AC. Complete the parallelogram AEFD; the diagonal AF is the force resulting from the combined action of the forces AB, AC, and AE. In like manner, completing the parallelogram AGHF, the diagonal AH is the force resulting from the combined action of AB, AC, AE, and AG, and so on of any number of forces.

This resulting force and the resulting motion may be much more expeditiously determined, in any degree of composition, by drawing lines in the proportion and direction of the forces in succession, each from the end of the preceding. Thus, draw AB, BD, DF, FH, and join AH; AH is the resulting force. The demonstration is evident.

60. It is to be noticed here, that in the composition of more than two forces, we are not limited to one plane. The force AD is in the same plane with AB and AC; but AE may be elevated above this plane, and AG may lead below it. AF is in the plane of AD and AE, and AH is in the plane of AF and AG.

Complete the parallelograms ABLE, ACKE, ELFK. It is evident that ABLFKCD is a parallelopiped, and that AF is one of its diagonals. Hence we derive a more general theorem of great use.

Three forces having the proportion and direction of the three sides of a parallelopiped, compose a force having the proportion and direction of the diagonal.

61. Any number of forces acting together on one particle of matter are balanced by a force that is equal and opposite to their resulting force; for this force would balance their resulting force which is equivalent to them in action. When this is duly considered, we perceive that each force is then in equilibrio with the equivalent of all the others; for a force can balance only what is equal and opposite to it. It appears very readily by the geometrical construction. If, instead of the circuit A, B, D, F, H, we take B, D, F, H, A, we have BA for the equivalent of the forces AC, AE, AG; but AB is equal and opposite to BA. Therefore the force AB is in equilibrio with the equivalent of all the others.

62. When any number of forces act on one particle of matter, and are in equilibrio, if they be considered as acting in parcels, the equivalents of these parcels are in equilibrio; for let the forces AB, AC, AE, AG, A*h*, be in equilibrio, and let them be considered in the two parcels AB, AC, and AE, AG, A*h*; then AD is the equivalent

of AB, BD (or AC), and DA is the equivalent of DF, FH, HA (or A*h*): now AD and DA balance each other. This corollary enables us to simplify many intricate complications of force; it also enables us to draw accurate conclusions from very imperfect observations. In most of our practical discussions we know, or at least we attend to, a part only of the forces which are acting on a material particle; and in such cases we reason as if we saw the whole: yet is our mathematical reasoning good with respect to the equivalent of all the parcels which we are contemplating, and the equivalents of the smaller parcels of which it consists; and the neglected force, or parcel of forces, induces no error on our conclusions.

63. In the spontaneous phenomena of nature, the investigation and discovery of our ultimate object of search is frequently very difficult, on account of the multiplicity of directions and intensities of the operating forces or motions. We may generally facilitate the process, by substituting equivalent forces or motions acting in convenient directions. It is in this way that the navigator computes the ship's place with very little trouble, by substituting equivalent motions in the meridional and equatorial directions for the real oblique courses of the ship. Instead of setting down ten miles on a course, S. 36. 52. W. he supposes that the ship has sailed eight miles due south, and six miles due west, which brings her near to the same place. Then, instead of fourteen miles south-west, he sets down ten miles south and ten miles west; and he proceeds in the same way for every other course and distance. He does this expeditiously by means of a traverse table, in which are ready calculated the meridional and equatorial sides of right angled triangles, corresponding to every course and distance. Having done this for the course of a whole day, he adds all the southings into one sum, and all the westings into another: he considers these as forming the sides of a right angled triangle; he looks for them,

paired together, in his traverse table, and then notices what angle and what distance corresponds to this pair. This gives him the position and magnitude of the straight line joining the beginning and end of his day's work.

The miner proceeds in the same way when he takes the plan of subterraneous workings, measuring, as he goes along, and noticing the bearing of each line by the compass, and setting down, from his traverse table, the northing or southing, and the easting or westing, for each oblique line: but there is another circumstance which he must attend to, namely, the slope of the various drifts, galleries, and other workings. This he does by noting the rise or the dip of each sloping line. He adds all these into two sums, and taking the risings from the dips, he obtains the whole dip. Thus he learns how far the workings proceed to the north, how far to the east, and how far to the dip.

The reflecting reader will perceive, that the line joining the two extremities of this progression will form the diagonal of a rectangular parallelopiped; one of whose sides lies north and south, the other lies east and west, and the third is right up and down.

The mechanician proceeds in the very same way in the investigation of the very complicated phenomena which frequently engage his attention. He considers every motion as compounded of three motions in some convenient directions, at right angles to each other. He also considers every force as resulting from the joint action of three forces, at right angles to each other, and takes the sum or difference of these in the same or opposite directions. From this process he obtains the three sides of a parallelopiped, and from these computes the position and magnitude of the diagonal. This is the motion or force resulting from the composition of all the partial ones.

This procedure is called the Estimation or Reduction of motions and forces.

64. A motion or force AB (fig. 9.) is said to be *estimated* in the direction EF, or to be *reduced* to this direction when it is conceived as compounded of the motions or forces AC, AD, one of which AC is parallel to EF, and the other AD is perpendicular to it. This expression is abundantly significant; for it is plain that the motion AD neither promotes nor hinders the progress along EF, and that AC expresses the whole progress in this direction.

65. In like manner, a force AB (fig. 10.) is said to be *estimated in*, or *reduced to*, a given plane EFGH, when it is conceived as resulting from the joint action of two forces AC, AD, one of which is parallel to a line *a b* drawn in that plane, and the other AD is perpendicular to it. The position of the line *a b* is determined by letting fall B *b* perpendicular to the plane, and drawing *b* P to the point P, in which BA meets the plane; then A *a* being drawn parallel to B *b*, will cut off *b a*, which is the reduction of the motion AB to the plane. Drawing AC parallel to *a b*, and completing the parallelogram ACBD, it is evident that the motion AB is equivalent to AD and AC, which is parallel to *a b*, and the three forces AB, AC, AD, are, as they should be, in one plane perpendicular to the plane EG.

66. If three forces AB, AC, AD, (fig. 11.), are in equilibrio, and are reduced to any one direction *d* A *l*, or to one plane EFGH, the reduced forces are also in equilibrio.

First, Let them be reduced to one direction *d l* by drawing the perpendiculars B *b*, C *c*, D *d*; make AL equal to AD, and join BL, CL, and draw the perpendiculars L *l*, C *c*; then, because the forces AB, AC, AD, are in equilibrio, ABLC must be a parallelogram, and AL is the force equivalent to AB and AC combined; then, because the lines D *d*, B *b*, C *c*, L *l*, are parallel, *d* A is equal to A *l*, and A *b* to C *o*, or to *c l*; therefore A *l* is equal to the sum of A *b* and A *c*, which are the reductions of AB and AC; therefore dA is equal to the same sum, and in equilibrio with them.

Secondly, Let them be reduced to one plane EFGH, and let $\alpha\beta$, $\alpha\kappa$, $\alpha\delta$, be the reduced forces. The lines D δ, A α, B β, C κ, L λ, are all parallel, being perpendicular to the plane; therefore the planes AB $\beta\alpha$ and CL $\lambda\kappa$ are parallel, and $\alpha\beta$, $\kappa\lambda$, are parallel. For similar reasons, $\beta\lambda$, $\alpha\kappa$, are parallel; therefore $\alpha\beta\lambda\kappa$ is a parallelogram. Also, because the lines D δ, A α, L λ, are parallel, and DA is equal to AL; therefore $\delta\alpha$ is equal to $\alpha\lambda$. But because $\alpha\beta\lambda\kappa$ is a parallelogram, the forces $\alpha\beta$, $\alpha\kappa$, are equivalent to $\alpha\lambda$; and $\alpha\delta$ is equal and opposite to $\kappa\lambda$, and will balance it; and therefore will balance $\alpha\beta$ and $\alpha\kappa$, which are the reductions of AB and AC to the plane EFGH, while $\alpha\delta$ is the reduction of AD; therefore the proposition is demonstrated.

The most usual and the most useful mode of reduction, is to estimate all forces in the directions of three lines drawn from one point, at right angles to each other, like the three plane angles of a rectangular chest, forming the length, the breadth, and the depth of the chest. These are commonly called the three co-ordinates. The resulting force will be the diagonal of this parallelopiped. This process occurs in all disquisitions in which the mutual action of solids and fluids is considered, and when the oscillation or rotation of detached free bodies is the subjeet of diseussion.

67. The only other general theorem that remains to be deduced from this law of motion is, that if a number of bodies are moving in any manner whatever, and an equal force act on every partiele of matter in the same or parallel directions, their relative motions will suffer no change; for the motion of any body A (fig. 12.) relative to another body B, which is also in motion, is compounded of the real motion of A, and the opposite to the real motion of B; for let A move uniformly from A to C, while B de-

scribes BD uniformly, draw AB, also draw AE equal and parallel to BD, join EC, DC, ED. The motion of A, relative to B, consists in its change of position and distance. Had A described AE, while B described BD, there would have been no change of relative place or distance; but A is now at C, and DC is its new direction and distance. The relative or apparent motion of A therefore is EC. Complete the parallelogram ACFE; it is plain that the motion EC is compounded of EF, which is equal and parallel to AC, the real motion of A, and of EA, the equal and opposite to BD, the real motion of B.

Now let the motions of A and B sustain the same change; let the equal and parallel motions AG, BH, be compounded with the motions AC and BD; or let forces act at once on A and B, in the parallel directions AG, BH, and with equal intensities; in either supposition, the resulting motions will be A *c*, B *d*, the diagonals of the parallelograms A G *c* C, and B H *d* D. Construct the figure as before, and we see that the relative motion is now *e c*, and that it is the same with EC both in respect of magnitude and position.

Here we still see the constant analogy between the composition of motions and the composition of forces. In the first case, the relative motions of things are not changed, whatever common motion be compounded with them all; or, as it is usually, but inaccurately, expressed, although the space in which they move be carried along with any motion whatever. In the second case, the relative motions and actions are not changed by any external force, however great, when equally exerted on every particle in parallel directions.

Thus it is that the evolutions of a fleet in a uniform current are the same, and produced by the same means, as in still water. Thus it is that we walk about on the surface of this globe in the same manner as if it neither revolved round the sun, nor turned round its axis. Thus it is that the same strength of a bow will communicate a certain velocity to an arrow, whether it is shot east, or west, or north, or south. Thus it is that the mutual actions of sublunary bodies are the same, in whatever directions they are exerted, and notwithstanding the very great changes in their velocities by reason of the earth's rotation and orbital revolution. The real velocity of a body on the earth's equator is about 3000 feet per second greater at midnight than at midday. For at midnight the motion of rotation nearly conspires with the orbital motion, and at midday it nearly opposes it. The difference between the velocities at the beginning of January and the beginning of July is vastly greater. And at other times of the day, and other seasons of the year, both motions of the earth are transversely compounded with the easterly or westerly motion of an arrow or cannon bullet. Yet we can observe no change in the effects of the mutual actions of bodies.

68. This is an important observation; because it proves that forces are to be measured by no other scale than by the motions which they produce. We have had repeated occasions to mention the very different estimation of moving forces by Mr Leibnitz; and have shewn how, by a very partial consideration of the action of those natural powers called *pressures*, he has attempted to prove, that moving forces are proportional to the squares of the velocities; and we shewed briefly, in what manner a right consideration of what passes when motion is produced by measurable pressures, proves that the forces really exerted are as the velocities produced. But the most copious proof is had from the present observation, that, in fact, the mutual actions of bodies depend on their relative motions alone.

69. The Leibnitzian measure of moving force is altogether incompatible with the universal fact now mention-

ed, viz. that the relative motions of bodies, resulting from their mutual actions, are not affected by any common motion, or the action of any equal and parallel force on both bodies: for this universal fact imports, that when two bodies are moving with equal velocities in the same direction, a force applied to one of them, so as to increase its velocity, gives it the same motion relative to the other, as if both bodies had been at rest. Here it is plain, that the space described by the body in consequence of the primitive force, and of the force now added, is the sum of the spaces which each of them would generate in a body at rest. Therefore the forces are proportional to the velocities or changes of motion which they produce, and not to the squares of those velocities. This measure of forces, or the position that a force makes the same change on any velocity whatever, and the independence of the relative motions on any motion that is the same on all the bodies of a system, are counterparts of each other. Since this independence is a matter of observation in all terrestrial bodies, we are entitled to say, that the powers which the Author of Nature has imparted to natural bodies are no way different from what are competent to matter once called into existence. And it also follows from this, that we must always remain ignorant of the absolute motions of bodies. The fact, that it has required the unremitted study of ages to discover even the relative motions of our solar system, is an argument to prove that the influence of this mechanical principle extends far beyond the limits of this sublunary world; nor has any phenomenon yet been exhibited which should lead us to imagine that it is not universal.

When we have made use of these arguments with some zealous partizans of Mr Leibnitz's doctrine, they have answered, that if indeed this independence of the relative motions of terrestrial bodies were observed to obtain exactly, it would be a conclusive argument. But the motion with which all is carried along is so great in comparison with the motions which we can produce in our experiments, that the small additions or diminutions that we can make to the velocity of this common motion must observe very nearly the proportions of the additions or diminutions of their squares. The differences of the squares of 2, 3, and 4, are very unequal; but the differences of the squares of 9, 10, 11, are much nearer to the ratio of equality; and the difference of the squares of 1000001, 1000002, 1000003, do not sensibly deviate from this ratio. But it is not fact that we cannot produce motions which have a very sensible proportion to the common motion. The motion of a cannon ball, discharged with one-third of its weight of powder, is nearly equal to that of the rotation of the earth's equator. When, therefore, we discharge the ball eastward, we double its motion; when to the westward, we destroy it. Therefore, according to Leibnitz, the action in the first case is three times the action in the second. In the first case it changes the square of the velocity (which we may call 1) from 1 to 4; and, in the second, it changes it from 1 to 0. But say the Leibnitzians, the velocity of rotation is but $\frac{1}{63}$ of the orbital velocity of the earth, and our observations of the velocities of cannon bullets are not sufficiently exact to ensure us against an error of $\frac{1}{31\frac{1}{2}}$. But the latter observations on the peculiar motions of the fixed stars concur in shewing, that the sun, with his attending planets, are carried along with a very great motion, which, in all probability, has a sensible ratio to the orbital motion of the earth. This must make a prodigious change on the earth's absolute motion, according as her orbital motion conspires with, opposes, or crosses, this other motion: the earth may even be at absolute rest in some points of its orbit. Thus will the composition with the motions produced in our experiments be so varied, that cases *must* occur when the difference of the results of the two measures of force will be very sensible.

But, farther, they have not attended to the agreement of our experiments, when the discharges of cannon are made in a direction transverse to that of the common motion. Here the immensity of the common motion, and the minuteness of our experimental velocities, can have no effect in diminishing the difference of the results of the two doctrines. This will appear distinctly to every reader who is much conversant in disquisitions of this kind; and it is in these more moderate motions that the complete independence of the relative motions on the common motions most accurately appears. Pendulum clocks and watches have been often executed which do not deviate from perfect equability of motion one part in 86400. This could not be obtained in all directions of the oscillations, if the forces deviated from the ratio of the velocities one part in 86400.

On the whole, we may consider it as established on the surest foundation, that the action of those powers of natural bodies which we call *pressures*, such as the force of springs, the exertions of animals, the cohesion of bodies, as well as the action of those other incitements to motion which we call *attractions* and *repulsions*, such as gravitation, magnetism, and electricity—is proportional to the change of velocity produced by it. And we must observe here, that this is not a mere mode of conception, the result of the laws of human thought, which cannot conceive a natural power as the cause of motion otherwise than by its producing motion, and which cannot conceive any degree of *moving* power different from the degree of the motion. This is the abstract doctrine, and it is true whether the pressures are proportional to the velocities or to the squares of the velocities. But we see farther, that whatever is the pressure of a spring (for example) on a quiescent body, yet the pressure actually exerted in producing a double velocity is only double, and not quadruple, as our first imperfect observations make us imagine.

70. Sir Isaac Newton has added another proposition to the number of laws of motion; namely, that *every action is accompanied by an equal and contrary reaction.* But in affirming this to be a law of nature, he only means that it is an universal fact: And he makes this affirmation on the authority of what he conceives to be a law of human thought; namely, that those qualities which we find in all bodies on which we can make experiments and observations, are to be considered as universal qualities of body. But we have limited the term *law of motion* to those consequences that necessarily flow from our notions of motion, of the causes of its production and changes. Now this third Newtonian proposition is not such a result. A magnet is said to act on a piece of iron when, and only when, the vicinity of the magnet is observed to be accompanied by certain motions of the iron. But it by no means follows from this observation, that the presence of the iron shall be accompanied by any motion, or any change of state whatever of the magnet, or any appearance that can suggest the notion that the iron acts on the magnet. When this was observed, it was accounted a discovery. Newton *discovered*, that the sun acts on the planets, and that the earth acts on the moon; and Kepler *discovered*, that the moon reacts on the earth. Newton had observed, that the iron reacts on the magnet; that the actions of electrified bodies were mutual; and that every action of sublunary bodies was, in fact, accompanied by an equal and contrary reaction. On the authority of his rule of philosophizing, he affirmed, that the planets react on the sun, and that the sun is not at rest, but is continually agitated by a small motion round the general centre of gravitation. He pointed out several consequences of this reaction. Astronomers examined the celestial motions more narrowly, and found that those consequences do

really obtain, and disturb all the planetary motions. It is now found that this reciprocity of action obtains throughout the solar system with the utmost precision, and that the third Newtonian proposition is really a law of nature, although it is not a law of human thought. It is a discovery. The contrary involves no absurdity or contradiction. It would indeed be contrary to experience; but things might have been otherwise. It is conceivable, and possible, that a ball A shall strike another equal ball B, and carry it along with it, without any diminution of its velocity. The fact, that the velocity of A is reduced to one-half, is the indication of a force residing in B, which force changes the motion of A ; and the intensity of this force is learned from the change which it produces. This *is found* to be equal to the change produced by A on B, and thus the reaction of B is *discovered* to be equal to the action of A.

It is highly probable, that this universality and equality of reaction to action is the consequence of some general principle, which we may in time discover; meanwhile we are entitled to suppose it universal, and to reason from this topic in our disquisitions about the actions of bodies on each other.

These propositions might have completed the doctrines of dynamics; but it appears that, in order to the production of a material universe which should accomplish the purposes of the Creator, it was necessary that there be certain characteristic differences between the forces inherent in the various collections of matter which compose this universe. The facts or physical laws (for the above-mentioned laws are metaphysical) of motion may be different from those which would have been observed had matter been left entirely to itself. This difference may have introduced other laws of motion as necessarily resulting from the nature of the forces. We have occasionally mentioned some instances where this appears to obtain, but gave good reasons for affirming, that a due examination of all circumstances which may be observed in the production or variation of motion by those forces, has demonstrated, that there are no such deviations from the two laws of motion already determined, but that all the mechanical powers of bodies, when considered merely as causes of motion, act agreeably to the same laws. Careful examination was, however, said to be necessary.

This examination must consist in distinctly noticing the circumstances that occur in the production of motion by any force whatever. It is by no means enough to state simply the intensity of the force and the direction of its exertion. If a force continue to act, it continues to vary the motion already produced. Should the force change its intensity or direction while it is acting, these circumstances must induce still farther changes in the motion; and it is not till all action has ceased that the motion is brought to its ostensible state, in which it is the object of our attention and our future discussions. Instances of the effects of such continued and such varied actions are to be seen in most of the phenomena of nature or art. The communication of motion by impulse is perhaps the only instance (very frequent indeed) that can be produced where this is not necessary: Nay, we shall perhaps find reason to conclude, that this instance is not an exception, and that even the communication of motion from one billiard ball to another is brought about by an action continued for some time, and greatly varied during that time. Much preparation is therefore necessary before we can apply the general laws of motion to the solution of most of the questions which come before us in the course even of our elementary disquisitions. We must lay down some general propositions which determine the results of the continued, and perhaps varied, actions of moving forces; and we must mark the different effects of the simple continuation of action, and also those of the variations in this continued action, both in respect of intensity and direction. The effect of a mere continuance of action must be an acceleration of the motion ; or a retardation of it, if the force continue to act in the opposite direction. The effect of the continued action of a transverse force must be a continual deflection, that is, a curvilineal motion. These must therefore now occupy our attention in their order.

OF ACCELERATED AND RETARDED MOTIONS.

71. All men can perceive, that a stone dropped from the hand, or sliding down an uniform slope, has its motion continually accelerated, and that the motion of an arrow rising perpendicularly through the air is continually retarded; and they feel no difficulty in conceiving these changes of motion as the effects of the continual operation of their weight or heaviness. The falling stone is in a different condition in respect of motion in the beginning and the end of its fall. In what respect do these states of the body differ ? Only in respect to what we call its *velocity*. This is an affection of motion ; it is an expression of the relation between the two notions or ideas which concur to form the idea of motion, namely, the space and the time. These are all the circumstances that we observe in a motion. Time elapses, and during its currency a space is described. The term *velocity* expresses the magnitude of the space which corresponds to some unit of time. Thus, the rate of a ship's motion is determined, when we say that it is nine miles in an hour, or nine miles per hour. We sometimes say (but aukwardly) " The motion is at the rate, or with the velocity, of a mile in three days." It is most conveniently expressed by a number of some given units of length, which completely make up the line described during this unit of time. But the mechanicians

express it in a way more general by a fraction, of which the numerator is a number of inches, feet, yards, fathoms, or miles, and the denominator is the number of seconds, minutes, or hours, employed in moving along this line. This is a very proper expression; for when we speak of any velocity, and continue to reason from it, we conceive ourselves to speak of something that remains the same, in the different occasions of using the term. Now if the velocity be constant, it is indifferent how long the line may be; because the time of its description will be lengthened in the same proportion. Thus if 48 feet be described in 12 seconds, 36 feet will be described in 9 seconds, 16 feet will described in 4 seconds, &c. Now $\frac{48}{12}$, $\frac{36}{9}$, and $\frac{16}{4}$, are fractions of equal value, being equal to $\frac{4}{1}$, or 4, that is, to the velocity of 4 feet per second. The value of this fraction, or the quotient of the number of the units of length, divided by the number of units of time, is the number of those units of length described uniformly in one unit of time.

But how shall we determine the velocity in any instant or in any point of a motion that is continually changing? Suppose that a body has fallen 144 feet, and that we would ascertain its velocity in that point of its fall, or the velocity which it has in passing through that point? In the next second the body falls 112 feet farther. This cannot be the measure of the velocity at the beginning of the fourth or the end of the third second. It is too great. The fall during the preceding second was 80 feet. This is too small. The mean of these two, or $\frac{80+112}{2}$, $=\frac{192}{2}$, $=96$, is probably more exact. Due attention to the nature of this motion shews us, that 96 is the proper measure, or that the motion at that instant is at the rate of 96 feet per second. But it is peculiar to this kind of motion that the half sum of the spaces described in two succeeding equal moments is the measure of the velocity in the middle instant. Therefore this method will not generally give an accurate measure. Yet it is indispensably necessary to obtain some accurate measure; for it is in this particular alone that the state of the body differs from its similar state in another instant. The difference of place makes no distinction; for if a body continue its motion unchanged, its condition in every different instant of time, or point of space, is unchanged or the same. The change of place is not a change of motion, but is involved in the very conception of the continuation of the motion. The change of condition consists, therefore, in the change of velocity: Therefore the change of velocity is the only indication, and the only measure of the action (perhaps accumulated) of the changing force. It is therefore the chief object of our search; and accurate measures of velocity are absolutely necessary.

When the velocity changes continually, there can be no *actual* measure of it. In what then does the magnitude of a velocity consist, when there is no actual measure of it? It is a certain undescribable DETERMINATION; by which, if not changed, a certain space *would* be uniformly described in a given unit of time. Thus we know, that if, when a stone has fallen 16 feet, its motion be directed along a horizontal plane, without diminution, it will move on for ever at the rate of 32 feet per second. The space which would be thus described is not the velocity, but the measure of the velocity. But the proportions of those spaces, being the proportions of those measures, are the proportions of the velocities themselves. We may discover these proportions in the following manner:

72. Let ACG (fig. 13.) be a line described by a body with a motion anyhow continually, but gradually, varied; and let it be required to determine the proportion of the velocity in any point C to the velocity in any other point F.

AXIOM.—If A be to B in a ratio that is greater than any ratio less than that of C to D, but less than any ratio greater than that of C to D, then A is to B as C to D.

Take the straight line *a c g* to represent the time of the body's motion along ACG, so that the points, *a*, *c*, *f*. *g*, may represent the instants of time in which the body passes through the points A, C, F, G; and the portions *a c*, *c f*, *f g*, of the line *a g*, may represent the times employed in describing the portions AC, CF, FG; and therefore *a c* is to *a f* as the time of describing AC to the time of describing AF.

Moreover, let *h k n o* be a line so related to the straight line *a c f g*, by the perpendicular ordinates *a h*, *c k*, *f n*, *g o*, that the areas *a c k h*, *a f n h*, *a g o h*, may be proportional to the portions AC, AF, AG, of the line described by the moving body; and let this relation be true with respect to every point B, D, E, &c. and the corresponding points *b*, *d*, *e*, &c

Then it is affirmed, that *the velocity in the point* C *is to the velocity in the point* F *as* c k *is to* f n.

Let the equal lines *b c*, *c d*, *e f*, *f g*, represent equal moments of time, and let B, D, E, G, be the points through which the body is passing at the instants *b*, *d*, *e*, *g*. Then the areas *b i k c*, *c k l d*, *e m n f*, *f n o g*, will represent, and be proportional to, the spaces BC, CD, EF, FG, which are described during the moments *b c*, *c d*, *e f*, *f g*.

Draw *t p* parallel to *a g*, so as to make the rectangle *b t p c* equal to the trapezium *b i k c*; and draw the lines *q v*, *u r*, *s x*, in the same manner, so that each rectangle may be equal to its corresponding trapezium.

If the motions had been uniform during the moments *b c* and *f g*, that is, if the spaces BC and FG had been uniformly described, then the velocity in the point C would have been to the velocity in the point F as *c p* to *f s*: For since the rectangles *b t p c* and *f s x g* are respectively equal to the trapeziums *b i k c* and *f n o g*; and since *b i k c* is to *f n o g* as BC is to FG, the rectangle *b t p c* is to the rectangle *f s x g* as BC to FG. But because those two rectangles have equal altitudes *b c* and *f g*, they are to each other in the proportion of their bases *c p* and *g x*, or *c p* and *f s*. Therefore BC is to FG as *c p* to *f s*. But if BC and FG are uniformly described in equal times, they are proportional to the velocities of those uniform motions. Therefore *c p* is to *f s* as the velocity with which BC is uniformly described to the velocity with which FG is uniformly described in an equal time.

But the motion expressed by the figure is not uniform, because the line *h l o* recedes from the axis *a g*, and the areas, cut off by the parallel ordinates, increase in a greater proportion than the corresponding parts of the axis; that is, the spaces increase faster than the times: for the moments *b c*, *c d*, *e f*, *f g*, being all equal, it is evident that the corresponding slips of the area continually augment. The motion is swifter at the instant *c* than at the instant *b*, and the velocity at the instant *c* is *greater* than that with which the space BC would be uniformly described in the same time. For the same reason, the velocity at the instant *f* is *less* than that with which the space FG would be uniformly described in the same time. Therefore the velocity at the instant *c* is to the velocity at the instant *f* in a greater ratio than that of *c p* to *f s*. In the very same manner, it will appear, by comparing the motion during the moment *c d* with the motion during the moment *e f*, that the velocity at the instant *c* is to the velocity at the instant *f* in a less ratio than that *c q* to *f r*.

Therefore the velocity in the point C is to the velocity in the point F in a greater ratio than that of *c p* to *f s*, but in a less ratio than that of *c q* to *f r*.

But by continually diminishing the equal moments *b c*, *c d*, *e f*, *f g*, it is evident that *c p* and *c q* continually approach to equality with *c k*; and *f r* and *f s* continually approach to equality with *f n*, that when *c p* is less than *c k*, *f s* is greater than *f n*, and when *c q* is greater than *c h*, *f r* is less than *f n*.

Therefore the velocity in the point C is to the velocity

in the point F in a ratio that is greater than the ratio of *any* line less than $c\ k$ to *any* line greater than $f\ n$, but which is less than the ratio of any line greater than $c\ k$ to any line less than $f\ n$. Therefore the ratio of the velocity in C to the velocity in F is greater than any ratio that is less than that of $c\ k$ to $f\ n$; but it is less than any ratio that is greater than that of $c\ k$ to $f\ n$. Therefore the velocity in the point C is to the velocity in the point F as $c\ k$ to $f\ n$.

This important theorem may be expressed in more general terms as follows:

If the abscissa a g *of a line* h k o *represent the time of any motion, and if the areas bounded by parallel ordinates be proportional to the spaces described, the ordinates are proportional to the velocities.*

Remark. The propriety or aptitude of expressing the time by the portions of the axis $a\ c\ g$, will, perhaps, appear more clearly in the following manner.

Let $a\ c\ g$ be any straight line, and let $h'\ k\ v$ be another line, straight or curved. Let the straight line $a\ h\ z$, perpendicular to $a\ g$, be carried uniformly down along this line, keeping always perpendicular to it, and therefore always parallel to its first position $a\ h\ z$. In its various situations $c\ k\ z$, $e\ m\ z$, &c. it will cut off areas $a\ c\ k\ h$, $a\ e\ m\ h$, &c. bounded by the axis by the ordinates $a\ h$ and $c\ k$, or by the ordinates $a\ h$ and $e\ m$, &c. and by the line $h\ k\ g$. By this motion the moveable ordinate is said, in the language of modern geometry, to *generate* the areas $a\ c\ k\ h$, $a\ e\ m\ h$, &c. At the same time, let a point A move along the line ACG, setting out from A at the instant when the line $a\ z$ sets out from a; and let the motion of the point A be so regulated, that the spaces AB, AC, AD, &c. generated by this motion, may increase at the same rate with the areas $a\ b$, $i\ h$, $a\ c\ k\ h$, $a\ d\ l\ h$, &c. or such that we shall have AB to AC as $a\ b\ i\ h$ to $a\ c\ k\ h$, &c. It is plain, that the motion along AG is the same with that described in the enunciation of the proposition: for because the motion of the ordinate $a\,z$, along the axis $a\,g$, is supposed to be uniform, the spaces $a\,b$, $a\,c$, $a\,d$, &c. are proportional to the times in which they are described, and may therefore be taken to measure or to represent those times.

73. Cor. 1. *In a motion continually varied, the velocities in the different points of the path are to each other in the limiting or ultimate ratio of the spaces described in equal times,* those times being supposed to diminish continually: for it is evident, that if the equal moments $b\,c$, $c\,d$, $e\,f$, $f\,g$, are supposed to diminish continually, till the instants b and d coalesce with c, and the instants e and g coalesce with f; then the ratio of $c\,k$ to $f\,n$ is the limit of the continually increasing ratio of $c\,p$ to $f\,s$, or of the continually diminishing ratio of $c\,q$ to $f\,r$. Sir Isaac Newton calls this the ultimate ratio of $c\,p$ to $f\,s$, or of $c\,q$ to $f\,r$. Now the ratio of $c\,p$ to $f\,s$ is, by construction, the same with the ratio of the rectangle $b\,t\,p\,c$ to the rectangle $f\,s\,x\,g$, and the ratio of $c\,q$ to $f\,r$ is the same with the ratio of the rectangle $c\,q\,v\,d$ to the rectangle $e\,u\,r\,f$. But the ratio of the rectangle $b\,t\,p\,c$ to the rectangle $f\,s\,x\,g$, is the same with the ratio of the space $b\,i\,k\,c$ to the space $f\,n\,o\,g$; that is (by hypothesis), the same with the ratio of the space BC to the space FG; and the ratio of the rectangles $c\,q\,v\,d$ and $e\,u\,r\,f$ is the same with that of the spaces CD and EF. Therefore the ratio of the velocity at C to the velocity at F is the same with the ultimate ratio of the small increments BC, FG, or CD, EF of the spaces generated in very small and equal times.

It is also evident, that because the ratio of $c\,k$ to $f\,n$ is the limit both of the ratio of $c\,p$ to $f\,s$ and of the ratio of $c\,q$ to $f\,r$, these ultimate ratios are the same, and that we may say that the velocity in C is to the velocity in F in the ultimate ratio of BC to EF, or in the ultimate ratio of CD to FG.

We also can easily perceive, that the ratio of the area

$b\,i\,k\,c$ to the area $e\,m\,n\,f$ approaches more near to the ratio of $c\,k$ to $f\,n$ as we take the moments $b\,c$ and $e\,f$ smaller. Therefore, in many cases of practice, where it may be easy to measure the spaces described in the different small moments of the motion, but difficult to ascertain their ultimate ratio, so as to obtain accurate measures of the proportions of the velocities, we may reduce the errors of measurement to something very insignificant, by taking these moments extremely small; and we shall diminish the error still more, by taking the proportion of the half sum of BC and CD to the half sum of EF and FG for the proportion of the velocities in C and F.

It often happens that we have it not in our power to compare the spaces described in small moments which are precisely equal. Still we can find the exact proportion of the velocities, if we can ascertain the ultimate ratio of the increments of the spaces, and the ultimate ratio of the moments of time in which these increments are described; for it is plain, by considering the gradual approach of the points p and r to the points k and n, that the ratio of $c\,k$ to $f\,n$ is still the ultimate ratio of the bases of rectangles equal to the mixtilineal areas, whether the altitudes (representing the moments) are equal or not. Now the bases of two rectangles are in the proportion of the rectangles directly, and of their altitudes inversely. But the ultimate ratio of the altitudes is the ultimate ratio of the moments, and the ultimate ratio of the rectangles is the ultimate ratio of the spaces described in those unequal moments. Therefore, in such cases, we have,

74. Cor. 2. *The velocities are in the ratio compounded of the direct ultimate ratio of the momentary increments of the spaces, and the inverse ultimate ratio of the increments (or moments) of the times in which these increments of the spaces are made.*

If s, v, and t, are taken to represent the magnitudes of the spaces, velocities, and times, and if $\dot{s}$, $\dot{v}$, and $\dot{t}$, are taken always in the limiting or ultimate ratio of their momentary increments, we shall have v always in the proportion of $\dot{s}$ directly, and of $\dot{t}$ inversely. We express this by the proportional equation $v \doteqdot \frac{\dot{s}}{\dot{t}}$, which is equivalent to the analogy $V : v = \frac{\dot{S}}{\dot{T}} : \frac{\dot{s}}{\dot{t}}$, or $V : v = \dot{S}\,\dot{t} : \dot{s}\,\dot{T}$.

75. *N. B.* Here observe, that this is not the only way of stating the relation of space and time—the abscissa may be made the time and the ordinate the space; then the velocity $= \frac{\dot{y}}{\dot{x}}$.

The converse of this proposition may be thus expressed:

76. *If the axis* a g *of the line* h k o *represent the time of a varied motion along the line* AG, *and if the ordinates* a h, b i, c k, *&c. be as the velocities in the instants* a, b, c, *or in the points* A, B, C; *then the areas* a b i h, a c k h, a d l h, *&c. are proportional to the spaces* AB, AC, AD, *&c.*

This may be demonstrated in the same way with the former; but the indirect demonstration is more brief, and equally strict.

If the spaces AC, AF, &c. are not proportional to the areas *a c k h*, *a f n h*, &c. they are proportional to some other areas *a c k h'*, *a f n' h'*, &c. which are bounded by the same ordinates, and by another line *h' k n'*. But because the areas *a c k h'*, *a f n' h'*, &c. are always proportional to the spaces AC, AF, &c. described on the line AG, the velocity in the point C is to the velocity in the point F as the ordinate *c k* is to the ordinate *f n'*. But, by hypothesis, the velocity in C is to the velocity in F as *c k* to *f n*, and *f n'* is equal to *f n*; which is absurd. Therefore the spaces AC, AF, are not proportional to any other areas, &c.

77. Cor. *The ultimate ratio of the momentary increments of the spaces is compounded of the ratio of the velocities, and the ultimate ratio of the increments of the times:* for when the moments $b\,c$, $e\,f$, are equal, it is evident, that the ultimate ratio of the rectangles $b\,c\,p\,t$, $e\,f\,r\,u$ is the same with the ultimate ratio of the increments of the spaces. But the ultimate ratio of these rectangles is the same with that of their bases $c\,p$ and $f\,r$; that is, the ratio of $c\,k$ to $f\,n$, that is, the ratio of the velocities. And when the moments are unequal, the ratio of the rectangles is compounded of the ratio of their bases and the ratio of their altitudes; that is, compounded of the ratio of the velocities and the ultimate ratio of the moments of time.

We have, therefore, $\dot{S} : \dot{s} = V\dot{T} : v\dot{t}$, and $\dot{s} \doteqdot v\dot{t}$.

It most commonly happens, that we can only observe the accumulated results of varied motions; and in them we only observe a space passed over, and a certain portion of time that has elapsed during the motion. But being able to distinguish the portions of the whole space which are described in known portions of the whole time, and having made such observations in several parts of the motion, we discover the general law that the motion affects, and we affirm this law to hold universally, even though we have not observed it in *every* point. We do this with a degree of probability and confidence proportioned to the frequency of our observation. It is not till we have done this, that we can make use of the first of these two propositions, which enables us to ascertain the velocity of the motion in its different moments. Thus if we observe, that a stone in falling descends one foot in the quarter of a second, 16 feet in a second, 64 feet in two seconds, and 144 feet in three seconds; the general law immediately observed is, "that the spaces described are as the squares of the times;" for 1 is to 16 as the square of $\frac{1}{4}$ to the square of 1. Again, 16 is to 64 as 1^2 to 2^2;

and 16 is to 144 as 1^2 to 3^2. Hence we infer, with great probability, that the stone would fall 36 feet in a second and a half; for 16 is to 36 as 1^2 to $1\frac{1}{2}^2$; and we conclude in the same way for all other parts of the motion.

78. This immediate observation of the analogy between the spaces and the squares of the times suggests an easy determination of the velocity in this particular kind of motion; and it merits particular notice, being very often referred to. We can take $a\,g$ to represent the time; and then, because the areas which are to represent the spaces described must be proportioned to the quares of the portions of $a\,g$, we perceive that the line which comes in place of $h\,k\,o$ must be a straight line drawn from a. For example, the straight line $a\,\delta\,\gamma$. For this is the only boundary which will give areas $a\,b\,\beta$, $a\,c\,\varkappa$, $a\,d\,\delta$, &c. proportional to $a\,b^2$, $a\,c^2$, $a\,d^2$, &c. And we perceive, that *any* straight line drawn from a will have this property.

Having thus got our representations of the times and the spaces, we say, on the authority of our theorem, that the velocity at the instant b is to the velocity at the instant d as $b\,\beta$ to $d\,\delta$, &c. And now we begin to make inferences, purely geometrical, and express our discovery of the velocities in a very general and simple manner. We remark, that $b\,\beta$ is to $d\,\delta$ as $a\,b$ is to $a\,d$; and we make the same affirmation concerning the magnitudes represented by these lines. We say that the velocity at the instant b is to the velocity at the instant d as the time $a\,b$ is to the time $a\,d$. We say, in terms still more general, that the velocities are proportional to the times from the beginning of the motion. We moreover perceive, that the spaces are also proportional to the squares of the acquired velocities; or the velocities are as the square roots of the spaces.

We can farther infer, from the properties of the triangle, that the momentary increments of the spaces are proportional to the momentary increments of the squares of the times, or of the squares of the velocities.

We also observe, that not only the whole acquired velocities are proportional to the whole elapsed times, but that the increments of the velocities are proportional to the times in which they are acquired; for $\pi\varkappa$ is to $\rho\varphi$ as $b\,c$ to $d\,f$, &c. Equal increments of velocity are therefore acquired in equal times. Therefore such a motion may, in great propriety of language, be denominated a UNIFORMLY ACCELERATED MOTION; that is, *a motion in which we observe the spaces proportioned to the squares of the times, is a motion uniformly accelerated;* and spaces in the duplicate ratio of the times form the ostensible characteristic of an uniformly accelerated motion.

79. Lastly, if we draw $\iota\,\lambda$ parallel to the axis $a\,b$, we perceive that the rectangle $a\,e\,\iota\,\lambda$ is double of the triangle $a\,e\,\iota$. Now, because $a\,e$ represents the time of the motion, and $e\,\iota$ represents the acquired velocity, the rectangle $a\,e\,\iota\,\lambda$ will represent the space which would be uniformly described with the velocity $e\,\iota$ during the time $a\,e$. But the triangle $a\,e\,\iota$ represents the space really described with the uniformly accelerated motion during the same time. Hence we infer, that the space that is described in any time, with a motion increasing uniformly from nothing, is one half of the space which would be uniformly described during the same time with the final velocity.

These are but a part of the inferences which we may draw from the geometrical properties of those representations which we had selected of the different measurable affections of motion. We may affirm, with respect to the motions themselves, all the inferences which relate to magnitude and proportion, and thus improve our knowledge of the motions.

We took the opportunity of this very simple and perspicuous example, to give our young readers a just conception of the *mathematical method* of prosecuting mechanical knowledge, and to make them sensible of the unquestionable authority for every theorem deduced in this manner.

One of the most important is, to discover the accumulated result of a motion of which we only observe the momentary increments. This is to be done by finding the area, or portions of the area, of the mixtilineal space $agoh$; and it is evidently analogous to the inverse method of fluxions, or the integral calculus.

In most cases, we must avail ourselves of the corollary $\dot{s} \doteq v\,\dot{t}$, and we obtain the solution of our question only in the cases where our knowledge of the quantities $\dot{s}$, $\dot{t}$, and v (considered as geometrical magnitudes, that is, as lines and surfaces), enables us to discover s and t.

OF ACCELERATING AND RETARDING FORCES.

80. HAVING thus discovered the proportions of the velocities in motions varying in any manner whatever, we can observe the variations which happen in them. These variations are the effects, and the only marks and measures of the changing forces. They are the characteristics of their kinds (considered merely as moving forces); that is, the indications of the directions in which they act; for this is the only difference *in kind* of which they are susceptible in this general point of view. If they increase the velocity, their direction must be conceived as the same with that of the previous motion; because the result of the action of a force is equivalent to the composition of the motion which that force would produce in a quiescent body with the motion already existing; and an increase of velocity is equivalent to the composition of a motion in the same direction.

Having no other mark of the force but the acceleration, we have no other name for it in the abstract doctrines of dynamics, and we call it an ACCELERATING FORCE. Had it retarded the motion, we should have called it a RETARDING FORCE.

In like manner, we have no measure of the *magnitude* or *intensity* of an accelerating force, but the acceleration which it produces. In order therefore to investigate the powers which produce all the changes of motion, we must endeavour to obtain measures of the cceleration.

A continual increase of velocity is the effect of the continued action of accelerating forces. If equal increments of velocity are produced in every succeeding equal moment of time, we cannot conceive that there is any change in the accelerating force. Therefore a uniformly accelerated motion is the mark of the unvaried action of an accelerating force, that is, of the continued action of a constant force; of a force whose intensity is always the same. When therefore we observe a body describe spaces proportional to the squares of the times, we must infer that it is urged forward by a force whose intensity does not change; and, on the other hand, a constant force must produce a uniformly accelerated motion by its continued action. And if any previous circumstances assure us of this continued action of an invaried force, we may make all the inferences which were mentioned under the article of uniformly accelerated motion.

That force must surely be accounted double which produces a double increment of velocity in the same time by its uniform action, we can form no other estimation of its magnitude. And, in general, *accelerating forces must be accounted proportional to the increments of velocity which they produce, by acting uniformly during the same or equal times.*

Supposing them to act on a body at rest. Then the velocity produced is itself the increment; and we must say, that accelerating forces are proportional to the velocities which they generate in a body in equal times. And because we found (No. 79.) that the space described with a uniformly accelerated motion is half the space which would be uniformly described in the same time with the final velocity, which space is the direct measure of this velocity, and because halves have the same proportion with the wholes—we may say, that *accelerating forces are proportional to the spaces through which they impel a body from rest in equal times by their uniform action.*

This is an important remark; because it gives us an easy measure of the force, without the trouble of first computing the velocities. It also gives us the only distinct notion that we have of the measurement of forces by the motions which they produce. When speaking of the composition of forces, we distinguished or denominated them by the sides and diagonal of a parallelogram. These lines must be conceived as proportional to the spaces through which the forces urge the body *uniformly* during the small and insensible time of their action, which time is supposed to be the same for both forces; for the sides of the parallelogram are supposed to be separately described in equal times, and therefore to be proportional to the velocities generated by the constituent forces. If indeed the forces do not act uniformly, nor similarly, nor during equal times, we cannot say (without farther investigation) what is the proportion of the intensity of the forces, nor can we infer the composition of their action. We must at least suppose, that in every instant of this very small time of their joint action, their direction remains unchanged, and that their intensities are in the same ratio. We shall see by and by, that with these conditions the sides of the parallelogram are still proportional to the velocities generated. In the mean time, we may take the spaces through which a body is uniformly impelled from rest (that is, with a uniformly accelerated motion) as the measures of the forces; yet these spaces are but the halves of the measures of the velocities. Then, if a body be moving with the velocity of 32 feet per second, and an accelerating force acts on it during a second, and if this force be such that it would impel the body (from a state of rest) 16 feet, it will add to the body a velocity of 32 feet per second. Accordingly, this is the effect of gravity—the weight of a pound of lead may be considered as a force which does not vary in its intensity. We know that it will cause the lead to fall 16 feet in a second; but if the body has already fallen 16 feet, we know that it is then moving with the velocity of 32 feet per second. And the fact is, that it will fall 48 feet farther in the next second, and will have acquired the velocity of 64 feet per second. It has therefore received an augmentation of 32 feet of velocity by the action of gravity during the second second; and gravity is in fact a constant force, causing equal increments of velocity in equal times, however great the velocities may be. It does not act like a stream of fluid, whose impulse or action diminishes as the solid body withdraws from it by yielding.

But supposing that we have not compared the increments of velocity uniformly acquired during equal times, in what manner shall we measure the accelerating forces? In such a case, that force must be accounted double which generates the same velocity, by acting uniformly during half the time; for when the force is supposed invariable, the changes of velocity which it produces are proportional to the times of its action; therefore if it produces an equal velocity in half the time, it will produce a double velocity in an equal time, and is therefore a double force. The same may be said of every proportion of time in which an equal change of velocity is produced by the uniform action of an accelerating force. The force must be accounted greater in the same proportion that the time required for the production of a given velocity in a body is less. Hence we infer, that *accelerating forces are in-*

versely proportional to the time in which a given change of velocity is produced by their uniform action.

By combining these two propositions we establish this general theorem;

82. *Accelerating forces are proportional to the changes of velocity which they produce in a body by their uniform action directly, and to the times in which these changes are produced inversely.*

If, therefore, A and a are the forces, V′ and v' the changes of velocity, and T′ and t' the proportions of time in which they are uniformly produced, we have

$$A : a = V' t' : v'T', = \frac{V'}{T'} : \frac{v'}{t'}$$

$$\text{And } a \doteqdot \frac{v'}{t'}.$$

The formula $a \doteqdot \frac{v'}{t'}$ is not restricted to any particular magnitude of v' and t'. It is true, therefore, when the portion of time is diminished without end; for since the action is supposed uniform, the increment of velocity is lessened in the same proportion, and the value of the fraction $\frac{v'}{t'}$ remains the same. The characters or symbols v' and t' are commonly used to express *finite* portions of v and t. The symbols v and t are used by Newton to express the same things taken in the ultimate or limiting ratio. They are usually considered as *indefinitely small* portions of v and t. We small abide by the formula

$$a \doteqdot \frac{v}{t}.$$

83. It must always be kept in mind, that v and t are abstract numbers; and that v refers to some unit of space, such as a foot, an inch, a yard; and that t refers to some unit of time, such as an hour, a minute, a second; and especially that a is the number of the same units of space, which will be uniformly described in *one* unit of the time

with the velocity generated, by the force acting uniformly during *that* unit. It is twice the space actually described by the body during that unit when impelled from rest by the accelerating force. It is necessary to keep hold of these clear ideas of the quantities expressed by the symbols.

On the other hand, when the measure of the accelerating force is previously known, we employ the theorem $a\,t' = v'$; that is, the addition made to the velocity during the whole, or any part, of the time of the action of the force, is obtained by multiplying the acceleration of one unit of time by the number of such units contained in t'.

These are evidently leading theorems in dynamics; because all the mechanical powers of nature come under the predicament of accelerating or retarding forces. It is the collection of these in any subject, and the manner in which they accompany, or are inherent in it, which determine the mechanical character of that subject; and therefore the phenomena by which they are brought into view are the characteristic phenomena. Nay, it may even be questioned, whether the phenomena bring any thing more into view. This force, of which we speak so familiarly, is no object of distinct contemplation; it is merely a something that is proportional to $\frac{\dot{v}}{\dot{t}}$. And when we observe, that the $\frac{\dot{V}}{\dot{T}}$, found in the motions that result from the vicinity of a body A, is double of the $\frac{\dot{v}}{\dot{t}}$, which results from the vicinity of another body B; we say that a force resides in A, and that it is double of the force residing in B. The accelerations are the things immediately and truly expressed by these symbols. And the whole science of dynamics may be completely taught without once employing the word *force*, or the conception which we imagine that we form of it. It is of no use till we come to study the mechanical history of bodies. Then, indeed, we must have some way of expressing the fact, that an acceleration $= \frac{32 \text{ feet}}{1''}$ is observed in every thing on the surface of this globe; and that an acceleration $= \frac{418 \text{ feet}}{1''}$ is observed over all the surface of the sun. These facts are characteristic of this earth and of the sun; and we express them shortly by saying, that such and such forces reside in the earth and in the sun. It will preserve us from many mistakes and puzzling doubts, if we resolutely adhere to this meaning of the term *force*; and this will carry mathematical evidence through the whole of our investigations.

84. As velocity is not an immediate object of contemplation, and all that we observe of motion is a space and a time, it may be proper to give an expression of this measure of accelerating force which involves no other idea. Supposing the body to have been previously at rest, we have $a \doteqdot \frac{v}{t}$. Multiply both parts of the fraction by t, which does not change its value, and we have $a \doteqdot \frac{vt}{t^2}$. But $vt = s$; and therefore $a \doteqdot \frac{s}{t^2}$.

The formula $a = \frac{s}{t^2}$ is equivalent to the proportion $t^2 : 1 = s : a$; and a would then be the space through which the accelerating force would impel the body in one unit of the time t. But this is only half of the measure of the velocity which the accelerating force generates during that unit of time. For this reason we did not express the accelerating force by an ordinary equation, but used the symbol $\doteqdot$. In this case, therefore, of uniform action, we may express the accelerating force by $a = \frac{2s}{t^2}$:

The following theorem is of still more extensive use in all dynamical disquisitions.

85. *Accelerating forces are proportional to the momentary increments of the squares of the velocities directly, and to the spaces along which they are uniformly acquired inversely.*

Let A′B, A′C, and AD (fig. 14.), be three lines, described in the same or equal times by the uniform action of accelerating forces; the motions along these lines will be uniformly accelerated, and the lines themselves will be proportional to the forces, and may be employed as their measures. On the greatest of them AD, describe the semicircle ABCD, and apply the other two lines A′B, A′C as chords AB, AC. Draw EB, FC perpendicular to AD. Take any small portions B*b*, C*c* of AB and AC, and draw *b e*, *c f* perpendicular to AD, and E*h* and F*k* parallel to AB and AC.

Then, because the triangles DAB and BAE are similar, we have $AD : AE = AD^2 : AB^2$. And because AD is to AB as the velocity generated at D is to the velocity generated at B (the times being equal), we have AD to AE as the square of the velocity at D to the square of the velocity at B; which we may express thus:

$$AD : AE = V^2, D : V^2, B.$$

For the same reasons we have also

$$AD : AF = V^2, D : V^2, C. \text{ Therefore}$$

$$AE : AF = V^2, B : V^2, C.$$

But because in any uniformly accelerated motion, the spaces are as the squares of the acquired velocities, we have also

$$AE : Ae = V^2, B : V^2\, b, \text{ and}$$

$$AF : Af = V^2, B : V^2\, c.$$

Therefore E*e* is to F*f* as the increment of the square of the velocity acquired in the motion along B*b* to the increment of the square of the velocity acquired along C*c*.

But, by similarity of the triangles ABD and E e h, we have

$$AB : AD = E\,e : E\,h;$$ and, in like manner,
$$AD : AC = F\,k : F\,f.$$ Therefore
$$AB : AC = E\,e \times F\,k : F\,f \times E\,h.$$

Now AB and AC are proportional to the forces which accelerate the body along the lines A B and A C; E e and F f are proportional to the increments of the squares of the velocities acquired in the motions along the portions B b and C c; and E h and F k are equal to those portions respectively. The ratio of AB to AC is compounded of the direct ratio of E e to F f; and the inverse ratio of E h to F k. The proposition is therefore demonstrated.

The proportion may be expressed thus:

$AB : AC = \frac{E\,e}{E\,h} : \frac{F\,f}{F\,k}$, and may be expressed by the proportional equation $AB \doteqdot \frac{E\,e}{E\,h}$ or, symbolically, $a \doteqdot \frac{(\dot{v^2})}{\dot{s}}$.

Remark. Because the motion along any of these three lines is uniformly accelerated, the relation between spaces, times, and velocities, may be represented by means of the triangle ABC (fig. 15.); where AB represents the time, BC the velocity, and ABC the space. If BC be taken equal to AB, the triangle is half of the square ABCF of the velocity BC; and the triangle ADE is half of the square ADEG of the velocity DE. Let D d and B b be two moments of time, equal or unequal. Then D d e E and B b c C are half the increments of the squares of the velocities DE and BC, acquired during the moments D d and B b. It was demonstrated, that the ratio of the area D d e E to the area B b c C is compounded of the ratio of DE to BC, and the ultimate ratio of D d to B b. But D d and B b are respectively equal to ι e and $\varkappa$ c. Therefore $D\,d\,e\,E$ is to $B\,b\,c\,C$, in the ratio compounded of the ratio of DE to BC, and the ultimate ratio of $\iota\, e$ to $\varkappa\, c$. If we represent DE and BC by V and v, the $\iota\, e$ and $\varkappa\, c$ must be represented by V' and v' the increments of V and v; and then the compound ratio will be the ratio of VV to $v\,v'$; and if we take the ultimate ratio of the moments, and consequently the ultimate ratio of the increments of the velocities, we have the ratio of $V\dot{V}$ to $v\,\dot{v}$. If, therefore, V^2 and v^2 represent the squares of the velocities, $V\dot{V}$ and $v\,v$ will represent, not the increments of those squares, but half the increments of them.

We may now represent this proposition concerning accelerating forces by the proportional equation $a \doteqdot \frac{v\dot{v}}{\dot{s}}$; and we must consider this as equivalent with $a = \frac{V^2 - v^2}{2\,(S - s)}$; keeping always in mind, that a, V, and v, relate to the same units of time and space, and that a is that number of units of the scale on which S and s are measured, which is run over in *one* unit of time.

This will be more clearly conceived by taking an example. Let us ascertain the accelerative power of gravity, supposing it to act uniformly on a body. Let the spaces be measured in feet and the time in seconds. It is a matter of observation, that when a body has fallen 64 feet, it has acquired a velocity of 64 feet per second: and that when it has fallen 144 feet, it has acquired the velocity of 96 feet per second. We want to determine what velocity gravity communicated to it by acting on it during one second. We have $V^2 = 9216$, and $v^2 = 4096$; and therefore $V^2 - v^2 = 5120$. $S = 144$, and $s = 64$, and $S - s = 80$, and $2\,(S - s) = 160$. Now $a = \frac{5120}{160}, = 32$. Therefore gravity has generated the velocity 32 feet per second by acting uniformly during one second.

86. *The augmentation of the square of the velocity is proportional to the force and to the space jointly.* For because

$a = \frac{v\dot{v}}{\dot{s}}$, we have $a\dot{s} = v\dot{v}$.

Thus we learn, that a given force, acting uniformly on a body along a given space, produces the same increment of the square of the velocity, whatever the previous velocity may have been. Also, in the same manner as we formerly found that the augmentation of the velocity was proportioned to the time during which the force has acted, so the augmentation of the square of the velocity is proportional to the space along which it has acted.

It is pretty plain, that all that we have said of the uniform action of an *accelerating* force may be affirmed of a *retarding force*, taking a diminution or decrement of velocity in place of an increment A uniformly retarded motion is that in which the decrements of velocity in equal times are equal, and the whole decrements are proportional to the whole times of action. Such a motion is the indication of a constant or invariable force acting in a direction opposite to that of the motion. We conceive this to be the case when an arrow is shot perpendicularly upwards; its weight is conceived as a force continually pressing it perpendicularly downwards.

In such motions, however great the initial velocity may be, the body will come to rest: because a certain determined velocity will be taken from the body in each equal successive moment, and some multiple of this will exceed the initial velocity. Therefore the velocity will be extinguished before the end of a time that is the same multiple of the time in which the velocity was diminished by the quantity above mentioned. It is no less evident, that the time in which any velocity will be extinguished by an opposing or retarding force is equal to the time in which the same force would generate this velocity in the body previously at rest. Therefore,

87. *1st.* The times in which different initial velocities will be extinguished by the same opposing force are proportional to the initial velocities.

88. *2d.* The distances to which the body will go till the extinction of its velocity are as the squares of the initial velocities.

3d. They are also as the squares of the times elapsed.

89. *4th.* The distance to which a body, projected with any velocity, will go till its motion be extinguished by the uniform action of a retarding force, is one half of the space which it would describe uniformly during the same time with the initial velocity.

It very rarely happens, that the force which accelerates the body acts uniformly, or with an unvaried intensity. The attraction of a magnet, for example, increases as the iron approaches it. The pressure of a spring diminishes as it unbends. The impulse of a stream of water or wind diminishes as the impelled surface retires from it by yielding. Therefore the effects of accelerating forces are very imperfectly explained, till we have shewn what motions result from any given variation of force, and how to discover the variation of force from the observed motion. This last question is perhaps the most important in the study of mechanical nature. It is only thus that we learn what is usually called *the nature of a mechanical force.* This chiefly consists in the relation subsisting between the intensity of the force and the distance of the substance in which it resides. Thus the nature of that power which produces all the planetary motions, is considered as ascertained when we have demonstrated that its pressure or intensity is inversely as the square of the distance from the body in which it is supposed to reside.

Acceleration expresses some relation of the velocity and time. This relation may be geometrically expressed in a variety of ways. In figure 13. the uniform acceleration or the universal relation between the velocity and the time is very aptly expressed by the constant ratio of the ordinates and abscisses of the triangle $a\,g\,\gamma$. The ratio of $d\,\delta$ to $a\,d$ is the same with that of $e\,\iota$ to $a\,e$, or that of $f\,\phi$ to $a\,f$, &c.; or the ratio of the increment of velocity $\pi\,\kappa$ to the increment of the time $\beta\,\pi$ or $b\,c$, or that of $i\,\phi$ to $\iota\,i$, &c. This ratio $\pi\,\kappa : \pi\,\beta$ is equivalent to the symbol $\frac{\dot{v}}{\dot{t}}$.

But when the spaces described in a varied motion are represented by the areas bounded by a curve line $h\,k\,o$, we no longer have that constant ratio of the increments of the ordinates and abscisses.

90. Therefore, in order to obtain measures of the accelerating forces, or at least of their proportions, let the abscissa $a\,e\,g$ (fig. 13.) of the line $h\,k\,o$ again represent the time of a motion. But let the areas bounded by parallel ordinates now represent the velocities, that is, let the whole area increase during the time $a\,g$ at the same rate with the velocities of the motion along the line AG. In this case the ordinates $b\,i$, $c\,k$, $d\,l$, &c. will be as the accelerations at the instants b, c, d, &c. or in the points B, C, D, &c.

This is demonstrated in the same way as the former proposition (No. 72.) If the accelerating force be supposed constant during any two equal moments $b\,c$ and $f\,g$, the rectangles $b\,c\,p\,t$ and $f\,g\,x\,s$ would express the increments of velocity uniformly acquired in equal times, and their bases $c\,p$ and $f\,s$ would have the ratio of the accelerations, or of the accelerating forces. But as the velocities expressed by the figure increase faster than the times during every moment, the force at the instant c is to the force at the instant f in a greater ratio than that of $c\,p$ to $f\,s$; but, for similar reasons, it is in a less ratio than that of $c\,q$ to $f\,r$; and therefore (as in the other proposition) the force at the instant c is to the force at the instant f as $c\,k$ to $f\,n$.

91. *Cor.* Because $c\,p$ is to $f\,s$ in the ratio compounded of the direct ratio of the rectangle $c\,p\,t\,b$ to the rectangle $f\,s\,x\,g$, and the inverse ratio of the altitude $b\,c$ to the altitude $f\,g$; and because these rectangles are proportional to the increments of velocity, and the ultimate ratio of the altitudes is the ultimate ratio of the moments or increments of the time—we must say, that *the accelerating forces (that is, their intensities or pressures producing acceleration) are directly as the increments of velocity, and inversely as the increments of the times:* Which proposition may be expressed, in regard to two accelerations A and a, by this analogy:

$$A : a = \frac{\dot{V}}{\dot{T}} : \frac{\dot{v}}{\dot{t}}$$

Or by the proportional equation $a \doteq \frac{\dot{v}}{\dot{t}}$. Also $a\,\dot{t} \doteq \dot{v}$, and $\int a\,\dot{t} = v'$. And thus do these theorems extend even to the cases where there cannot be observed an immediate measure, either of velocity or of acceleration; because neither the space nor the velocity increases uniformly.

The theorem $a \doteq \frac{\dot{v}}{\dot{t}}$ is employed when we would discover the variation in the intensity of some natural power. We observe the motion, and represent it by a figure analogous to fig. 13. where the abscissa represents the times, and the area is made to increase at the same rate with the spaces described. Then the ordinates will represent the velocities, or have the proportion of the velocities. Then we may draw a second curve on the other side of the same abscissa, such that the areas of this last curve shall be proportional to the ordinates of the first. The ordinates of this last curve are proportional to the accelerating forces.*

92. On the other hand, when we know from other circumstances that a force, varying according to some known law, acts on a body, we can determine its motion. The intensity of the force in every instant being known, we can draw a line so related to another line representing the time that the ordinates shall be proportional to the forces: The areas will be proportional to the velocities. We can draw another curve to the same absciss, such that the ordinates of this shall be proportional to the areas of the other, that is, to the velocities of the motion. The areas of this second curve will be proportional to the spaces described.

93. We must now observe, that all that has been said concerning the effects of accelerating forces continually varying, relates to *changes* of motion, independent of what the absolute motions may be. The areas of the line whose ordinates represent the velocities do not necessarily represent the spaces described, but the change made on the spaces described in the same time; not the motions, but the changes of motion. If, indeed, the body be supposed to be at rest when the forces begin to act, these areas represent the very spaces that are passed over, and the ordinates are the very velocities. In every case, however, the accelerations are the real increments of the velocities.

This circumstance gives a great extension to our theorems, and enables us to ascertain the disturbances of any species of regular motion, apart from the motions themselves, and thus avoid a complication which would fre-

* See Barrow's *Lect. Geometr.* passim.

quently be inextricable in any other way. And this process, which is merely mathematical, is perfectly conformable to mechanical principles. It is in fact an application of the doctrine of the composition of motion; a doctrine rigidly demonstrated when we measure a mechanical force by the *change* of motion which it produces. Acceleration is the continual composition of a new motion with the motion already produced.

We may learn from this investigation of the value of an accelerating force, that no finite change of velocity is effected in an instant by the action of an accelerating force. When the fig 13. is used for the scale of accelerations, and they are represented by the ordinates of the line *h k o*, the increment of velocity is represented by an area, that is, by a slip of the whole area; which slip must have some altitude, or must occupy some portion of the abscissa which represents time. Some portion of time, however small it may be, must elapse before any measurable addition can be made to the velocity. The velocity must change *continually*. As no motion can be conceived as instantaneous, because this would be to conceive, that in one instant the moving particle is in every point of its momentary path; so no velocity can change, by a finite quantity, in one instant; because this would be to conceive, that in that instant the particle had all the intervening velocities. The instant of change is at once the last instant of the preceding velocity, and the first of the succeeding, and therefore must belong to both. This cannot be conceived, or is absurd. As a body, in passing from one part of space to another, must pass in succession through all the intermediate places; so, in passing from one velocity to another, it must in succession have all the intermediate velocities. It must be *continually* accelerated; we must not say *gradually*, however small the steps.

94. But to return from this digression:

The most frequent cases which come under examination do not shew us the relation between the forces and times, but the relation between forces and spaces. Thus, when a piece of iron is in the neighbourhood of a magnet, or a planet is considered in the neighbourhood of the sun, a force is acting on it in every point of its path, and we have discovered that the intensity of this force varies in a certain proportion. Thus, a spring varies in its pressure as it unbends; gunpowder presses less violently as it expands, &c. &c.

Our knowledge is generally confined to some such effect as this. We know, that while a body is moving along a line ADE (fig. 16.), it is urged forward by a force, of which the intensity varies in the proportion of the ordinates BF, CG, DH, EI, &c. of the line FGHI.

To investigate the motion or change of motion produced by the action of this force, let CD be supposed a very small portion of the space s, which we may express by s'. Draw GK perpendicular to DH. Then, if we suppose that the force acts with the unvaried intensity CG through the whole space CD, the rectangle CDKG will express half of the increment of the square of the velocity (No. 85.) We may suppose that the force acts uniformly along the adjoining small space Dr with the intensity DH. The rectangle DH$o\,r$ will in like manner express another half increment of the square of the velocity. And in like manner we may obtain a succession of such increments. The aggregate or sum of them all will be half the difference between the square of the velocity at B and the square of the velocity at E.

If we employ f to express the indetermined or variable intensity of the accelerating force, and v to express the variable velocity, and v' its increment *uniformly* acquired; then the rectangle CDKG will be expressed by $f\,s'$. We have seen that this is equal to $v\ v'$. Therefore, in every case where we can tell the aggregate of all the quantities $f\ s$, it is plain that we will obtain half the difference between the squares of the velocities in B and E, on the supposition that the intensity of the force was constant along each little space, and varied by starts. Then, by increasing the number, and diminishing the magnitude, of those little portions of the space without end, it is evident that we terminate in the expression of the real state of the case, *i. e.* of a force varying continually; and that in this case the aggregate of these rectangles occupies the whole area AEIF, and is equivalent to the fluent of $f\ s$, or to the symbol $\int f\ \dot{s}$ used by the foreign mathematicians to express this fluent, which they indeed conceive as an aggregate of small rectangles $f\ s'$. And we see that this area expresses half of the augmentation of the square of the velocity. Therefore,

If the abscissa AE (fig. 16.) *of a line* FGI *is the path along which a body is urged by any accelerating force, and if the ordinates* BF, CG, DH, *&c. are proportional to the forces acting in the points* B, C. D, *&c. the intercepted areas* BCGF, BEIF, *&c. are proportional to the augmentations of the square of the velocity.* *

Observe that the areas BCGF and DEIH are also proportional to the augmentations made on the squares of the velocities in B and in D.

Observe also, that it is indifferent what may have been the original velocity. The action of the forces represented by the ordinates make always the same addition to its square; and this addition is half the square of the velocity which those forces would generate in the body by impelling it from rest in the point A.

---

* See Newton's *Principia*, i. 39.

96. Lastly, on this head, observe, that we can state what constant or variable force will make the same augmentation of the square of the velocity by impelling the body uniformly along the same space BE; or along what space a given force must impel the body, in order to produce the same increase of the square of its velocity. In the first case, we have only to make a rectangle BEN φ, equal to the area BEIF, and then B φ is the intensity of the constant force wanted. In the second case, in which the force EO is given, we must make the rectangle A α OE equal to the area BEIF, and AE is the space required.

97 The converse of this proposition, *viz. If the areas are as the increments of the square of the velocity, the ordinates are, as the forces*, is easily demonstrated in the same way; for if the elementary areas CDKG and EIM *e* represent increments of the squares of the velocity, the accelerating forces are in the ratio compounded of the direct ratio of these rectangles, and the inverse ratio of their altitudes, because these altitudes are the increments of the space (No 85.) Now the base CG of the rectangle CDKG, is to the base EI of the rectangle EIM *e* in the same compounded ratio; therefore the force in C is to the force in E as CG to EI.

98. The line *h k o* (fig. 13.) was called by Dr Barrow, who first introduced this extensive employment of motion into geometry), the SCALE *of velocities;* and the line FHL (fig. 16.) was named by him the *scale of accelerations*. Hermann, in his *Phoronomia*, calls it the *scale of forces*. We shall retain this name, and we may call *h k o* of fig. 13. the *scale of* accelerations, when the areas represent the velocities. Newton added another scale of very great use, *viz*. a scale of times. It is constructed as follow.

99. Let ABE (fig. 16.) be the line along which a body is accelerated, and let FHI be the scale of forces, that is, having its ordinates FB, HD, IE, &c. proportional to the

forces acting at B, D, E, F, &c.; let $f\,h\,i$ be another line so related to ABE, that C g is to E i in the inverse subduplicate ratio of the area BFGC to the area BFIE; or, to express it more generally, let the squares of the ordinates to the line $f\,g\,i$ be inversely, as the areas of the line FHI intercepted between these ordinates and the first ordinate drawn through B; then the times of the bodies moving from a state of rest in B are as the intercepted areas of the curve $f\,g\,i$.

For let CD and E e be two very small portions of the space described equal times. They will be ultimately as the velocities in C and E. The area FBCG is to the area FBEI as the square of E i to the square of C g (by construction); but the area FBCG is to FBEI as the square of the velocity at C to the square of the velocity at E (by the proposition;) therefore the square of the velocity at C is to the square of the velocity at E as the square of E i to the square of C g; therefore E i is to C g as the velocity at C to the velocity at E, that is, as CD to E e: but since E i : C g = CD : E e, we have E i × E e = C g. × C D, and the elementary rectangles C $g\,k$ D and E $i\,m\,e$ are equal, and may represent the equal moments of time in which CD and E e were described. Thus the areas of the line $f\,g\,l$ will represent or express the times of describing the corresponding portions of the abscissa.

We may express the nature of this scale more briefly thus. Let B E be the space described with any varied motion, and $f\,g\,l$ a curve, such that its ordinates are inversely as the velocities in the different points of the abscissa, then the area will be as the times of describing the corresponding portions of the abscissa.

100. In all the cases where our mathematical knowledge enables us to assign the values of the ordinates of the figure 16, we can obtain the law of action of the forces, or the nature of the force; and where we can assign the value of the areas from our knowledge of the pro-

portions of the ordinates or forces, we can ascertain the velocities of the motion. We shall give an example or two, which will shew the way in which we avail ourselves of the geometrical properties of figure in order to ascertain the effects of mechanical forces.

1*st*. Let the accelerating force which impels the body along the line AB be constant, and let the body be previously at rest in B; the line which bounds the ordinates that represent the forces must be some line φ HN parallel to AB. The area BDH φ is to the area BEN φ as the square of the velocity at D to the square of the velocity at E. These areas, having equal bases DH and EN, are as their altitudes BD and BE. That is, the spaces described are as the squares of the acquired velocities. And we see that this characteristic mark of uniformly accelerated motion is included in this general proposition.

101. 2*d*. Let us suppose that the body is impelled from A (fig 17.) towards the point C, by a force proportional to its distance from that point. This force may be represented by the ordinates DA, EB, $e\, b$, &c. to the straight line DC We may take any magnitude of these ordinates; that is, the line DC may make any angle with AC. It will simplify the investigation if we make the first force AD = AC. About C describe the circle AH a, cutting the ordinate EB in F; let $e\,b$ be another ordinate, cutting the circle in f very near to F; draw CH perpendicular to AC, and make the arch Hh = fF, and draw $h\,c$ parallel to HC; join FC and DH, and draw F g perpendicular to fb. Let IMLK be another ordinate.

The area DABE is to the area DAKL as the square of the velocity at B to the square of the velocity at K. But DABE is the excess of the triangle ADC above the triangle EBC, or it is half of the excess of the square of CA or CF above the square of CB, that is, half the square of BF. In like manner, the area DAKL is equal to half the square of KM; but halves have the same ratio as the integers; therefore the square of BF is to the square of KM as the square of the velocity at B to the square of the velocity at K; therefore the velocity at B is to the velocity at K as BF is to KM. The velocities are proportional to the sines of the arches of the quadrant AFH described on AC.

Cor. 1. The final velocity with which the body arrives at C, is to the velocity in any other point B as radius to the sine of the arch AF.

Cor. 2. The final velocity is to the velocity which the body would acquire by the uniform action of the initial force at A as 1 to $\sqrt{2}$; for the rectangle DACH expresses the square of the velocity acquired by the uniform action of the force DA; and this is double of the triangle DAC; therefore the squares of these velocities are as 1 and 2, and the velocities are as $\sqrt{1}$ and $\sqrt{2}$, or as 1 to $\sqrt{2}$.

Cor. 3. The time of describing AB is to the time of describing AC as the arch AF to the quadrant AFH.

102. For when the arch Ff is diminished continually, it is plain that the triangle fiF is ultimately similar to CFB, by reason of the equal angles Cib (or CFB) and fiF, and the right angles CBF and fFi; therefore the triangles fg F and CBF are also similar. Moreover, Bb, is equal to Fg, Ff is equal to h H, which is ultimately equal to c C; therefore since the triangles fg F and CFB are similar, we have Fg : Ff = FB : FC = FB : HC; therefore Bb is to c C as FB to HC, that is, as the velocity at B to the velocity at C; therefore Bb and c C are described in equal moments when indefinitely small; consequently equal portions Ff, h H, of the quadrant correspond to equal moments of the accelerated motion along the radius AC; and the arches AF, FM, MH, &c. are proportional to the times of describing AB, BK, KC, &c.

Cor. 4. The time of describing AC with the unequally accelerated motion, is to the time of describing it uniformly with the final velocity, as the quadrantal arch is to the radius of a circle; for if a point move in the quadrantal arch so as to be in F, f, M, H, &c. when the body is in B, b, K, C, it will be moving uniformly, because the arches are proportional to the times of describing those portions of AC; and it will be moving with the velocity with which the body arrives at C, because the arch h H is ultimately = C c. Now if two bodies move uniformly with this velocity, one in the arch AFH, and the other in the radius AC, the times will be proportional to the spaces uniformly described; but the time of describing AFH is equal to the time of the accelerated motion along AC; therefore the proposition is manifest.

103. *Cor.* 5. If the body proceed in the line C a, and be retarded in the same manner that it was accelerated along AC, the time of describing AC uniformly with the velocity which it acquires in C is to the time of describing AC a with the varied motion, as the diameter of a circle to the circumference; for because the momentary retardations at K′, B′, &c. are equal to the accelerations at K and B, &c. the time of describing AC a is the same with that of describing AH a uniformly with the greatest velocity; that is, to the time of describing AC uniformly as AH a to AC, or as the circumference of a circle to the diameter. Therefore, &c. *N. B.* In this case of retarding forces it is convenient to represent them by ordinates K′L, B′E, a D′, lying on the other side of the axis AC a; and to consider the areas bounded by these ordinates as subtractive from the others. Thus the square of the velocity at K′ is expressed by the whole area DACK′L′C, the part CK′L′, being negative in respect of the point DAC. This observation is general.

Cor. 6. The time of moving along KC, the half of AC, by the uniform action of the force at A, is to that of describing AC a by the varied action of the force directed to C, and proportional to the distance from it, as the diameter of a circle to the circumference; for when the body is uniformly impelled along KC by the constant force IK, the square of the velocity acquired at C is represented by half the rectangle IKCH, and therefore it is equal to the velocity which the variable force generates by impelling it along AC (by the way, an important observation). The body will describe AC uniformly with this velocity in the same time that it is uniformly accelerated along KC. Therefore by *Cor.* 5. the proposition is manifest.

Cor. 7. If two bodies describe AC and KC by the action of forces which are every where proportional to the distances from C, their final velocities will be proportional to the distances run over, and the times will be equal.

For the squares of the final velocities are proportional to the triangles ADC, LKC, that is, to AC^2, KC , and therefore the velocities are as AC, KC. The times of describing AC and KC uniformly, with velocities proportional to AC and KC, must be equal; and these times are in the same ratio (*viz.* that of radius to $\frac{1}{4}$ of the circumference) to the times of describing AC and KC with the accelerated motion. Therefore, &c.

Thus by availing ourselves of the properties of the circle, we have discovered all the properties or characters of a motion produced by a force always directed to a fixed point, and proportional to the distance from it. Some of these are remarkable, such as the last corollary; and they are all important; for there are innumerable cases where this law of action obtains in Nature. It is nearly the law of action of a bowstring, and of all elastic bodies, when their change of figure during their mutual action is moderate; and it has been by the help of this proposition, first demonstrated in a particular case by Lord Brouncker and Huyghens, that we have been able to obtain precise measures of time, and consequently of actual motions, and consequently of any of the mechanical powers of Nature. It is for this reason, as well as for the easy and perspicuous employment of the mathematical method of proceeding that we have selected it.

Instead of giving any more particular cases, we may observe in general, that if the intensity of the force be proportional to any power whose index is $n-1$ of the distance, and if a be the distance from the fixed point at which the body begins to be accelerated, and x its distance from that point in any part of the motion, the velocity will be $\doteqdot \sqrt{a^n - x^n}$. This is very plain, because the increment CGHD of the area of fig. 16. which is also the increment of the square of the velocity, is $\doteqdot x^{n-1}\dot{x}$, and the area is $\doteqdot x^n$; and the whole area, corresponding to the distance a, is a^n. Therefore the portion of the area lying beyond the distance x is $a^n - x^n$. This is as the square of the velocity, and therefore the velocity is as the square root $\sqrt{a^n - x^n}$ of this quantity.

This proposition, $f\dot{s} \doteqdot v\dot{v}$, or $f \doteqdot \frac{v\dot{v}}{\dot{s}}$, is the 39th of the first book of Newton's *Principia*, and is perhaps the most important in the whole doctrine of dynamics, whether employed for the investigation of forces or for the explanation of motions. It furnishes the most immediate data for both purposes, but more especially for the last. By its help Newton was able to point out the numerous disturbances of the planetary motions, and to separate them from each other; thus unravelling, as it were, that most intricate motion in which all are blended together. He has given a most wonderful specimen of its application in his Lunar Theory.

We now are able to explain all the puzzling facts which were adduced by Leibnitz and his partisans in support of their measure of the forces of bodies in motion. We see why four springs, equally bent, communicate but a double velocity, and nine springs but a triple velocity;

why a bullet moving twice as fast will penetrate an earthen rampart to a quadruple depth, &c. &c.

This theorem also gives a most perspicuous explanation of the famous doctrine called *conservatio virium vivarum*. When perfectly elastic bodies act on each other, it is found that the sum of the masses multiplied by the squares of the velocities is always the same. This has been substituted, with great encomiums, by the German philosophers in place of Des Cartes's principle, that the quantity of motion in the universe, estimated in one direction, remains always the same. They are obliged, however, to acknowledge, that in the actions of perfectly hard bodies, there is always a loss of *vis viva*, and therefore have denied the existence of such bodies. But there is the same loss in the mutual actions of all soft or ductile or even imperfectly elastic bodies; and they are miserably puzzled how to explain the fact: but both the *conservatio* and the *amissio* are necessary consequences of this theorem.

In the collision of elastic bodies, the whole change of motion is produced during the short time that the bodies are compressed, and while they regain their figure. When this is completed, the bodies are at the same distance from each other as when the mutual action began. Therefore the preceding body has been accelerated, and the following body has been retarded, along equal spaces; and in every point of this space the accelerating and the retarding force has been equal. Consequently the same area of fig. 17. expresses the change made on the square of the velocity of both bodies. Therefore, if V and U are the velocities before collision, and v and u the velocities after collision, of the two bodies A and B, we must have $A \times \overline{V^2 - v^2} = B \times \overline{u^2 - V^2}$, and therefore $A \times V^2 + B \times U^2 \times = A \times v^2 + B \times u^2$.

But in the other class of bodies, which do not completely regain their figure, but remain compressed, they are nearer to each other when their mutual action is ended than when it began. The foremost body has been accelerated along a shorter space than that along which the other has been retarded. The mutual forces have, in every instant, been equal and opposite. Therefore the area which expresses the diminution of the square of the velocity, must exceed the area expressing the augmentation by a quantity that is always the same when the permanent compression is the same; that is, when the relative motion is the same. $A \times \overline{V^2 - v^2}$ must exceed $B \times \overline{u^2 - U^2}$, and $A \times V^2 + B \times U^2$ must exceed $A \times v^2 + B \times u^2$.

This same theorem is of the most extensive use in all practical questions in the mechanical arts; and without it mechanics can go no farther than the mere statement of equilibrium.

Hermann, professor of mathematics at Pavia, one of the ornaments of the mathematical class of philosophers, has given a pretty demonstration of this valuable proposition in the *Acta Eruditorum Lipsiæ* for 1709; and says, that having searched the writings of the mathematicians with great care, he found himself warranted to say, that Newton was the undoubted author, and boasts of his own as the first synthetical demonstration. The purpose of this assertion was not very apparent at the time; but long after, in 1746, when Hermann's papers, preserved in the town-house of Pavia, were examined, in order to determine a dispute between Maupertuis and Koenig about the claim to the discovery of *the principle of least action*, letters of Leibnitz's were found, requesting Hermann to search for any traces of this proposition in the writings of the mathematicians of Europe. Leibnitz was by this time the envious detractor from Newton's reputation; and could not but perceive, that all his contorted arguments for his doctrine received a clear explanation by means of this proposition, in perfect conformity to the usual measure of moving forces. Newton had discovered this theorem long before the publication of the *Principia*, and even before the discovery of the chief proposition of that book in 1666: for in his Optical Lectures, the materials of which were in his possession in 1664, he makes frequent use of a proposition founded on this (see No. 42.) We may here remark, that Hermann's demonstration is, in every step, the same with Dr Barrow's demonstration of it as a theorem merely geometrical, without speaking of moving forces (see *Lect. Geometr.* xi. p. 85. edit. 16.), but giving it as an instance of the transformation of curves, which he calls SCALES of velocity, of time, of acceleration, &c.

The two fundamental theorems $f\dot{t} = \dot{v}$, and $f\dot{s} = v\dot{v}$, enable us to solve every question of motion accelerated or retarded by the action of the mechanical powers of nature. But the employment of them may be greatly expedited and simplified by noticing two or three general cases which occur very frequently.

104. *These may be called similar instants of time, and similar points of space which divide given portions of time, and of space in the same ratio.* Thus the middle is a similar instant of an hour or of a day, and is the similarly situated point of a foot or of a yard. The beginning of the 21st minute, and of the 9th hour, are similar instants of an hour and of a day. The beginning of the 5th inch, and of the 2d foot, are similar points of a foot and of a yard.

105. *Forces may be said to act similarly when their intensities in similar instants of time, or in similar points of space, are in a constant ratio.* Thus in fig. 17. when one body is impelled towards C from A, and another from K, each with a force proportional to the distance of every point of its motion from C, these forces may be said to act similarly along the spaces AC and KC, or during the times represented by the quadrantal arches AFH, KNO. The following propositions on similar actions will be found

very useful on many occasions; but we must premise a geometrical lemma.

106. If there be two lines EFGH (fig. 18), *efgh*, so related to their abscisses AD, *ad*, that the ordinates IK, *ik*, drawn from similar points I and *i* of the abscisses, are in the constant ratio of AE to *ae*; then the area ADHE is to the area *adhe* as the rectangle of AD × AE to the rectangle *ad* × *ae*.

For let each abscissa be divided into the same number of equal and very small parts, of which let CD and *cd* be one in each. Inscribe the rectangles CGID, *cgid*. Then because the number of parts in each axis is the same, the lengths of the portions CD and *cd* will be proportional to the whole abscisses AD and *ad*. And because C and *c* are similar points, CG is to *cg* as AE is to *ae*. Therefore CD × CG : *cd* × *cg* = AD AE : *ad* × *ae*. This is true of each pair of corresponding rectangles; and therefore it is true of their sums. But when the number of these rectangles is increased, and their breadth diminished without end, it is evident that the ultimate ratio of the sum of all the rectangles, such as CDHG to the sum of all the rectangles *cdhg* is the same with that of the area ADHE to the area *adhe*, and the proposition is manifest.

107. *If two particles of matter are similarly impelled during given times, the changes of velocity are as the times and as the forces jointly.*

Let the times be represented by the straight lines ABC (fig. 19.) and *abc*, and the forces by the ordinates AD, BE, CF, and *ad*, *bc*, *cf*. Then if B and *b* are similar instants (suppose the middles) of the whole times, we have BE : *be* = AD : *ad*. Therefore, by the lemma, the area ACFD is to *acfd* as AC × AD to *ac* × *ad*. But these areas are proportional to the velocities (No. 72.), and the proposition is demonstrated. For the same reason, the change of velocity during the time AB is to the change during ab as $AB \times AD$ to $ab \times ad$.

Cor. 1. If the times and forces are reciprocally proportional, the changes of velocity are equal; and if the forces are inversely as the times, the changes of velocity are equal.

108. *If two particles be similarly urged along given spaces, the changes made on the squares of the velocities are as the forces and spaces jointly.*

For if AC (fig. 19.) and $a\,c$ are the spaces along which the particles are impelled, and the forces are as the ordinates AD and $a\,d$, the areas ACFD and $acfd$ are as the changes on the squares of the velocities. But these areas are as $AC \times AD$, and $ac \times ad$. Therefore, &c.

Cor. 2. If the spaces are inversely as the forces, the changes of the squares of the velocities are equal; and if these are equal, the spaces are inversely as the forces.

Cor. 3. If the spaces, along which the particles have been impelled from a previous state of rest, are directly as the forces, the velocities are also as the forces. For, because the changes of the squares of the velocities are as the spaces and forces jointly, they are in this case as the squares of the forces or of the spaces; but the changes of the squares of the velocities are in this case the whole squares of the velocities; therefore the squares of the velocities are as the squares of the forces, and the velocities are as the forces. *N. B.* This includes the motions represented in fig. 17.

109. *If two particles be similarly impelled along given spaces, from a state of rest, the squares of the times are proportional to the spaces directly, and to the forces inversely.*

Let ABC (fig. 19.) $a\,b\,c$ be the spaces described, and AD, ad, the accelerating forces at A and a. Let V, B express the velocity at B, and v, b the velocity at b.

Let GHK and $g\,h\,k$ be curves whose ordinates are inversely as the velocities at the corresponding points of the abscissa. These curves are therefore exponents of the times (No. 99.) Then, because the forces act similarly, we have, by the last theorem, AC × AD : $a c \times a d =$ V^2, B : v^2, b, $= h'b^2$: HB^2. Therefore HB : $h b =$ $\sqrt{ac \times ad} : \sqrt{AC \times AD}$, and therefore in a constant ratio. Call this the ratio of m to n. But, since the ordinates of the lines GHK, $g h k$ are inversely as the velocities, the areas are as the times (No. 99.); and since these ordinates are in the constant ratio of m to n, the areas are in the ratio of AC × m to $a c \times n$. Therefore (calling the times of the motions T and t) we have

$$T : t = m\,AC : nac; \text{ and therefore}$$
$$T^2 : t^2 = m^2 \times AC^2 : n^2 \times ac^2. \text{ But}$$
$$m^2 : n^2 = ac \times ad : AC \times AD. \text{ Therefore}$$
$$T^2 : t^2 = ac \times ad \times AC^2 : AC \times AD \times ac^2,$$
$$\text{Or } T^2 : t^2 = ad \times AC : AD \times ac.$$
$$\text{Or } T^2 : t^2 = \frac{AC}{AD} : \frac{ac}{ad}.$$

The attentive reader will observe, that these three propositions give a great extension to the theorems which were formerly deduced from the nature of uniformly accelerated motion, or of uniform action of the forces, and were afterwards demonstrated to obtain in the momentary action of forces any how variable.

The first of the three propositions, $V : v = F \times T : f \times t$, is the extension of the theorem $f \times \dot{t} = \dot{v}$. The second, $V_2 : v^2 = F \times S : f \times s$, is the extension of the theorem $f \times \dot{s} = v \dot{v}$. And the third, $T^2 : t^2 = \frac{S}{F} : \frac{s}{f}$, is the extension of $f = \frac{\dot{s}}{(t^2)^{\cdot}}$, or of $f \times (t^2)^{\cdot} = \dot{s}$. These theorems hold true of all *similar* actions; and only for this reason, are true of uniformly accelerated motions, or uniform actions.

There remains one thing more to be said concerning the action of accelerating forces. Their magnitude is ascertained by their effect. Therefore that is to be considered as a double force which produces a double quantity of motion. Consequently when a body A contains twice the number of equal atoms of matter, and acquires the same velocity from the action of the force F that another body a, containing half the number of atoms, acquires from the action of a force f, we conceive F to be double of f. That this is a legitimate inference appears clearly from this, that we conceive the sensible weight of a body, or that pressure which it exerts on its supports, as the aggregate of the equal pressure, of every atom, accumulated perhaps on one point; as when the body hangs by a thread, and, by its intervention, pulls at some machine. Without inquiring in what manner, or by what intervention, this accumulation of pressure is brought about, we see clearly that it results from the equal accelerating force of gravity acting immediately on each atom. When this weight is thus employed to move another body by the intervention of the thread, which is attached to one point perhaps of that body, it puts the whole into motion, generating a certain velocity v in every atom, by acting uniformly during the time t. We conceive each atom to have sustained the action of an equal accelerating force, whose measure is $\frac{v}{t}$. Without considering how this force is exerted on each atom, or by what it is immediately exerted, or how it is diffused through the body from the point to which the weight of the other body is applied by means of the thread; we still consider it as the aggregate of the action of gravity on each atom of that other body. Moreover, attending only to the motion produced by it, and perhaps not knowing the weight of the impelling body, we measure it, as a moving force, by considering it as the aggregate of the forces propagated to each atom of the impelled body, and measured by $\frac{v}{t}$. If we know that the impelled body contains the number m of atoms, the aggregate of forces is $m\frac{v}{t}$, or $\frac{mv}{t}$.

But since we measure forces by the quantity of motion which they produce, we must conceive, that when the same force is applied to a body which consists of n particles, and produces the velocity u, by acting uniformly during the same time t, the force $n\frac{u}{t}$ is equal to the force $m\frac{v}{t}$.

110. Newton found it absolutely necessary to keep this circumstance of acceleration clear of all notions of quantity of matter, or other considerations, and to contemplate the affections of motion only. He therefore considered $\frac{v}{t}$ as the true original measure of accelerating force, and $m\frac{v}{t}$ as an aggregate. He therefore calls the aggregate a *vis motrix*, a *moving force*, measured by the *quantity of motion* that it generates. And he confines the term *accelerating force* to the quantity $\frac{v}{t}$, measured by the *acceleration* or *velocity* only. It would be convenient, therefore, also to confine the symbol f to $m\frac{v}{t}$, and to retain the symbol a for expressing the accelerating force $\frac{v}{t}$.

This appellation of *motive force* is perfectly just and simple; for we may conceive it as the same with the accelerating force which produces the velocity m times v in one particle, by acting on it uniformly during the time t. This motion of one particle having the velocity $m\,v$, is the same with that of m particles having each the velocity v.

If therefore a motive force f act on a body consisting of m particles, the accelerating force a is $=\frac{f}{m}\frac{v}{t}$

Therefore the three last propositions concerning the similar, the uniform, or the momentary actions of *moving* forces, when expressed in the most general terms, are,

$$v' \doteq \frac{f}{m\,t'}.$$

$$v^{2\prime} \doteq \frac{f\,s'}{m}, \text{ or } v\,\dot{v} = \frac{f\,\dot{s}}{m}.$$

$$t^2 \doteq \frac{m\,s'}{f}.$$

OF DEFLECTING FORCES.

111. WHEN we observe the direction of a body to change, we unavoidably infer the agency of a force which acts in a direction that does not coincide with that of the body's motion; and we may distinguish this circumstance by calling it a DEFLECTING FORCE. We have already shewn how to estimate and measure this deflecting force, by considering it as competent to the production of that motion which when compounded with the former motion, will produce the new motion. No. 44. Now, as all changes of motion are really compositions of motions or forces, it is evident that we shall explain the action of deflecting forces when we shew this composition.

We may almost venture to say *a priori*, that all deflections must be continual, or exhibit curvilineal motions; for as no finite velocity, or change of velocity, can be produced in an instant by the action of an accelerating force, no polygonal or angular deflection can be produced; because this is the composition of a finite velocity produced in an instant. Deflective motions are all produced by the composition of the former motion, having a finite velocity, with a transverse motion continually accelerated from a state of rest. Of this we can form a very distinct notion, by taking the simplest case of such accelerated motion, namely, an uniformly accelerated motion.

112. Let a body be moving in the direction AC (fig. 20.) with any constant velocity, and when it comes to A, let it be exposed to the action of an accelerating force, acting uniformly in any other direction AE. This alone would cause the body to describe AE with a uniformly accelerated motion, so that the spaces AD, AE would be as the squares of the times in which they are described. Therefore, if AB be the space which it would have described uniformly in the time that it describes AD by the action of the accelerating force, and AC the space which it would have described uniformly while it describes AE by the action of the accelerating force—nothing more is wanted for ascertaining the real motion of the body but to compound the uniform motion in the direction AC with the uniformly accelerated motion in the direction AE. AD is to AE as the square of the time of describing AD to the square of the time of describing AE; that is, as the square of the time of describing AB to the square of the time of describing AC; that is, as AB^2 to AC^2 (by reason of the uniform motion in AC). This composition is performed by taking the simultaneous points B, D, and the simultaneous points C, E, and completing the parallelograms ABFD, ACGE. The body will be found in the points F and G in the instants in which it would have been found at B and C by the uniform motion, or in D and E by the accelerated motion. In the same manner may be found as many points of the real path as we please. It is plain that these points will be in a line AFG, so related to AE that $AD : AE = DF^2 : EG^2$; or so related to the original motion AC, that $AB^2 : AC^2 = BF : CG$, &c. This line is therefore a parabola of which AE is a diameter, DF and EG are ordinates, and which touches AC in A.

Having thus ascertained the path of the body, we can also ascertain the motion in that path; that is, the velocity in any point of it. We know that the velocity in the point G is to the velocity of the uniform motion in the direction AC as the tangent TG is to the ordinate EG; because this is the ultimate ratio of the momentary increment of the arch AFG to the momentary increment of the ordinate EG. Thus is the velocity in every point of the curve determined. We have taken it for granted, that the line of projection touches the path, and that the direction in every point is that of the tangent. To suppose that the curve, in any portion of it, coincides with the tangent, is to suppose that the body is not deflected; that is, is not acted on by a transverse accelerating force: And to suppose that the tangent makes a finite angle with any part of the path, is to suppose that the deflection is not continual, but by starts—both of which are contrary to the conditions of the case. No straight line can be drawn between the direction of the body and the succeeding portion of the path, otherwise we must again suppose, that the deflection is subsultory, and the motion angular.

113. But while the investigation is so easy when the direction and intensity of the deflecting force in every point of the curve are known, the investigation of the deflecting force from the observed motion is by no means easy. The observed curvilineal motion always arises from a composition of a uniform motion in the tangent with some transverse motion. But the same curvilineal motion may be produced by compounding the uniform motion in the tangent with an infinity of transverse motions; and the law of action will be different in these transverse motions according as their directions differ. We must learn, not only the intensity of the deflecting force, and the law of its variation, but also its direction in every point of the curve. It is not easy to find general rules for discovering the direction of the transverse force; most commonly this is indicated by extrinsic circumstances. The deflecting force is frequently observed to reside in, or to accompany some other body. It may

be presumed, therefore, that it acts in the direction of the line drawn to or from that body; yet even this is uncertain. The most general rule for this investigation is to observe the place of the body at several intervals of time before and after its passing through the point of the curve, where we are interested to find its precise direction. We then draw lines, joining those places with the places of the tangent where the body would have been by the uniform motion only. We shall perhaps observe these lines of junction keep in parallel positions: we may be assured, that the direction of the transverse force is the same with that of any of these lines. This is the case in the example just now given of a parabolic motion. But when these lines change position, they will change it gradually; and their position in the point of contact is that to which their positions on both sides of it gradually approximate.

But all this is destitute of the precision requisite in philosophical discussion. We are indebted to Newton for a theorem which ascertains the direction of the transverse force with all exactness, in the cases in which we most of all wish to attain mathematical accuracy, and which not only opened the access to those discoveries which have immortalised his name, but also pointed out to him the path he was to follow, and even marked his first steps. It therefore merits a very particular treatment.

114. If a body describes a curve line ABCDEF (fig. 21.) lying in one plane, and if there be a point S so situated in this plane that the line joining it with the body describes areas ASB, ASC, ASD, &c. proportional to the times in which the body describes the arches AB, AC, AD, &c. the force which deflects the body from rectilineal motion is continually directed to the fixed point S.

Let us first suppose that the body describes the polygon ABCDEF, &c. formed of the chords AB, BC, CD, DE, EF, &c. of this curve: and (for greater simplicity of argument) let us consider areas described in equal successive times; that is, let us suppose that the triangles ASB, BSC, CSD, &c. are equal, and described in equal times. Make B c = AB, and draw c S.

Had the motion AB suffered no change in the point B, the body would have described B c in the equal moment succeeding the first: but it describes BC. The body has therefore been deflected by an external force; and BC is the diagonal of a parallelogram (No. 45, 46.), of which B c is one side, and c C is another. The deflecting force will be discovered, both in respect of direction and intensity, by completing the parallelogram B c C b. B b is the space which the deflecting force would have caused the body to describe in the time that it describes B c or BC. Because B c is equal to BA, the triangles BS c, BSA are equal. But (by the nature of the motion) BSA is equal to BSC. Therefore the triangles BSC and BS c are equal. They are also on the same base BS; therefore they lie between the same parallels, and C c is parallel to SB. But c C is parallel to B b. Therefore B b coincides with BS, and the deflecting force at B is directed toward S. By the same argument, the deflecting force at the angles D, E, F, &c. is directed to S.

Now, let the sides of the polygon be diminished, and their number increased without end. The demonstration remains the same; and continues, when the polygon finally coalesces with the curve, and the deflection is continual.

When areas are described proportional to the times, equal areas are described in equal times; and therefore the deflection is always directed to S. Q. E. D.

The point S may, with great propriety of language, be called the CENTRE OF DEFLECTION, OR THE CENTRE OF FORCES; and forces which are thus continually directed to one fixed point, may be distinguished from other deflecting forces by the name CENTRAL FORCES.

The line joining the centre of forces with the body, and which may be conceived as a stiff line, carrying the body round, is usually named the RADIUS VECTOR.

115. The converse of this proposition, *viz.* that if the deflecting forces be always directed to S, the motion is performed in one plane, in which S is situated, and areas are described proportional to the times—is easily demonstrated by reversing the steps of this demonstration. The motion will be in the plane of the lines SB and B *c*; because the diagonal BC of the parallelogram of forces is in the plane of the sides. Areas are described proportional to the times; for C *c* being parallel to SB, the triangles SCB and S *c* B are equal; and therefore SCB and SAB are equal, &c. &c.

116. *Cor.* 1. When a body describes areas round S proportional to the times, or when it is continually deflected toward S, or acted on by a transverse force directed to S, the velocities in the different points A and E of the curve are inversely proportional to the perpendiculars S *r* and S *t*, drawn from the centre of forces to the tangents in those points; that is, to the perpendiculars from the centre on the momentary directions of the motion: For since the triangles ASB, ESF are equal, their bases AB, EF are inversely as their altitudes S *r*, S *t*. But these bases, being described in equal times, are as the velocities; and they ultimately coincide with the tangents at A and E.

117. *Cor.* 2. If $B\alpha$ and $F\iota$ be drawn perpendicular to SA and SE, we have $SA \times B\alpha = SE \times F\iota$, and $SA : SE = F\iota : B\alpha$: For $SA \times B\alpha$ is double of the triangle BSA, and $SE \times F\iota$ is double of the equal triangle SFE.

118. *Cor.* 3. The angular velocity round S, that is, the magnitude of the angle described in equal times by the radius vector, is inversely proportional to the square of

the distance from S. For when the arches AB, EF are diminished continually, the perpendiculars $B\alpha$ and $F\iota$ will ultimately coincide with arches described round S with the radii SB and SF. Now the magnitude of an angle is proportional to the length of the arch which measures it directly, and to the radius of the arch inversely. In any circle, an arch of two inches long measures twice as many degrees as an arch one inch long; and an arch an inch long contains twice as many degrees of a circle whose radius is twice as short. Therefore, ultimately, the angle ASB is to the angle ESF as $B\alpha$ to $F\iota$, and as SF to SB jointly; that is, as $B\alpha \times SF$ to $F\iota \times SB$. But $B\alpha : F\iota = SE : SA$ (*Cor.* 2.). Therefore $ASB : ESF = SE \times SF : SB \times SA$, $=$ ultimately $SE^2 : SB^2$.

This corollary gives us an ostensible mark, in many very important cases, of the action of a deflecting force being always directed to a fixed point. We are often able to measure the angular motion when we cannot measure the real velocities.

Having thus discovered the chief circumstances which enable us to ascertain the direction of the deflecting force, we proceed to investigate the quantity of this deflective determination in the different points of a curvilineal motion. This is a more difficult task. The momentary effect of the deflecting force is a small deviation from the tangent; and this deviation is made with an accelerated motion. The law of this acceleration regulates the curvature of the path, and is to be determined by it. We may be allowed to observe by the way, that it appears clearly from the form in which Newton has presented all his dynamical theorems, that we are indebted to these problems for the immense improvement which he has made in geometry by his invention of fluxions. The purposes he had in view suggested to his penetrating mind the means for attaining them; and the connection between dynamics and geometry is so intimate, that the same theorems are in a manner common to both. This is particularly the case in all that relates to curvature. We have seen how the curvature of a parabola is produced by a force acting uniformly. The *momentary* action of all finite forces may be considered as uniform; and therefore the curvature will be that of some portion of some parabola; but it will be difficult to determine the precise degree without some farther help. We are best acquainted with the properties of the circle, and will have the clearest notions of the curvature of other curves by comparing them with circles.

The curvature of a circular arch of given length is so much greater as its radius is shorter; for it will contain so many more degrees in the same length; and therefore the change of direction of its extremities is so much greater. Curvatures may always be measured by the length of the arch directly and the radius inversely.

119. Suppose a thread made fast at one end of a material curve ABCD (fig. 22.) and applied to it in its whole length. Taking hold of its extremity D, unfold it gradually from the curve DCBA; the extremity D will describe another curve D *c* *b* *a*. This geometrical operation is called the EVOLUTION of curves, and D *c* *b* *a* is called the EVOLUTE of DCBA, which is called the INVOLUTE of D *c* *b* *a*. Perhaps this denomination has been given from the genesis of the area or surface contained by the two lines, which is folded up and unfolded somewhat like a fan. When the describing point is in *b*, the thread *b* B is, undoubtedly, the momentary radius of a circle *e* *b* *f*, whose centre is B, the point of the involute which it is just going to quit. The momentary motion of *b* is the same, whether it is describing an arch of the evolute passing through *b*, or an arch of a circle round the centre B. The same line *b* *t*, perpendicular to the thread *b* B, touches the circle *e* *b* *f* and the curve D *b* *a* in the point *b*. This

circle $e\,b\,f$ must lie within the curve D b a on the side of b B toward a; because on this side the momentary radius is continually increasing. For similar reasons, the circle e b f lies without the curve on the other side of b B. Therefore the circle e b f both touches and cuts the curve D b a in the point b. Moreover, because every portion of the curve between b and D is described with radii that are shorter than b B, it must be more incurvated than any portion of the circle e b f. For similar reasons, every portion of the curve between b and a must be less incurvated than this circle; therefore the circle has that precise degree of curvature that belongs to the curve in the point b; it is therefore called the EQUICURVE CIRCLE, or the CIRCLE OF CURVATURE, and B is called the centre, and B b the RADIUS OF CURVATURE. It is easy to perceive that no circle can be described which shall touch the curve in b, and come between it and the circle e b f; for its centre must be in some point i of the radius b B. If i b be less than B b, it must fall within the curve on both sides of b, and if i b is greater than B b, the circle must fall without the curve on both sides of B b. The circle e b f lies closer to the curve, has closer contact with it, than any other, and has therefore got the whimsical name of OSCULATING CIRCLE; and this sort of contact was called OSCULATION.

This view of the genesis of curve lines is of particular use in dynamical discussions. It exhibits to the eye the perfect sameness of the momentary motion, and therefore of the momentary deflection, in the curve and in the equicurve circle, and leaves the mind without a doubt but that the forces which produce the one will produce the other. A great variety of curves may be described in this way. If perpendiculars be drawn to the curve D b a in every point, they will intersect each other, each its immediate neighbour, in the circumference of the curve DBA: and geometry teaches us how to find the curve DBA which shall produce the curve D b a by evolution.

It is a matter worthy of remark, that the path of a body that is deflected from rectilineal motion by a *finite* force, varying according to any law whatever, may always be described by evolution. This includes almost every case of the action of deflecting forces; none being excepted but when, by the opposite action of different forces, the body is in equilibrio in one single point of its path.

Our task is now brought within a very narrow compass, namely, to measure the deflection in the arch of a circle.

Had the motion represented in fig. 21. been polygonal, it is plain that the deflecting force in the point B is to that in the point E as the diagonal B b of the parallelogram ABC b to the diagonal E i of the parallelogram DEF i; therefore let ABCZY be a circle passing through the points A, B, and C, and let the radius vector BS cut the circumference in Z; draw AZ, CZ, and the diagonal AC, which necessarily bisects and is bisected by the diagonal B b. The triangles b BC and CBZ are similar; for the angle C b B is equal to the alternate angle A B b or ABZ, which is equal to ACZ, standing on the same chord AZ. And the angle CB b, or CBZ, is equal to CAZ, standing on the same chord CZ; therefore the remaining angle b CB is equal to the remaining angle AZC; therefore ZA is to AC as BC to B b, and B b $= \frac{AC \times BC.}{AZ}$ In like manner E $i = \frac{DF \times EF}{E\,z}$

Now let the points A and C continually approach, and ultimately coalesce with B; it is evident that the circle ABCZY is ultimately the equicurve or coinciding circle at the point B, and that AS ultimately coalesces with, and is equal to, BS, and that AC×BC is ultimately $2BC^2$; therefore ultimately

$$B\,b : E\,i = \frac{2BC^2}{BZ} : \frac{2EF^2}{E\,z}, \text{ or} = \frac{BC^2}{\frac{1}{2}BZ} : \frac{EF^2}{\frac{1}{2}E\,z}.$$

Now BC and EF, being described in equal times, are as the velocities. B b and E i are the measures of the velocities which the deflective forces at B and E would generate in the time that the body describes BC or EF, and are therefore the measures of those forces. *They are as the squares of the velocities directly, and inversely as those chords of the equicurve circles which have the directions of the deflection.*

Observe, that B b or E i is the third proportional to half of the chord and the arch described; for

$$Bb : BC = BC : \frac{BZ}{2}.$$

It is evident, that as the arches AB, BC, continually diminish, AC is ultimately parallel to the tangent B r, and BO is equal to the actual deflection from the tangent. The triangles BOC and AOZ are similar, and $BO = \frac{OC^2}{OZ}$, or ultimately $= \frac{BC^2}{BZ}$. We may measure the forces by the actual deflections, because they are the halves of the measures of the generated velocities; and we may say that

120. *The actual momentary deflection from the tangent is a third proportional to the deflective chord of the equicurve circle and the arch described during the moment.*

Either of these measures may be taken, but we must take care not to confound them. The first is the most proper, because the change produced on the body (which is the *immediate* effect and measure of the force) is the determination, left inherent in it, to move with a certain velocity. This is the measure also which we obtain by means of the differential or fluxionary calculus; but the other measure must be obtained when our immediate object is to mark the actual path of the body.

Thus have we obtained a measure of deflecting force, and, in the most important cases, a method of discovering its direction. It only remains to point out the relation between the intensity of the force, the curvature of the path, and the velocity of the motion. These three circumstances have a necessary connection; for we see that the intensity is expressed by certain values of the other two in the formula $f = \frac{\text{Arch}^2}{\frac{1}{2}\text{ Chord}}$, or $f = \frac{2\,BC}{BZ}$. The deflective velocity B b is acquired in the time that the body describes BC; therefore the deflective velocity is to the velocity in the curve as B b to BC. The velocity B b is acquired by an accelerated motion along BO; for while, by progressive motion, the body describes BC, it deflects from the tangent through a space equal to the half of B b, because the momentary action of the deflecting force may be considered as uniform. The progressive velocity BC may be generated by the same force, uniformly acting through a space greater than BC; call this space x. The spaces along which a body must be uniformly impelled in order to acquire different velocities, are as the squares of those velocities; therefore $Bb : BC = Bo : x$; but $Bb : BC = BC : \frac{1}{2}BZ$; therefore $Bb : BC^2 = Bb : \frac{1}{2}BZ$, and $Bb : \frac{1}{2}BZ = Bo : x$, and $Bb : Bo = \frac{1}{2}BZ : x$; but B o is $\frac{1}{2}$ of B b; therefore x is $\frac{1}{4}$ of BZ; that is,

121. *The velocity in any point of a curvilineal path, is that which the deflecting forces in that point would generate in the body by impelling it uniformly along one fourth part of the deflective chord of the equicurve circle.* If the velocity increase, the chord of the equicurve circle must increase; that is, the path becomes less incurvated. If the force be increased, the curvature will also increase, for the chord of curvature will be less.

There is another general observation to be made on the velocity of a curvilineal motion, which greatly assists us in our investigations.

122. *If a body describes a curve by the action of a force always directed to a fixed point, and varying according to any proportion whatever of the distances from that point, and if another body, acted on by the same centripetal force, move toward the centre in a straight line, and if in any one case of equal distances from the centre of force the two bodies have equal velocities, they will have equal velocities in every other case of equal distances from the centre.*

Let one body be impelled from A (fig. 23.) toward C along the straight line AVDEC, and let another be deflected along the curve line VIKk. About the centre C describe concentric arches ID, KE, very near to each other, and cutting the curve in I and K, and the line AC in D and E; draw IC, cutting KE in N, and draw NT perpendicular to the arch IK of the curve, and complete the parallelogram ITNO. Let the bodies be supposed to have equal velocities at I and at D.

Then, because the centripetal forces are supposed to be the same for both bodies when they are at equal distances, the accelerating forces at D and I may be represented by the equal lines DE and IN; but the force IN is not wholly employed in accelerating the body along the arch IK, but, acting transversely, it is partly employed in incurvating the path. It is equivalent to the two forces IO and IT of which only IT accelerates the body. Now IKN is a right-angled triangle, as is also the triangle INT; and they are similar; therefore $IN : IT = IK : IN$, or $DE : IT = IK : DE$; that is, the force which accelerates the body along DE is to the force which accelerates the body along IK as the space IK is to the space DE; therefore (No. 86.) the increment of the square of the velocity acquired along DE is equal to the increment of the square of the velocity acquired along IK. But the velocities at D and I were equal, and consequently their squares were equal; and these having received equal increments, therefore the squares of the velocities at E and K are equal, and the velocities themselves are equal. And since this is the case in all the corresponding points of the line AC and the curve VIK, the velocities at all equal distances from C will be equal.

It is evident that the conclusion will be the same, if the bodies, instead of being accelerated by approaching the centre in the straight line AC, and in the curve VIK, are moving in the opposite directions from E to A, or from I to V, and are therefore retarded by the centripetal force.

123. *Cor.* Hence it follows, that if a body be projected from any point, such as V, of the curve, in a line tending straight from the centre, with the velocity which it had in that point of the curve, it would go to a distance VA, such, that if it were impelled along AV by the centripetal force, it would acquire its former velocity in the point V; also in any point between V and A it will have the same velocity in its recess from the centre that it has there in its approach to the centre.

The line BLFG, whose ordinates are as the intensities of the centripetal force in A, V, D, E, or in A, V, I, K, may be called the SCALE or EXPONENT of force; the areas bounded by the ordinates AB, VL, DF, EG, &c. drawn from any two points of the axis, are as the squares of the velocity acquired by acceleration along the intercepted part of the axis, or in any curvilineal path, while the body approaches the centre, or which are lost while the body retires from it. When we can compute these areas we obtain the velocities (see No. 102.)

We are now in a condition to solve the chief problem in the science of dynamics, to which the whole of it is, in a great measure, subservient. The problem is this.

Let a body be projected with a known velocity from a given point and in a given direction, and let it be under the influence of a mechanical force, whose direction, intensity, and variation, are all known: it is required to determine its path, and its motion in this path, for any given time?

124. This problem is susceptible of three distinct classes of conditions, which require a different investigation.

1. The force may act in one constant direction; that is, in parallel lines.

2. The force may be always directed to a fixed point.

3. It may be directed to a point which is continually changing its place.

1st, When the force acts in parallel lines, the problem is solved by compounding the rectilineal accelerated motion which the force would produce in its own direction with the uniform motion which the projection alone would have produced. The motion must be curvilineal, when the accelerating force is transverse, in any degree whatever, to the projectile motion; and the curvilineal path must be concave on that side to which the deflecting force tends; for the force is supposed to act incessantly. The place of the body will be had for any time, by finding where the body would have been at the end of that time by each force acting alone, and by completing the parallelogram. Thus, suppose a body projected along AB (fig. 20.) while it is continually acted on by a force whose direction is AD. Let D and B be the places where the body would be at the end of a given time. Then the body will, at the end of that time, be in F, the opposite angle of the parallelogram ABFD. But it has not described the diagonal AF, because its motion has been curvilineal, as we shall find by determining its place at other instants of this time.

The velocity in any point F is found by first determining the velocity at D, and making DT to DF as the velocity at D to the velocity at B (that is, the velocity of projection, because the motion along AB is uniform.)

Then draw TF. Then AB is to TF as the constant velocity of projection to the velocity at F. We have seen already (No. 112—119.) that TF is a tangent to the curve in F. Hence we may determine the velocity at F in another way. Having determined the form of the path in the way already described, by finding its different points, draw the tangent F *d*, cutting the line DA in *d*. Then the velocity at A is to that at F as AB to *d* F. Hence also we see, that the velocities in every point of the curve are proportional to the portion of the tangents at those points which are intercepted between *any* two lines parallel to AD.

Either of these methods for ascertaining the velocity, in this case of parallel deflections, will in general be easier than the general method in No. 121. by the equicurve circle.

It was thus that Galileo discovered the parabolic motion of heavy bodies.

2*d*. We must consider the motions of bodies affected by centripetal or centrifugal forces, always tending to one fixed point. This is the celebrated *inverse problem of centripetal forces*, and is the 42d proposition of the first book of Newton's *Principia*. We shall give the solution after the manner of its illustrious author; because it is elementary, in the purest sense of the word, keeping in view the two leading circumstances, and these only, namely, the motion of approach and recess from the centre, and the motion of revolution. By this judicious process, it becomes a pattern by which more refined, and, in some respects, better solutions should be modelled. At the same time we shall supply some steps of the investigation which his elegant conciseness has made him omit.

125. Let a body, which tends to C (fig. 24) with a force proportional to the ordinates of the exponent BLFG, having the axis CA, be projected from V in the direction

VQ, with the velocity which the centripetal force would generate in it by accelerating it along AV. It is required to determine the path or orbit VIK of the body, and its place I in this orbit, at the end of the assigned time T?

Suppose the thing done, and that I is the place of the body. About the centre C, with the distances CV and CI, describe the circles YV and ID. Draw CIX to the circumference, and draw the ordinate DF of the exponent of forces, producing it toward x, and produce the ordinate VL toward a. Let V t be the distance to which the body would go along the tangent VQ in the time T, and join t C. Let this be supposed done for every point of the curve. Let $a\,i\,k$ and $a\,x\,y$ be two curves so related to the curve VIK, that the ordinate DF cuts off an area V$a\,i$D equal to the orbital sector VCI, and an area V $a\,x$D equal to the circular sector VCX.

Then, because the velocity of projection is given, the distance V t is known, and the area of the triangle VC t. But this is equal to the area VCI, by the laws of central forces (No. 115.) Therefore the area V$a\,i$D is given. Also, because the area VCI increases in the proportion of the time, the area V a i D increases at the same rate. Therefore, having these subsidiary curves $a\,i\,k$, $a\,x\,y$, the problem is solved as follows:

Draw an ordinate D i, cutting off an area V $a\,i$ D proportional to the time, and describe a circle DIR. Then draw a line CX, cutting off a sector VCX, equal to the area V $a\,x$ D cut off by the ordinate D $i\,x$. This line will cut the circle DR in the point I, which is the point of the orbit that was demanded.

But the chief difficulty of the problem consists in the description of the two subsidiary curves $a\,i\,k$ and $a\,x\,y$, into which the lines VIK and VXY are transformed. We attain this construction by resolving the motion in the arch of the orbit into two motions, one of which is in the direction of the transverse force, or of the radius vector, and the other is in the direction of revolution, or perpendicular to the radius.

Let V k and IK be two very small arches described in equal moments, and therefore ultimately in the ratio of the velocities in V and I (No. 73.) Describe the circle KE, cutting IC in N. Draw KC and k C, and $k\,n$ perpendicular to VC.

The element ICK of the orbit is $= \frac{\text{IC} \times \text{KN}}{2}$, or to $\frac{1}{2}$ IC × KN. This is equal to the element D $i\,k$ E of the area V $a\,i$ D, or to D i × DE, or to D i × IN. Therefore IN : KN $= \frac{1}{2}$ IC : D i, or 2 IN : KN = IC : D i, and D $i = \frac{\text{IC}\quad\text{KN}}{2\ \text{IN}}$.

Now let A $l\,f\,g$ be the exponent of the velocities, that is (N° 86.), let V l^2 be to D f^2 as ABLV to ABFD, or V l : D $f = \sqrt{\text{ABLV}} : \sqrt{\text{ABFD}}$. Make V v and I i in the tangents respectively equal to V l and D f. Draw $v\,u$ and $i\,o$ perpendicular to VC and IC, and $v\,m$ perpendicular to LV produced. Let $m\,r\,z$ be an equilateral hyperbola, having VC, ZC, for its asymptotes, and cutting FD produced in r. Then the ordinates V m, D r, are inversely proportional to CV, CD, or V m : D r = CD : CV, = CI : CV. But because the momentary sectors VC k and ICK are equal, $k\,n$: KN = CI : CV. Therefore,

	V m : D r = $k\,n$: KN
but	V v : V m = V k : $k\,n$
and	I i (or D f) : V v = IK : V k
therefore	I i : D r = IK : KN
but	I i : $i\,o$ = IK : KN, by similar triangles.

Therefore D $r = i\,o$, and $i\,o$: V m = VC : CI.

Also, by similarity of triangles, I o : $i\,o$ = IN : KN, and 2 I o : $i\,o$ = 2 IN : KN.

Now it was shewn, that in order that the space D $i\,k$ E may be equal to the space ICK, we must have

$$2\,IN : KN = IC : Di$$

or $$2\,Io : io = IC : Di$$

but $$io : Vm = VC : IC$$

therefore $$2\,Io : Vm = VC : Di$$

and $Di = \frac{VC \times Vm}{2\,Io}$.

Having obtained Di, we easily get Dx; for the ultimate ratio of ICK to XCY is that of IC^2 to VC^2. Therefore make

$$IC^2 : VC^2 = Di : Dx.$$

Thus are the points of the two subsidiary curves aik, axy, determined.

The rectangle $VC \times Vm$ is a constant magnitude; and is given because VC is given, and Vm is the given velocity Vl, diminished in the ratio of radius to the sine of the given angle CVQ.

But the line $2\,Io$ is of variable magnitude, but it is also given, by means of known quantities. Io^2 is $= Ii^2 - io^2$, $= Df^2 - Dr^2$, and $Io = \sqrt{Df^2 - Dr^2}$. Moreover, $Df^2 =$ ABFD, and $Dr^2 = \frac{VC^2 \times Vm^2}{IC^2}$. Therefore

$2\,Io = 2\sqrt{ABFD - \frac{VC^2 \times Vm^2}{IC^2}}$, expressed in known quantities, because ABFD is known from the nature of the centripetal force.

Let the indeterminate distance CI or CD be $= x$, and let the ordinate DF, expressing the force, be y. Let VC be a, and Vm be c, and let ab be a rectangle equal to the whole area of the exponent of forces lying between the ordinate AB and the ordinate CZ, so that $ab - \int y\dot{x}$ may represent the indeterminate area ABFD.

We have $Di = \frac{ac}{2\sqrt{ab - \int y\dot{x} - \frac{a^2c^2}{x^2}}}$

and $Dx = \frac{a^3c}{2x^2\sqrt{ab - \int y\dot{x} - \frac{a^2c^2}{x^2}}}$.

REMARK. We have hitherto supposed that the velocity of projection is acquired by acceleration along AV. But this was merely for greater simplicity of argument, and that the final values of D i and D x might be easier conceived. In whatever way the velocity is acquired, it will still be true, that when in any point V we make V l to V m as the momentary increment V k of the arch is to the perpendicular $k\,n$ on the radius vector, we shall have in every other point, such as I, the line D f to the line D r as the increment IK of the arch to KN. And in the final equation D f will still be expressed by $\sqrt{a\,b - \int y\,x}$.

126. *Cor.* 1. The angle which the path of the projectile makes with the radius vector is determined by this solution; for I i is to $i\,o$ as radius to the sine of this angle; which sine is therefore $= \frac{a\,c}{x\sqrt{a\,b - \int y\,x}}$

127. *Cor.* 2. When the magnitude $\frac{a\,c}{x}$ is equal to $\sqrt{a\,b - \int y\,x}$, the path is perpendicular to the radius vector, and the body is a tone of the apsides of its orbit, and begins to recede from the centre after having approached to it, or begins to approach after having receded.

128. *Cor.* 3. The curvature of the orbit VIK is also determined in every point; for the curvature of any line is inversely as the radius of the equicurve circle, and this is to the chord which passes through C as radius to the sine of the angle CI i. Because the velocity in any point I is $= \sqrt{\text{ABFD}}$, and is equal to what the centripetal force at I would produce, by impelling the body along $\frac{1}{4}$th of the deflective chord of the equicurve circle, we have this chord $= 4\,\frac{\text{ABFD}}{\text{DF}}$. Or we obtain it by taking a third proportional to the momentary deflection and the momentary arch of the curve, or by other processes of the higher

geometry, all proceeding on the quantities furnished in this investigation.

129. Such is the solution of this celebrated problem given by Newton, who may justly be called the inventor of the science of which it is the chief result, as well as of the geometry, by help of which it is prosecuted. We cannot give this glory to Galileo; for his simple problem of the motion of bodies affected by uniform and parallel gravity, however just and elegant his solution may be, was peculiar; and the same must be said of Huyghens's doctrine of centrifugal forces. Besides, these theorems had been investigated by Newton several years before, as corollaries which he could not pass unnoticed, from his general method. This is proved by letters from Huyghens.

130. Whether we consider this problem as a piece of mere mathematical speculation, or attend to its consequences, which include the whole of the celestial motions in all their extent and complication, we must allow it to be highly interesting, and likely to engage much attention in the period of ardent inquiry which closed the last century. Accordingly, it was no sooner known, by the publication of the *Mathematical Principles of Natural Philosophy* in 1686, than it occupied the talents of the most eminent mathematicians; and many solutions were published, some of which differ considerably from Newton's; some are more expeditious, and better fitted for computation. Of these, the most remarkable for originality and ingenuity are those of de Moivre, Hermann, Keill, and Stewart. The last differs most from the methods pursued by others. M'Laurin's propositions on this subject, and in that part of his fluxions which treats of curvature, are highly valuable, classing the chief affections of curvilineal motions geometrically, as they are suggested by the fluxionary method; and then shewing, in a very instructive manner, the connection between these mathematical affections of motion and the powers of nature which produce them.

This part of his excellent work is a fine example of the real nature of all inquiries in dynamics; shewing that it differs from geometry little more than in the language, in which the word *force* is substituted for *acceleration*, *retardation*, or *deflection*. We recommend the careful perusal of these propositions to all who wish to have clear conceptions of the subject. Dr John Keill and Dr Horsley (bishop of Rochester) have given particular treatises on the motions of bodies deflected by centripetal forces inversely proportional to the cubes of the distances; induced by the singular motions which result from this law of action, and the multitude of beautiful propositions which they suggest to the mathematician. Newton, indeed, first perceived both of these peculiarities, and has begun this branch of the general problem. He first demonstrated the description of the logarithmic and hyperbolic spirals, and indicated a variety of curious recurring elliptical spirals, which would be described by means of this force, and shewing that they are all susceptible of accurate quadrature. Several of those authors affect to consider their solutions as more perfect than Newton's, and as more immediately indicating the remarkable properties of such motions; and also affect to have deduced them from different and original principles. But we cannot help saying, that their claims to superiority are very ill founded; there is not a principle made use of in their solutions which was not pointed out by Newton, and employed by him. The appearance of originality arises from their having taken a more particular concern in some general property of curvilineal motions; such as the curvature, the centrifugal force, &c. and the making that the leading step of their process. But Newton's is still the best; because it is strictly elementary, aiming at the two leading circumstances, the motion to or from the centre, and the motion of revolution round that centre. To these two purposes he adapted his two subsidiary curves.

131. Is it not surprising, that 25 years after the publication of Newton's *Principia*, a mathematician on the continent should publish a solution in the Memoirs of the French Academy, and boast that he had given the first demonstration of it? Yet John Bernoulli did this in 1710. Is it not more remarkable that this should be precisely the solution given by Newton, beginning from the same theorem, the 40th. I. Prin. following Newton in every step, and using the same subsidiary lines? Yet so it is. Bernoulli actually reduces the whole to two functions; namely, $\frac{ac}{\sqrt{ab - \int \varphi \dot{x} - \frac{a^2 c^2}{x^2}}}$ and $\frac{a^2 c}{\sqrt{a b x^4 - \int \varphi x^4 \dot{x} - a^2 c^2 x^2}}$; which last is plainly the same with Newton's $\frac{Q \times CX^2}{A^2 \sqrt{ABDF - Z^2}}$; because Newton's $\frac{Q}{A}$ is the same with $\frac{ac}{x}$ and Newton's $A^2\sqrt{ABFD - Z^2}$ is the same with $x^2 \sqrt{ab - \int \varphi \dot{x} - \frac{a^2 c^2}{x^2}}$, which Bernoulli has changed into $\sqrt{a b x^4 - \int \varphi x^4 x - a^2 c^2 x^2}$.

Bernoulli's chief boast in this dissertation is, that *now* philosophers may be assured that the planets will always describe conic sections; a truth of which they had not as yet received any proof; because, says he, Newton's argument for it in the corollary of the 13th proposition is inconclusive, and because he had not been able to accommodate his demonstration of the 41st and 42d proposition to the particular case of the planetary gravitation. Newton's demonstration in the corollary of the 13th proposition is just, founded on the principle on which the very demonstration of the 42d, adopted by Bernoulli, proceeds, and without which that demonstration is of no force; namely, that a body, in given circumstances of situation, velocity, direction, and centripetal force, can describe no other figure than what it really describes. Newton did not accommodate the demonstration of the 42d proposition to the planetary motions, because he had already demonstrated the nature of their orbits; but mentions the case of a force proportional to the reciprocal of the cubes of the distance; not as a deduction from the 42d, but because it *was not* a deduction from it, and admitted of a very singular and beautiful investigation by methods totally and essentially different.

132. It cannot be expected that we can proceed to consider the various departments of this celebrated problem. We are only giving the outlines of the general doctrines of dynamics; and we have bestowed more time on those which are purely elementary than some readers may think they deserve. We were anxious to give just conceptions of the fundamental principles of dynamics; because we know that nothing else can entitle it to the name of a demonstrative science, and because we see much indistinctness and uncertainty, and a general vagueness or want of precision, in several elementary works which are put into the hands of persons entering on the study. This leads to errors of more consequence than a person is apt to think; because they affect our leading thoughts of mechanism itself, and our notions of the intimate nature of the visible universe.

133. Many very general doctrines of dynamics remain untouched; all, namely, that relate to the rotative motion of rigid bodies, and all that relate to the mutual action of bodies on each other in the way of impulse. These subjects will be resumed in another part of this work.

Notwithstanding these omissions, we must observe that no new principle remains to be considered. We have given all that are necessary; and there is no question that occurs in the cases omitted, which cannot be completely answered by means of the propositions already established. We have taught how to discover the existence and agency of a mechanical force, to measure and characterise it, and then to state what will be its various effects, according to the circumstances of the case.

134. Proceeding by these principles, men have discovered an universal fact, that every ACTION of one body on another is accompanied by an equal REACTION of that other on the first, in the opposite direction; that is, to express it in the language of dynamics, "all the phenomena which "make us infer that the body A possesses a force by "which it changes the motion of the body B, shew, at "the same time, that B possesses a force by which it "makes an equal and opposite alteration in the motion "of A." This, however, is not a doctrine of abstract dynamics: it does not flow from our idea of force; therefore it was not included in our list of the LAWS OF MOTION. It is a part of the mechanical history of nature, just as the law of universal gravitation is; and it might be called the law of UNIVERSAL REACTION. Newton has, in our apprehension, deviated from his accustomed logical accuracy, when he admits, as a third *axiom* or law of motion, that reaction is always equal and contrary to action. It is a physical law, in as far as it is *observed* to obtain through the whole extent of the solar system. But Newton himself did not, in the subsequent part of his work, treat it as a logical axiom; that is, as a law of human thought with respect to motion: for he labours to prove, by *fact and observation*, that it really obtains through the whole extent of the solar system; and it is in this discovery that his chief claim to unequalled penetration and discernment appears.

135. Availing ourselves of this fact, we, with very little

trouble, state all the laws of impulsion. The body A, for example, moving to the westward at the rate of eight feet per minute, overtakes the double body B, moving at the rate of four feet per minute. What must be the consequence of their mutual impenetrability, and of the equality and contrariety of action and reaction? Their motions must be such that both sustain equal and opposite changes. They must give, in some way or other, *this* indication of possessing equal and opposite forces. This will be the case if, when the changes are completed, A and B move on in contact at the rate of four feet per minute: for here A has produced in each half of B a change of motion two; and therefore a totality of change equal to four. This is the effect, the mark, the measure, of the *impulsive* force of A; for it is the whole *impulsion*. B has produced in A a change of motion four, equal to the former, and in the opposite direction. This is the effect, mark, and measure, of the *repulsive* force of A; for it is the whole *repulsion*. And this is all that we observe in the collision of two lumps of clay; and the observation is one of the facts on which the reality of the physical law of equal action and reaction is founded: and we can make no farther inference from *this* fact.

But the event might have been very different. A and B may be two magnets floating on corks upon water, with their north poles fronting each other. We know, by other means, that they really possess forces by which they equally repel each other. The dynamical principles already established tell us also what must happen in this case. That both conditions of equal reaction and sensible repulsion may be fulfilled, A must come to rest, and B must move forward at the rate of four feet per minute. The same thing must happen in the meeting of perfectly elastic bodies, such as billiard balls. If elastics are known to be imperfect in any degree, our dynamical principles will still state the effect of their collision, in conformity to the law of equal reaction.

136. In like manner, all the motions of rotation are explained or predicted by means of the same principles of dynamics applied to the force of cohesion. This is considered as a moving force, because, when the attraction of a magnet acts on a bit of iron attached to one end of a long lath floating on water, the whole lath is moved, although the magnet does not act on it at all: some other force acts on it; it is its cohesion; which is therefore a moving force, and the subject of dynamical discussion.

137. And thus it appears that these subjects do not come necessarily, nor, perhaps, with scientific propriety, under the category of dynamics, but are parts of the mechanical history of nature. Yet, did a work like ours give room in this place, the study of mechanical nature might be considerably improved, by giving a system of such *general* doctrines as involve no other notions but those of force and its measures, and the hypothesis of equal reaction. Some very general, nay universal, consequences of this combination might be established, which would greatly assist the mechanician in the solution of difficult and complicated problems. Such is the proposition, that *the mutual actions of bodies depend on their relative motions only, and require no knowledge of their real motions.* This principle simplifies in a wonderful manner the most difficult and the most frequent cases of action which nature presents to our view; but at the same time gives a severe blow to human vanity, by forcing us to acknowledge that we know nothing of the real motion of any thing in the universe, and never shall know any thing of it, till our intellectual constitution, or our opportunities of observation, are completely changed.

138. M. D'Alembert has made this principle still more serviceable for extricating ourselves from the immense complication of actions that occurs in all the spontaneous

phenomena of nature, by presenting it to us in a different form, which more distinctly expresses what may be called the *elements* of the actions of bodies on each other. His proposition is as follows (*see his Dynamique*, p. 73.):

"In whatever manner a number of bodies change their motions, if we suppose that the motion which each body would have in the following moment, if it were perfectly free, is decomposed into two others, one of which is the motion which it really takes in consequence of their mutual actions, the other will be such, that if each body were impressed by this force alone (that is, by the force which would produce this motion) the whole system of bodies would be in equilibrio."

This is almost self evident: for if these second constituent forces be not such as would put the system in equilibrio, the other constituent motions could not be those which the bodies really take by the mutual action, but would be changed by the first.

For example, let there be three bodies P, Q, R, and let the forces A, B, C, act on them, such as would give them the velocities p, q, r, in any directions whatever, producing the momenta, or quantities of motion, $P \times p$, $Q \times q$, $R \times r$, which we may call A, B, C, because they are the proper measures of the moving force. Let us moreover suppose, that, by striking each other, or by being any how connected with each other, they cannot take these motions A, B, and C, but really take the motions a, b, and c. It is plain that we may conceive the motion A impressed on the body P, to be composed of the motion a, which it really takes, and of another motion α. In like manner, B may be resolved into b, which it takes, and another β; and C into c and $\varkappa$. The motions will be the same, whether we act on P with the force A, or with the two forces a and α; whether we act on Q with the force B, or with b and β; and on R with the force C, or with c and $\varkappa$. Now by the supposition, the bodies actually take the motions a, b, and c; therefore the motions α, β, and $\varkappa$ must be such as will not derange the motions a, b, and c; that is to say, that if the bodies had only the motions α, β, and $\varkappa$ impressed on them, they would destroy each other, and the system would remain at rest.

Mr D'Alembert has applied this proposition with great address and success to the very difficult questions that occur in the motions and actions of fluids, and many other most difficult problems, such as the precession of the equinoxes, &c. The cause of its utility is, that in most cases it is not difficult to find what forces will put a system in equilibrio; and combining these with the known extraneous forces whose effects we are interested to discover, we obtain the motions which really follow the mutual action of the bodies.

This is not, properly speaking, a principle: it is a form in which a general fact may be conceived. In the same way the celebrated mathematician De la Grange observed, that a system of bodies, acting on each other in any way, is in equilibrio, if there be impressed on its parts forces in the inverse proportion of the velocities which each body takes in consequence of their action or connection; and he expresses this universal fact by a very simple formula; and, calling this also a principle, he solves every question with ease and neatness, by reducing it to the investigation of those velocities. In this way he has written a complete system of dynamics, to which he gives the title of *Mechanique Analytique*, full of the most ingenious and elegant solutions of very interesting and difficult problems; and all this without drawing a line or figure, but accomplishing the whole by algebraic operations.

139. But this is not teaching mechanical philosophy; it is merely employing the reader in algebraic operations, each

of which he perfectly understands in its quality of an algebraic or arithmetical operation, and where he may have the fullest conviction of the justness of his procedure. But all this may be (and, in the hands of an expert algebraist, it generally is,) without any notions, distinct or indistinct, of the things, or the processes of reasoning that are represented by the symbols made use of.

OF PROJECTILES,

OR THE

MOTION OF SUBLUNARY BODIES IN FREE SPACE.

140. In the process of our examination of the motions in the solar system, it appears that terrestrial gravity, or the heaviness of common sublunary bodies, is only a particular case of the mutual tendency of all matter towards all matter. It further appears, that a body on the surface of our globe gravitates in a line that is directed very nearly to the centre of the earth; and that the intensity of this gravitation is inversely proportional to the square of its distance from this centre.

Bodies let fall, or projected in any direction on the surface of this earth, move under the influence of this force, and their motions may be computed from the general doctrines of dynamics in the same manner as we computed the motions of the planets. They will either fall in the direction of gravity, or will rise in the opposite direction, or will describe a curve line concave toward the earth, which will be an ellipsis, parabolic, hyperbolic, or a circle, according as the velocity and direction of the projection may have been combined.

But, in the greatest projections that we can make, the force of gravity is so nearly the same in every point of the path, that we may suppose it to be accurately so, without any sensible error, were it ten times greater than it is. Therefore in all disquisitions about projectiles, it would be useless affectation to embarrass ourselves with the variations. None of our projectiles rise a mile in the air, which is about $\frac{1}{3900}$ of the mean radius of the earth, and will occasion a diminution of gravity nearly equal to $\frac{1}{1950}$, a quantity altogether insignificant.

For the same reasons, although the directions of gravity in the different points of the projectile's flight, are lines converging nearly to the centre of the earth, we may consider them as all parallel, because none of our projectiles fly four miles, which produces a convergency of nearly four minutes, a deviation from parallelism which needs not be regarded.

In general, therefore, we may consider all such projectiles as under the influence of equal gravity acting in lines parallel to the vertical or plumb-line drawn through the place of projection. This reduces the theory of projectiles to a great degree of simplicity.

Accordingly, this is the first department of mechanical philosophy which first received improvement by the application of mathematical knowledge. We are indebted for this fortunate introduction of mathematics into the doctrines of motion, to the celebrated Florentine Galileo Galilei. This excellent philosopher read his discourses on local motion, about the beginning of the 17th century. Those lectures contain the whole of this doctrine, nearly in the state in which it continued till about the middle of last century. There is no branch of natural philosophy that has met with so much assistance and encouragement, it having been considered in all nations as the foundation of the art of gunnery; an art unfortunately too much connected with the security of every nation. It has therefore been patronised by princes and magistrates—most costly establishments have been made for its cultivation; the mathematicians have occupied themselves with its problems, and more numerous and expensive volumes have been published on this than on any other part of mechanical philosophy. Yet there is none in which so little improvement has been made. Galileo's lessons contain every thing that has been done in a scientific way, till M. Robins in 1746 gave it a form altogether new.

141. We shall first consider the perpendicular ascents and descents of heavy bodies; and in the next place their curvilineal motion when projected in directions deviating from the vertical.

142. The motion of a falling body is uniformly accelerated, and that of a body thrown straight upward, is uniformly retarded.

For the accelerating or retarding force is constant, and therefore the motions are such as were considered in articles 80, &c.

143. All the characteristic phenomena of these motions having already been sufficiently considered, all that is wanted for the application to this class of mechanical phenomena is merely one experimental determination of the accelerative power of gravity, that is, the velocity, or increment of velocity which gravity will generate in a body by acting on it uniformly during some given time. Galileo, who first demonstrated that an invariable gravity must produce a uniformly accelerated motion, was also among the first who appealed to experiment in all inquiries. We now think lightly of this, and wonder that a man shall think of another argument who has this in his power. But when Galileo began to communicate his knowledge to the world, this was the last support that a philosopher would think of. They had received a parcel of topics from their master, which had been handed down in the schools during many ages; and from these was every

thing accounted for or explained. Aristotle, or his immediate pupils, had said that the velocities of falling bodies increased with their weights; Galileo's doctrine was incompatible with this, and he thought himself obliged to use arguments in his support. He said that if Aristotle's doctrine be true, two crown pieces must fall faster when sticking together than when unconnected, which, said he, is contrary to common experience. Not doubting that he had convinced his audience, he described the experiments which he was to exhibit next day, shewing that in a double time, a body would fall four times as far, &c. The experiments were performed in the dome of the great church, before a vast concourse of people, and succeeded most perfectly. Yet so little were the philosophers moved by this kind of argument, that they represented Galileo as a dangerous person, unfriendly to the state, and he was obliged to leave his native city in a few days, and take shelter in Padua. It is very remarkable that Baliani, one of the first geometers and mathematicians of that age, and who perfectly understood Galileo's speculations on this subject, should teach another doctrine, reviving, or supporting an old scholastic assertion that the velocity of a falling body might be as the space fallen through, calling this motion also a uniformly accelerated motion.

144. Galileo found more difficulty than one should expect in his endeavours to obtain an exact measure of the power of gravity, and indeed could not obtain one that was satisfactory. But the difficulty of the task, and his struggle to accomplish it, were big with advantages to science. A body falls so fast, that a considerable error in the conclusion arises from a very small error in estimating the time; and the great difficulty was how to estimate the time. It was in this casting about for a measure of a small portion of time that Galileo first thought of the pendulum. His penetrating and sagacious mind enabled him to see that there must be a fixed proportion

between the time of a vibration and that of falling through its length, although his mathematical knowledge did not yet enable him to find it out; he saw an immediate consequence of this if true, namely, that the vibrations of two pendulums should be in the subduplicate ratio of the lengths, because this must be the proportions of the times of falling through those lengths (55.) This he would try; and he found that it was so. Delighted with this success, he immediately compared the time of falling from the top of the great dome with that of a pendulous vibration, by making a pendulum of such a length that it peformed precisely one vibration in the time of the fall. In this time, the body, moving with the final velocity, would describe a space double of that fallen through. He then counted with patience the number of vibrations made by his pendulum in an interval of time, measured by the transit of two stars. Thus he obtained the time, and the velocity generated in that time by the uniform action of gravity. Galileo made this to be about 31 feet of our measure in a second, and said that it was certainly somewhat more; because his experiments on falling bodies convinced him that their motion is retarded by the air.

These efforts and resources of an ingenious mind are worthy of record, and are instructive to others. But Galileo did not attain the accuracy in this measure that we now possess. The honour of the accurate statement of the time of a pendulous oscillation, and that of the fall through its length, was reserved for Mr Huyghens. This proportion was determined by him by a most ingenious and elegant physico-mathematical process. He also gave us the pendulum clock, by which time can be measured with as much accuracy as a line can be divided.

Aided by these inventions, we have now obtained the most precise measure of the accelerating powers of gravity; and we can now say that its intensity is such in the latitude of London, that by acting uniformly on a body for one second of time, it generates in it the velocity of 32 feet two inches per second, and a heavy body falls 16 feet one inch in that time.

These are standard numbers, of continual use in all mechanical discussions, and should be carefully kept in remembrance. Not only so, but we should acquire distinct notions of them in this respect, viz. as standard numbers. Gravity is known to us in two ways; our most familiar acquaintance with it is as a pressure, which we feel when we carry a heavy body. With this we can compare the pressure of a spring, the exertion of an animal, the pressure of a stream of water or wind, the intensity of an attraction, &c. by setting them in opposition and equilibrium. The philosopher, and especially the physical astronomer, and cultivator of the Newtonian philosophy, is well acquainted with gravity as an accelerating and a moving force, capable of accelerating, retarding, or deflecting the body in which it inheres, or on whose intimate particles it acts without intermedium. He can compare the gravity of a stone with that of the moon, or of Jupiter, or with the force that produces the precession of the equinoxes. The general mechanician, observing that all other pressures, such as that of a spring, of an animal, &c. are also moving forces, by combining those two aspects of gravity, makes a most important use of it by comparing other forces with weights, and thence inferring the motions which those forces will produce. Thus, knowing that an arrow $\frac{7}{8}$ oz. weight, by falling 18 inches acquires the velocity of $10\frac{1}{2}$ feet per second, he infers, that when drawn to the head by a bow of 62 pounds, it will be discharged with the velocity of 233 feet per second.

We shall therefore, in future, compare every force with gravity, and express the accelerative power of this standard by 32, meaning that by acting on every particle of a body for a second, it will generate the velocity of 32 feet per second, and cause the body to describe 16 feet with a motion uniformly accelerated. We may find it convenient, on some occasions, to use the numbers 386, and 193, which are the inches in $32\frac{2}{12}$ and $16\frac{1}{12}$ feet.

145. The questions that interest us at present are those concerning the relations between the time, t, of any fall, the height, h, of that fall, and the velocity, v, that is uniformly acquired in falling; so that when any one of those things is given, the others may be found out.

I. Since the variations of velocity are proportional to the times in which they are produced, we have

$$1': t'' = 32 : 32\, t''$$

$$\text{and } v' = 32\, t''$$

$$\text{and } t'' = \frac{v'}{32}.$$

N. B. The time t is always supposed to be a number of seconds, and the height h a number of feet, and the velocity v a number of feet uniformly moved over in one second.

A falling body therefore acquires an increment of 32 feet per second in every second of its fall, and an ascending body has its velocity lessened as much during every second of its rise. A body *falling* during four seconds acquires the velocity of 128 feet per second.

But if the body has been projected downward, with the velocity of 100 feet per second, then, at the end of 4″, it is moving at the rate of 228 feet per second.

A body projected straight upwards with the velocity of 160 feet per second, will at the end of the first second of its rise, have the velocity 128. At the end of 2″ it will be moving at the rate of 96 feet per second. Its velocity at the end of the third second will be 64. At the end of the fourth second it will be 32, and at the end of five seconds it will stop, and begin to fall.

The times of the rise and the subsequent fall are equal.

146. II. Since the heights are as the squares of the times of the fall or ascent, we have

$$1'': t^2 = 16 : 16\,t^2$$
$$\text{and } h = 16\,t^2 \text{ and } \sqrt{h} = 4\,t$$
$$\text{also } t^2 = \frac{h}{16}, \text{ and } t = \frac{\sqrt{h}}{4}.$$

A heavy body, falling during four seconds, falls 256 feet.

A body rising straight upwards 144 feet employs 3 seconds in its ascent.

147. III. Because the heights fallen through are also proportional to the squares of the velocities acquired at the end of the fall, we have

$$32^2 : v^2 = 16 : h$$
$$\text{and } h = \frac{16}{32^2}\,v^2, \text{ and } \sqrt{h} = \frac{4}{32}\,v, = \frac{v}{8}$$

and, conversely, $v = 8\,\sqrt{h}$, and $v^2 = 64\,h$.

148. All questions concerning the perpendicular ascents and descents of heavy bodies may be solved by means of the two equations

$$v = 32\,t = g\,t$$
$$h = 16\,t^2, = \tfrac{1}{2}g\,t^2.$$

An easy mode of extempore computations is had, by remarking that since a heavy body falls 16 feet in a second, and acquires the velocity 32, it falls 1 foot in $\frac{1}{4}$th of a second, and acquires the velocity D.

In every second of the fall, the velocity is increased by 32—and in every foot of the fall, the square of the velocity is increased by 64.

In many questions, particularly in hydraulics, it is convenient to have the measures in inches.

149. Now, $\sqrt{193} : \sqrt{1} = 386 : 27{,}785$. Therefore a heavy body by falling one inch acquires the velocity 27,785 inches, or nearly $27\frac{3}{4}$ inches per second.

150. Did gravity impel a body uniformly along a space equal to the radius of the earth, it would generate the velocity, which would enable the body to describe a parabola, having the centre of the earth for its focus. If projected straight upwards with this velocity, it would never return.

151. Now $\sqrt{16} : \sqrt{\text{Earth's rad.}} = 32 : 36{,}680$ feet. This is the velocity now spoken of. Suppose the earth uniformly dense, and a pit to the centre. A heavy body would acquire, by falling down this pit the velocity 25,866. Greater velocities than either of these can be produced by forces which we know. *Aurum fulminans* expands with the velocity of at least 42 miles per second.

It does not seem necessary to insist further on the rectilineal ascents and descents of heavy bodies, and therefore we proceed to consider their curvilineal motions, when projected in any direction that deviates from the perpendicular. These are the motions which are understood to form what is called PROJECTILES.

152. These motions are not only interesting to the philosophical mechanist, as examples of a constant deflecting force, and a uniform deflection in parallel lines, but also to the artillerist; because the motion of shot and shells are cases of this question, which comprehend the whole of his art. It has therefore been very much cultivated; and there is no branch of mechanical philosophy on which so much has been written, or so many experiments made for its improvement. The experimental cultivation of this branch could scarcely be prosecuted by private persons; but, in all the states of Europe, there are public establishments for this purpose, and no expence has been spared for bringing to perfection an art in which the fate of nations has unfortunately much dependence.

But, notwithstanding this liberal encouragement, and the numberless volumes which have been published on the subject, it cannot be said to have improved much as a science since it came out of the hands of its inventor, and his immediate pupil Tartaglia; and we shall be greatly disappointed if we look for that nice agreement between the results of the most approved theory and what we observe in the flight of great shot and shells. The theory, however, is unexceptionable; and the enormous deviations that we see in the actual performance of artillery, is owing to the resistance of the air. This was long considered as insignificant, even after Newton had given us sufficient information to the contrary. But the gentlemen of the profession made little account of the speculations of a private philosopher, and continued to regulate their theories by notions of their own. They have been at last convinced of their mistake by the curious experiments and discoveries of Mr Robins, and are improving their practice in some measure. But we now find, that the theory of the motion of heavy bodies through a resisting fluid, is one of the most abstruse and difficult tasks that the mechanician can take in hand.

At present, we are about to consider this subject merely as a particular case of motions regulated by gravitation, reserving the particular consideration of the modifications of these motions by the resistance of the air, till we shall have made ourselves acquainted with the general laws of such resistance.

153. Let a body (Plate II. fig. 1.) be projected in any direction AB, which deviates from the vertical AW. Then it would move on in this direction, and in equal succeeding moments would describe the equal spaces AB, BH, HI, IK, KL, &c. But suppose, that when the body is at B it receives an instantaneous impulse in the direction of the vertical BB′, such that by this impulse it would describe the line B b uniformly in the same time that it would have continued its motion along BH. Or, to speak more accurately, let the motion or velocity B b be compounded with the motion BH. The body must describe the dia-

gonal BC of a parallelogram B b CH, and, at the end of this second moment, it must be in C, in the vertical line HCC′, and moving with the velocity BC. Therefore, in the third moment it would describe CN, equal to BC. But let another impulse in the direction of the vertical CC′ generate the velocity C c, equal to B b. By the composition of this with the motion CN, the body will describe the diagonal CD of the parallelogram C c DN, and at the end of the third moment must be in D, moving in the direction and with the velocity CD. It would describe DO equal to CD in the fourth moment. Another impulse of gravity D d, in the vertical, and equal to either of the former impulses, will make the body describe DE; and an equal impulse E e will deflect the body into EF; and another impulse F f will deflect it into FG, &c.

Thus it is plain that the body, by the composition of these equal and parallel impulses, will describe the polygonal figure ABCDEFG, all in one vertical plane, and in every instant or point, such as E, will be found in the vertical line KE, drawn from the point at which it would have arrived in that instant by the primitive projection.

Now, let the interval between these impulses be diminished, and their number be increased, without end. It is evident that this polygonal motion will ultimately coincide with the motion in a path of continued curvation, by the continual and unvaried action of gravity.

The line described by the body has evidently the following properties.

154. 1*st.* If a number of equidistant vertical lines BB′, HCC′, IDD′, KEE′, &c. be drawn, cutting the curve in B, C, D, E, &c.; and if the chords AB, BC, CD, DE, &c. drawn through the points of intersection, be produced till they cut the verticals in H, N, O, P, &c. the intercepted portions HC, ND, OE, PF, &c. are all equal.

155. 2*d.* The curve is a parabola, in which the verticals

BB′, CC′, &c. are diameters. The property mentioned in the last paragraph belongs exclusively to the parabola. As the circle is the curve of uniform deflection in the direction of the radius, so the parabola is the curve of uniform deflection in the direction of the diameter. That the curve in which the chords drawn through the intersection of equidistant verticals cut off equal portions of these verticals is a parabola, is easily proved in a variety of ways. Since Bb, Cc, Dd, Ee, are all equal, and the verticals are equidistant, $BcdE$ must be a straight line. So must $CdeF$; BE must be parallel to CD, and CF to DE. Therefore BF and CE are parallel, and are bisected in m and o by the vertical DD′. Also, if FC be produced till it meet the next vertical in i, iB is equal to Dm. All this is very plain. Hence

$$iB, \text{ or } Dm : dm = BF : mF, = mF : oE;$$
$$\text{but } \quad dm : Do = mF : oE;$$
$$\text{therefore } Dm : Do = mF^2 : oE^2;$$

and D, E, F, are in a parabola, of which Dm is a diameter, and oE, mF are semiordinates. We should prove, in the same manner, that BG is parallel to CF, and AG to BF, and $Dm : DD' = mF^2 : D'G^2$, and the points D, F, G, in the same parabola.

156. Thus we have demonstrated, that the equal and parallel impulse of gravity produces a motion in a parabola whose diameters are perpendicular to the horizon. This was the great discovery of Galileo, and the finest example of his genius. His discoveries in the heavens have indeed attracted more notice, and he is oftener spoken of as the first person who shewed the mountains in the moon, the phases of Venus, the satellites of Jupiter, &c. But in all these he was obliged to his telescope; and another person who had common curiosity would have seen the same things. But, in the present discovery, every step was an effort of judgment and reasoning, and the whole investigation was altogether novel. No attempt

had been made, since the first dawn of mechanical science, to explain a curvilineal motion of any kind; and even the law of the composition of motion, though faintly seen by the ancients, had never been applied to any use (except by Stevinus) till this sagacious philosopher saw its immense importance, and brought it into constant service.

157. The process employed by Galileo in this investigation, and which has been copied by almost all the writers on the subject, is considerably different from the one now gone through. Galileo supposes the heavy body to fall in the vertical BB′ with a uniformly accelerated motion, describing spaces as the squares of the times. He supposes this motion to be compounded with the uniform motion in the direction of the tangent BR. Then, supposing that B t and BT are fallen through while B r and BR are described by the motion of projection, it follows, that because B r is to BR as the time of describing B r to the time of describing BR, we shall have $Bt : BT = Br^2 : BR^2$. Therefore, completing the parallelograms B t C r, BTSR, we have $Bt : BT = tC : TS$, and the points B, C, S, are in a parabola, whose diameter is BT, and has BR a tangent in B.

No doubt, the result of these suppositions agrees perfectly with the phenomena, and gives a very easy and elegant solution of the question. But, in the first place, it is more difficult, or takes more discourse, to prove this continued composition of motion (almost peculiar to the case) than to demonstrate the parabolic figure: and, secondly, it is not a just narration of the fact of the procedure of nature. There is no composition of such motions as are here supposed. When the body is at C, there is not a motion in the direction parallel to B r, compounding itself with a motion in the vertical, having the velocity which the falling body would have as it passes through the point t. The body is really moving in the direction CS of the tangent to the parabola, and it there receives the same infinitesimal impulse of gravity that it received at B. Its deflection, therefore, from the line of its motion, does not make any finite angle with that motion. Therefore, although Galileo's demonstration does very well for a mere mathematical process, like the navigators calculation of the ship's place by tables of difference of latitude and departure, it by no means answers the purpose of the philosophical investigation of a natural phenomenon. The method we have followed is a bare narration of the facts—considering the motion of the body in every instant as it really is, and stating the force then really affecting its motion.

We have not scrupled to make use of the method employed by Newton in the demonstration of his fundamental proposition on curvilineal motions, first conceiving the action of gravity to be subsultory, and the motion to be polygonal, and then inferring a similar result from the uninterrupted action of gravity. But if any person is so fastidious as to object to this, (as John Bernoulli has done to Newton's method,) he may remark, that the motion B *b* which we compared with BH, in order to produce the motion BC, is just double of the space B *t*, through which the body falls during the motion along BH. Therefore the figure will be such that the curvilineal deflection will be one half of B *b*, or of HC, and the tangent to the curve, whatever it is, will bisect HC. Then, during the next moment, since the deflective action of gravity is supposed the same, the body will be as much deflected from its path in C, that is, from the new tangent CS, whatever direction that tangent may have, as it was in the preceding moment. This gives us *s* D equal to *r* C, and this obtains throughout. Without entering on any discussion on the progress of the deflection in the different points of the arch BC or CD, it is enough for our purpose to shew that the curve described is such that when equidistant verticals are drawn, and tangents drawn through their intersections with the curve, the portions of the verticals cut off by the tangents are everywhere equal. This also is a property of the parabola exclusively. That BCD is a parabola, of which BT is a diameter, and BR a tangent, is easily seen. For, drawing D u parallel to BR, it is plain that v N $= 2\ r$ C, and ND $= 2\ s$ D, $= 2\ r$ C. Therefore v D $= 4\ r$ C, and B $u = 4$ B t, and B t : B $u = t\ C^2 : u\ D^2$. And we should prove, in the same manner, that y E $= 9\ r$ C, &c.

Having thus ascertained the general nature of the path of a projectile, we must now examine its motion in this path, determining its velocity in the different points, and the time employed in the description of the arches. For this purpose we must first ascertain the precise parabola described under the conditions of the projection, that is, depending on its direction and velocity. To do this in a way naturally connected with the acting forces, we shall consider the velocity of projection as having been generated by falling through some determinate height.

158. Let us therefore suppose that the body is projected from B, in the direction BR, with the velocity acquired by falling through the vertical VB. Make BT equal to VB, and BR equal to VT or 2 VB, and, lastly, draw TS parallel to BR, meeting the parabola in S.

It is plain that BR is the space which would be uniformly described with the velocity of projection in the time of falling through VB. Also B r is the space that would be uniformly described, with the same velocity, in the time of falling through B t. Therefore BR is to B r as the time of falling through VB to that of falling through B t. But, since BT is equal to VB, B r is to BR as the time of falling through B t to the time of falling through BT. Therefore BR is to B r as the time of falling through VB to that of falling through B t. But, since BT is equal to VB, B r is to BR as the time of falling through Bt to the time of falling through BT. Therefore we have Bt : BT $=$ Br^2 : BR^2. But, in the parabola, we have Bt : BT $= t$C^2 : TS^2, $=$ Br^2 : TS^2. Therefore TS is equal to BR or to twice VB or BT. Therefore TS^2 $=$ 4 BT^2, $=$ 4 BT $\times$ BV, $=$ BT $\times$ 4 BV. But, in a parabola, the square of any ordinate TS is equal to the rectangle of the absciss BT and the parameter of that diameter. Therefore 4 VB is the parameter of the diameter BT, and VB is the fourth part of that parameter.

If, therefore, the horizontal line VZ be drawn, it is the directrix of the parabola described by a body projected from B in any direction, with the velocity acquired by falling from V.

159. *Cor.* 1. As this is true for any other point, C, D, &c. it follows that the velocity in any point of the path is that which a heavy body would acquire by falling from the directrix to that point.—*N. B.* This agrees with the determination given in Art. 121. in the most general terms, for curvilineal motions of every kind. For it is well known that the equicurve circle passing through the vertex of any diameter of a parabola, cuts off a chord equal to its parameter. Now this is evidently the deflective chord in the present case, because the diameters are all vertical lines, in the direction of gravity. It agrees equally with the determination given in Art. 122.

160. *Cor.* 2. Hence also we learn that the velocities in any two points, such as B and D, are proportional to the portions $v\,y$ and D$\,t$ of the tangents through those points which are intercepted by the same diameters. Thus, $v\,y$ is a portion of the tangent B$\,y$, intercepted by the diameters DD and EE′, which also intercept a portion of the tangent D$\,t$. For these portions of tangents are in the subduplicate ratio of the lines VB and ZD. Now the velocities acquired by falling through VB and ZD are in this subduplicate ratio of the spaces fallen through.

161. Such is the Galilean Theory of the parabolic motion of projectiles, a doctrine valuable for its intrinsic excellence, and which will always be respectable among philosophers, as the first example of a problem in the higher department of Mechanical Philosophy.

We are now to consider it as the foundation of the art of gunnery. But it may be affirmed, at setting out, that the theory is of very little use for directing the practice of cannonading. Here it is necessary to approach as near as possible to the object, and the hurry of service allows no time for geometrical methods of pointing the piece after each discharge. When the gun is within 300 yards of the object, the gunner points it straight on it, or rather a little above, to compensate for the small deflection which obtains, even at this small distance. Sometimes the piece is elevated at a small angle, and the shot, discharged with a very moderate velocity, drops on the ground, and bounds along destroying the enemy's troops. But, in all these cases, the gunner is directed entirely by practice, and it cannot be said that the parabolic theory is of any service to him.

Its principal use is for directing the bombardier in the throwing of shells. With these it is proposed to destroy buildings, to break through the roofs of magazines, to destroy troops, by bursting among them, &c. Such objects, being generally under cover of the works of a place, cannot be hit by a direct shot, and therefore the shells are thrown with such elevated directions that they get ever the works, and produce their effect. These shells are of great weight, sometimes exceeding 200 lb. The mortar from which they are discharged must be exceedingly strong, that it may resist the explosion of the powder able to impel this vast mass to a great distance. They are therefore most unwieldy, and it is found most convenient to have them almost solid, and unchangeable in their position. The shell is thrown to the intended distance by employing a proper quantity of powder. This is found incomparably easier than to vary the elevation of the mortar. We shall also find, that when a proper elevation has been selected, a small deviation from it, unavoidable in such service, is much less detrimental than if another elevation had been chosen. Mortars, therefore, are frequently cast in one piece with their bed or carriage, having an elevation that is not far from being the best on all ordinary occasions, and the rest is done by repeated trials with different charges of powder.

Still, however, in this practice, the parabolic motion must be understood, that the bombardier may avail himself of any occasional circumstance that may be of advantage to him. We shall therefore consider the chief problems that the artillerist has to resolve, but with the utmost brevity; and the reader will soon see, that more minute discussion would be of very little service.

162. The velocity of projection is measured by the fall that is necessary for acquiring it. It has generally been called the force, or IMPETUS; we shall distinguish it by the symbol f. Thus, in Plate II. fig. 2, 3, 4, FA is the height through which the body is supposed to fall, in order to acquire the velocity with which it is projected from A.

The distance AB between the piece of ordnance and the object, is called the AMPLITUDE, and also the *range* $=r$.

Let the angle EAB contained between the vertical and the direction of the object be called the ANGLE OF POSITION, $=p$.

And let the angle DAB contained between that direction and the axis of the piece, be called the direction of the mortar $=d$, and let z express the zenith distance or angle EAD, contained between the axis of the mortar and the vertical line AE.

The leading problem, from which almost all the others may be derived, is the following.

163. Let a shell be thrown from A (fig. 3, 4), with the

velocity required by falling through the vertical FA so as to hit an object B. Required the direction AD of the projection.

Let AH be a horizontal line, and AB the line of position of the object. In the vertical AF, take AE = 4 AF, and on EA describe an arch of a circle ED d A, which shall touch the line of position AB. Draw through the object the vertical line BD, cutting the circle in D and d, and join AD and A d. I say that AD or A d are the directions required. Join ED and E d.

For, because AB touches the circle in A, the angle ADE is equal to the exterior angle EA a, or DBA, and the alternate angles EAD, ADB are equal. The triangles ADB and EAD are therefore similar, and DB : DA = DA : AE, and $DA^2 = DB \times EA$. Therefore B is in a parabola, of which the vertical AI is a diameter, AD a tangent in A, and AE the parameter of that diameter. If, therefore, the body be projected from A in the direction AD, with the velocity acquired by falling through FA, the fourth part of this parameter, it will describe a parabola AVB which passes through B.

By the same reasoning, it is demonstrated that the body will hit the mark B, if projected in the direction A d with the same velocity, describing the parabola A v B.

From this very simple construction, we may draw several very instructive corollaries.

164. *Cor.* 1. When the vertical line passing through B cuts the circle EDA, it always cuts it in two points D and d, giving two directions AD and A d, either of which will solve the problem.

Cor. 2. But if the vertical through b only touch the circle, as it touches it in one point only, it gives but one direction, along which the body must be projected to hit the mark b. This direction is AG.

165. *Cor.* 3. The direction AG evidently bisects the

angle EAB, and the directions AD and A *d* are equidistant from the middle direction AG.

166. *Cor.* 4. If the vertical passing through B do not meet the circle described on AE, according to the conditions specified, the object is too remote to be struck by a body projected from A with the velocity acquired by falling from F. There is no direction that will enable it to go so far on the line AB. The distance A *b* is the greatest possible with this velocity, and it is attained by taking the elevation AG which bisects the angle EAB. We may therefore call A *b* the *maximum range* on the line AB, and AG the *middle direction*.

167. *Cor.* 5. The distances on a given line of position to which a body will be projected in a given direction AD, are proportional to the squares of the velocities of projection. For the figure being similar, the range AB has the same proportion to AF, the fall necessary for acquiring the velocity. Now the falls are in the duplicate ratio of the velocities required by falling. Therefore, &c.

The converse of this problem is solved with the same facility of construction.

168. Let a body be projected in the direction AD, with the velocity acquired by falling through FA, it is required to find to what distance it will reach on the line AB.

Describe, as before, on AE, $= 4$ AF, the circle EDA, touching AB, and cutting AD in D. Through D draw the vertical DB, cutting AB in B. Then B is the point to which the projectile will reach. The proof is too evident to need discussion.

Lastly, suppose the object B to be given, and also the line of direction AD (which is a very common case, seeing that our mortars are often so fixed in their beds that their elevation can be very little altered) it is required to determine the velocity that must be given to the projectile.

Draw through the object the vertical BD, meeting the direction in D. Draw the vertical AE, and make it a third proportional to DB and DA, that is, make AE $= \frac{DA^2}{DB}$, and take FA $= \frac{EA}{4}$. Then FA is the fall which will generate the velocity required for the projection. The demonstration of this is also very evident.

169. Notwithstanding the great simplicity of the construction of these problems, we cannot obtain numerical solutions for practice with equal simplicity, except when the line of position is horizontal, as in fig. 2. This indeed is the most general case, and there are few situations so abrupt as to deviate very far from this case, the greatest height of a fortress commonly bearing but a small proportion to the distance of the mortar.

When AB is a horizontal plane, as in fig. 2. the arch EDA is a semicircle.

In this case the maximum range A*b* is equal to AC, the radius of the circle, and equal to twice the height FA necessary for acquiring the velocity of the projection.

This greatest range is obtained by elevating the mortar 45 degrees from the horizon.

170. The ranges, with different directions, are proportional to the sines of twice the angles of elevation For, drawing GC, DL, *d l*, perpendicular to EA, and drawing the radii CD and C *d*, we have CG equal to the range A *b* and *l d*, equal to the range AB. Now CG is the sine of the angle ACG, which is double of GAB, and *l d* is the sine of AC *d*; which is double of AE *d*; which is equal to the elevation *d* AB; and the same is true of all other elevations. We may always employ this analogy as radius to the sine of twice the angle of elevation, so is twice the height necessary for acquiring the velocity to the range of the projection on a horizontal plane.

171. The height to which the projectile rises above the

horizontal plane is as the square of the sine of elevation. For OV the axis of the parabola is $\frac{1}{4}$th of DB or LA;—and FA, the height to which the projectile would rise straight upward, is $\frac{1}{4}$th of EA. Now EA : LA $=$ EA^2 : AD^2 $=$ rad.^{2} : sin.^{4} AED, $=$ rad.^{2} : sin.^{2} elevation. Therefore FA : VO $=$ rad.^{4} : sin.^{2} elevation,—also VO : v O $=$ sin.^{2} DAB : sin.^{2}, d AB, &c.

172. The times of the flights are as the sines of the elevation. For the velocities in the directions AD, A d, being the same, the times of describing AD and A d uniformly will be as AD and A d. Now AD and A d are as the sines of the angles AED and AE d, which are equal to the angles DAB and d AB. Now the times of describing AD and A d uniformly with the velocity of projection are the same with the times of describing the parabolas AVB and A v B.

173. When the object to be struck is on an inclined plane AB, ascending, as in fig. 3. the arch EDA is less than a semicircle; and when it is on a descending plane, as in fig. 4. EDA is greater than a semicircle. This considerably embarrasses the process for obtaining the direction, when the impetus and the object are given, or conversely. It has been much canvassed by the many authors who deliver theories of gunnery, and the parabola affords many very pretty methods of solving the problem. Dr Halley's, in the Philosophical Transactions, No. 179. is peculiarly elegant. Mr Thomas Simpson's also, in No. 486. is extremely ingenious and comprehensive, and has been reduced to a very elegant simplicity by Frisius in his Cosmographia. But neither of these methods shew so distinctly the connection between the different circumstances of the motions, or keep the general principle so much in view, as the one here given; and all the arithmetical operations which finally result from them, are precisely similar to those deduced from our construction.

174. The following method, suggested by the simple construction now given, is probably as easy and expeditious as any.

Draw the horizontal line HA a, fig. 3. and 4. cutting the vertical drawn through B in K, let C be the centre of the circular arch EDA. Join AC, and draw GC, cutting the verticals through A and B in the points f and g. Also draw CD and C d. Let p represent the angle of position EAB, and d the angle of direction DAB, which the axis of the piece makes with the line of position AB. Also let z be the angle EAD which the axis makes with the vertical. Let r express the range AB, and f the fall FA necessary for communicating the velocity of projection. Then the parameter of the parabola at the point of projection is $4f, = \text{AE}$, and using S to express the sine.

We have $\text{AB} : \text{DB} = \text{S, EAD} : \text{S, DAB}, = \text{S}, z : \text{S}, d.$

$\text{DB} : \text{DA} = \text{S, DAB} : \text{S, DBA}, = \text{S}, d : \text{S}, p.$

$\text{DA} : \text{AE} = \text{S, DEA} : \text{S, EDA}, = \text{S}, d : \text{S}, p.$

Therefore $\text{AB} : \text{AE} = \text{S}, z \times \text{S}, d : \text{S}^2, p.$

That is $r : 4f = \text{S}, z \times \text{S}, d : \text{S}^2, p.$

And $r \times \text{S}^2, p = 4f \times \text{S}, z \times \text{S}, d.$

Hence are derived formulæ which solve all the questions contained in the problem.

175. I. $r = \dfrac{4f \times \text{S}, z \times \text{S}, d.}{\text{S}^2, p.}$

176. II. $f = \dfrac{r \times \text{S}^2, p.}{4\,\text{S}, z \times \text{S}, d.}$

177. III. $\text{S}, d = \dfrac{r \times \text{S}^2, p.}{4f \times \text{S}, z.}$

The answers to the questions expressed in the two first cases are obtained by a single operation. In the first case, the maximum value of r, which corresponds with the elevation AG, is a third proportional to AE and AD, and will be had by the analogy $\sin.^2 p : \sin.\ \frac{1}{2} p = 4f : r.$

We also may remark that the ranges made with the

same velocity, and on the same declivity, are as the products of the sines of d and of z.

178. But these formula do not afford so ready an answer, when d is the thing wanted, as one would expect from their simplicity. When d is unknown, z is also unknown. In this case we must remark, that S, $z \times$ S, d is equal to $\frac{\cos.\ \overline{z \backsim d} - \cos.\ \overline{z+d}}{2}$, and that $z + d = p$.

This changes our formula into $r \times \sin.^2 p = 4 f \times \frac{\cos.\ \overline{z \backsim d} - \cos.\ \overline{z+d}}{2}$, $= \overline{2 f \times \cos.\ z \backsim d - \cos.\ p}$, $= 2 f \times \cos.\ \overline{z \backsim d} - 2 f \times \cos.\ p$. Therefore we have

$r \times \sin.\ p + 2 f \times \cos.\ p = 2 f \times \cos.\ \overline{z \backsim d}$.

Having obtained the arch $z \backsim d$, and having $z + d = p$, we easily obtain d, it being $= \frac{p + z \backsim d}{2}$. The process is much expedited by the help of a table of natural sines. We must remember that when the projection is made on an ascending plane, the quantity $2 f \times \cos.\ p$, is to be added to $r \times \sin.^2 p$; but that it is to be subtracted from it if the projection is made on a declivity.

179. But a plainer method may be taken, although not so obviously deduced from the general principle. The position of the object B being known, its horizontal distance AK is known. Call this h. The middle direction AG is also known. The line f A is also known, being $= 2 f$. Now f C $= 2 f \times$ tan. fAC, $= 2 f \times$ cotan p. Call this b. Then C g is $= h + b$, or $= h - b$, according as the projection is made on an ascending or a descending plane. Now we have

$$f\text{A} : \text{C}g = \cos.\ p : \cos.\ \overline{d - z} \text{ or}$$
$$2 f : h \pm b = \cos.\ p : \cos.\ \overline{d - z}.$$

Then to $\frac{1}{2} p$ ($= \frac{1}{2}\ \overline{d + z}$) add $\frac{1}{2}\ \overline{d - z}$, and we obtain d.

This is, in fact, the process to which we are ultimately

led by every method that is taken for the solution of this case of the problem.

The construction suggests another process, which may be more acceptable to some readers. The angle fAG is $\frac{1}{2}$ EAB. Therefore $f\,G = 2\,f \times \tan. \frac{1}{2}\,p$, and $2\,f \times \tan. \frac{1}{2}\,p - AK = g\,G$, = the versed sine of $z \backsim d$, CA, or $\frac{2f}{\sin. p}$ being radius.

There are two questions more that must be solved before the artillerist can have all the information he requires. In the throwing of shells, it is of peculiar importance that the fuse of the shell burn during the whole time of the flight, but no longer; and it would be best of all were it ended when the shell is about six feet from the ground. This requires an exact knowledge of the time of the flight.

180. The time of the flight is the same with that of falling through DB. We must therefore calculate DB in feet. Then (146) $t = \frac{\sqrt{DB}}{4}$; $t^2 = \frac{DB}{16}, = \frac{r \times \sin. d}{16 \sin. z}$. From the sum of the logarithms of the range (measured in feet) and the sine of the direction, take the sum of the logarithm of 16 and the sine of the zenith distance, and half the remainder is the time of the flight, measured in seconds.

If the best or middle direction had been chosen, which is generally not far from being the case, DB is equal to BA or r. Therefore in this case we have $t = \frac{\sqrt{r}}{4}$.

Lastly, with respect to the velocity and momentum with which the projectile makes its stroke, this is easily deduced from the property of the parabolic motion. We know the velocity of projection, or the velocity at A, namely, that which is acquired by falling through FA. In like manner, the velocity at B is that acquired by falling through F α, (B α being drawn parallel to the horizon.) Therefore, $\sqrt{FA} : \sqrt{F\alpha}$ = velocity at A : velocity at B.

181. Such are the practical problems with their more useful corollaries which have been deduced from the Galilean theory of projectiles, and it only remains to compare the results of this theory with observation. As this is the simplest case of the curvilineal motion of bodies in free space, and we have seen such exquisite coincidence in the celestial motions, in which the complication is incomparably greater, we are led to expect an equal coincidence in the case of terrestrial projectiles. But it is not easy to institute the comparison. The planets describing orbits which return through the same points from which they set out, their revolutions are repeated, and we can easily state the moment of their being in any given point of their orbits. But in our projections the whole flight is over in a very short time, and every time we see them they are new motions, independent on, and unconnected with any prior or posterior projection. There is, however, one case, in which the comparison seems even preferable to any we can make in the case of the planetary motions, namely, the motion of a jet of water or other fluid issuing through a small orifice,—here, as every drop follows the same path, we have the whole path exhibited to us at once. We see, in one moment, a particle in every point of the path. We can measure this path, and state its accurate form. The only other example that affords an opportunity for examination is the flight of a bomb-shell in the night time. We see this by the light of its fuse ; but the appearance is too transitory to give us any accurate information.

When the jet of water is very moderate, as when it issues with a velocity not exceeeding 20 or 30 feet per second, the curve formed by the jet is found to coincide with a parabola. And when experiments are made with jets so limited, especially if the run of water is not too small, the correspondence with the theory is as exact as we can wish. The furthest jet is made on a horizontal plane with the elevation of 45°—and the elevations which give the smaller ranges are observed to have the due proportion, viz. that of the sines of twice the angles of elevation.

But when the velocity of efflux is so great that the water is seen to spread as it issues, and soon divided into spray, we then observe a great deviation. When the jet is very oblique, rising in an angle of 45° for example, we can plainly see that the curve deviates greatly from a parabola with its axis perpendicular to the horizon. The remote branch of the curve is seen to be much less sloping than the rising branch,—and in the very great jets which are to be seen in some great water-works, the falling branch is almost perpendicular at its remote extremity; and the highest point of the curve is far from being in the middle between the spout and the place where the water falls. This unequal division of the curve by its highest point may also be observed in the flight of a bomb-shell, and of an arrow.

The time of rising to the top of the curve should be equal to the time of its descent on the other side. But it is sensibly greater when the elevations are small, and sensibly less when the projection is made with a great elevation.

The greatest horizontal range should be made when the elevation is 45°; and this is generally understood to be the case. But in every experiment that has been made with considerable velocities, the elevation which gives the greatest range is considerably below that of 45 degrees. A strong bow will send the arrow much farther with an elevation of 36° or 38°, than with 45°; and a piece of ordnance does the same.

182. The horizontal ranges should be equal that are

made with elevations equidistant from 45°. The following experiments were made at La Fere by the Chevalier Borda, with a 24-pounder brass cannon, with the same charge of powder in each experiment. Three discharges were made with each elevation, and the medium range is here set down:

Elevation.	*Range.*
15°	1950
30	2235
45	2108
60	1700
75	950

Here it appears that the range at 75°, which should have been the same with that at 15°, wanted 25 of being one half of it. The range at 60° is but $\frac{3}{4}$ths of the range at 30°. And the range at 45° falls considerably short of that at 30°.

The same gentleman made a similar experiment with a brass field six-pounder, with two pounds of powder. It ranged 1590 yards when elevated 45°, and 1700 when elevated 30°.

A 12-pounder ranged at	19°	672
	10	495
It should have ranged		362
Another piece at	8	690
at	4	600
It should have ranged		350

The range with an elevation of 45° should be twice the height through which a body must fall in order to acquire the velocity of the projection.

This comparison, which is of main importance, can be pretty well made with jets of water, because the velocity of efflux can be deduced from the quantity which issues in a given time. In these experiments a defalcation may be observed, even when the velocity is very moderate, and when it is considerable, the defalcation is very great indeed.

183. It is not easy to institute such a comparison in the case of the prodigious velocities of military projectiles, and we were long ignorant in this matter and had notions so erroneous, that our conclusions misled us still farther from the truth. At last, however, a method was discovered of measuring these great velocities. Notwithstanding the magnificent establishments for the improvement of the art in all the governments of Europe, and the numberless experiments which have been made, no improvement has been made till very lately. The experiments are of a kind that professional men alone, aided by the establishment, can undertake—and they have not been spared—but have been so injudicious that they did more harm than good. Yet they appear unexceptionable. Thus, from a medium of 18 discharges with the same quantity of powder, and only three degrees of elevation, that the flight might be almost a straight line; it was found that 510 yards were described in 2½ seconds, from which it was inferred that the velocity was 204 yards per second or 612 feet. Calculations being made for longer flights from this standard were found so discordant, and led to such inconsistencies, that it was plain that the first principle was erroneous.

184. At last, Mr Benjamin Robins, a private person, but eminent for mathematical and philosophical knowledge, accomplished this difficult task, in a manner that gives complete satisfaction. The third law of motion teaches us that when a body A makes any change in the motion of another body B, it sustains an equal change in the opposite direction. If, by striking B, it gives it a quantity of motion 1, then it loses the same quantity of its own motion. If therefore A, moving with the velocity 604 feet per second, strike on a body B a hundred times bigger than itself, it will give it about the 100th part of this velo-

city, nearly 6 feet per second, which is very moderate. Mr Robins therefore discharged musket balls against a block of wood hanging like a pendulum, and so constructed that he could tell, with great exactness, what velocity it had acquired from the blow. We shall see in due time how this may be inferred, with great accuracy, from the vibration which it will cause this suspended block of timber to make.

Proceeding in this way, Mr Robins found that a musket ball, discharged with the ordinary service allotment of powder, issues from the muzzle of the piece with a velocity between 1600 and 1700 feet in a second. A body falling 4 feet acquires a velocity of 16 feet per second; therefore, to give it the velocity of 1600 feet requires a fall of 40,000 feet, or 13,333 yards, $= 7\frac{6}{10}$ miles.

Such a musket, therefore, elevated 45°, should send the ball almost 16 miles—but, when the trial is made, it is rarely found much to exceed $\frac{1}{2}$ a mile, deviating from the theory 31 parts out of 32. A 24 pound shot, when discharged with the usual service of powder, has nearly the same velocity. Yet such a ball will very seldom go three miles, which is but $\frac{1}{5}$th of what the theory requires.

This is an enormous deviation from the theory —It is not so great in more moderate velocities. Thus Mr Robins found that a musket ball with the velocity 400, ranged only 450 yards when elevated 20°. We shall find that this velocity with this elevation, should have given a range of 1050 yards. The experiment gave but $\frac{3}{7}$ths of this.

Another, having the velocity 700 elevated 8°, ranged 690 instead of 1400, which it should have ranged.

185. Thus it appears, that the actual motions of those projectiles differ so monstrously from the theory, that it becomes of no use for directing the practice, except perhaps in bombardment, where the velocities are vastly more moderate, and the observed deviations from theory are much smaller, and so regular that an equation can be adapted

to them, whereas the deviations of cannon shot and of small arms seem unsusceptible of any such reduction. I may almost venture to say, that the numerous and splendid volumes which have been written on the parabolic theory of projectiles, are little more than ingenious amusements for mathematicians, but deserve little attention from the gunner or bombardier. All that can be done for his instruction, is to make a collection of experiments, with every variety of gun, elevation, quantity of powder, and manner of disposing it in the piece. In this collection, no conclusion should be drawn except by a medium of several discharges in similar circumstances. By far the greatest part of the experiments already published have little value, being injudiciously made—and they differ more from one another than from any theory—ranges with small elevations are of no use for establishing the theory, because the smallest deviation from the direction in the vertical produces a great difference in the range, insomuch that the ranges with 4° of elevation are frequently observed to exceed those made with the same powder and 6° of elevation. Such deviations are unavoidable when the ball has much room in the piece.

186. Such being the state of this article of mechanical philosophy, it is our business to inquire into the cause of the deviation from results so simple, and which seem so firmly established on the first principles of mechanics. The cause is by no means abstruse to any person that recollects that this globe is surrounded by a material atmosphere. A body cannot move through this atmosphere without pushing the air out of its place, that is, without giving motion to a quantity of matter. It is a fact, without exception, that when a body in motion displaces another, the moving body loses as much motion as the other body acquires. Although we are far from knowing with precision what motion is thus communicated to the air, we can draw several conclusions from the general fact of its

being put in motion, which must be very nearly true. A ball displaces the air nearly in the same manner with whatever velocity it moves. Therefore, the quantity of aërial motion which it generates, will be nearly proportioned to the quantity of matter thus put in motion, and the motion thus induced on each particle, that is, the velocity with which it is pushed aside. The product of the quantity of air displaced, multiplied by the velocity of the removal, will express the quantity of motion generated in the air, and extinguished in the ball. The quantity displaced in a second must be proportional to the space passed over by the moving ball. And the motion given to the same parcel of air must also be proportional to the velocity of the ball. The product is proportional to the square of the velocity of the ball.

Therefore, the motion lost by the ball, or its momentary diminution of velocity, must be in this proportion.

Suppose another ball, of the same weight, but having twice the diameter, moving with the same velocity. It must displace a greater quantity of air, and this in proportion as its surface is greater, that is in proportion as the square of its diameter is greater. The momentary loss of motion, therefore, will be as the square of the velocity, and as the square of the diameter jointly.

Suppose a ball of equal diameter with the first, but of a greater density. If moving with the same velocity, it generates and loses the same quantity of motion; therefore its momentary diminution of velocity must be so much the less, as its quantity of matter is greater. Thus should a ball of 10 pounds, and one of 20, having the same diameter, move with the same velocity 3, the quantities of motion are 30 and 60. Now should the first lose in any time the velocity 1, it loses the quantity of motion 10. The other loses the quantity of motion 10, and its remaining quantity of motion is 50. Therefore its velocity is $2\frac{1}{2}$, and it has lost the velocity $\frac{1}{2}$. Thus it is that we see a soap bubble descend very slowly, while a cork ball of the same size falls much faster, and a leaden ball still faster.

These things will be more particularly considered afterwards, and are only mentioned just now, to help us to form a notion of the more remarkable circumstances of this retardation. We see, in general, that a denser body sustains a smaller diminution of its velocity when it sustains the same diminution of its momentum, and that, with respect to bodies of the same density, the smaller will be most retarded, because their momenta are in a smaller proportion to the momenta of the larger bodies than their surfaces are, to which the retardations are (*cæteris paribus*) proportional. In general, the momentary diminution of velocity is proportional to $\frac{d^2 v^2}{d^3}$, that is to $\frac{v^2}{d}$, where v is the velocity, and d the diameter.

187. Philosophers have allowed themselves to be misled by the great disproportion between the density of the air and that of an iron or leaden ball, which is from 9,000 to 11,000 times heavier, and have therefore thought that the retardation occasioned by so rare a fluid must be insignificant. They have been confirmed in this by the result of chamber experiments, with a jet of water or mercury, in which there is a sufficient coincidence with experiment. Yet a little reflection might have convinced them, that what impels our ships, and overturns buildings, breaks down trees, and produces other ravages, cannot be insignificant in such rapid motions.

It is really surprising that so little has been learned in a busy period, and on an interesting subject, especially when we think on the immense establishments in most nations for the cultivation of this art. Or, while mere military men prosecuted this subject in their own way, with their moderate stock of mathematical knowledge, it is very surprising that men of science have not improved the art, especially since the elements have been so plainly and distinctly laid down a century ago by Sir Isaac Newton.

Newton was the first who considered the mechanical action of fluids in motion, and he has given us the elements of all the science we yet possess on this subject. His theory of what is now called the resistance of fluids proceeds on the acknowledged principles of mechanics, and has nothing in it that is not strictly demonstrated, except the assumption of the particular constitution of a fluid. This he announces *as a hypothesis.* He gives two forms of this hypothesis, one of which corresponds to the only distinct notions we can form of the mechanical affections of common air. He shortly demonstrates, that the diminution of motion produced by it is the same with what would be produced by the weight of a column of this fluid, having the resisted surface for its base, and having for its height the height necessary for acquiring the velocity of the motion. Thus, a square in its moving in water 32 feet per second, sustains a resistance equal to the weight of a prism of water 16 feet long—this will be about seven pounds. Air being 840 times lighter, the resistance is but the 840th of this, or $1\frac{1}{3}$ ounces. If the surface move 50 times faster than this, or 1600 feet in a second, the resistance will be $50 \times 50 \times 1\frac{1}{3}$ ounces, nearly 21 pounds on every square inch. The resistance to a musket ball should be 72 times its weight, by Newton's principles.

The inattention of artillerists to all this is still the more surprising, as Newton supported his theory by the most unexceptionable experiments, made by him with the utmost care. One chief reason seems to have been the nature of the experiments. Newton's measures of the actual resistance of the air were inferred from observing the diminution which it caused in the vibrations of pendulums. These inferences required such a load of calculation, that scarcely any person but Newton had patience to submit to it. And, as his observed resistances were very small, they were disregarded.

Artillerists were also greatly mistaken in their notions of the initial velocities of cannon shot. They had inferred these measures from the time a cannon ball flew a certain number of yards; not suspecting, that in a second of time the velocity is reduced to little more than one half. A remarkable experiment was made at Leith in 1783, which is very instructive in this respect. A ball of 103 pounds was discharged by eleven pounds of powder, at an elevation of 15°. A 24 pound ball was discharged with eight pounds of powder, with the same elevation. They were discharged almost at the same instant. They alighted within one-fourth of a second of each other, and not above sixty yards asunder. From such an experiment, a person might be led to think, that the two balls moved with nearly the same velocity; yet the smallest reflection must convince us, that this cannot be nearly the case. By the best judgment that can be formed of the action of gunpowder, and by comparing these discharges with such as have been measured, it appears that the initial velocity of the great ball was nearly 800 feet per second, and that of the 24 pound ball was 1450. In two seconds the 24 pound ball was reduced to the velocity 807. In 5½ seconds both were reduced to 520; the smaller ball having gone 4300 feet, and the large one 3500. In 22 seconds the great ball had just passed the other, having gone 9500 feet, and still retaining the velocity 270. The 24 pounder had the velocity 210. It was necessary to calculate these motions for every tenth part of a second.

We see, from this experiment, how difficult it is to learn the velocities. We are therefore under great obligations to Mr Robins for his discovery of an unexceptionable method of acquiring this knowledge. He begins his examination of this question by first finding the resistance to slow motions—this he did in a way that was equally accurate and original. The surface, of which he examined the resistance, was made to move round an axis at the distance of six or eight feet, at a certain determined rate, and the force requisite for this purpose was measured by weights. In this manner, he found that a ball $4\frac{1}{2}$ inches diameter (which is the size of a 12 pound shot) moving 25 feet per second, sustained a resistance of 0,72 ounce avoirdupoise; therefore, if moving $29\frac{1}{2}$ feet per second, it sustains a resistance of one ounce, for he found the resistances accurately proportional to the square of the velocity. Now, suppose this ball to move with the velocity with which it is discharged with four pounds of powder, that is, 1600 feet per second. The resistance would be 64×64 times as much, or about 184 pounds, or $15\frac{1}{2}$ times the weight of the ball. Now, the weight of the ball would diminish its velocity 32 feet in the first second, that is, while it flies the 1600 feet. The resistance would diminish it $15\frac{1}{2}$ times as much, or 488 feet, reducing its velocity to 1172 feet, before it has gone 500 yards.

188. So inattentive have we been to the truths furnished us by Newton. But Robins did not rest satisfied with this deduction from the resistances observed in slow motions. He contrived a still more beautiful and original method of determining directly the initial velocities of military projectiles. This completed the discovery, and gave a new form to the science of artillery;—I should rather say, that he made it altogether a new science.

Mr Robins discharged the bullet against a block of timber hanging like a pendulum, and the stroke set it into vibration. We shall learn in due time how the velocity, communicated to the pendulum, may be inferred with great accuracy from the extent of the vibration. As the bullet stuck fast in the wood, its velocity was reduced to the velocity of the pendulum; and, as the quantity of motion remains unchanged, as much as the block of timber exceeds the ball in weight, so much did the velocity of the ball exceed that of the pendulum. Thus, by making the pendulum sufficiently heavy, the velocity may be made as moderate as we please.

With this apparatus, Robins discovered that a musket ball weighing $\frac{1}{12}$th of a pound, when fired from a barrel 45 inches long with half its weight of powder, issues with the velocity of between 1600 and 1700 feet in a second.

This was a most precious invention; for it not only enabled Mr Robins to determine the present important question, but also to examine the whole practice of artillery, shewing what proportion of powder, and what form of chamber, and what disposition of the loading, produced the best effect, and what is the effect of varying any one of all those circumstances. All the received nostrums in artillery adopted by men ignorant of principle, and sanctioned by long authority, were now open to a fair, easy, and accurate trial; and it was now put in our power to establish fixed principles and maxims of practice.

Mr Robins also suggested another method of determining the velocities by means of the recoil of the piece, which he also suspended like a pendulum, and observed the extent of its vibration. He found, that the part of the recoil that was produced by the powder only was the same, whether the piece was discharged with or without a ball. This method is much easier than the other, especially with great guns.

189. These inventions of Mr Robins have been prosecuted with great zeal and success by Sir R. Thomson, afterwards Count Rumford, with small arms, and Professor Hutton at Woolwich has extended the experiments to balls of 3 pounds. It would be of great service to have two or three trials with great ordnance; for as we find that a considerable modification was necessary for connecting the trials in small arms with those on shot of 3

pounds, we may certainly conclude that as great a modification will be necessary for connecting any of the experiments yet made with the performance of a piece of great ordnance. The ingenious discoverer would certainly have prosecuted his invention with superior address, being a person of uncommon sagacity and acuteness, and a most excellent mathematician. But soon after he was appointed chief engineer to the East India Company, in which station he would have had the finest opportunities. This excellent philosopher died of a fever, and the world was deprived of all the knowledge which his experience would have enabled him to communicate.

His dissertations on this subject were published by his friend Mr Wilson, in 2 vols. They have been commented on by Euler, Borda, and other mathematicians of the first eminence. The experiments recorded in them, and those by Count Rumford, and still more those by Professor Hutton, form a most excellent collection of facts, from which almost every thing that is important in the art may be deduced.

The fertile genius of Mr Robins did not stop at the discovery of the initial velocities of bullets. His invention furnished him with a method of determining by direct experiments the assistance sustained by musket balls. By the medium of many shots fired against his pendulum at the distance of 25 feet, he obtained the medium velocity with which the ball struck the pendulum. He then removed it 50 feet further off, and repeated the experiment. The medium velocity was smaller. This diminution was produced by the resistance of this additional 50 feet of air. And thus its measure, or rather its effect, is given in the most unexceptionable manner. This is perhaps one of the most ingenious inventions of the last century, for scarcely any thing seems more removed from all means of determination. It is therefore worth while to record some of Mr Robins's experiments, which will give us standard numbers, to which we can have recourse on all occasions.

190. The pendulum apparatus was set up at 25 feet distance from the muzzle of the musket, which was firmly fixed to an immoveable mass. After firing several shots into it, it was removed 50 feet further off, and the experiment was repeated. It was then placed 50 feet still further off, and a third trial was made. The results were as follow, a mean being taken of each set.

8 Shot at	25 feet struck with the velocity			1670
		50 interval	Diff.	123
8	75		. .	1547
		50 interval	Diff.	122
8	125		. .	1425

Here we see that a musket ball, moving at the rate of 1670 feet per second lost 123 feet of velocity per second, in passing through fifty feet of air, and that, when moving at the rate of 1547 feet per second, it lost 122 feet of its velocity in passing through the next 50 feet. Now, as the ball began the passage through the first 50 with the velocity 1670, and ended with the velocity 1547, we may take 1608 for the medium velocity of this passage. It therefore passed through these 50 feet of air in $\frac{1}{32}$ part of a second, and it was during this small moment of time that the air's resistance retarded its motion. This enables us to form a very clear and distinct notion of this resistance, by comparing its retarding effect with the retardation that the weight of the ball would have produced in the same time. Now we know that in $_{32}$ part of a second, the weight of the ball would have diminished its velocity exactly one foot. The air's resistance has diminished it 123. It is therefore so much greater than the weight of the ball, or it is equivalent to a contrary pressure of about 10 pounds and a quarter. The ball is retarded in the same manner that it would have been in it been fired straight up, and had a weight of 10 pounds dragged after it by a wire.

This is surely a very remarkable phenomenon, and what no person would have thought possible, that a fluid so rare, and so yielding as air, should oppose such a resistance. Yet had the military gentlemen who had the science or the art committed to their hands for improvement, condescended to study the propositions published by Newton more than 50 years before, they would have seen that this resistance could not be less than 5 pounds. For this magnitude results from Newton's theorems by the most simple calculations. But, as has been already observed, they did not imagine that the velocity of a musket ball was nearly so great as Mr Robins found it.

In another experiment Mr Robins found

at 25 feet distance the velocity	1690
175	1300
difference	390

This retardation, when considered in the same manner, shews a resistance equal to 11½ pounds, or 138 times the weight of the ball.

In another experiment, in which the velocity was 1100 feet per second, Mr Robins found the resistance only 33½ times the weight of the ball.

Here is a new field of experiment opened to those who wish to improve this art, or the science of the resistance of elastic fluids. It is surprising that it has not been occupied by any philosopher or artillerist, and that they have been contented with the experiments of the ingenious discoverer. Mr Robins has indeed made a great number—but few of them have been published. His own curiosity was greatly excited by the curiosity of the discoveries that he had already made, and their importance for establishing some general laws on this subject. His death has therefore been an irreparable loss.

By comparing a great variety of experiments made as now described, Mr Robins found that when he compared the resistance to slow motions with those having the velocity of 1600 or 1700 feet per second, on the supposition that they increased in the duplicate ratio of the velocity, the resistance to those rapid motions deviated from this proportion in the ratio of 3 to 1 nearly, or were nearly thrice as great as on this supposition.

When the velocity was only 1100 feet per second, the deviation from the duplicate ratio of the velocity was in the proportion of 11 to 7, still very great, but far from the former.

In all cases Mr Robins found the resistances to exceed those of the Newtonian theory pretty considerably, even in the slowest motions, and that they deviated more and more from the duplicate ratio of the velocity as the velocities increased. It is surprising that this assertion of Mr Robins has not excited some of the eminent mathematicians to a repetition of the experiments. Some of them have published most elaborate dissertations on the theory of resistance in elastic fluids. These theories seem incompatible with this fact, unless some elements be taken in of which they make no mention. At any rate, a collection of experiments made in the manner we have described, on all velocities from 100 feet per second up to 2000, would be of more service both to mechanical science and to the art of artillery than any thing that has yet been done, and till we get such a collection, all our theories must be very imperfect, and probably very erroneous.

191. Mr Robins, in the course of those experiments on the resistance of the air made a curious observation. In very moderate velocities the retardations were nearly as their squares. As the velocities were increased the resistances increased at a somewhat greater rate, but with a certain observable regularity, till the velocity exceeded

1100 feet per second. But when the velocity is increased from 1100 to 1200, the increase of resistance is prodigious. After this the resistance goes on increasing nearly with its former regularity.

This is a singular fact, that there should be such a jump in the variation of the resistances. Mr Robins guesses at the cause of this, with great probability of being right. As the ball rushes through the air, the air falls in behind it, being pressed in by the weight of the surrounding air. The ball may move so rapidly, that the air cannot fill up the place left by the ball. In this case, the ball is retarded, not only by the resistance of the air which it displaces, but also by the statical pressure of the atmosphere. This last is nearly equal to the pressure of 12 pounds.—Now we can tell the velocity with which the air can fall in behind the ball. This must be the same with which it can rush into a void. This, by calculation, is about 1100 or 1200 feet per second.

It is remarkable that this is also nearly the velocity of sound, and there is an observation of this kind which seems to have some connection with the mechanical fact observed by Mr Robins. If a person stand in such direction from a cannon when it is discharged that the ball may pass him at no great distance, he will hear the noise made by the ball rushing through the air all the time of its flight, and as the ball approaches him, the noise should become more audible. But he will hear the noise loudest at the very first, immediately following the report of the gun; and after about two seconds he may observe the sound change all at once, and not only become more faint, but even change its kind, after which the sound increases as the ball comes nearer. It seems highly probable that this abrupt alteration in the sound takes place just at the time that the resistance undergoes such a change; and that it is owing to the difference in the nature of the undulations when there is a void behind the ball, and when there is not. That there is such a difference in the primitive agitation of the air, is confirmed by observing the way in which water falls in behind a stick drawn through it. When the motion of the stick is rapid, a void is left behind it, to a considerable depth, and the agitation of the water is extremely different from what may be observed in slow motions.

192. From the preceding observations, Mr Robins infers that great charges of powder are absolutely useless in the service of artillery, especially when the distance of the object is considerable. If we would increase the velocity from 1600 to 2000, we must employ half as much more powder, with a hazardous strain to the gun, and the velocity will be reduced to 1300 before the ball has proceeded 500 yards. Had it been discharged with the velocity 1600, it would have retained 1292 at that distance, less than the other by 8 feet only. The velocity 1000 requires only one-third of the powder, and at the same distance the remaining velocity is very nearly 900. We may certainly say that a velocity exceeding 1100 should not be aimed at, because it is so rapidly reduced by the great resistance. But this subject will come better before us when we are considering the action of expanding fluids, and we shall then take occasion to distinguish those cases of service which render the increase of velocity useful, as in the battering in breach, in order to cut through the revetement of a fortification at the bottom. Here a deep penetration is necessary, which depends on a great velocity, although the shot be not so heavy. Whereas, in the subsequent battering down, a great momentum with a heavy ball, though moving more slowly, is found more effective in shaking down the undermined mass.

193. Mr Robins made another curious and original observation in the course of his experiments. He found that shot not only deflected downward by the continual

action of gravity, but went frequently to the right or to the left of the vertical plane passing through the axis of the gun. He discovered that this was owing to the knocking of the ball on the side of the chace or barrel as it went along, so that at leaving the muzzle of the gun, it had generally a very rapid rotation round an axis. This rotation was altogether uncertain, both as to its direction and its velocity. The effect, however, was very manifest; one side of the ball was moving in the direction of the projection, and the other side was moving in the opposite direction. From this must result a force that continually deflects the ball toward that side that is moving backwards. Mr Robins proved this by discharges from a crooked barrel, which caused the ball always to whirl in one way, and he found the deflection to be always toward that side from which the barrel bent away, that is, toward that side on which the line of the barrel is convex.

Great care should therefore be taken to prevent this whirling motion of shot. No more windage should be allowed than what is sufficient for passing the drawing ladle round the shot to bring it out. When the exact aim is matter of peculiar consequence, it would be advisable to use oval bullets, which cannot turn in the barrel of the piece, or to case them in canvas, or otherwise, so that they shall need to be rammed down.

The surer aim that may be taken with a rifled piece, depends on the rotation of the ball being secured. It must be round the line of its motion; for the effect of the rifles in the barrel is to make the ball make half a turn or a whole turn round the axis of the barrel as it goes along it, and it continues this motion very nearly in the direction of its flight. This steadying effect of a rotation round the line of motion, is plainly seen in a common shuttlecock, which flies quite wild unless it twist briskly round its axis. In like manner, there is a great nicety in putting on the feathers of an arrow. They should be so placed, that in looking along any of the feathers, the point of the arrow should appear about half an inch to the left hand of the direction of the feather. Such an arrow, let fall with its point down, will sensibly turn round its axis. If it turn briskly, the feathers are too oblique, and will greatly impede its progressive motion. But it will fly more true. Small pieces of ordnance have been made with rifled barrels, and the balls cast with protuberances to fit them. Their motion has been extremely steady, but their force is so much diminished by this great increase of resistance, that the contrivance has been given up as unserviceable.

194. It has been thought indispensably necessary to make the axis of the trunnions of a cannon considerably below the axis of the chace. It is thought that if made otherwise, the trunnions have not sufficient connection with the metal of the gun. But the effect of this is hurtful to the certainty of the aim, and has other inconveniences. For the axis of the barrel being above the axis of the gun's motion, the pressure of the powder on the breech causes the breech to press strongly, and with a jerk on the bed or quoins which support it, by which the muzzle is made to spring up in the instant that the ball is going along the barrel, and it is thus deranged from the aim given it by the gunner. It is also this violent thump given by the breech of the gun to the carriage that causes the gun to spring up again. This is extremely troublesome. It is most remarkable in guns upon field carriages, owing to the greater elasticity of the long sides of the carriage. But it is very troublesome also on garrison and ships carriages, particularly in the latter, causing the gun to strain or disorder the breeching, and even sometimes to strike the deck above.

195. All this would be avoided by placing the axis of the trunnions on the line crossing the axis of the barrel at right angles. This was done with complete effect in a

set of guns made in 1759, for a fine privateer frigate, built by the late admiral Sir Charles Knowles. These guns, having the breech supported (for the experiment) by a thin chip band-box were repeatedly discharged without making any impression on it. A nine-pounder of the ordinary form was tried along with them, its breech being supported by a little oaken box almost an inch thick. The first discharge crushed it to pieces.

196. The same effect is observed in small arms. The line of the barrel, produced backward, passes considerably above that part of the but end which is applied to the shoulder, and when the piece is discharged, it never fails to rise above the direction intended for the shot. It is still worse in pistols. These are made almost in the most unserviceable form, the line of the barrel passing so far above that part of the but end which is grasped in the hand, that the pistol starts wide of the direction in which we think to discharge it, and it is only by practice that we can learn to take a tolerably good aim with a pistol. They would be greatly improved by making their but end exactly in the form of a joiner's smallest saw, which I think is called a tenon saw. If thus shaped, with the barrel pointing backward to the convexity which fills the grasp of the hand, they would be perfectly steady in the firing.

OF CORPUSCULAR FORCES IN GENERAL.

197. We have explained in sufficient detail, the chief phenomena which result from the heaviness of bodies in motion. The exquisite agreement between the legitimate deduction from the premised principles of dynamics with these phenomena, has had the double effect of giving us the grounds of a complete confidence in the truth of those principles, and in *their* competency to the explanation of the appearances of nature.

The effects of the heaviness or gravity of bodies will be found to mix with those of every other property, which observation shall shew us to be found in all matter, and therefore we must carefully separate the effects of gravity from the other circumstances of the complicated phenomenon, before we presume to make any inference in support of the existence and the nature of any other property which we may suppose the subject to possess.

Proceeding according to the method proposed, we must now consider the phenomena which indicate a mechanical property of existing matter, that seems next, in respect of its generality or extent, to gravity.

198. Motion is continued or preserved in bodies by the perseverance or inertia of matter. It is necessarily communicated to other bodies by this sensible impenetrability. It is modified by the action of gravity, so as to produce all the variety of mechanical changes that we observe. But this conservation, communication, and modification of motion, are susceptible of infinite varieties by circumstances which seem to characterise different classes of bodies, and make them appear of very different natures. A piece of glass lies still on a table, or slides down an inclined plane. It sinks, if laid on the surface of water—it rises to the surface, if plunged into quicksilver—it makes a pit, if it fall on clay, but rebounds if it fall on glass, and is sometimes shivered into pieces—the ball which hits another, drives it before it, if it be elastic—but only carries it along, if unelastic, or goes through it if soft. A vessel filled with grain will allow it to heap above the brim. Water will not heap in this manner. If the vessel have a leak, a little of the grain will *fall* out, with the velocity with which any heavy body begins to fall, and the greatest part will remain behind. If it be filled with water, the water begins to *flow* out, with a much greater velocity, and it runs all out. If one part of a piece of wood be gently struck, the whole is moved together. If one part of a parcel of sand be struck ever so gently, it is moved alone, and the rest remain still. If the air be comprised in a vessel, and an opening be made, it rushes out with violence and will drive a ball before it, as we see in the air gun. But should we compress water in it with triple or quadruple force, no such violent efflux is observed. A bit of silver will lie for ever at the bottom of a glass of water. It will soon rise, and be uniformly disseminated through the whole of a glass of aquafortis. It will remain in this diffused state, if we add a little brandy—it will all fall to the bottom, if we add a little spirit of hartshorn.

It appears, therefore, that our knowledge, even of the *sensible* changes of motion, is incomplete, till we know its modifications in consequence of the hardness, softness, fluidity, or aerial form of bodies; and that our knowledge of those internal motions which produce the sensible changes of form in the phenomena of chemistry and physiology, is really nothing till we understand the circumstances on which those changes depend. We might indeed consider hardness, softness, elasticity, and fluidity as phenomena, describing exactly those forms of aggregation, and then proceed to consider the mechanical consequences deducible from those forms. But it will be more suitable to the nature of philosophical research, and more satisfactory to the curious mind, if we can discover some principle of connection, or some circumstance of resemblance, among those forms, the modification of which produces those distinguishing appearances, and which *immediately* operates, even in the mechanical phenomena that we observe, and which it is *our peculiar province* to study. Should we be so fortunate as at the same time to acquire such a knowledge of this connecting principle as will tend to explain those differences of form, and regulate the effects of the different modifications of tangible matter, it is plain that we gain a considerable extension of our knowledge, and that we have a better chance of being able to comprehend the nature of those operations of natural causes which are the subjects of chemical and physiological science.

199. I shall therefore think our attention very properly bestowed on a few pages, in which I shall endeavour to explain, or at least to describe, the chief appearances of that property of matter by which many particles are united in one mass, having different degrees of connection, from the greatest hardness to the most perfect fluidity. It is usually called the *Power* of COHESION, the power by which the particles of tangible matter co-

here. This name is perhaps as proper as any, and, like the term Gravity, is merely the expression of what we conceive to be its effect. It may indeed be objected to this denomination, that it does not seem to include that mode of aggregation that we call aerial or gaseous, in which the particles, instead of cohering, or manifesting any tendency to unite in one mass, shew, on the contrary, a tendency to separate farther from one another. For all such expansive fluids seem to require external compression to prevent this separation. The objection is not improper, and were it possible to find a word in our language which would, in its common use, include this relation, manifested in many cases between the particles of matter, it should be preferred. But, unless we multiply the denominations of the properties of existing matter, almost without end, we shall not find words sufficiently precise. It is too precipitant to ascribe the phenomenon to a different power of nature, unless our knowledge of the other is so distinct and clear that we are certain that this phenomenon cannot be explained by any of them. Besides, the most coherent bodies are perhaps similar in this respect to the expansive fluids. That external pressure forces them into smaller room we are certain, by experience. It is reasonable, therefore, to believe, that when the pressure of the atmosphere is removed, they occupy larger dimensions, although the difference is too small for our perception. And, on the other hand, there is, in all probability, a certain distance between the particles of air, at which their evasive or repulsive tendency ceases. This must be the case at the top of the atmosphere, for there gravity has no tendency to compress the superficial particles.

200. It will, on the whole, be better to consider this property which modifies and regulates the sensible forms and mechanical relations of tangible matter more generally. And, as its distinctive effects are immediately re-

lative to the insensible particles, and more remotely to the masses composed of them, we may call it the CORPUSCULAR FORCE, and the corpuscular action of matter.

201. I have called it a force, meaning by this term whatever is the immediate cause of a change of motion. That property by which the particles of matter cohere, has undoubtedly a title to this denomination. Our very first conceptions of it state it in opposition to mechanical force, and as balancing, to a certain degree, that force; for the most obvious appearance of it is the resistance made to our attempts to separate the cohering particles. We know that the particles of wood cohere with force, because it requires force to separate them; and we give to cohesion different degrees of force, measured by the forces which are just not able to separate the parts of the coherent body. Cohesion presents itself to us as a force in the most familiar and palpable conception of the term, an obstacle that resists our strain exerted to separate the parts of the body, in the very same way, and exciting the same feeling, as if a man were pulling against us. But cohesion has also every *philosophical* title to the appellation of Force—it is a moving force;—for when I push or pull one part of a firm body, the remote parts of this body are put in motion, although I have not exerted any of my force on them. How is this effected? Solely by what I have called corpuscular action. We cannot conceive an atom of matter as moved from its place unless a moving force acts on it. There must therefore be something in the parts to which I apply my force, which occasions a moving force to be excited in, or exerted on, every particle that is moved. Did I apply the same exertion to part of a parcel of sand, I should move what I acted on, but no more; and I should infer that the particles of sand do *not* cohere. The cohesion is invariably inferred from the motion, and cohesion is conceived as a moving force. In strict language the word Cohesion, like

Gravitation, is only a term for the fact, that the parts cohere; but we have used it also to express that power, that force, which produces the cohesion.

We see plainly that the force of cohesion is susceptible of degrees, and that these are measurable, whether it be considered as a mere strain, or as a moving force. The cohesion of a string may not only balance a strain equal to two pounds, while it yields to a strain of three pounds, but we find also that it is sufficient to twitch suddenly a pound ball into a moderate motion, but will be broken if we attempt, by a more violent twitch, to force the ball into a more rapid motion, or to force a larger ball into an equal motion. In these last instances we consider cohesion as a moving force.

The consideration which now excites the curiosity of the philosopher is the manner, or the means by which a force applied to one part of an assemblage of particles mutually related by this corpuscular force, affects all the rest of the assemblage. An ordinary observer finds no difficulty in the matter; the parts, says he, cohere, and therefore the distant parts are moved. But, in like manner, he would find no difficulty in accounting for an egg's being crushed by a man standing on it; the man, he says, is heavy, and therefore his weight must crush the egg. We have seen, however, that this question has greatly agitated the philosophers, and that they are not yet agreed about the cause of this pressure of gravity. They cannot explain it any other way than by calling it a property of matter; and this is acknowledged to be no explanation at all. So has the question concerning the cohesion of matter; and many and various opinions have been entertained about the nature of the force of cohesion and all its modifications. But it is most probable that our attempts to explain it, that is, to deduce it from more simple or original principles, so as to be able to state cohesion as an effect, will be as fruitless as they have been in the case of gravity, and that it will be more prudent in us to abstain from the research, and content ourselves with collecting the laws by which the operations of this force, whatever it is, are regulated. By this procedure, becoming the modesty of limited intellects, Newton has enriched science with an immense body of curious and useful knowledge, of which a brief account has been given already.

I shall not therefore employ any time in stating and refuting many different hypotheses which have been proposed for explaining cohesion. They may be seen in several performances of the 17th century, particularly in Muschenbroek's posthumous System of Natural Philosophy.

Our search after the laws of cohesion will not indeed lead us into such magnificent scenes of contemplation as filled the admiring mind of Newton. But we shall find it full of very interesting and unexpected objects, objects in which the beautiful symmetry of nature, and the transcendent wisdom of its author, are written in the most legible characters. Examples of manifest subserviency to the most useful purposes will be seen in many quarters of this varied scene.

202. The phenomena, which result from the operation of those natural powers which produce the cohesion of tangible matter, are so various and even dissimilar, that it is not easy to form a plan by which our inquiries may proceed. I think that it will greatly assist us in this respect, if we set out with an observation, or rather a conjecture of Sir Isaac Newton, which he formed very early in life, and seems to have been more and more thoroughly confirmed in as his observations of nature were multiplied and extended. Sir Isaac Newton says, in the first edition of his Principles of Natural Philosophy, that as the distant bodies of the visible universe are connected by gravitation, so the particles of which those bodies consist are connected by mechanical forces, which

are exerted only at those small and imperceptible distances which determine their peculiar densities; and, as the regular motions of the planets are distinctly explained by considering them as instances of mutual deflections towards one another, by the action of a force inversely proportional to the square of their distances, so, says he, *all* the changes which are observed in common tangible matter may be explained by shewing them to be examples of mutual inflections or tendencies to inflection, by the action of a force varying according to some law of the distances, which we may perhaps be able to determine by means of the phenomena. He says that he has actually observed that some bodies act on others without coming into mathematical contact, in a way similar to the action of magnetical and electrical bodies, that is, they somehow cause those other bodies to approach them, or to recede from them, which phenomena we commonly call Attraction * and Repulsion. He has observed, that when the distance between them is less than a certain very minute quantity (about the 800) of an inch, the bodies are uniformly attracted,—at a distance a little greater they are repelled, and at a distance still greater, but within certain limits, they are again attracted. He has observed several alternations of this kind in the motion of light, as it passes near to any solid body, such

* Should I, in imitation of Newton, call this force attraction, I do not mean to *explain* the phenomena by attraction as a physical cause, but merely employ a word, perhaps not very well chosen, as a general term for a very general phenomenon. The endeavour or effort to approach, and the resistance to separation, are phenomena as much alike as the acceleration of a falling, and the retardation of a rising body. They are so inseparably connected that we cannot but consider them as phenomena of the same kind, indicating the same cause or power. These tendencies are of the same nature, whether the distances of the parts so related be great or small. If attraction be a suitable denomination for the mutual approach of distant bodies, it is no less suitable to the resistance which the touching particles of a body manifest to separation. I do not mean to decide as to the cause of either, and merely consider them as similar phenomena.

as a pin, or the edge of a ruler or the like. All the attempts of philosophers to explain the motions of the planets, or of magnetical and electrical bodies, by the mechanical impulsion of æthers, magnetical and electric fluids, failed of this purpose when examined by the severe laws of mechanism, and nothing seemed to remain but to acknowledge the agency of centripetal forces, whether intrinsic in the central or in the attracted body, or in both. Sir Isaac, many years after this first intimation of his conjecture, when his experiments in every branch of natural knowledge had given him much more information, shewing him this corpuscular action in a vast variety of forms, says, in his Optics, published in 1706, that he is much disposed to think that the phenomena of cohesion, and all the hidden motions in chemical changes, are produced in the same way, that is, by attracting and repelling forces, which are exerted only at those small and insensible distances that intervene between the particles of tangible matter.

203. That this was not a transient conjecture, but that it had the full assent of his understanding, is evident from the extensive application that he makes of it to the explanation of the most remarkable phenomena of chemistry, such as solution, precipitation, crystallisation, fermentation, &c. occupying many quarto pages, and the frequent allusion to those reflections in other parts of his writings; and, to define more precisely what he meant by all this, the last phenomenon which he attempts to explain by it is the reflection and refraction of light. Now this had been explained in his Principia, on principles precisely similar to those employed in explaining the phenomena of the solar system, viz. by the action of deflecting forces.

Since Newton considers the cohesion of tangible matter as effected in the same way, it is plain that he conceived all such masses as consisting of a number of discrete particles or atoms, which are held in their places by the balanced action of those corpuscular forces. In this respect, therefore, Newton may be said to have revived the ancient atomistical doctrine, but with this capital distinction, that the atoms are not contiguous, but separated from one another.

I do not see how this notion of the intimate constitution of tangible matter can be entertained, without, at the same time, conceiving the forces which connect them as inherent in them, or somehow belonging to them, like the ποιοτητες of the ancient philosophers, by which one kind of substance was distinguished from another.

I do not speak of what a philosopher thinks himself warranted or obliged to do, after a scrupulous consideration of the subject, but of what he actually does, in the familiar and unguarded conceptions which he rapidly forms of things. I can only say, that I think that such cohering powers belong to matter, in the same manner, as I conceive weight or heaviness to belong to it.

This is, unquestionably, a curious subject, and of great importance. Besides the chemical phenomena to which it has a most immediate relation, it is very interesting to the mechanical philosopher. The strength of solids, the communication of motion by impulsion, the nature of fluid and aerial union, and the modifications of mechanical action by this union, are elementary articles of his science; and it would be a most pleasing acquisition to be able to deduce all these from the acknowledged laws of mechanism, including in this term the action of central forces.

204. That the particles or ultimate atoms of bodies are actuated by forces *e distanti*, is rendered probable by many well known facts. All bodies are susceptible of compression and dilatation. This seems incompatible with the absolute contact of the ultimate atoms; nearer than contact is inconceivable. This compression or dilatation requires force to produce it, and to continue it.

If this force is removed, the parts resume their former distances, and the mass its former bulk. This cannot be explained by the fibrous and contorted structure of bodies, allowing compression by the bending of the fibres, as we observe in a piece of sponge. For, in all such cases, we come to an ultimate fibre: if this be bent, the atoms of which it consists are brought nearer to one another on the concave side of the bend, and are removed farther from one another on the convex side, than when the fibre regains its natural shape.

All bodies expand by heat, and contract by cold; and this change takes place in all their dimensions of length, breadth, and thickness.

205. Fluidity under a great degree of sensible compression seems mechanically impossible, if the particles of the uncompressed fluids are supposed to have been in absolute contact. Air may easily be reduced into one half of its ordinary bulk. If this be supposed to be effected by the compression of its elastic particles, which may be supposed spherical, and to touch one another only in single points, we must then suppose that these spheres are compressed on each other into perfect cubes, (because a sphere is nearly one half of a cube of the same height). Now such a mass could no more have fluidity than a box of blown bladders when equally compressed. The particles must now be in contact, not in single points, but each with the whole of its six square sides—yet is air perfectly fluid in this state—so is water, in any degree of compression. Its particles have never indeed been compressed into cubes, touching one another all over; because water has never been compressed into one half of its natural bulk. But they must be mutually dimpled, and the touching surfaces exactly fitting each other like so many joggles. Such an assemblage could not be fluid.

But we have much more distinct proofs of mutual forces

connecting the particles of tangible matter, and acting on particles at a distance.

206. Fluids, in falling, gather into spherical drops—rain drops are most accurate spheres, as may be demonstrated by the phenomena of the rainbow. The angle which it makes with the line from the sun to the spectator, shews that the light is reflected from an exact sphere. Oil rises through water in spherical drops, and spreads on the surface in an exact circular disk. All figures would be indifferent for these collections of particles if mere contact were sufficient; but the prominent particles are drawn aside, or drawn down, just as the top of a wave is drawn aside on the surface of the ocean by the action of gravity. We need not, however, suppose that the attraction between the particles of a drop extends to the distance of the radius of the drop—but *some* distance is necessary for producing this change of form, and *any*, the most minute, will suffice. If we take a glass tube of the shape of Plate II. fig 5. of $\frac{1}{10}$th of an inch in diameter, and fill it with water, it will retain it, when held horizontal. But if we touch the surface of standing water with the orifice A, the whole, or very nearly the whole, will *run* out. Mere *adhesion* will not do this. If, after touching the water with the end A, we gently drew away the tube, mere adhesion would account for the water quitting the tube; but the effect now mentioned indicates a force exerted by the particles of the water in the vessel, which draws off laterally the water in the tube.

The mutual adhesion of solid and fluid bodies exhibits many appearances in confirmation of this assertion.

207. If we take a slender glass pipe of a tapering form, as in fig. 6., and after thoroughly wetting the inside, by causing the water to run through it, we shake all the water out of it, and then put in a drop or two, holding the wide end downwards, the drop of water will remain at the

lower or wide end. If we now hold the tube horizontal, the water will gradually move from the wide to the narrow end of the tube. This can never arise from mere adhesion. There is some force, acting from a distance, however small that distance may be—and the force is attractive. It may be understood in this way. Suppose that glass and water mutually attract each other, and that this action extends to the small distance *a b* (fig. 6.) The glass of the ring, whose breadth is *a b*, acts on the water, urging it in the direction *a b*. But the ring *d e* at the other end of the drop of water urges it in the opposite direction *d e*. The force at the end *a* is to that at the end *d* as the circumference of the ring *a b* to the circumference of the ring *d e*. Now it is demonstrated in hydraulics, that in order that two opposite forces may be in equilibrio when acting on the two ends of such a column of fluid, they must be proportional to the surfaces to which they are applied. But this is not the case here. The forces are as the acting circumferences, whereas the surfaces are as the squares of those circumferences. The force at the end *d* is not enough for balancing the force at *a*, and the latter must prevail—therefore the water will move in the direction *d a*. We see, by the way, the force is attractive—had it been repulsive, the water would have moved in the direction *a d*.

If we consider this fact carefully, we shall see, that by going towards the smaller end of the tube, the length of the little column of fluid is increased. The surface of glass in contact with the water is also increased. For if it is in a part where the diameter is one half of its former magnitude, the length of the column must be quadrupled, and the touching surface doubled.

208. We observe water and other liquids rise up all round the circumference of the vessel, forming a concave curve. If a glass plate AB (fig. 7.) be wetted and dipped into still water, we observe it to heap up on both

sides of the plate, forming a curve *a b c*, of a very sensible height. We are not to imagine that the attraction of glass for water extends to so great adistance as *a c*, or even as the perpendicular height *a e* of the curve. We shall soon be convinced that the extent of this action is too small to be measured by us. It is probable that only a very small portion, perhaps the portion *d a*, is the sole agent. As the water climbs up the side of the plate, it comes within the sphere of action of a portion still higher, and is drawn up by it, and the water thus raised drags more after it, by the mutual adhesion of the particles of water, till the weight of the whole is just a balance for the attraction of the acting portion *d a* of glass still above it. We shall see, however, by and by, that *a d* is not the acting portion of the glass.

If the plate extends horizontally to any distance, the water rises along the whole intersection of the plate with the surface of the still water. Thus it rises all round in a tea-cup. The quantity raised will therefore be proportional to the horizontal extent of the intersection.

Should we bring another plate CD, so near the plate AB, that the curve *f g h*, caused by AB, interferes with the curve *i k l*, caused by the plate CD, the water will rise still higher between the plates. It will rise till the whole weight of water raised just balances the doubled action of the glass.

Unless we know the law by which the attraction is diminished by an increase of distance, we cannot say what proportion of height will result from different distances between the plates. As far as experiment can be made with precision, it would seem that the height of water between the plates is inversely proportional to their distance. This shews that the greatest distance at which the mutual attraction is exerted is very small, and incomparable with any measurable distance between the plates. For, were it otherwise, the water would rise higher, because there would then be some of the water acted on by both plates, and *its weight* doubly diminished. Were the heights of the water accurately proportional to the distance between the plates inversely, we must infer that there is no sensible portion of the whole water immediately acted on by both plates. Or, in other words, the distance between the plates is incomparably greater than the distance at which water and glass attract each other.

209. Dr Brook Taylor describes, in No. 336. of the Philosophical Transactions, a very beautiful experiment, which gives us a more accurate mean of judging in this matter, and is indeed the source of all our precise ideas of this subject.

A pair of mirror plates ABCD, (fig. 8.) are set in water. They join in the vertical line AB, and are separated a very small matter at the other end CD, thus forming two vertical planes inclined to one another in a very small angle. The water rises between them, and its upper surface forms *a very exact hyperbola k a m*, of which AB and BC are the two assymptotes. Let *c d* and *g h* be equal small portions of BC, and then *a b c d* and *e g h f* will be two columns of water supported by the attraction. Let the angle formed by the plates be represented by the plan KBC. The sections $\varkappa\, c\, d\, \delta$ and $\gamma\, g\, h\, \chi$ of the supported columns, are evidently proportional to their distances from B. By the nature of a hyperbola the heights *a c* and *e g* are in the inverse proportion of B *c* and B *g*. Therefore the quantities of water in the two columns are equal, and their weights are equal.

This is a valuable experiment, because it states with so much accuracy the relation between the heights of the columns and the distances of the plates. We also learn from it, that the whole touching surfaces *a c d b* and *e g h f* are not acting in the elevation of the water. For the weight raised is the true measure of the force in action, because it precisely balances it. Now the weight is the same in these two columns, while the surfaces in contact with the water are unequal in any degree. We must grant that, over the whole touching surface, there is the simple *adhesion*, and *this* may be proportional to the surface. But we see that this is not the proportion of the action; for the actions are equal, since the balancing forces, the weights of the columns, are equal.

210. There seems to be but one way of accounting for this. If we suppose that only a *very minute* portion of the surfaces *a c d b* and *e g h f* act on the water, either a small parallelogram adjoining to *c d* and *g h*, or adjoining to *a b* and *e f*, then those being equal, their actions will be equal, and will balance equal weights of water, as is actually the case. We certainly cannot take this acting portion in any intermediate part of the cohesion, because there is no part distinguishable from another by any circumstance of its situation, except the extremities of the column. Dr Jurin, for reasons that will soon be mentioned, supposes that the acting portions are adjoining to *a b* and *e f*, (see his Dissertations in the Philosophical Transactions, No. 355, &c.) And he has been followed by the greatest part of those who have considered this subject. But we shall find it much more probable that it is at the very bottom of the column that the acting portions of the plates are situated. Indeed the greater probability of this may be inferred from the great obliquity of the hyperbola in the vicinity of the assymptote AB. The line *a b* is not nearly equal to *e f*, a circumstance which seems necessary, if the acting portion were immediately adjoining to the surface of the water. But we shall presently see much more cogent reasons.

If we make the breadth *g h* of the section of a column equal to *g γ*, the distance between the plates at that place, the section *γ g h χ* may be considered as a square. Suppose this to be the section of a square pipe of glass.

If it be true that it is only a minute portion of the touching glass which supports the water, and if this be incomparably less than the distance between the plates, then, whether this be at the top or bottom of the supported column, we should expect the water to stand twice as high in the square pipe as between the plates where the section is the same. For, in case of the plates, the acting surfaces are only the two sides $\gamma\ x$ and $g\ h$; but in the pipe, we have the whole periphery $\gamma g\ h x$ in action. This being double, should support a double weight of water, and therefore a double height. This experiment cannot easily be made, as such a pipe can scarcely be had. But a cylindrical pipe will do quite as well. For if the diameter of the round pipe be equal to the side of the square one, the proportion of the two peripheries is precisely the same with that of the two sections.

I have made this experiment with the greatest care, and I find that this expectation is completely answered. The water is supported twice as high. This is a momentous experiment. It has been long observed, that water is raised in slender pipes to heights which are inversely as the diameters, so far agreeing with the plates. This experiment completes the agreement.

211. Here then, in slender pipes, we have another means of measuring the natural force employed in supporting the liquid. The heights of the liquid are inversely as the diameter of the pipes. The sections of the pipes are directly as the squares of the diameters. Therefore the quantity of water, or the weights supported, are directly as the diameters. The surfaces in contact with the water, and therefore the simple adhesion, are equal in them all. Here, therefore, we have another incontestible proof that the whole surfaces are not acting in supporting the water. In the plates, the weights supported were equal in all the columns, and the surfaces were unequal in any degree.

In the pipes, the adhering surfaces are equal in them all, but the weights supported are unequal in any degree.

All will be reconciled, if we suppose that it is only a small portion of the touching surfaces that is effectual in supporting the water—a narrow parallelogram adjoining to *a b* or *c d*, perhaps not the 100th of an inch in breadth. This being equal to *g h*, equal weights are supported. This in the pipes, being a narrow ring at the top or bottom, is proportional to the diameter; and weights of water in this very proportion are supported.

I hoped to find some comparability between the distance at which this attraction is exerted and the distance between the plates, by the form of the curve *k a m*. I found it to be a very accurate hyperbola as far as the distance between the plates exceeded $\frac{1}{300}$th of an inch. This was inferred from the distance from AB where the curve began to deviate from the hyperbola, and the distance of the plates at the other end DC. I ascribed the deviation from the hyperbola in smaller distances to the action of the attractive force of the two plates interfering and being exerted on the same particles of water. But I was mistaken. The deviations in these small intervals were quite anomalous, and were greatly changed by the most trifling changes of distance at DC, and were uniformly greater in some places than in others. I found at last, that the cause of deviation from the hyperbolic curve was the want of perfect flatness of the plates—the inequalities of their surfaces bore a sensible proportion to the 1900th part of an inch. Employing two plates made by Dollond, and finished in the most perfect manner, I found the hyperbola strictly preserved, as far as I could perceive the coloured liquor that was between the plates. In these very small intervals, it is perceived with difficulty.

It deserves remark, that in making this experiment, the plates *seem* to attract one another with considerable force. When the thin wedge which separates the edges DC is drawn out, we find that it is strongly pressed by them, and they close with each other as it is drawn out, and it is difficult to get another wedge introduced. This is undoubtedly owing to the mutual attraction of the glass and the fluid.

212. This phenomenon of the rise of water, or other liquids, in slender pipes, has long engaged the attention of philosophers under the denomination of CAPILLARY ATTRACTION; being so named, because it is sensible only in pipes whose bore is extremely small, little exceeding the diameter of a hair. It has been much studied, both because it is curious, and because it has a very extensive and multifarious influence on many important operations of nature in our sublunary world. To this it is owing, that the rains which fall in the high grounds do not immediately run down to the sea with continually increasing velocity, but are retained by the soil, and, slowly filtering down through it, are delivered in the springs and fountains at the foot of the hills, so as to afford a constant and nearly uniform supply of moisture. By capillary attraction does the oil or melted tallow rise slowly through the wick of a lamp, where it is converted into steam by the heat of the surrounding flame, and, blowing out in all directions from the wick, is set on fire when it reaches the surrounding air. There *only* is it inflamed, although we generally suppose the whole space luminous. But it is only a luminous film. By capillary attraction the juices of the soil are taken up by plants and carried to the remotest leaf, where they are partly evaporated, as the oil is dissipated from the wick of the lamp. By capillary attraction is the lymph and other fluids taken up and transported to all parts of the animal body. No doubt the organical structure, both of plants and animals, is also concerned in the motion of the sap and the animal juices; but we are certain, by numerous observations,

that much of this is effected by *mere capillary attraction.* It is solely by this that a piece of dry wood absorbs a great deal of moisture, and swells, with almost irresistible force.

213. Such influence in the operations of nature justify the attempts of the philosophers to acquire a more precise and confident knowledge of this operation. I have already observed, that Dr Jurin was the first who gave any information on which an opinion can be formed. The ingenious Dr Hooke, indeed, had, before this time, speculated on the subject, and attempted to explain the phenomena by a principle which he called *congruity.* But men were not very familiar then with the accurate logic that is required in all philosophical disquisition, and Dr Hooke had little success in his endeavours. Dr Jurin reasoned more consequentially, from the beautiful experiment of Dr Brook Taylor, and its correspondence with the rise of water in capillary tubes. (See Phil. Trans. vol. xxx.)

Suppose two tubes AB, AB (fig. 9. No. 1. & 2.) which support the water at heights inversely as their diameters. Dr Jurin maintains, that the water is supported entirely by the action of the ring B *c* immediately above the surface of the water, B *c* being supposed the greatest distance at which the mutual attraction is exerted. For, says he, this is the only part of the tube from which the water must recede upon its subsiding, and therefore the only one which, by the force of its cohesion or attraction, opposes the descent of the water. To make this more manifest, he bids us suppose the tube to be divided into squares B *d*, *d e*, *e f*, &c. equal to B *c*. The glass in B *c* attracts the water in B *d* upwards. It is attracted downwards by the water in *d e*. This being inferior to the action of the glass above it, the water is supported. In every other part of the tube, such as *e f*, the water is equally attracted both upwards and downwards, and is not sensibly affected by those actions. The whole weight is therefore carried by the ring B *o*, aided by the mutual adhesion of the particles of the water. It will therefore be supported at any height, provided that this mutual adhesion be equal to the weight of the column.

In support of this reasoning, he observes, that the height at which the water will be supported depends entirely on the diameter of the *upper* orifice, whatever may be the width of the tube below. Thus, in fig. 9. No. 3. the water is supported at the same height as in No. 1. although the lower part of the tube is smaller. In No. 4. the water is *supported* at the same height as in No. 2. although it would have been *raised* only to *b* by the lower part of the tube. It is remarkable, that in a tube formed like No. 4. the water stands at the same height, although the lower part of the tube be of any width, even several inches, provided that the upper orifice at B is of the proper diameter; and this obtains, even though the apparatus is placed under the receiver of an air pump. The water, therefore, is not supported, as was asserted by some, by the pressure of the air. Others maintained that the water was supported in the wide part of the tube, by its adhesion to the arch immediately below the slender tube. But a *curious* experiment of Dr Jurin's seems to overturn this account. If the wide part of the apparatus No. 5. be filled with water up to D, very near the arch, and kept there by shutting the tube at the bottom. Then let a single drop of water enter at the top, and set the apparatus into the water in the cistern, and open the tube below. The drop let into the upper orifice will occupy the whole of the slender tube, and the water will remain at the height to which it was filled, though it do not come in contact with the arch. The whole seems still to be supported by the action of the upper orifice.

Such is the reasoning by which Dr Jurin's theory is supported. It is undoubtedly very specious, and it obtained the ready acquiescence of all the naturalists. Mr Hawksbee has varied the experiments in many ways, and a multitude of others have been added by Bulfinger in two long dissertations, in the Petersburgh Commentaries. Mr Achard also, of the Berlin Academy, has added many more, and philosophers were satisfied.

214. Yet the theory is certainly erroneous in its first principles. Granting that the particles of water in the uppermost ring *d* B are attracted upward by the glass in the ring B *c* above them, it must be as much attracted downward by the glass in the ring *d e*. The attraction of this ring cannot be supposed to cease when the water has passed it. If the whole tubes from end to end be divided into portions equal to B *c*, the water is in equilibrio (in respect of this attraction by the glass) in every part of the tube, except the lowest portion *m n*. For every other portion has an equal ring of glass above and below it. Therefore, if there be any truth in this mutual action of glass and water, this theory of capillary attraction is in consistence with it. Let us consider the consequences of this action a little more particularly.

215. Let C F (fig. 10.) be a glass tube, lying horizontally, and containing water, or any liquid that it attracts, between A and B, and let us suppose that the mutual attractions of glass and water extend to the distance AC or BD. The water in this tube will continue at rest, being as much attracted in the direction BA by the glass in AC, as it is attracted in the direction AB by the glass in BD.

Suppose now that the tube is cut through at A as in No. 2., by this the equilibrium of the water is destroyed, there being no force now to balance the attraction of BD. The water will therefore be carried in that direction, and will be again in equilibrio, when it gets into the situation ED. It will then be equally acted on by AE and DF. In this manner the water may be made to remove, in the direction AB, as far as we please by continually cutting off the attracting ring at one end.

This attraction has a limited force, and may be balanced by any opposite force. It may be balanced by the force of gravity, by holding the pipe upright, as in No. 3. or better, as in No. 5. by making the tube bend upward. The tube being cut off at A, and the water having retreated into the space ED, we may suppose water added by the other end. This will make no change in EA till the level part of the tube be filled,—then the addition will begin to push the water from E toward A, and will reach A when the weight of the column GH is equal to the attraction. In this condition of the apparatus, the whole is supported (agreeably to Dr Jurin's opinion) by the attraction of HI, unbalanced by any attraction at A. If the weight of the water in AB, No. 3. is just equal to the attraction of AC or BD, then, when the tube is held upright the water will descend, till it arrives at the mouth of the pipe, occupying the space C *d*. The weight of the water is then in equilibrio with the attraction of the glass in B *d*. But it is not the attraction of B *d* that supports it; for B *d* acts only on the water in *d e*; for the water in *d e* is as much attracted downward by the glass in B *d*. And if we examine every portion of the tube, we find the water in the same state of equilibrium, excepting the lowest portion AC. This is attracted upward by the glass in AE, without any opposite attraction to balance it. It is balanced, only by the weight of the water between *c* and *d*. This weight is supported by the attraction of the glass in EA for the water in AC. We may suppose this little portion of water in AC to become ice, without adhering to the tube, and without any change in the attracting forces. It will then perform the office of a moveable plug, loaded with the water in A *d*, and supported by the attraction of the glass in AE.

When things have been thus adjusted, bring the lower

orifice of the pipe into contact with the surface GH of standing water, as in No. 4. The water below it, which was formerly in hydrostatic equilibrio, and at rest, is now attracted upwards by the glass in CA, as much as the water in CA is attracted by the glass in AE. Thus there are now two portions of water, viz. ACC′A′, and CIKC′, which are attracted upwards by the glass in CA and AE. Let g represent the attraction of glass for water. There is now a force $2\,g$, acting upwards, in the place of the force g which operated before. This should support twice as much weight, and the water should now stand twice as high as before.

But there is now brought into action another force, which did not exist before. The water in the portion AC of the tube is attracted downwards by the water CIKC′ in the cistern. Let w express the force with which the water in AC is attracted by the water CIKC′ in the cistern. The whole force which now acts on the water in the tube is the force $2\,g$, acting upwards, and the force w acting downwards. Therefore the height to which the water will be raised will be such that its weight is $= 2g - w$.

216. Hence it plainly follows, that if the mutual attraction of water for water be *any thing less than twice the attraction of water for glass*, it will stand higher within the tube than without it. If w be merely equal to $2\,g$, the water will be on the same level both within and without the tube. We learn here, that we must not infer, from the rise of fluids in slender pipes, that their particles attract each other more feebly than they attract the particles of the pipe, which is the general opinion. We are entitled only to infer, that the attraction of the particles of the fluid is not twice as great as their attraction for those of the pipe.

Such are the genuine consequences of a mutual attraction between the particles of a liquid and those of a solid.

This attraction is unquestionable, and therefore this theory of the rise of water in capillary tubes must be admitted, in preference to Dr Jurin's, which is incompatible with the very principle by which the author thought it established.

217. This explanation of capillary attraction was first proposed by the celebrated Mr Clairaut, in his dissertation on the Figure of the Earth, published in the year 1743. But it lay there unnoticed by naturalists not much conversant in mathematical disquisitions. Indeed, it comes in so incidentally, and is so slightly mentioned, that it does not seem to have arrested the notice even of mathematicians. Lalande first took particular notice of it, and published it in a small duodecimo pamphlet in 1770. It does not seem, even yet, to be very generally known.

218. Both Mr Clairaut and Mr Lalande consider capillary attraction as a case of universal gravitation; but this is certainly a mistake. Were it gravitation, a fluid would stand above the level of the cistern only when it is less dense than the pipe. Now, we know that water will stand higher in a small quill than the water in the cistern. Were it gravitation, a glass pipe would support different fluids at heights inversely proportional to their density. This is far from being the case. A very slender pipe supported the different fluids as follows: *

	Inch.
Oil of turpentine	1,35
Spirits of wine	2,5
Water	5,5
Caustic vol. alkali	6,25
Solution of sal ammoniac	8,07

This attraction must therefore be considered as a *speci-*

* See Note I. at the end of the volume.

fic mechanical affinity analogous to chemical affinities or elective attractions;—It is different in different substances for the same kind of glass, and in different kinds of glass for the same fluid.

219. If a glass pipe be plunged into mercury, the fluid is lower within the pipe than without it. This shews that the attraction of the particles of mercury for each other is more than twice as great as their attraction for the glass. When two glass plates are set in mercury, it forms a curve convex upwards, lower than the surface of the mercury in the cistern.

220. It will naturally be asked here, how it happens, as in Dr Jurin's experiments, that the height at which the water is supported depends so expressly on the size of the upper orifice, and not at all on that of the lower orifice?

It has been already explained how this attraction causes a drop of water to move from the wide to the narrow end of a tapering tube. Now, it is a matter of observation, that if we take a glass syphon ABC (fig. 11.) having the legs of different calibres, but both capillary, and put water into it, the water will stand so in the two legs, that the elevation of the surface C in the smaller leg above the surface A in the larger, is precisely equal to the difference between the heights at which each of the legs would have supported the water above the level of the cistern. Suppose that the smaller leg would have raised water to a height m, and the larger to the height n, then the elevation Ca will be $m - n$. This experiment, which has been often repeated, may be received as a sufficient proof, that a contraction of the diameter in any part of the capillary canal is always accompanied by an excess of attraction directed to the narrow part of the canal. Here we have a measure of that excess, and it corresponds precisely with the measure that was deduced from the principles of hydraulics. To apply this to the present question, suppose two capillary tubes a and b (fig. 12.)

and that *a* supports water to the height AM, which we shall call *m*, while the wider tube *b* supports it only to the height AN, which we shall call *n*. Let the tube C have its lower part AB of the same bore with the tube *b*, and its upper part BM of the same bore with the tube *a*. Let the fourth tube *d* have its lower calibre equal to that of *a*, and its upper have the calibre *b*.

At the lower orifice of the tube *c* there is a force *n* acting upwards. At B there is a force $m - n$ also acting upwards. Add these two together, and their *sum* is the compound force, and measures the height to which the water will be raised. Now the sum of *n* and $m - n$ is *m*, and therefore the water will be raised to the same height by *c* and by *a*.*

At the lower orifice of the tube *d* there is a force *m* acting upwards. At the point B there is a force $m - n$ acting downwards. *The difference* of those two forces is the measure of the compound force, and of the height to which the water will be raised. Now the difference of *m* and $m - n$ is *n*, and the water will be raised to the same height in the tubes *d* and *b*.

221. Thus it is manifest that the phenomena which gave such authority to the theory of Dr Jurin are perfectly consistent with Mr Clairaut's theory. Dr Jurin was also mistaken, I think, in saying that the water was not sustained in the wide vessel terminating above in a capillary tube by the adhesion to the arch of the wide part. It is very true that his experiments succeeds, as he says, where a small part just adjoining to the arch is left empty. But this happens only under the pressure of the

---

* Here it must be noted that in order to *raise* the water by this tube to the height of AM, it is necessary that AB do not exceed AN of the tube *b*. Wherever the contraction B is, the water will be *supported* at the height AM, provided that the tube be previously filled with water. It will sink till the height of the remaining column be AM. But the empty tube will *raise* it no higher than AN of the tube *b*.

air. If the apparatus be in an exhausted receiver, the water will descend in the wide part, which it will not do if the apparatus be full of liquor, so as to be in contact with the whole of the arch. It will not, indeed, even then, remain long suspended. For, *in vacuo*, a quantity of air, which had been chemically united with the water is detached from it, and gets to the top. As soon as a bubble gets to the lower orifice of the capillary tube, the water in the wide part of the apparatus descends. The adhesion of the top of a lateral column of water to the arch is abundantly able to carry its weight, for it is almost immensely greater than the attraction of the infinitesimal ring of the capillary tube in the middle, which, according to Dr Jurin, supports the whole. If air were not disengaged in the manner now mentioned, I have no doubt but that the water would be supported *in vacuo* to the height of many feet. We see mercury supported to the height of 70 inches.

222. This subject seemed to merit a particular discussion on account of its manifold influence on natural operations, as well as because it is a very good example of the corpuscular forces, susceptible of measure, and of mathematical consideration. It rectifies our judgments concerning the intensity or magnitude of the attractive force of the particles of water and other liquids. We are disposed to consider this as extremely small, because we find it so easy to separate one parcel of water from another. But this separation does not give us the smallest information on this point, and the mutual cohesion of two particles of water may be as great as that of two particles of steel, for any thing that we know to the contrary. The phenomena of capillary attraction shew it to be very considerable.

A pipe of one-tenth of an inch in diameter supports a column of nearly half an inch in height. This weighs one grain*. This is supported by the action of a ring of glass, whose breadth cannot exceed the 2000th part of an inch, (nay it is probable that it is almost incomparably less) and which is one-eighth of an inch in circumference nearly. The surface of it is therefore nearly $\frac{1}{1333333}$th of an inch, and the adhesion of water to a square inch would be 1,333,333 grains, or 190 pounds avoirdupois. But this is far from being the just measure of the mutual attraction of the parts of water. It is only *the difference between this attraction and twice the attraction of water for glass.* It may be but a very minute portion of the mutual attraction of the parts of water, and I am much disposed to think that this is the case. When the almost absolute incompressibility of water is compared with the other phenomena of cohering matter, and the whole considered in the manner that will be unfolded by degrees, I apprehend that we are well warranted, nay obliged, to draw this conclusion.

223. The parts of water are easily separated, because they are easily made to slide over each other, as will be particularly explained, when we come to consider the mechanism of liquid aggregation. This mutual attraction extends to a very minute distance, and, when the particles are so far separated, they attract no more. Suppose gravity to extend only a foot from the surface of the earth, and suppose that it is covered with balls of a foot diameter, which attract each other with *any enormous* force. If we take one of these balls in our hand and raise it, another will stick to it and follow it, a third will adhere to that, and a fourth to the third, and thus we shall lift all the balls from the earth, and will feel only

* A cylinder of water one-tenth of an inch in diameter, and twelve inches high, weighs 23.862 grains, or very nearly 24 grains; therefore very nearly 1 grain when half an inch in height. A cylinder one inch diameter, therefore, weighs 2386, very nearly 2400, or 5 ounces troy. This is a convenient number to keep in mind, = 5 oz. + 200 gr. avoirdupois.

the weight of one, because as soon as a ball is a foot from the ground, it is no longer heavy. It is in this way that we may separate the particles of water from the mass.

224. The measures which we have acquired of capillary attraction lead to several useful inferences. It was already observed, that it is owing to this that water is retained by the soil: not only so, but it will be raised by it. If we fill a hollow tube with dry sand, and set it in water, we shall observe that the water will soak the sand to a considerable height,—so much greater as the sand is finer; in fine impalpable clay, the water will rise to almost any height.

It is in this way that oil rises in the wick of a lamp. The wick is a bundle of fibres, with interstices between them. The fluid rises in those interstices, and will rise almost twelve times higher in the interstices of cylindrical fibres than through pipes of the same diameter, as will easily appear by calculating the area of the interstice, and its acting periphery.

I already took notice of the swelling of wood by moisture. The force exerted in this way is sometimes very great. In raising grindstones from the quarry, it is a usual practice to hew a long cylinder, of the diameter required, and then grooves are cut round it, at the distance required for the thickness of a grindstone. The grooves are filled with wedges of dry wood. Water is thrown on them, and in an hour or two the whole column is found neatly separated into grindstones. It is in the same way that sponge swells with water.

225. These are not cases of mere absorption. The water is *drawn* into the cavities by an active force, acting at some small distance, otherwise the effect of mere adhesion would be completed when the cavities are merely filled. But more is still attracted, and this causes the flexible cavities to enlarge. It would require a pretty long discus-

sion to explain how this effect is produced. But a very simple instance of the fact will conduce to a better comprehension of the mechanism in more complicated cases. Scatter a few drops of mercury on a plate of mirror glass lying on a table, and let the drops be as nearly as possible of a size. Lay on the plate so sprinkled another very light plate of mirror glass. If the drops of mercury are numerous, and pretty uniformly distributed, the upper plate will not greatly flatten them. Now lay weights on the upper plate. The drops of mercury will be more and more flattened as the weights are increased. Now let the weights be gradually and equably lifted off. The flattened drops, which are now spread out to a considerable breadth, will gradually contract, and will regain the breadths which they acquired by the first laying on of the plate.

The inference from these phenomena is incontrovertible. The particles in each drop of mercury are not in corpuscular equilibrium except when in a spherical form, and an external force is necessary for giving them any other shape. Their own gravity flattens the drops a little, even without the upper plate. The weight of the plate flattens them still more. But, as it lessened, the drops become more and more spherical. This cannot be, unless the upper plate remove farther from the lower one —this it naturally does. In like manner, in the foiling of looking-glasses, it is necessary to lay on a very great weight to squeeze out all the mercury that is not absolutely necessary for saturating the tin foil, and if these are taken off, a great quantity of mercury is again sucked in. I found that a plate of 18 inches by 14 sucked in $5\frac{1}{2}$ ounces, which must have raised the plate about $\frac{1}{20}$th of an inch. Therefore, in these facts it appears, that the drops tend to a spherical form *with force*, for they lift up the plate, even when loaded with weights.

226. It is evident, from the different heights to which

the same glass will raise different liquors, and different kinds of glass raise the same liquor, that it is a specific attraction, depending on the constitution of the particles, both of the solid and the fluid. Sir Isaac Newton observed a phenomenon of the same kind with respect to the action of bodies on the rays of light. He observed, that when a beam of the sun's light passed by the edge of a knife, or any other solid body, the rays of one colour were more deflected than those of another which passed by at the very same distance. Also, a blue ray was as much deflected when it passed at the distance of $\frac{1}{480}$th of an inch as a red ray which passed at the distance of $\frac{1}{333}$th of an inch. In short, he observed, in those delicate experiments, that the forces exerted between the light and the solid body varied by a variation of distance, just as gravity does, (though not in the same proportion), and was different in relation to the differently coloured rays, therefore specific. Having also demonstrated that reflection and refraction are performed by means of the same forces which produce the inflections and deflections now mentioned, Newton justly inferred, from the different refracting qualities of different bodies, that those forces are also specific in respect of different substances, and that, in particular, the particles of inflammable bodies exert stronger forces than others. From this, his sagacious mind conjectured that diamond is an inflammable body, a conjecture which has been completely verified within these few years. Mr Tennant's experiments prove it to be the poorest specimen of carbon.

227. When we reflect on this specific nature of the forces which produce the phenomena of capillary attraction, and the share which these phenomena have in many natural operations, several considerations are suggested which seem deserving of notice. I am persuaded that much of the operation of the absorbent system in animals and vegetables depends upon mere capillary attraction, or,

at least, upon mere mechanical corpuscular force. I imagine that it is chiefly in this way that the nutritious and otherwise useful parts of our food are taken up by the lacteals from the intestinal canal, and conveyed to other parts of the body. If so, much of the operation depends on the orifice by which a lacteal communicates with the intestine. The operation will be affected by a change of mere aperture—it will be affected by every change in the constitution of that orifice, as also of the fluid taken up. By the difference of its substance, a vessel of the same diameter with another, and opening into the same canal, may abstract very different juices, and in this way various secretions may arise—a change of composition, whether in the fluid or the vessel, may accelerate or obstruct the operation *. Capillary attraction, therefore, seems to deserve the attention of medical men. The poor inhabitants of the island of Ormus, who cannot afford to bring fresh water from the Continent, are said to quench their thirst by lying down in cisterns of sea water. I saw an experiment of the same kind, made in 1761, at the Royal Academy at Portsmouth. A man who had abstained from drinking till his mouth was parched, went into a tub of sea water, and in a quarter of an hour his thirst was almost gone, and he could spit freely. His skin had absorbed a very considerable quantity of fresh water.

228. We know that in like manner, a plant that languishes for want of water will be completely refreshed by immersion, although none gets at the roots. At the same time it is probable that, both in the animal and vegetable

* Some experiments, by a very intelligent and ingenious gentleman, on the rise of the oil and tallow in the wick of a lamp, shew remarkable differences in this respect. The publication of these experiments might produce great improvement in the manufacture of candles, and the preparation of inflammable matter. I hope that the ingenious author will not withhold them from public service.

economy, organical structure and mechanism are concerned in these operations. Although the greatest part of the juices rise through the plant by mere capillary attraction, it is very likely that there are also contrivances in the roots which may be said to force or pump it up, and valves to prevent its improper return.

Having employed so much time in describing the phenomena of corpuscular action between the particles of solids and fluids, I shall be more brief in my account of such as appear in the action of solids on solids. A small number will suffice.

229. When naturalists intimated their acquiescence in the conjecture of Newton, that cohesion was effected by means of mutual forces acting beyond the particles, it was said that were this the case, nothing more is wanted for reuniting two pieces of broken glass but putting them together again. But this is altogether contrary, say they, to experience. But we should not expect this, as the law of corpuscular action is unknown (at least as yet) we cannot tell but that at a distance somewhat greater than the distance of cohesion, the particles may repel each other. This is not absurd, but it is not even improbable. We see that magnets having their similar poles fronting each other, repel at all considerable distances, but when brought very near, they generally, though not always, attract each other. What should hinder *particles* from having a similar relation. It may require a great force to bring the parts of a broken body near enough for giving occasion for the exertion of attracting forces. Besides, when we press together the surfaces of the fracture, it is perhaps only in a few points that they are near enough for attraction. Accordingly, it is known that if two pieces of metal are *scraped* very clean, a severe blow will make them cohere so as to be inseparable. It is thus that flowers of gold and silver are fixed on steel and other metals. The steel is first scraped clean, and a thin bit

of gold or silver is laid on it, and then the die is applied by a strong blow with a hammer. It is remarkable that they will not adhere with such firmness, if they adhere at all, when the surfaces have been polished in the usual way, with fine powders, &c. This is always done with the help of greasy matters. Some of this probably remains, and prevents that *specific* action that is necessary. I am disposed to think that the scraping the surfaces also operates in another way, viz. by filling the surface with scratches, that is, ridges and furrows. These allow the air to escape as the pieces come together by the blow. If the mere blow were sufficient, a coin would adhere fast to the die. But, in coining, the flat face of the die first closes with the piece of metal, and effectually confines the air which fills the hollow that is to form the relief of the coin. This air must be compressed to a prodigious degree, and, in this state, it is still between the die and the coin. We may say that the impression on the coin is really formed by this included air; for the metal, in this part of the coin, is never in contact with the die. I know of two cases which greatly confirm this conjecture. The dies chanced to crack in the highest part of the relief, and after this, were thrown aside, (although in one, for a common coin, the crack was quite insignificant), because the coin could seldom be parted from them.

230. Air seems to adhere to most bodies. When cold distilled water is poured into a glass which has been previously cleared from all dampness, numerous air-bubbles are observed to form all over the glass; and, by tapping it with any thing hard, they are disengaged, and others are formed in their places. The air seems to be disengaged by the superior affinity of the water. Now, this air may be an obstacle to the approximation of the bodies that is required for bringing the attractive force of cohesion into action. In ductile bodies, whose constitution ap-

proaches to softness, the due approximation may be easily effected. Thus, two pieces of lead (and, I am well informed, of pure gold) may be forced into perfect cohesion by a strong pressure, combined with a sliding motion, when they are made very clean. Two pieces of the gum caoutchouc, fresh cut with a clean knife, when pressed moderately together, adhere as strongly as any other part. I apprehend, however, that it is a very difficult thing to bring the surfaces of two bodies as near as the natural distances between their particles. The grounds of my opinion will soon appear.

231. There is a precious experiment by Mr Huyghens in No. 86. of the Philosophical Transactions. A piece of mirror glass being laid on the table, and another, to which a handle was cemented on one surface, being gently pressed on it, with a little of a sliding motion, the two adhered, and the one lifted the other. Lest this should have been produced by the pressure of the atmosphere, Mr Huyghens repeated the experiment in an exhausted receiver, with the same success.

Here, then, is a most palpable demonstration that the adhesion is the effect of corpuscular attracting forces. It may be ascribed to the attraction of gravitation, with as much propriety as the attractions in Mr Cavendish's experiment mentioned in § 475. But another valuable part of Mr Huyghens's experiment overturns this supposition.

He found that one plate carried the other, although they were not in mathematical contact, but had a very sensible distance between them. He found this by wrapping round one of the plates a single fibre of silk drawn off from the cocoon. The adhesion was vastly weaker than before, but still sufficient for carrying the lower plate.

Here then is a most evident and incontrovertible example of a mutual attraction acting at a distance. Mr

Huyghens found that if, in wrapping the fibre round the glass, he made it cross a fibre already wrapped round it, there was no sensible attraction. In this case, the glasses were separated by a distance equal to twice the diameter of a fibre of silk.

232. I said that this experiment shewed that it was not the attraction of gravitation that produced the adhesion. I have repeated the experiment with the most scrupulous care, measuring the distance of the glasses (the diameter of a silk fibre) and the weight supported. I find this, in all cases, to be nearly $14\frac{1}{2}$ times the action of gravity. The calculation is obvious and easy. I tried it in distances considerably different, according to the diameter of the fibre. I must inform the person who would derive his information from his own experiments, that there are many circumstances to be attended to which are not obvious, and which materially affect the result. The silk fibres are not round, but very flat, one diameter being almost double of the other. The 2400th part of an inch may be considered as the average smaller diameter of a fibre. A magnifying glass must be used, and great patience in wrapping the fibre round the glass so that it may not be twisted. A flaxen fibre is much preferable, when gotten single, and fine enough, for it is a perfect cylinder. I must also inform him, that no regularity will be had in experiments with bits of ordinary mirror; these are neither flat enough, nor well enough polished. We must employ the square pieces which are made and finished by a *very few London artists* for the specula of the best Hadley's quadrants. These must be most carefully cleared of all dust or damp. Yet this must not be done by wiping them with a clean cloth; this infallibly deranges every thing, by rendering the plate electrical. I succeeded best by keeping them in a glass jar, in which a piece of moist cloth was lying, but not touching the glasses. When wanted, the glasses are taken out with a pair of

tongs, and held a little while before the fire, which dissipates the damp which had adhered to them, and which prevented all electricity. With these precautions, and a careful measurement of the diameter of the silk fibre, the experiments will rarely differ among themselves one part in ten.

There is another circumstance to be particularly attended to. Although one plate will lift the other by means of a very gentle pressure, it will not long support it without the utmost attention. The lower piece always slides to one side, and drops off. Hence some have ascribed the transitory adhesion to the pressure of the air, which is at an end as soon as enough of air has insinuated by the sides. But a little reflection will convince any one versant in hydrostatics and pneumatics, that this supposition is altogether erroneous; without a mutual attraction, the glasses would not adhere for a moment, having air already between them.

The glass drops off by the sliding to one side, because it is next to impossible to hold the glasses perfectly level. If in the least inclined to the horizon, the lower glass, having no obstruction from any thing like friction, glides along the inclined plane with *perfect* freedom. If the plates have been hard pressed, with a sliding or grinding motion, the adhesion is then either very strong, or nothing at all; when they do adhere, it seems to be another stage or alternation of the force, as will be explained by and by But they rarely adhere, owing to fragments torn off by the grinding. The glasses will be scratched by it.

I thought this capital experiment worthy of a very minute description, it being that which gives us the means of mathematical and dynamical treatment in the greatest perfection.

233. We are not to imagine that the corpuscular force which we have just now considered is the attraction of cohesion. This would be to suppose that $\frac{1}{2400}$th of an

inch is nearly the natural distance of the particles of coherent bodies. But it certainly exceeds that distance many thousand times. In the manufacture of gold lace, a silver wire is covered with a certain quantity of gold leaf, and then drawn to an extreme fineness. We are certain that the covering of gold is not more than one fourteen millionth part of an inch thick. Yet this film is so continuous and compact, that the microscope can discover no pores in it, and it completely defends the inclosed silver from the action of aquafortis.

It is now time to take notice of a corpuscular action extremely different from those hitherto mentioned. Every one has observed certain insects run about on the surface of the water without sinking or wetting their feet. If one of them be carefully viewed with a magnifying glass, a pit will be observed on the surface of the water all round each foot, resembling what is made in a mattress or feather bed when a person stands on it. If the feet of those insects be examined with a microscope, they will be seen to consist of five or six spreading hairs, ranged as the rays of a star are usually drawn. When on the water, each fibre is surrounded by a pit much broader than the fibre.

By these extensive pits, a quantity of water is displaced, equal in weight to the insect, and the insect is supported. Yet the feet are not wetted. There is something which keeps the water from coming into contact, which keeps it at a distance, and thus forms the pit: perfectly similar to this is the common experiment of making a clean and polished needle swim on water. It forms just such a pit, and is buoyed up in the same way.

The feet of the insect, and the needle, *repel* the water, exerting corpuscular forces directly opposite to attraction. We may call them *repulsions*, without pretending to explain their means of acting, merely to distinguish the effect. Of such forces there is a vast variety. It is owing to such repulsions that it is so difficult to wet many powders, such as that of the puff-ball, the lycopodium. It is this which makes the down and the fur of aquatic birds and beasts almost impenetrable to water.

234. It is owing to this that we see a dew-drop lie so beautifully brilliant on the leaves of plants in the morning. We have the fullest assurance from the laws of optics that the part of its surface where the reflection is so brilliant is not in contact with the leaf. The dew-drop may sometimes be made to roll along the leaf, retaining all its brilliancy. In this case it makes no watery trace on it. One may see the effect of contact very plainly by taking a little drop or ball of crystal, and, attending to the vivacity of the reflection from its posterior surface, touch the surface of water with that part which reflects with such vivacity; the vivid reflection is destroyed in that instant, and scarcely any reflection remains. I have often observed the large drops of a warm summer's shower roll about for a second or two on the surface of still water, and some of the small drops which they have dashed up have fallen down and rolled about for a long while, allowing me to drive them about by fanning them. They are remarkably brilliant while thus rolling about, thus shewing incontestibly that they are not in contact with the water.

We shall soon have abundant evidence that the distance at which one body in this manner supports another, giving rise to *physical* contact, and giving us the sensation of touch or feeling, is considerably less than the distance between the glasses in the Huyghenian experiment. It is reasonable therefore to conclude, that in some intermediate distance between the diameter of a silk fibre and the distance of a dew-drop from the leaf, the bodies neither attract nor repel. They will repel if pushed nearer, and attract if separated a little. Such alternation is observed in some magnetical experiments, and we can generally put the magnets at that intermediate distance in which they neither attract nor repel. But when, as in the case of the drops that we have been speaking of, one body is carrying the weight of another, they are somewhat nearer than the distance of inactivity. They are so much nearer, that a mutual repulsion is exerted, equal to the weight of the drop; and when, in Huyghens's experiment, one glass plate suspends another, they are at a distance somewhat greater than the distance of inaction: they are so much more distant that both exert an attraction equal to the weight of the suspended glass. These cases resemble what happens when a body is supported by a spring. When laid on the spring, it compresses it till an elasticity is exerted equal to the weight of the body. When suspended from the spring, it stretches it till a sufficient elasticity is exerted. But while the spring is of its natural shape, no elasticity, either attractive or repulsive, is exerted.

235. I have already mentioned the observations of Newton, in which it appeared that when the rays of light pass by the edge of a solid body at a certain distance, they are inflected towards it; if at another distance, they are deflected from it; and if at a third distance, they are inflected towards it; and if at a fourth, they are deflected, &c. &c. But these observations may be thought peculiar to the action of bodies on light, and even to depend on a cer ain theory concerning the nature of light. But there is another observation of Newton's more directly and unequivocally to our purpose, shewing that solid bodies act on one another at a measurable distance.

236. If we take the object glass of a long telescope, having its radius of convexity not less than 15 or 20 feet, and lay it on a piece of very flat and well finished mirror, or a piece of finely polished metal, or black marble, we shall not observe any thing remarkable in the reflection of the light, unless the object lens is heavy, or is pressed down by the hand. But if we press it down, we shall ob-

serve in the place of contact of the convex lens and flat surface, something like a greasy spot, having the uncertain pearly colour of a fish scale. Pressing harder, we observe a coloured ring form round the point of contact, reflecting white light in the middle of its breadth, with blue on its inner edge, and red on its outer edge. Pressing still harder, the ring enlarges in diameter, and a new spot forms in the centre, which, by greater pressure, becomes a ring, and is succeeded by another central spot. And this succession may be continued till a great number of rings are thus produced. A more careful examination shews us that it is not a succession of spots which occupies the centre, but of simple colours, which follow one another in a determined order, so as to form rings such as have been described blue on the inner edge, and ruddy on the outer. When the pressure is made very strong, a bright silvery spot appears in the centre, ruddy all round; and the next increase of pressure produces a black spot in the middle of the silvery spot, as if a ragged hole were made in a round bit of silver leaf.

If we now gradually diminish the pressure, we shall see the rings contract, and vanish in the centre in succession.

237. On this observation Newton erected a most ingenious fabric of optical doctrines, which will be considered in their proper place. He immediately began to consider every thing mathematically. He measured the diameters of the rings, and, knowing the convexity of the lens, he calculated the distances between them in the brightest part of each rings, and he found that the distance at each ring exceeded the distance at the ring immediately within it by $\frac{1}{89000}$th of an inch nearly. He also found that these successive differences were all equal. For the diameters of the rings were as the square roots of the numbers 1, 3, 5, 7, &c. and the diameters of the dark spaces between them as the square roots of the numbers 2, 4, 6, 8, &c.

Newton's mind being completely occupied by the important optical inference which he drew from this phenomena, he paid no attention to the circumstance in which *we* are now chiefly interested, namely, the pressure that must be employed to produce these rings. He counted in favourable circumstances upwards of 20 rings, which he had followed with his eye from their first formation in the centre till their greatest expansion. Suppose that he observed 20 in this manner. He concluded, that when the black spot appeared in the centre the glasses were then in contact, and not before. On this supposition, the distance between the glasses at the 20th ring is $\frac{1}{4450}$th of an inch. Very rarely, however, would he perceive a ring at so great a thickness of the glass without some optical assistance which he describes.

238. Now it is a matter of fact that, unless the lens be heavy, or pressed down, no coloured spot appears in the centre. The glasses therefore are not in contact, but are distant from one another at least $\frac{1}{4450}$th of an inch. At this distance the lower glass supports the weight of the upper glass. If pressure be added to this weight, the glasses are brought nearer, and this approximation is indicated by the colour which appears in the centre. Every time that we produce the red in the centre, by increasing the pressure, we must conclude that the glasses are $\frac{1}{29000}$th of an inch nearer than they were at its preceding appearance; and thus a relation may be discovered between the distances and the intensities of the resisting forces, in the same manner as Newton inferred the law of gravity from the deflections produced by it.

Nothing can be better established than the conclusion from this experiment, viz. that those bodies act on each other at a distance, in the same sense of the words as when we say that the sun and planets act on one another. We may say that they repel one another, in the same way as we say that two magnets, or two electrified bodies, repel one another. We do not mean by this abbreviation of language to assign the mode of action, but merely the appearance of the effect.

239. I have repeated this experiment of Newton with great care, and I find it most accurately described by their illustrious author. I hoped to ascertain the law of action by means of the pressures employed. But I found myself unable to express the relation between the distance and force of the acting particles by an equation. We may easily see that this must be difficult; for it is not the central particles alone that are acting, but also those all around, to an unknown distance, were acting with different forces. The experiment requires very well finished glasses; ordinary low priced lenses are good for nothing in this research; their errors of figure bearing a considerable proportion to $\frac{1}{29000}$th of an inch. But they shew the rings very well, if the convexity be sufficiently small. Even the lens of a pair of spectacles will do this, if it is of 20 or 30 inches focal distance, but the rings will be exceedingly small, and scarcely distinguishable.

Newton supposed that the glasses were in absolute mathematical contact where the black spot appeared, for which reason he thought it was that they allowed the light to pass through as if they had been one continuous mass of glass. But it may be asked what authority Newton had for this supposition. We are indebted to him for a complete answer to this question, but an answer very different from his declared opinion. This we obtain by means of another precious experiment of Newton's. The beautiful colours of a soap bubble, which had often amused him in his boyish years, now recurred to his memory with strong persuasions of their importance in his present speculations, and he immediately examined them with the attention of a philosopher.

240. Having blown a soap bubble of a small size, he

covered it with a bell glass, to ward off the disturbance by the air, and carefully noted the appearances of colour in the bubble. He saw, after some little time, like a pearl-coloured spot in the very uppermost point or zenith of the soap bubble. This in a little spread itself into a round spot, ruddy in the centre, and blue round the circumference. This widened into a ring, ruddy within and purple without. Another spot formed in the centre; this also became a ring, the other ring enlarging in the mean time. The second spot became a ring, and was succeeded by a third, which underwent the same change. In short, coloured rings formed and enlarged on the upper hemisphere of the soap bubble, in the same succession, and having the same colours as those seen between the object glasses; but incomparably larger, more distant, and more brilliant. He could sometimes count 50 of them. After some time, there formed in the zenith a bright silver-coloured spot, in which there soon appeared a ragged hole, which sometimes enlarged itself to the breadth of one third of an inch or more, and then the bubble burst. This central dark spot seemed at first altogether without reflection; but more careful inspection shewed that it still reflected a minute quantity of light.

Newton had already inferred from the experiment with the glasses, that the different colours reflected at different distances from the common centre of the rings depended on the different distances of the glasses. Here, in the experiment of the soap bubble, we have the same succession of colours. Newton explains them by observing that the bubble grows gradually thinner at top, by the subsidence of the clammy liquid, and that the different colours depend on the thickness of the film where they appear. He proved this immediately after this discovery, by splitting talc till it produced permanent colours. Even glass may be blown so thin as to exhibit them. But those optical inferences are not our proper object at present. It is enough that we have seen that the film may be so very thin as to give no vivid reflection; yet it has some thickness, for the bubble exhibits the spot for some little time before it bursts.

241. Here then we have abundant evidence that the appearance of the black spot between the glasses did *not prove* that they were in *mathematical contact* in that place, but only that the distance between them was too small for producing any sensible reflection of light. There is no doubt of the spot between the glasses, and the spot in the soap bubble being corresponding phenomena. They are remarkably distinguished from all the coloured spots that successively formed them. The silver-like rings which surrounded this black spot are quite unlike all the coloured rings, being incomparably more vivid. But the chief distinction is the abrupt, irregular, and ragged inner edge of the silvery ring which surrounds the spot. It is exactly like a hole carelessly torn in the middle of a bit of silver leaf, whereas the edges of all the coloured rings, and even the *outer* edge of this silver ring, are undefined, like the edges of the rainbow.

The conclusion seems therefore unquestionable, that we have no proof from the black spot between the glasses that they are in mathematical contact in that place.

We know, by the first experiment, that a very considerable force is necessary for producing the black spot. A greater pressure makes it broader, and in all probability this is partly by the mutual yielding of the glasses. I found that before a spot, whose surface is a square inch, can be produced, a force exceeding 1000 pounds must be employed. When the experiment is made with thin glasses, they are often broken before any black spot is produced.

242. What is it that we properly, and without any figure of speech, call a pressure? It is something that we are informed of solely by our sense of touch. What do

we feel by means of this sense, when the upper lens lies in our hand? It is not the matter of this lens, for we now see that there is some measurable distance between the lens and the hand; it is this repulsion. Give a blind man a strong magnet in his hand, and let another person approach the north pole of a similar magnet to its north pole. The blind man will think that the other has pushed away the magnet he holds in his hand with something that is soft. In the same manner, if the blind man be electrified, another person passes the open palm of his hand to and fro near the blind man's cheek; he will say that cobwebs are driven across his cheek. All this is owing to the electrified hand repelling every hair or down of the face, and causing it to bend this way and that way, as if it were really touched by a cobweb.

There is therefore an essential difference between *mathematical and physical contact*; between the absolute annihilation of distance, and the actual pressure of adjoining bodies. We must grant that two pieces of glass are not in mathematical contact till they are exerting a mutual pressure not less than 1000 pounds per inch. For we must not conclude that they are in contact till the black spot appears; and even then we dare not positively affirm it. My own decided opinion is, that the glasses not only are not in mathematical contact in the black spot, but the distance between them is vastly greater than the 89000th part of an inch, the difference of the distances at two successive rings. My reasons for thinking so cannot be laid before you till you have acquired some optical knowledge.

Now we have complete explanation of the curious facts mentioned above; the free motion of the insects on the surface of water. Its brushy feet are in physical, but not in mathematical contact with the water, and by repelling it, depress so much of it that they are supported. And here we have an instructive piece of information. If the water be pretty warm, or if we mix a small portion of spirits with it, the insect can no longer walk on it, but sinks on it up to the belly. There is, therefore, in this case, a *specific* law of corpuscular action, suited to the purposes of this insect, but different perhaps from the more general repulsion that takes place between all bodies. It is also explained how the dew-drop rolls along a cabbage leaf, of sparkling brilliancy, and does not leave a trace.

243. Mr Saussure mentions a thing of the same kind, which I had often observed, without reflecting on its curiosity. The particles of a fog or mist, as they pass by, rebound from any thing that they are driven against by the wind, like a tennis ball. Another example mentioned by Saussure is very amusing. If a dish of warm coffee without cream be set in the sunshine, and sheltered from any stream of air, the vesicles of vapour which rise from it often fall down again, and roll about on the surface of the coffee, most brilliant and sparkling. Therefore they are not in contact with the liquor, because in that case you would have no brilliancy. In the same manner may the rain-drops of a warm summer shower be often seen to roll about on the surface of water, brilliant like a dew-drop, and for the same reason, because they are not in contact with the surface. Electricity supplies us with facts to the same purpose. If the discharge of the coated phial be made through a chain lying loosely on a table, or on a glass plate, it is rendered sparkling all over. If the chain be hanging in the air, forming a bight, it will not be nearly so luminous by the discharge. If a great weight be hung on its middle, no light will be observed. The explanation is easy. A spark is produced at every link, when they are not in mathematical contact. When this is almost completely produced by the weight, the light must cease. If a chain be part of a galvanic circle, the shock is not transmitted by it, unless it be well stretched.

244. I flatter myself that the experiments of Huyghens

and Newton, and the spontaneous phenomena of nature which have been mentioned, shew, in a manner still more distinct than the phenomena of capillary attraction, that the particles of tangible matter act on each other with moving forces, at certain small and measurable distances, in the same manner that the sun and planets mutually act on one another. But the distances now under consideration are greater, almost incomparably greater, than what should be considered as the natural cohering distance of the particles even of the rarest substance. Therefore we must not consider the phenomena which have now been described as examples of the action of those forces which produce the phenomena of cohesion, in all its modifications of elasticity, ductility, softness, viscidity, and fluidity, whether liquid or aerial.

Yet these examples are of the greatest use in our attempts to investigate the intimate constitution of tangible matter; because they shew us that there really exist in nature mechanical or moving forces, acting, like gravitation, at a distance, but clearly distinguishable from it, by their law of variation by a change of distance. While gravity produces sensible effects at the utmost boundary of the solar system, these other forces seem limited in their exertion to a small fraction of an inch, perhaps not exceeding $\frac{1}{2000}$th part in any instance; and in this narrow bounds we observe great diversity in the intensity, although we have not yet been able to ascertain the law of variation. What is of peculiar moment, we have seen that those corpuscular forces even change their kind by a change of distance, producing, at one distance, the mutual approach, and at another distance the mutual separation of the acting corpuscles, from being attractive, becoming repulsive. Now when an attractive force, by a gradual variation of distance, becomes repulsive, we cannot avoid thinking that, if we could hit on the exact distance, we should find that the particles neither attract nor repel.

We even observe a phenomenon which greatly resembles this. When two magnets of a soft temper are placed with their north poles fronting each other, and are at any considerable distance from each other, they invariably recede. If we push them gradually nearer, we find this tendency to recede gradually increase, as the distance diminishes, till the repulsion acquire a maximum of intensity, after which it rapidly diminishes, and at a certain distance of the magnets, it vanishes entirely; and when we bring them still nearer, they evidently attract each other, and this attraction increases till they come into contact.

Here is a very distinct analogy with what we have discovered on a much smaller scale. Nothing hinders us from supposing that the force by which cohesion is effected has a similar law of action. From this supposition we can deduce certain distinct consequences, which we can compare with the phenomena of cohesion. We shall find them extremely conformable, as will be shewn by and by; and thus we can form to ourselves mechanical notions of the intimate constitution of tangible matter, and of the procedure of nature in operating many changes which we see its masses undergo. We can do all this with a degree of confidence which we should never have had without those experiments.

Let us therefore consider a little the train of conclusions which we are entitled to draw concerning the change in the corpuscular forces, occasioned by a change of distance between the particles.

245. At all considerable distances, bodies attract each other by gravitation, but at certain very small distances, they repel one another, and at other very small distances, they attract.

246. (*a*) The distance at which one glass plate attracts another, in the Huyghenian experiment, is greater than the distance at which they repel one another, exhibiting coloured rings.

For, while the one suspends the other, with a silk fibre interposed, no colours appear between them; take away the silk fibre, and press them strongly together, and colours are produced: these vanish when the pressure is removed, and in this state the plates again attract.

247. (*b*) The distance at which glass plates repel, exhibiting colours, in Newton's experiment, exceeds that in which glass attracts water, in the phenomenon of simple humefaction, or capillary attraction.

For, when water is admitted between the glasses, in which case it is attracted by, and adheres to them, the coloured rings appear between the glasses as before, only the thickness exhibiting any particular colour is diminished in the proportion of 4 to 3. This shews plainly that the adhesive distance is contained (perhaps many thousand times) in the colorific distance. As a farther confirmation of this, it may be remarked, that when a film of water evaporates from a glass, the same colours appear in the vanishing film, just before it disappears. A drop of oil of turpentine, by spreading out on the surface of water, exhibits those colours when it has become thin enough, and they change as its thinness advances, by its diffusion on the surface of the water. The thickness of this transparent film of oil of turpentine may be estimated by comparing the diameter of the drop with the extent of its diffusion. It will be found to have a very sensible proportion to the diameter of a capillary tube which supports oil of turpentine at a certain height; and this shews that this diameter is incomparably greater than the sphere of capillary attraction.

248. (*c*) There is the greatest reason for believing that this small distance is vastly greater than the natural or cohesive distance between the particles of glass or of water. If it were not, I scarcely can conceive how it can cause the water to rise in the pipe. It would require a long discussion to make this clear, but we are not reduced

to this as the sole argument. We have now acquired considerable information concerning the law of cohesive force, within certain limits. We see that a change in their mutual distance is accompanied by a change of force. When liquids gather into drops, it is because the particles are otherwise at such distances from one another on their different sides that they are not in equilibrio, and it is only by changing those distances that the balance can be effected. This motion is a clear indication of a change of intensity by a change of distance. It is perhaps more distinctly perceived by compressing the fluid. Here we see that a greater force is required for producing, and for maintaining a greater compression. Fluids which gather into drops are not susceptible of great compression. But aerial fluids exhibit it almost without limit. Thus air is seen in some experiments to occupy a thousand times as much space as in some others, and it will expand still more if permitted. In this expanded state its particles are ten times more distant from one another than in its denser state; therefore, in its denser state, the mutual repulsion must be conceived as reaching to the tenth particle. This may perhaps constitute the mechanical difference between the aggregation of liquids and that of airs, gasses, or vapours. The action of liquids may perhaps extend only to the adjoining particle, while that of the expansive fluids may extend over many. There are great difficulties attending both of these hypotheses.

249. The change of cohesive intensity by a change of distance between the particles is as distinctly, or more distinctly seen in the cohesion of solid bodies. All that we are acquainted with are susceptible (in various degrees) of compression and dilatation. A greater force is required for producing, and for maintaining a greater change of bulk, and when the force is withdrawn, the body resumes its natural bulk, if the change has not been

too great. We must conclude from these facts, that the particles of a solid body, when at their natural distances, neither attract nor repel. When forced nearer to one another they repel, and when drawn asunder they attract.

250. In our experiments of this kind, a fact is observed that is perhaps universal. If the compression or dilatation has been very moderate, so that the change of distance between two adjoining particles is but a minute portion of that distance, it is found that it is proportional to the attractive or repulsive corpuscular force which is excited by the change. A double, triple, quadruple force is required for making a double, triple, or quadruple change of distance between the particles. Dr Robert Hooke made this discovery with respect to springs and all elastic bodies, in 1660, and expressed it by the phrase *ut tensio sic vis*. It was this observation that suggested to him his noble improvement on pocket watches, by putting a spiral spring on the axis of the balance. The balance assumes a certain quiescent position. If any force can turn it 10 degrees from this position, in opposition to the spring, a double force will turn it 20 degrees, a triple force will turn it 30 degrees, &c. Such a balance, therefore, vibrating by the action of this spring, will perform its vibrations in the same time, whether they extend 10, 20, or 30 degrees on each side of the quiescent position. Mr J. Bernoulli investigated the curve into which an elastic rod will be bent, on the supposition that the attractive and repulsive forces which are brought into action by the bending, are proportional to the change produced in the distances between the particles. When this curve was compared with experiment, they coincided most perfectly. Mr Coulomb of the French academy suspended bodies by long wires, and then, twisting the wire a certain number of turns, he let it go, and observed the oscillations. He found them perfectly isochronous,

whether the wire was twisted once, or ten, or twenty times. This proves incontrovertibly that the forces exerted between the particles are exactly as the changes of distance. Now when this is the case, we may say with confidence that those compressions are but very minute portions of the natural distances of the particles. The truth of this inference will appear very clearly when we consider Mr Boscovich's attempt to investigate the laws of corpuscular action.

Thus we see that the analogical reasoning concerning the force of cohesion, founded on the actual observation of alternate attractive and repulsive corpuscular forces altogether different from cohesion, are fully supported by the phenomena of cohesion itself.

251. From a collective view of all those facts, we must conclude that the forces by which the particles of tangible matter cohere in its various forms of aggregation, are, like gravity, forces which act at a distance, and that they vary, both in quantity and direction, by a variation of distance. We must also conclude that the distance at which one body suspends another (for it is not confined to the Huyghenian experiment with glass) is greater than that of cohesion or capillary attraction; and, since we see that enormous pressures are necessary, in the Newtonian experiment, to bring the nearest parts of two bodies within the 89000th part of an inch of each other, we must conclude that at these greater distances the particles of bodies act on one another, and that this repulsive action is probably the immediate cause of physical contact, exciting the sensation of touch, and the feeling of pressure, and that this is the immediate cause of all the mutual pressures which we observe bodies exert on each other. It is therefore the immediate cause of all the motions and changes of motion which are produced by those pressures. All those may and must happen without any real mathematical contact of the bodies.

252. The deflection of a planet from the tangent of its orbit, the deflection of a cannon ball from its parabolic path, the approach of a piece of iron to a magnet, the similar motion of electrified bodies, the suspension of one plate of glass by another, the repulsion between two object glasses, the motion of one body when it is struck by another (arising from the same repulsion), the motion of water in a capillary tube, and the motion of the particles of a springy body when it is bending or unbending, are all facts of one kind. They are all equally the effects of natural powers which act between distant particles. Of this I apprehend every intelligent person will be persuaded, if he steadily confines his attention to the particles *really* in action.

253. If a vortex, or a stream of fluid of any kind, be considered as necessary for explaining the deflection of a planet, it is equally necessary for explaining the motion of a billiard ball when struck by another. Nay, it is as necessary for explaining the action of this vortex or stream of fluid. For, since the appearance of the black spot between the glasses does not entitle us to say that they are in mathematical contact, every particle of this vortex must be granted to repel at some minute distance, or else we must suppose another vortex belonging to each particle of the first to render it impulsive. In short, this view of the subject cuts off at once all explanations by the help of invisible impelling fluids, æthers, atmospheres, or by whatever name they may be called. All changes of motion have for their immediate causes those powers of nature which we have called accelerating forces, accompanying the particles, and brought into action, or excited, by the mere distance and situation of the related particles. There is no *mechanical* difference between them. The only difference is the distance in which the force is exerted, and the variation of intensity by a variation of distance. If we know

these two circumstances, we have all the knowledge that can be of any use. How gravity or any other power produces its effect is of no other use but the gratification of curiosity; and if it were gratified, we should be equally curious to find out the cause of every step of this process of efficiency

254. Planetary action extends to the utmost bounds of the solar system, and is in the inverse duplicate ratio of the distances. Magnetism also extends to a great distance, as we learn by the motions of the mariner's needle. The law of action seems to resemble that of gravitation. Electricity is also extensive, and has the same law. Physical contact, or PRESSURE, becomes *sensible* at the distance of the 5000th part of an inch nearly, and decreases much faster than in the inverse duplicate ratio of the distances. I could infer this from my experiments with the glasses with great confidence, although I could not assign the precise law. Cohesion, with all its modifications, has a much more limited range, perhaps not the millionth of the millionth of an inch. This may be inferred from the perfect continuity of the gold or the silver wire employed in the manufacture of gold lace, where it is the fourteen millionth of an inch in thickness. Yet even in this minute scale, we see by the experiments with springy bodies, that this minute distance may be subdivided into many portions, and that each distance has a peculiar intensity of cohesion belonging to it. Its law of variation, taking it generally, is unknown. For the observation of Hooke, *ut tensio sic vis*, is true only when the changes of distance are very small in comparison with the whole natural distance between the particles. We shall consider this more particularly afterwards, and shall learn why Hooke's observation is so generally true. The modifications of cohesion are innumerable, producing an endless variety of sensible forms, solid, fluid, vaporous, in each of which the law of action between the corpuscles is pro-

bably different. Also, in each of these forms we have subordinate varieties, which make bodies hard, soft, elastic, unelastic, plastic, ductile, viscid; and lastly, there are other modifications of the corpuscular force, which produce the phenomena of solution, precipitation, crystallisation, &c. &c. &c. All and each of these are ultimately mechanical forces, producing local motion and changes of motion.

255. In this range of observation there are two extremes. On the one hand, enlarging our scale, we have electricity, magnetism, and gravitation. This last leads us to the bounds of the solar system. Nay, there are appearances which render it probable that it extends at least to some of the fixed stars. But we have not sufficient authority for extending it to all. Gravitation may cease at a certain distance; nay it may change to a repulsive force at greater distances, and the visible universe may consist of parcels which are in equilibrio with one another, as the particles of a common body are in equilibrio between a state of attractions and repulsions. Each parcel of connected stars, magnificent as it is in our eye, may thus constitute a portion or particle of the universe.

256. Our imaginations are lost in the contemplation of such a scene. But there is no absurdity in the thought. The ingenious Dr Halley proved from the law of gravitation, that there may be within this globe a scene of existence and habitation altogether undisturbed by the gravitation of external bodies, yet every thing having weight and stability nearly the same as on the surface of the earth. To its inhabitants the scene may appear as extensive as the heavens appear to us.

257. There is just as wonderful a series of connecting forces on the other hand, when we consider the *smaller* scales on which they are exerted. A pint of corrupted puddle is perhaps as great a universe to its countless inhabitants as the visible heavens are to us. It requires some

effort of imagination to subdivide the 480th part of an inch into the 50 different portions in which Newton observed the alternate inflection and deflection of light; or to subdivide the 4450th of an inch into the 20 portions where light was alternately reflected and transmitted. Yet far within the limit of one unit of this subdivision, viz. $\frac{1}{89000}$th of an inch, are exhibited all the alternations of attraction and repulsion which we observe in the compression and dilatation of bodies. In these minute scales of force there must be many differences, both in the magnitude of the scale, according as the body is dense or rare, and in the intensity of the force, according as the body is hard, soft, viscid, &c. &c. and in the law of its variation.

But the most wonderful variety is exhibited to our observation in the structure of regular bodies, such as crystals; and still more in the structure of the organized bodies of vegetables and animals. We have acquired some knowledge of the more obvious particulars of our own structure; we understand something of the *efficiency* of this structure, how it is capable of performing its office. We understand all this by means of the laws of mechanics, that is, the general facts observed in the agency of all pressures. We have now learned that pressures are forces acting at a distance, which are so small in comparison with ourselves that we cannot perceive them.

258. Now, in those incomparably smaller structures which the microscope has brought into view, we behold machines, which perfectly resemble, both in form and in function, the large animal machines which are more familiarly known to us. These little machines must be moved in the same way as the larger. The whale and the minnow, the minnow and its fry (not one tenth of an inch long) have the same veins, and arteries, and nerves, the same livers and muscles, and their functions in both are the same. The minnow is a giant when compared with the red ant; yet this little creature is constructed with a

variety of parts, and with a polished elegance in every member, that exceeds the most delicate piece of human art, beyond all imagination. It is really a wonder. One of its antinnæ, (not $\frac{1}{80}$th of an inch in length) has 12 joints, each consisting of a finely polished spherical ball, moving between two hemispherical sockets, all finished with the utmost attention to elegance of shape and fitness for angular motion. The most elaborate performance of human art being laid beside it, and viewed with the same glass, appears a piece of the rudest botching. On this beautiful little creature, Mr Southern discovered an acarus, a vermin to which the red ant is a world. Even this animal is formed with symmetry, and a great variety of parts.

Beyond this our eyes, with all the assistance of the microscope, can discern no more structure. We can only gaze and wonder at many still smaller animals, which the microscope exhibits, swimming about with great vivacity, and indicating by their motions that they are effecting the same purposes which are effected by the larger animals; this authorises us to infer that they are of similar mechanism, and that their operations are performed by the same principles of mechanical energy.

259. The conclusion which I think must be drawn from these observations, is that those alternations of attraction and repulsion which accompany the gradual diminution of distances, as we continue to subdivide the extent of a hair's breadth, must go beyond all our conceptions of minuteness. The distances at which *our* pressures are excited, and *our* sensations of touch, will no more answer the purposes which we see accomplished by those little creatures, than an axe and a sledge-hammer will suffice for making a repeating watch contained in the size of a seal ring. We must grant that there are scales still smaller, on which those necessary forces complete all their variation of intensity and direction.

And lastly, since all those little objects consist of parts extremely different, both in component ingredients and in structure, we are certain that the particles of each of those parts are still of a very compounded nature, far removed from the simplicity of a primary atom of matter, and therefore consisting of many atoms, connected by forces of a class still more removed from those we are acquainted with.

260. On the whole, we see that within the narrow limit of the four or five thousandth part of an inch, (the greatest distance of sensible pressure,) there actually are a numberless variety and alternation of attractive and repulsive forces exerted between the particles and atoms of tangible matter, and that it is to their immediate agency that we must ascribe all the diversities of connection, form, and distinguishing properties, which characterise the objects by which we are surrounded in this sublunary world.

In giving them the denomination of attracting and repelling forces, and in calling their effects attraction and repulsion, nothing is meant but a denomination. No explanation of the manner of acting is intended. But they are all supposed to be of the same kind, and similar to the gravity or heaviness of terrestrial matter. If any explication can be given of gravitation, the same must be applicable to those forces which connect the particles of tangible matter. All that we are entitled to say of them is that they vary in intensity and direction by every change of distance, from the utmost bounds of the solar system to the actual coalescence of the acting atoms.

What then must we conceive to be the ultimate action of atom on atom, as we diminish their distance without end, and just before the annihilation of all distance between them? It seems to me that it must be an inseparable repulsion. We see in the sensible masses a repulsion which is prodigiously great, before they can be brought to that vicinity at which no light is reflected by the glass plates; and we have no evidence that we can bring them into mathematical contact by any force whatever.

261. It is very surprising that the mathematicians and philosophers at the end of the 17th century, who were so arduously engaged in applying the newly discovered fluxionary mathematics to the explanation of the mechanical phenomena of nature, did not turn their attention to a department which promised the richest crops of discoveries to reward their labours. Newton had also given the philosophers a new system of mechanics, particularly fitted for this research, and had demonstrated its competency by the most successful examination of the great movements of the universe. He had also remarked some very encouraging analogies, which seemed to admit the same manner of treatment in the study of the corpuscular phenomena, and had even pointed out to them many phenomena of this class, which seemed to require this method alone for their explanation, and to reject it from every other quarter; and he gives us some account of his own conjectures on the subject in the queries subjoined to his Optics. Nothing however was done that was of any service, if we except the speculations of Bernoulli, Mariotte, and others, about the strength of solid bodies. Yet even these attempts were encouraging. The immense difficulty of the task was doubtless the great obstacle, and seems to have deterred even Newton from formally engaging in it. Dr John Keill indeed gave, in No. 315 of the Philosophical Transactions, a number of general theorems concerning the action of forces attracting or repelling according to various laws of the distance; and Dr James Keill and Dr Friend gave Theories

of Physiology and Chemistry, founded, as they said, on similar theorems. But these theories were filled with gratuitous assumptions of forms and motions, which rendered them altogether ridiculous. This brought the whole department of study into discredit, and it remained entirely neglected for many years.

262. The first who (so far as I know) attempted to revive this study in a serious manner, was the celebrated magnetical philosopher Dr Gowin Knight. He published, in 1748, his "Attempt to explain all the phenomena "of nature by means of two principles, attraction and "repulsion, shewing that gravitation, cohesion, magnet-"ism, electricity, &c. are all the operation of those two "principles."

Mr Knight supposes two species of material atoms, one of which attracts another atom of the same kind, and the atoms of the other species mutually repel each other. The atoms of different species probably attract one another; but of this he is not certain. He then shews how the attractive atoms coalesce into particles; and how those particles must be surrounded with the repelling atoms constipated on their surface, and even surrounding them like an atmosphere. From this combination arises another class of particles, which are mutually attractive or repulsive, according to the proportion and disposition of the atoms of the two spaces of which they consist.

Mr Knight then proceeds to explain from those premises the structure and sensible properties of bodies of various kinds. The whole is digested into formal propositions and corollaries, very distinctly expressed, and the law of action which he assumes (viz. a force inversely as the distance) is applied mathematically. There is very considerable ingenuity and great simplicity in the management of his principles; but it requires only a moderate attention to the unchangeable laws of dynamics to shew that almost every one of his propositions is false. At the same time it is not unsusceptible of improvement, and it deserves a serious perusal.

This study has never been resumed and prosecuted with method and perseverance. Yet the case does not seem desperate; although we may never be able to acquire such accurate knowledge of the corpuscular relations of tangible matter as we have of the simple laws of the planetary motions, and the ordinary doctrines of mechanics, hydrostatics, and pneumatics, it is highly probable that a steady and judicious prosecution of it would bring to light some *general* laws which might be of material service.

263. Father Boscovich, one of the first mathematicians of Europe, and of very extensive knowledge of the phenomena of nature, struck with the importance and probability of Sir Isaac Newton's conjectures, has endeavoured to revive this study, and published at Vienna in 1759 a most ingenious work, which he called *Theoria Philosophiæ Naturalis ad unicam legem virium, in Natura existentium redacta.* This is of a very different cast indeed from Dr Knight's, and is undoubtedly one of the most curious productions of the last century, filled with original and ingenious notions of natural things, and explanations of all the general appearances and mechanical relations of different kinds of matter. It richly deserves the serious perusal of every philosopher. Although the ingenious author is far from having attained the knowledge of this characteristic law of matter, or established a perfect theory, he has deduced many legitimate consequences from the phenomena, which are of very important service in all mechanical disquisitions; and I may even venture to say that, if we shall ever acquire the knowledge of a true theory, it will resemble Mr Boscovich's in many of its chief features. For which reason I shall give a slight sketch of its leading propositions, referring

the reader to the work itself for a full account of the theory *. The following are the elements of this theory.

264. *1st,* All matter consists of indivisible and inextended *atoms.*

265. *2d,* These atoms are endowed with attractive and repelling forces, varying, both in intensity and direction, by a change of distance, so that at one distance two atoms attract each other, and at another distance they repel.

266. *3d,* This law of variation is the same in all atoms. It is therefore mutual; for the distance of *a* from *b* being the same with that of *b* from *a,* if *a* attract or repel *b, b* must attract or repel *a* with precisely the same force.

267. *4th,* At all considerable or *sensible* distances this mutual force is an attraction, sensibly proportional to the square of the distance inversely. It is the attraction called *gravitation.*

268. *5th,* In the small and insensible distances in which sensible contact is observed, and which do not exceed the 1000th or 1500th part of an inch, there are many alternations of attraction and repulsion, according as the distance of the atoms is changed. Consequently, within this narrow limit, there are many situations in which the two atoms neither attract nor repel.

269. *6th,* The force which is exerted between the two atoms, when their distance is diminished without end, and is just vanishing, is an insuperable repulsion, so that no force whatever can press two atoms into mathematical contact.

Such, according to Boscovich, is the constitution of a

---

* See Note II. at the end of the Volume.

material atom, and it is the *whole* of its constitution, and the immediate efficient cause of *all* its properties.

270. Two or more atoms may be so situate, in respect of distance and position, as to constitute a PARTICLE of the first order. Two or more such particles may constitute a particle of the second order, and so on, to any degree of composition.

Mr Boscovich proceeds to deduce such consequences of this constitution as may be called elementary; and then to shew that these are sufficient for constituting a substance having all the sensible qualities, forms, and physical properties that we observe in tangible matter; and that, although all atoms of matter are precisely similar, there must result from their combination, and the joint action of their atomical forces, an inexhaustible variety of external form, mutual relations and actions, fully adequate to explain all the phenomena of the material universe.

271. All that can be done in the short while allowed for this discussion is merely to enable the *attentive* student to form a just notion of this most ingenious theory, and of its competency to the explanation of nature. I trust that he will see enough to incite him to a perusal and serious study of the work, and that he will find it full of curious and valuable information.

Mr Boscovich represents his law of atomical action in the Newtonian manner, by what he calls an experimental curve.

272. Let the distances of two atoms be estimated on the line CAC, Plate III. fig. 1 A being the situation of one of them, while the other is placed any where on this line. When placed at *i*, for example, we may suppose that it is attracted by the atom A, with a certain force. We can represent the intensity of this force by the length of a line *i l*, perpendicular to AC, and we can express the direction of it (namely the direction *i* A, because it is attrac-

tion) by placing *i l* above the axis or line of distances AC. Should the atom be at *z*, and be repelled, we can express the intensity of repulsion by *z t*, and its direction (A *z*) by placing *z t* below the axis.

This may be supposed done for every point of the axis, and then a line DEFGHIKLMNOPQRSTV may be drawn through the extremities of all the perpendicular ordinates. This is the exponential curve, or scale of forces, in the manner employed in DYNAMICS.

As there are supposed a great many alternations of attraction and repulsion, it follows that the exponential curve must consist of various branches, lying on different sides of the axis, and must therefore cross it in many points, such as E, G, I, L, N, P, R. All those are supposed to be contained within a very small fraction of an inch, not exceeding the distance of the glasses in the Huyghenian experiment. AR may represent this distance, magnified by a microscope.

273. The branch most distant from A is STV. This must be of such a form that its ordinates CV, *v* ♄, *o p*, &c. may be inversely as AC^2, $A\,v^2$, $A\,o^2$, &c. RC must be an assymptote to this branch; and the branch ED next to A, must have the perpendicular AB for its assymptote, because the ordinate, expressing repulsion, increases beyond all limit. The intermediate branches of the curve must be determined by means of the phenomena of cohesion, capillary attraction, the repulsion of the glasses in the Newtonian experiment, &c. &c.

It is plain that an atom situated in any of the points E, G, I, &c. where the curve crosses the axis, will neither be attracted nor repelled by the atom A. But there is a remarkable difference between the condition of an atom situate in E, and of one placed in G. If the atom E be pushed a little nearer to A, as to *b*, (A being held fast in its place), it will be repelled by A, and a force equal to the repulsion *b r* must be employed to keep it in *b*; and if this force be withdrawn, it will come back again to E. On the other hand, if E be drawn off to *d*, (A being held fast) it will be attracted by A, with the force *d h*, and if the extending force be withdrawn, it will go back to E.

Hence it appears that two atoms situated in A and E, will compose a sort of particle, which will have a certain degree of permanency of form. It may be compressed by an external force, or it may be distended, but will recover its bulk AE when the compressing or stretching force is removed.

274. An atom placed in G will be in a very different condition. It may remain there for ever, if no compressing or distending force is applied to it. But, if compressed in the smallest degree, for example to *i*, it is immediately attracted by A, and flies towards it with an accelerated motion, and will finally settle in E. If G be drawn off to *m*, it is immediately repelled with the force *m n*, and will fly farther off, and proceed to I, or perhaps farther. Hence it appears that the atoms A and G cannot compose a permanent particle, but the smallest disturbance will immediately destroy it irrecoverably.

275. It is plain that the points E, I, N, R, are situations of the first kind, and that G, L, and P, are of the second kind. Boscovich calls the first LIMITS OF COHESION, and the second LIMITS OF NON-COHESION. These last are better named LIMITS OF DISSOLUTION.

I said that particles, such as AE or AI, have a certain degree of permanent form, resisting compression or dilatation, and again recovering their natural bulk. But this is only when the disturbing force, and the change of bulk, have been moderate, the change of distance being very small in comparison of the whole distance between two intersections of the exponential curve with the axis. For if the particle AI be compressed into less room than AG, the atom will not return to I when the compressing force is withdrawn. For it will now be attracted by A, and the particle AI will collapse into the bulk AE. If AI be stretched beyond the bulk AL, it will not collapse into AI, but will take the form AN. The only particle that cannot be changed by any compression is AE, as is evident.

It is evident that the component atoms of particles so constituted are in a state of inactivity on each other, unless some external force be applied. The atoms A and I neither attract nor repel each other. But when the particle is actually compressed, the atoms being pushed nearer than their natural inactive distance, they immediately repel; in like manner they attract if the particle be stretched. The approximation or separation of the atoms gives occasion or opportunity to the exertion of the attractive or repulsive forces which are inherent in, or at least always accompany the atoms. Thus these forces may be said to be *excited* in them, or brought into action. The compression or dilatation is therefore the *occasion*, though not the *efficient cause* of the mutual attractions and repulsions. We must always keep this distinction in mind.

276. The magnitude or intensity of those atomical forces at the different distances determine the form of the exponential curve. If a very moderate force produces a sensible compression, the ordinates on each side of the limit must be short, expressive of those small repulsions. The exponential curve must therefore cross the axis very obliquely, that the ordinates expressing the attracting and repelling forces may increase slowly. But if it require a very great force to produce a very small compression, the small compression A *b* must have a great ordinate *b r*, and the exponential curve crosses the axis almost perpendicularly.

We may also remark (and the remark is important) that when the dilatation or compression bears a very small proportion to the natural distance of the atoms, the compression or dilatation will be very nearly proportional to the forces employed. For, when the compression E b is very small in comparison with AE, the arch r E differs little from a straight line, and therefore the ordinates $r\,b$, $r'\,b'$ are very nearly proportional to E b and E b'. The same may be affirmed of the distension E d. The arch E h differs little from a right line, and the ordinates are nearly as the abscissæ. But it is otherwise when the dilatation is considerable. When the dilatation E d is much greater, the ordinates $d\,h$ must now increase more slowly, and, when the dilatation has increased to a certain degree, the ordinates will even diminish again, and they are reduced to nothing in the neighbouring limit of dissolution. Many examples of this will occur in the phenomena of nature. On the other hand, when we observe the compression or dilatation proportional to the force employed, we may conclude that the compression or separation of the atoms is very small in comparison with their natural inactive distance.

277. We may now take the example of this simplest constitution of a particle to shew that, although the law of atomical action be the same in all matter, the action of particles composed of such atoms may be unspeakably various, according to the distance and position of the two component atoms. We have not room for much of this discussion, and must content ourselves with one or two of the most simple cases. But the attentive reader will find no difficulty in extending the inferences to more complex cases. We shall first consider the action of such a particle on an atom placed in the axis or line joining the component atoms.

Suppose then a particle XY (Plate III. fig. 2) composed of two atoms X and Y, which are placed in this *first* limit of cohesion A and E of fig. 1. In order to know the condition of an atom placed any where in the line AC, we may suppose the whole exponential curve shifted toward C, so as now to refer to the atom E as it formerly referred to the atom A. Thus, for any point in the line EC, where we suppose the third atom to be placed, there are two ordinates, one of them an ordinate to the curve belonging to A, and the other an ordinate to the curve belonging to E. If those ordinates are on the same side of the axis, then the third atom is either attracted by both A and E, or is repelled by both. We must therefore draw an ordinate in that point, equal to the sum of the two ordinates. This will express the force actıng on the third atom. But if the ordinates of the two curves lie on opposite sides of the axis, the third atom is attracted by one, and repelled by the other component atom. Therefore we must make an ordinate equal to the difference of the two, and place it above or below the axis, according as the repulsive or attractive forces prevail. Doing this for every point, we may draw a curve through the extremities of all the ordinates, and it will be the exponential curve for the particle AE or XY expressing its action on a third atom placed in the line AR. Accordingly, the curve in fig. 2. is constructed in this very way. By this we see that the law of action of the particle XY, differs greatly from that of the component atoms X and Y. The branch $d\,A\,d'$ between the two assymptotes $c\,b$ and $c'\,b'$, is easily understood. At A, the repulsions of the two atoms X and Y must exactly balance, and as the third atom comes nearer either to X or to Y, it is more strongly repelled by it than by Y or by X.

278. If an exponential curve had been constructed for a particle consisting of two atoms A and I, it would have turned out still more unlike the primitive exponential. This will easily be conceived by shifting the primitive curve from A to I. We shall then find that several of the attractive branches of one curve stand opposed to repulsive branches of the other, the consequence of which must be

that the particle AI will be almost inactive on a third atom, till it be moved almost to the furthest limit of the small scale of corpuscular action.

279. It is much more important that we learn the law of action on an atom placed out of the line AC. But we must still begin with a simple case. Let us take a particle composed of the two atoms A and I, situated in this second limit of cohesion, and let us consider the action of the particle AI in fig. 3. on an atom placed any where in the line BCB, which bisects AI at right angles. Suppose the third atom placed in u. Then join A u and I u, and taking Au in the compasses, transfer it from A in fig. 1. to u, and draw the ordinate $u\,y$. This is an ordinate of an attractive branch of the exponential. Therefore, in fig 3. set off from u, towards A, and towards I, the two lines $u\,y$, uy, equal to the ordinate $u\,y$ of fig. 1. The atom u is attracted by A and I with the forces $u\,y$, uy. Complete the parallelogram $u\,y\,v\,y\,u$, (which is a rhombus) and its diagonal $v\,u$ will express the force with which u is attracted in the direction u C. Draw $u\,y'$ perpendicular to u C, and make it equal to $u\,v$, placing it on the right hand of BC, if it represent an attractive force, and on the left as $z\,t'$, to represent a repulsive force. Let this be done for every point of the line BCB, and draw a curve line CGHIKLMNOPQRSTV through the extremities of all the ordinates. This will be the exponential of the action of the particle AI on an atom placed any where in the line BcB'.

We see that this line of action differs most remarkably from that of a single atom. It has indeed, like the single atom, many alternations of attraction and repulsion. But instead of an insuperable repulsion at the greatest vicinity, we see that the repulsion changes, at G, to attraction, which continues all the way to the centre, and that in the very centre C there is no force, either attractive or repulsive. Such a particle, therefore, will be easily penetrable, though composed of atoms exerting an insuperable repulsion.

It is scarcely necessary to advertise the reader that the exponential curve extends on both sides of the line AI, and that the attractive and repulsive branches below C are on the opposite sides of B C and B C. Thus GHI and G′ H′ I′ are repulsive branches. Observe also that all the branches on the right hand of B C B′ express a force which produces a motion of the third atom in the direction BB′, and all those on the left hand express forces which produce a motion in the direction B′B.

It is not unworthy of remark that our particle AI greatly resembles a magnet in its action. For it has been already observed that this particular construction of the particle renders it almost inactive on an atom placed in the line AI, while it acts very sensibly on an atom placed in the line BB In like manner a magnet acts strongly in the direction of its poles, and is without action in the direction of its equator. Moreover, since we observe that a parcel of small magnets, or of magnetical fragments of iron, floating on quicksilver, have a disposition to cluster together in a particular way, rather than in any other; and since this arises from that difference in action which we call polarity, we must ascribe similar tendencies to a collection of particles constituted like AI. They will possess polarity, and will cluster together in one way in preference to all others. Keeping this in mind will greatly aid us in conceiving some of the hidden operations of nature.

280. Thus then we see that, even in this simplest constitution that can be imagined for a particle, the action on another atom is susceptible of great variety, by the mere difference of position and distance between the two component atoms. But it must be farther remarked, that all these differences of action on a third atom are confined to the small and insensible distances which lie

within the limits of physical contact. At all considerable distances, we shall find nothing but the action of gravitation, inversely proportional to the square of the distances, and also proportional to the number of atoms which compose the acting particle. This will easily appear, when we consider the primitive exponential curve with a little attention. The whole distance AR (fig. 1.) does not exceed the thousandth part of an inch. Therefore supposing A and R to be the component atoms, (which constitution of the particle will occasion the greatest possible shifting of the exponential curve), then, even at the distance of only a quarter of an inch from the particle, the sum of the ordinates of the two curves will not differ one part in five hundred from the double of either; and at the distance of an inch, will not differ one part in a hundred thousand. Therefore, at all sensible distances, we may conclude that the forces are in the inverse duplicate ratio of the distances from the middle of the particle, and are double of the force tending to either of the component atoms. The slightest consideration will shew us that this is equally true of a particle composed of any number of atoms, the particle itself being of insensible magnitude. Its action on a distant atom will be inversely as the square of the distance, and proportional to the number of atoms in the particle.

281. This is equally, and even more accurately, true of its action on an atom situated in the line BB′ of fig 3. For, if we take even so small a distance as CR, which is within the limits of physical contact, the distances AR IR differ very little from CR. It is surely unnecessary to insist longer on what is so plain.

From these considerations, we see that all the varieties in the law of corpuscular action will be observed only in the small and insensible distances which lie within the limits of physical contact, while the masses of matter consisting of the atoms so combined will exhibit, in all sensible distances, the attraction of gravitation, proportional to the number of atoms, that is, to the quantity of matter in the mass.

282. We have hitherto considered only such particles as consist of two atoms. But it is evident that, if the three sides of a triangle are respectively equal to three limits of cohesion, and if an atom be placed in each of the three angles, they will be in a state of indifference or inactivity, in respect of one another. If one of these atoms be moved a little away from the other two, this increase of distance will be the occasion of attractive forces exerted mutually between them, so that the other two will be drawn after it. Or, if the atom be pushed towards the other two, so as to lessen its distance from them, repulsive forces are exerted, and the two remote atoms will also be pushed away. This is surely the character of a particle. But we must examine this connection more particularly.

We have already seen that if an atom be placed in the point u of the line BB′, fig. 3., it is attracted by the particle AI with a force $u\,y'$, and that if placed in z, it is repelled with the force $z\,t'$. In either case, therefore, it is impelled towards the situation N, where its distances NA or NI is the same with NA of the primitive exponential, fig. 1. This happens, not only when the atom is drawn away from its quiescent situation N in the direction NB, or pushed in the opposite direction NC, but also, when removed from N in *any* direction. Thus, in fig. 4. let the third atom be drawn laterally from N, into the situation n. By this removal, it is drawn a little farther from the component atom A. It is therefore attracted by it, we may suppose with the force $n\,o$. By the same removal, it is brought a little nearer to the atom I. It is therefore repelled by it, suppose with the force $n\,p$. By the joint action of the forces $n\,o$ and $n\,p$, the atom is impelled in the direction $n\,q$, and made to approach its quiescent situation N. It requires a force $q\,n$ to keep it in n, (A and I being supposed to be held fast by some means in their places) and when this force is withdrawn, it returns to N.

Thus we see that A, I, and N, will compose a particle, having both length and breadth, and that it retains both its bulk and its shape with a certain force. It has all the characters of a particle. It is surely needless to shew that all these things are true, although the form of the particle be not that of an isosceles triangle. All the reasoning would have been the same, although one side had been AI of the primitive curve, fig. 1. another AN, and the third AR. All that is required is that the three sides of the triangle be limits of cohesion.

283. It is of particular importance to attend to the lateral force $n\,q$ exerted by the atom N, fig. 4. when it is drawn a little aside from the line NC joining it with the centre of the particle AI. This circumstance gives us a clear notion how a number of atoms may be disposed and combined so as to compose a material surface. For, if they be all placed symmetrically, in the angles of triangles, so that each atom may be in a limit of cohesion with its adjoining atoms, as is represented by the points in fig. 5. such an assemblage will form a sort of material surface. If the atoms in the line AB are held fast in their places, and a small force be applied to the atom CD, tending to urge them towards E, it is plain that this must excite the corpuscular forces inherent in the atoms. Those in the side DB will be compressed, and will exert repulsive forces, while those in the side CA will be drawn farther from each other, and will exert attractive forces. We shall have a very distinct and a very just notion of this mechanism by supposing all those atoms to be connected by slender elastic spiral wires like corkscrews. Urge any one atom nearer to another, and we compress the interposed spring. It resists, and when we withdraw the compressing force, it unbends, and pushes the atom to its

former distance. If we draw an atom farther from its neighbour, the intervening spring, being stretched, exerts a contractile force, which will, in like manner, bring the atom back to its natural quiescent position. A little attention to this, (for without attention nothing can be learned) will make it evident that the flexion which the external force produces on the bodies so connected, and the resistance opposed by the bodies when thus bent, and the recovery of the original form when the external force is withdrawn, all proceed from the elasticity of the connecting wires. This elastic action obtains whenever a spring is compressed or stretched, *and not otherwise.* In like manner, the inherent corpuscular forces must be *exerted*, or an opportunity must be given for their exertion. This is done by changing the natural distances of the atoms, in which they exert no force, for greater or smaller distances, in which they exert attractive or repulsive forces. And we easily perceive that this excitation of the atomical forces takes place *over the whole assemblage.* For, when D fig. 5. is pulled in the direction DE, it drawn away from C, and is pressed a little towards 1. Therefore 1 resists, and is at the same time pressed towards 4. This side, therefore, becomes a sort of fulcrum or prop, by means of the exerted repulsions. C is drawn after D, and therefore is separated from 3, and attraction is excited between C and 3. The atom 3 is thus drawn away from 5, and attraction is exerted between them, by which 5 is drawn away from A. And thus all the atoms on the side DB are in a state of compression, exerting repulsive forces, while those on the side CA are stretched, and exerting attractive forces. We need only consider what spiral wires would be in a state of extension, and what in a state of compression; and we may be assured that they are exerting forces precisely similar to the atomical forces in the theory of Boscovich.

Before quitting the subject of the material surface, there remains another important observation.

284. Let A and B (fig. 6.) be two atoms, The circles described round them as centres with a full line, are supposed to have limits of cohesion for the radii. The dotted circles have for their radii limits of dissolution. Therefore an atom in C will remain there. So will an atom in N, D, E, G, F, K, or M. Suppose N drawn towards *a*, it tends to come back to N. For if it be situated in any part of the little quadrangle between N and *a*, it is attracted both by A and by B. In like manner, if it be in any part of the quadrangle between N and *b*, it is attracted by A, and repelled by B, and therefore tends towards N. In short, into whichever of the four quadrangles round N it be taken, the combined action of A and B tend to urge it towards N.

The atom will also remain at rest in *a*, *b*, *c*, *d*, *e*, *f*, or *g*. Suppose *a* to be drawn into any of the quadrangles which surround it, for example, into the quadrangle *a* C, it is repelled both by A and B Therefore it will not return to *a*, but will immediately go to C. If it be taken into the quadrangle *a* N, then, being attracted by A and B, it will immediately go to N. If taken into the quadrangle *a* D, being attracted by A and repelled by B, it will go to D. If taken into the quadrangle *a* E, it is attracted by B and repelled by A, and will go to E. In like manner, an atom placed in *b*, or in *c*, or in *f*, if disturbed in the smallest degree, will immediately leave that position, and will settle in one of the neighbouring intersections of limits of cohesion.

Thus it appears that in all the intersections of the full circles the third atom will combine with the other two, and compose a particle of three atoms, having a bulk and shape which resist a moderate changing force. But an atom situated in any interval of the dotted circles cannot form a particle, but will be forced by the smallest disturbing force to alter its situation, and assume another, in which it may compose a permanent particle. It is evident that if the atoms A and B have many limits of cohesion intervening, the number of intersections, such as C, D, E, N, &c. must be very great, and therefore the particle of three atoms is susceptible of a great variety of forms.

After shewing how two atoms may compose the simplest particle that is possible, it was shewn that the law of action of this particle on a third atom is extremely different from the primitive law which characterises an atom of matter, and that it was susceptible of a great variety. We might now proceed to shew that the action of a particle consisting of three atoms differs from both, and is susceptible of much greater variety. But this is so evident, and the detail of the differences would be so tedious and complicated, without terminating in any thing very general, and at the same time precise, that it is better to abstain from the discussion, and to proceed to something that will bring us more speedily to a correspondence with our observation of the phenomena of tangible matter.

285. Therefore, we shall end the whole of this elementary part of Boscovich's theory, by shewing how it will account for the appearance and sensible qualities of a mass of tangible matter. The attentive reader must have in some measure guessed at this already. Suppose the three atoms A, B, C, (fig. 7.) placed in limits of cohesion, composing a particle, having both length and breadth. To make the case as simple as possible, let AB, BC, and CD, be all equal. We can now suppose a fourth atom D, above the plane ABC, and so situated as to be in limits of cohesion with each of the other three atoms. We may still take the most simple case possible, and suppose each of the triangles ADB, BDC, CDA, to be equi-

lateral, and the whole to be sides of a regular tetrahedron or triangular pyramid.

Here we have a particle having length, breadth, and thickness, fit for being a particle of tangible matter. It is now evident that in the same manner that a surface was constituted in fig. 5, a mass of any size and shape, having length, breadth, and thickness, may be composed of atoms, all arranged in this way, each being in the angle of a regular tetrahedron. This would be a mass perfectly homogeneous. It will have a certain degree of firmness; that is, it will resist a certain compressing or dilating force, on the withdrawing of which it will recover its natural bulk and shape; or it may not be perfectly homogeneous, if the tetrahedrons are not all regular, yet still having their angles situated in limits of cohesion with each of the adjoining angles, or rather with the atoms situated in them; or the atoms may be in the angles of cubes, or indeed of any figures whose sides are limits of cohesion. It is plain, that if the distances between these atoms be so small as altogether to escape our observation, the assemblage will appear a continuous uninterrupted body. It will have the mechanical properties that we perceive in all bodies. It will gravitate in proportion to the number of atoms contained in it, that is, in proportion to its quantity of matter. It will exhibit all the appearances of coherent matter, because when brought near enough to another such collection of atoms, the nearest atoms of each will act on each other with that repulsion which is seen between the object glasses, producing sensible contact and pressure. One part of it being impelled, the whole will be put in motion, because the particles immediately impelled cannot move without either coming nearer to the others, or separating farther from them. Either of these events are the occasion of corpuscular forces being excited among the particles; and those must produce motion. We shall afterwards see, with complete evidence, that they must produce *the very motions that we observe.* We will be greatly assisted in our conceptions of all this internal and unseen mechanism by still comparing this assemblage of atoms, connected by attracting and repelling forces, with a number of bodies connected by spiral elastic wires. We may suppose those bodies (so many balls) arranged in the angles of regular or irregular solids, till they form a cluster of any size and shape. This may be considered as a magnified representation of the piece of tangible matter. A small bit of the pith of elder, when viewed through a microscope, will give us a very good notion of this structure. It consists of a number of dodecaedrons, the sides of which are so thin as scarcely to be visible; and there is in each angle a little knot, connected with its neighbouring knots by visible threads, which strengthen the angles of the planes. The whole is a piece of beautiful cage-work. The knots are analogous to our atoms, and the connecting threads and films, being elastic, exert contractile or protrusive forces, according as they are stretched or compressed.

Such a structure will lie on the table, will carry another, and be compressed a little by the weight. Such a collection will be moved by another hitting it, and the motion of the parts remote from the part that is hit is effected solely by the elasticity of the connecting springs, and this elasticity is excited by the compression or dilatation of the whole structure. The similitude in mechanic action is very accurate.

286. It is not necessary to prosecute the theory of Mr Boscovich much farther. It is very evident that what we have said of a mass composed of atoms acting on each other is equally applicable to a mass consisting of particles composed of such atoms. Such particles act on each other as the atoms do, but the law of action, that is, the change

of force by a change of distance, must be very different. Still, however, the general ostensible results will be similar; and, since all the actions are ultimately derived from the atomical forces, and arise from their combination, the actions of different particles must depend entirely on the manner in which their atoms are situated in respect of each other. Therefore, there may, and there must, be an almost infinite variety in the manner of acting of different substances, according to the structure of the particles; and all those differences will be confined to the small and insensible distances in which the atoms and particles are situated, while in all sensible distances we shall see nothing but the attraction of gravitation proportional to the quantities of matter. The phenomena of solid and liquid cohesion, with all its varieties of softness, hardness, ductility, viscidity, and perfect fluidity, must arise from the different manner in which the atoms are arranged in forming a particle. Thus, if they are so arranged that the action of a particle is the same in every direction, varying only with a variation of distance, and of any degree of intensity, however great, the situation of one particle among the rest must be perfectly indifferent; and this assemblage will have the ostensible qualities of a fluid. But if the action of a particle is greater in one part than in another, or greater in one direction than in another, such particles will not take *any* situation, among the others indifferently. They will coalesce in one way rather than another, just as we observe a parcel of small magnets floating on quicksilver to do. These will not cluster together except in their own way. Such particles will have a sort of polarity. This constitution, in a greater or less degree, must be supposed in the particles of all solid, and in a still smaller degree in viscid fluid bodies. The notions which the preceding observatious enable a reflecting mind to form of the construction of particles, point out this as likely to be the most general constitution of tangible matter; that is to say, the number of particles which act equally in every direction, is not probably so great as that of particles having inequalities of action on the different sides. For a certain *determinate* arrangement of atoms is necessary for the first, whereas the *possible* arrangements exceed all numeration. Perfect fluidity, therefore, devoid of all viscidity, will be comparatively rare, and the varieties of viscid, soft, and firm cohesion will be very great. All this is agreeable to what we observe.

287. It seems to result from the preceding observations that all bodies should be perfectly elastic. For, when a particle AI (fig. 1.) is compressed, and the compressing force suddenly withdrawn, the atom I, which was compressed to some point between I and G, will immediately fly back to I with an accelerated motion; and therefore will not stop there, but will pass on towards L, till its motion is gradually reduced to nothing by the attraction of A. It will come back to I with an accelerated motion, pass it as far as before, and thus will oscillate for ever on each side of I, and the whole particles will be made to vibrate.

But we have seen that if the disturbing force exceed a certain moderate quantity, the particle will not return to I but may go to E or to N, fig. 1. And we have seen (284) that when the limits of cohesion and dissolution are very numerous, the situations in which a third atom may rest in relation to a particle consisting of no more than two atoms, are still more numerous. Now, when we consider a body as composed of particles, each of which contains many atoms, it is evident that the situations where a particle may be at rest, are increased in number beyond all power of calculation.

288. What should be the ostensible effect of this constitution? If we examine the motion of the particles by the methods laid down in dynamics, we shall find that all bodies, when dilated or compressed by very small

forces, should recover their form like perfectly elastic bodies. But if the disturbing force has exceeded a certain quantity, which may be very different in different constructions of the particles, the restituent force must be less than the changing force—the form will not be completely restored, and the bodies must appear imperfectly elastic; and the sensible elasticity will be so much the less as the particle has been forced over a greater number of limits; because it will only return to the left of those it has passed over.

This is a consequence of the theory that is confirmed by the most extensive experience. Mr Coulomb was engaged (for a particular purpose) in a series of experiments on the oscillations of springs, particularly of twisted wires. He suspended a nicely turned ball or cylinder by a wire of a certain length, and fitted it with an index, which pointed out the degrees of the torsion. He found that when a wire of 20 inches long was twisted ten times, the index returned to its primitive position, if repeated a thousand times, and the oscillations were made in equal times, whether wide or narrow. But if it was twisted eleven times, the index did not return to its first place, but wanted nearly a whole turn of it. Here then the parts of the wire had taken new relative positions, in which they were again at rest. But what was most remarkable in Coulomb's experiments was this. He found that after the wire had taken this set, (as it is termed by the artisans) it exhibited the same elasticity as before. It allowed a torsion of ten turns, and when let go, it returned, and after its oscillations were finished, it rested in the position from which it had been taken. I was much struck with this experiment, and immediately repeated it on a great variety of substances, with the same result. The most unelastic substance that I know is soft clay. I got a thread made of fine clay at a pottery, by forcing it through a syringe. It was about $\frac{1}{12}$th of an inch in diameter, and eleven feet long. While quite soft, (and smeared with olive oil, to prevent its stiffening by the evaporation of its moisture,) I fastened it to the ceiling, and fixed a small weight and an index to its lower end. I found that it made 5½ turns a hundred times and more, without the smallest diminution of its elasticity, always recovering its first position. But when I gave it seven turns, it returned only 5½. Thus it took a set. In this new arrangement of its parts, I found that it again bore a twist of 5½ turns without taking any new set. And I repeated this several times. I then gave it ten turns, in the same direction with the first seven. It returned 5½ as before, and was again perfectly elastic within this limit.

I therefore look upon this consequence of Boscovich's theory as perfectly agreeable to what we observe in tangible matter. Mr Coulomb infers from his experiments that the ultimate particles of bodies are perfectly elastic, and that the imperfect elasticity that we observe is owing to the particles shifting their positions when forced too far from their present situations. This is a very familiar and easy way of conceiving it, and it is very evidently the case in such a body as clay, which we can twist and force into any mutual position of its parts. But this perfect elasticity of the ultimate particles can be conceived in no other way than that described by Boscovich. I may add that Coulomb is probably right in ascribing the shifting of position *solely to the particles*, and not to the atoms of which they are composed. For if the atoms of a particle shift their places in the particle, the nature and manner of acting of that particle will be changed. It is also *a demonstrable* consequence of Boscovich's theory, that the forces which resist a change of position among the atoms, must be incomparably greater than the similar forces of the particles; and also that the number of quiescent situations for particles must be incomparably greater than for the atoms of a particle.

289. Another appearance of tangible matter shews a most encouraging conformity to the theory. When bodies are very moderately compressed or dilated, the forces employed are proportional to the change of distance between the particles. This appears most exactly true in the experiments of Dr Hooke, on which he founded his theory of springs, expressed in the phrase *ut tensio sic vis*, and his noble improvement of pocket watches by applying a spiral spring to the axis of the balance, which, by its bending and unbending, produced a force proportional to the angle of the oscillations, and therefore made them isochronous, whether wide or narrow. It is also confirmed by the experiments of Coulomb on twisted wires; and by the form of the elastic curve, as determined by Bernoulli, on the supposition that the forces with which the particles attracted and repelled each other are proportional to their removal from their natural quiescent positions. But it is found that when the compression or dilatation is too much increased, the resistance does not increase so fast; that it comes to a maximum by still increasing the strain, then decreases, and the body takes a great set, or breaks. All this is perfectly analogous to the forces expressed by the ordinates of our exponential curve. In the immediate vicinity of the limits of cohesion, the ordinates increase nearly in the ratio of the abscissæ, then they increase more slowly, come to a maximum, decrease again, till we come to a limit of dissolution.

I have now said enough, I think, for giving a pretty distinct notion of this theory, and its competency to explain many of the mechanical phenomena of nature. The curious student will not content himself with this slight sketch, but will go to the work itself. There he will find a most ingenious application of these principles to the explanation of many abstruse and curious phenomena, some of which seem to throw unsurmountable obstacles in our way. Nothing can seem more difficult to explain than the free passage of light.

The observations which have occupied our attention are only such as relate to the more obvious mechanical changes of condition produced by the mutual action of bodies. Mr Boscovich's notion of the impenetrability of the masses are very curious, but are omitted, as not of much service in our particular views.

290. I may end this exhibition by saying, that it is really wonderful how fertile this theory appears when we apply it to the explanation of physical phenomena. All masses will have gravity proportional to $\frac{1}{x^2}$, and proportional to the quantity of matter in the mass toward which it is directed. So that on the surface of this earth, gold and a feather will fall equally fast; and the pressure which we call this weight will be (sensibly) as their quantity of matter; and (which is of prime importance) all the mechanical actions, contemplated merely as changes of motion, will be mutual, equal, and opposite. But, in all the properties which depend on the small insensible distances, such as those in which light is reflected, refracted, or deflected, the distances at which our fibres are stimulated mechanically, those of the palate for taste, those of the nostrils for smell, those of the ear for sound, those of the eye for vision,—the distance at which the particles of the air are exerted to tremor, and rest on each other so as to propagate this tremor—the distance at which the nerves of our skin are stimulated by roughness—the distance at which cohesions are effected in all their varieties—the distance at which secretions, fermentations, solutions, precipitations—the action and polarity of particles which operate in crystallisations, and all the wonders of chemistry which distinguish the different kinds of substances,—in all these distances there must be a diversity, fully adequate to ac-

count for all the variety observed in the world. There does not appear the smallest necessity for supposing that any one primary atom of matter differs in any respect whatever from every other atom. One and the same curve line is the exponential of the force of every atom. Not only does no other seem necessary, notwithstanding the immense variety that we observe, but it would seem that there cannot be another. It would not be agreeable to the universal fact of action and reaction—a fact to which we know of no exception.

291. It may now be asked why so much time has been occupied with this theory, however ingenious? Has it any foundation in nature? Will it assist us in our future researches? Can it be that tangible body is not that continuous uninterrupted substance that it is generally supposed to be, and that it is merely an assemblage of points, distant from one another, and as unlike a solid body as any thing we can conceive?

In answering such questions, we must always recollect that our ultimate judgments depend more immediately, or at least more certainly, on the nature of human intellect, than on their agreement with the state of external nature. Of that perhaps we have but little. But we can tell what notions we can form with distinctness, what are obscure, and what are incongruous, impossible, or absurd. We also know what kind of evidence produces belief, and what does peremptorily exclude it. It is certain that we account a piece of polished gold or of glass, to be a piece of continuous uninterrupted matter. There is no alternative but that it is interrupted. But to say that it is so is not enough. We must prove in some way or another that it is not continuous. We see no interruption, we feel none, and therefore may require them to be pointed out to us. As this has not been done with the evidence of sense, it has been flatly refused, and the greatest part of philosophers hold tangible matter to be something continuous, consisting of parts in absolute contact, and that one body acts on another by coming into contact with it.

There are cases, however, in which we have learned to distrust our first judgments concerning the information given by our senses. Sometimes the errors of opinion proceed from the imperfection of our senses. Thus a glass globe foiled in the usual way, appears uniformly resplendent by reflected light; but if held between the eye and the light, we perceive the metallic covering to be very partial, and that light comes through it as through a cobweb, but still we see no pores till we take a magnifying glass.

In other cases we observe our error, by its absolute inconsistency with judgment, formed on the most uexceptionable manner. Thus it is certain, that in many cases, where no interruption can be observed, there are nevertheless vacuities through which other matter may pass. The hardest marble is penetrable by oil, which it imbibes, and is discoloured by it. Gold and silver do, in like manner, imbibe mercury. The most specious proof is obtained by the free passage of light through transparent bodies. Something passes through a piece of glass which illuminates, which warms, which oxygenates metals and other inflammables; and which, in other circumstances, reduces metallic oxydes; which, in short, gives the strongest indication of its materiality.

292. From these, and many similar facts, philosophers thought themselves under the necessity of admitting interruption in the continuity of tangible matter, and were led to the atomistic theory, which Newton first enounced with some precision, and which Boscovich has reduced to a system. The philosophers of antiquity also had some imperfect notion of this kind. But they went no farther than the mere division, in imagination, of the visible masses, into smaller parts of the same kind, without taking any pains to shew how the sensible

qualities of bodies could result from the mere combination of such fragments. Even the chemical philosophers, who taught that there were different elementary particles, did nothing to shew in what those elements differed, or what was the principle of their connection, from fragments of the sensible masses. Newton attempted this; but though his notions have all the originality and acuteness that we should expect from Sir Isaac Newton, they are offered merely as miscellaneous thoughts, without any serious attempt to support this novel opinion. Indeed those observations of Newton are rather attempts to explain some of the chief discriminating relations of different substances, than to establish a general doctrince concerning the sensible forms of tangible matter, and the mechanical consequences of this doctrine.

Boscovich, on the other hand, endeavours at once to establish this doctrine. Therefore, he begins with the bold assertion that this atomodynamical constitution of tangible matter is not only a conceivable and possible thing, but that no other constitution is compatible with the acknowledged phenomena of existing matter. His chief argument is taken from what is called the LAW OF CONTINUITY. A moving point cannot pass from one point of space to another, without passing through all the intermediate points of its push in succession. In like manner, a body moving with the velocity 5 feet per second, cannot change this rate of motion to 4 feet per second, in an indivisible instant of time, as it is generally thought to do in collision, for that instant is the end of its motion with the velocity 5, and the beginning of its motion with the velocity 4. It is not a portion of time interposed between them, but the mutual boundary of both. Therefore, to suppose this instantaneous change, is to suppose that in that indivisible instant, the body has both the velocity 5 and the velocity 4, which is inconceivable or absurd.

293. There must therefore be some way by which this

impossibility is avoided. This will be accomplished only by means of a mutual force, acting on both bodies before they come into contact, in the manner in which we really observe all changing forces to act, that is, diminishing the velocity of the one, and increasing that of the other, by insensible degrees. We shall, in due time, have the fullest proof that this is the case *in fact*, in all the changes of motion observed in this created world; and we may limit our inquiries to this scene. All atoms of matter, therefore, must be endowed with an insuperable repulsive force, extending to some distance, which shall make the mathematical contact absolutely impossible.

It does not require much reflection to see that, without this supposition, the law of continuity must be violated in every collision of bodies.

But repulsive forces alone cannot produce a mass of coherent matter. Mutual attractions are equally necessary; and, as the law of continuity obliged Mr Boscovich to consider this repulsive force as acting at a small and insensible distance, it was very natural for him to consider the attractive forces as similar in this respect. Having admitted this, it did not require a great deal of consideration or ingenuity to contrive a constitution, such as has been described, where the atoms should be situated at such distances from one another, that when pushed nearer they should repel, and when drawn asunder they should attract, and when at their natural distance, they should neither attract nor repel. The rest of his study was in order to be able to shew, by the many alternations of attraction and repulsion, and by other accommodations, how, from one law of action for every atom, may be derived, by composition, all the varieties of form and of mutual action that we observe in the world around us.

294. All this was pure speculation on the part of Mr Boscovich; for he does not seem to have known the

many facts by which this corpuscular action *e distanti* is confirmed. He mentions indeed the experiments of Newton on the inflexion of light as it passes near solid bodies, and also by those by which it appears that light is reflected and refracted before it comes into mathematical contact. But he takes no notice of the still more precious experiments of Huyghens, and those of Newton with the object glasses. Nor does he consider the phenomena of capillary attraction as any thing different from mere adhesion, or as affording proofs of action *e distanti.* But, in the progress of his discussion, many things occur, which greatly illustrate and confirm his first notions of the discrete structure of tangible matter.

This theory of Mr Boscovich unquestionably receives great support from the various phenomena of corpuscular action, which I mentioned before giving an account of his doctrine. His great argument from the law of continuity will come to be considered and applied afterwards, and, I presume, will be found to be of great force. It was mentioned just now, merely to shew how the ingenious author was led to form his theory.

295. I shall just add to all Mr Boscovich's arguments for the discrete structure of body, that I cannot form to myself any conception of continuous matter that is compatible with a difference of density. If the space be completely filled, that is, if there be no part of it in which there is not matter, I cannot conceive that more matter can be put into that space, nor what can be meant by a difference in density. Yet nothing can be more convincingly ascertained than the different density of the most solid and continuous bodies. When we talk of pores in such a body as gold, or water, or air, *in which none can be discovered by the senses, however assisted,* I cannot see what authority we have for supposing that the ultimate particles touch one another in *any* points whatever. We are induced to suppose this, solely because our familiar ac-

quaintance with matter, from our earliest infancy, has been by means of our sense of touch. This is the only way in which we usually put matter in motion, and it is to this sense that we always appeal in any case of doubt about the materiality of a thing. We must *feel it*, or see it produce effects which would excite the sensation of touch in ourselves. We have allowed this habit to influence our speculative notions, and have indolently or carelessly adopted it as a principle, that contact is absolutely necessary for the cohesion of matter, and for the communication of motion by impulse. But surely who ever holds tangible matter to be absolutely continuous, cannot admit such a thing as the passage of light through it. There is indeed no absurdity in the conception of continuous matter. On the contrary, it is the clearest and the most simple. But it excludes all difference of density. For nothing can exceed the density of what completely fills the space. If again we explain the difference of density by pores, then since these pores are altogether unperceived, and are admitted merely to get rid of a difficulty, there is no reasonable objection remains against the opinion of Boscovich, that not only there are vacuities or pores, but that the particles are not in contact on *any* side. I need not repeat the arguments used in the beginning of this article against the mutual contact of the particles of compressed air, or compressed water. I hope that it is now sufficiently plain that physical contact, and all its consequences, and particularly the production of motion by impulse, may be produced by means of attractive and repulsive forces, without real mathematical contact; and that Newton's experiments with the glasses and the soap bubble shew that *such physical contact*, and the communication of motion, are *really so produced*.

296. I see no reason, therefore, for refusing the doctrine taught by Mr Boscovich, as far as relates to its

great principle, the discrete constitution of matter, and the non-existence of mathematical contact. These two great points are not more opposite to the suggestions of sense than the rotation and revolution of the earth, which we find ourselves *obliged* to admit, on account of the incompatibility of the common opinion with other phenomena. It is for such reasons only that the daily and annual motion of the sun, and the absolute rest of our own habitation, are considered as false judgments or prejudices. The present question is perfectly similar in this respect. The continuity and contact of the particles of tangible matter are to be accounted prejudices, *only because they are incompatible with other undoubted phenomena,* and because they are incompatible with the law of continuity, the violation of which is inconceivable.

With respect to the imputation of reviving the occult qualities of the peripatetics by the name of attraction and repulsion, the defence is nearly the same as in the case of the gravitation of the planets. We have seen instances in which the mutual adhesion, and the mutual separation of the component particles of tangible matter are incontestible facts,—are general laws in nature. Nothing is meant by attraction and repulsion but the immediate efficient causes of those phenomena. A philosopher may say that the deflection of a planet is effected by the action of an etherial vortex. He may and he must say the same thing here. But it may then be asked him how this ether produces the deflection? He will probably say that it is in the same way that a stream of water or wind, or even a stream of sand, would deflect the motion of any body that it strikes. But I apprehend that this answer of the philosophers, which has long been thought complete, is now rendered altogether invalid. For we have seen that it cannot be shewn that one body ever really strikes another, even when it puts it into motion. The ether is employed, because it is a fact that a body of sensible magnitude impels another before it comes into mathematical contact. Therefore, since we have no evidence that one body really hits another, we cannot say that a particle of this ether hits the body. Another ether is therefore necessary for explaining *its* impulse, if an ether be necessary for explaining the impulse of the sensible body; and we must have a third ether to account for the action of the second, &c. &c. &c. There is no end to this kind of mechanism. In short, active powers, different from the stroke of a moving body, must be admitted, whatever origin we assign to them, and we must add them to our notions of mechanism, and no longer limit the term *mechanical explanation* to an explanation by impulsion and contact. The manner in which both gravitation and impulsion are effected, will probably always remain unknown, but the approach and the separation are not occult qualities, but manifest phenomena; and both are accompanied by that *pressure* which we and all mankind have been accustomed to call a force, which is therefore another manifest phenomenon, and not an occult quality. If better names can be found for them let them be employed, but attraction and repulsion seem to distinguish the phenomena very clearly.

There seems no reasonable objection against extending the action of these forces to the particles, and even to the atoms, although it must be acknowledged that we do not so clearly see their operation on those particular subjects. But surely, when a mass of tangible matter is compressed or dilated, the actual exertion of forces, opposing the approach or separation of the particles, is as distinctly perceivable as in the similar experiments of Newton and Huyghens, although in this last case the actual approach and recess of the subjects are seen, and can be measured, while in the first these motions are altogether imperceptible in the subjects which are thus moved. The cases are so analogous that it is fastidious to refuse admitting the same mode of explanation.

297. Perhaps more solid objections may be made to one circumstance of Mr Boscovich's theory. He holds his atoms to be unextended mathematical points. I cannot form any notion of one of these as an individual subject, different from the notion I have of a mere point of space. The individuality of a point of space consists in its being *that particular* point of space, so and so situated with respect to certain other points of space. This individual point of space is therefore immovable. Surely, therefore, the active powers ascribed to matter are conceived as the endowments, or the attendants of something that is different from a mere point of space.

I am by no means certain that this inextension of an atom is indisputably requisite in a theory which maintains the discrete constitution of matter. Far less am I certain that it is necessary for the truth of this constitution, and for its utility in philosophical discussion, that I should perfectly understand the nature of this subject or substratum of the active powers of matter. But neither can I, on the other hand, conceive those powers to have no substance to which they belong, or to which they are related, either intrinsically or extrinsically. I cannot conceive one set of attractive and repelling powers putting another set in motion by impulse. These are words without ideas. It is not likely that human intellect will either acquire more adequate knowledge of the subject, or that it will ever be able to divest itself of the thought that there is a substance in which those powers are inherent.

298. Notwithstanding this limitation and imperfection of our knowledge of this subject, we shall find several consequences of this manner of considering the mechanical relation of the parts of tangible matter, which we may employ with confidence, to help us in our inquiries; and I trust that I may be able by this help to illustrate many

things which would otherwise have been very obscure. But, at the same time, I am aware that the most scrupulous caution must be continually observed in making use of this help. We must at all times avoid forming any general conclusions without firmly establishing, by mathematical reasoning, the individual fact which we make the basis of our inference; and we must never trust our imagination in its views of the *combined* motions or actions of many particles or atoms, except in cases where we can demonstrate the legitimacy of such procedure. I should even add that the difficulties in this study of corpuscular attraction and repulsion have been greatly augmented by some late discoveries.

The action which we are now considering is limited to the small and insensible distances in which the particles of tangible matter are situated in respect of one another; and the doctrine which I am now prosecuting considers the corpuscular force as depending, both as to kind and degree, solely on the distance and position of the acting particles. Now, in that numerous and important class of phenomena which are known by the name of galvanic, there occur corpuscular actions which seem altogether inexplicable on those principles. We see a wire, the remote end of which communicates with the positive pole of Volta's pile, attract to itself the particles of metal which are floating round it in the solution. Now let the remote end of this wire communicate with the other pole, of the pile. In an instant the action of the wire is totally changed; it no longer attracts the metal, and even seems to repel it. Yet there is no change in the distance and position of the acting particles. I know that magnetism presents phenomena which are extremely similar, and which we can explain in perfect consistency with the laws of attraction and repulsion; but the admission of a similar explanation in the present case, gives a complexity to the question which almost blasts all our hopes.

299. Notwithstanding this discouragement, this department of science is deserving of the most careful cultivation. All the changes of material nature, when analysed to the utmost, are examples of matter influenced by moving forces This is the case even when the sunshine discharges some gay vegetable colour; some matter is moved from its former place, and takes a new arrangement. The phenomenon would be completely explained, could we, from our knowledge of the law of action of the forces, tell the path of the removed atom, and its motion in that path, as we can do in the case of the moon.

We are not without encouragement, even in this perplexed research. We are obtaining now and then pretty distinct traces of general laws. We have found that the mutual action of the particles of such gasses as we have been able to confine and measure, is nearly proportional to the density of the gas in its different states of compression; and therefore it diminishes nearly in the proportion that the distance of the particles increases. We have also found that there is a great difference in the laws of action in different gasses in the moment of their production from a solid or liquid body. Thus the gas produced by the inflammation of gunpowder exhibits an elasticity nearly proportional to its density, through the whole range of its expansion. But the gas produced from fulminating gold, silver, or mercury, shews an action of the particle which diminishes much more by an increase of distance. Scarcely any barrel is able to withstand their explosion, yet they have not half of the force for impelling a ball by their *continual* expansion along the barrel.

300. I observed, on another occasion, that a certain sect of philosophers insist that all forces diffused from a centre decrease in the inverse duplicate ratio of the distances; and that if any other law is observed, it arises from the manner in which different atoms, acting by this

primordial law, are combined. But this will not agree with many phenomena of tangible matter. If the primordial force be proportional to $\frac{1}{x^2}$, the action of all compound particles must decrease more slowly, and be proportional to $\frac{1}{x-m}$. Now we are certain that there are many forces which decrease faster than in the ratio of $\frac{1}{x^2}$, and must be proportional to $\frac{1}{x+m}$. The phenomena of cohesion in hard and brittle bodies much more resemble the effects of a force proportional to $\frac{1}{x^3}$, for at the smallest sensible distance the cohesion vanishes entirely. The action of aurum or argentum fulminans undoubtedly decreases faster than $\frac{1}{x^4}$.

The observations contained in the first part of this article give sufficient authority for concluding, that the parts of what we have called tangible matter are related to each other in a way which greatly resembles the mechanism described by Mr Boscovich; so that this description may be of considerable service in our subsequent examinations of those changes of motion that are produced by the actions of sensible masses on one another. It directs us to a way of conceiving the connection of the distant parts of a body, without which we shall in vain look for a satisfactory explanation of some fundamental truths. Of this we shall soon have full evidence.

301. I cannot help taking notice of a performance which has been repeatedly announced in our periodical publications, as ready for the press, by the title of *Recherches Mathematiques sur la constitution intime des Corps. Par A. G. Bue'.* An index or precis of the work has also been some time in circulation. The author professes to ex-

plain the intimate structure and constitution of tangible matter. He builds his whole system on the following proposition, which ascertains the primordial property of matter, by which *alone* all the phenomena of the visible universe are effected and explained.

FUNDAMENTAL THEOREM.

Any two elements of matter will, in virtue of their mere existence, mutually approach each other in every instant, by a space which is inversely proportional to the square of their distance.

The author does not indeed say that they *must* so approach. His only meaning seems to be, that if there be any relation of approach or recess mutual between them, this *must* be its character. The momentary change of place must be inversely proportional to the square of their mutual distance. Having demonstrated this fundamental theorem, the very ingenious author proceeds to explain, that is, to account for all the phenomena of the material world, the phenomena of cohesion, in all its varieties of hardness, softness, ductility, elasticity, fluidity; and the phenomena of magnetism, electricity, chemical affinity, optics, &c. &c.

Mr Bue' is an expert and elegant algebraist, and has given a vast apparatus of the most refined equations, expressing the mechanical possibilities of nature. In particular, he gives *une equation finale*, which expresses the UNIVERSE "*touts les faits possibles.*"

He then applies these equations to particular cases, and says that this and that phenomenon are explained, that is, are shewn to be necessary results from this primitive essential property of matter. Then comes his chief question,—"*If, among all the possibilities expressed by his final* "*equation, a certain number tally exactly with the same num-*

" *ber of the observed phenomena of the universe, what is the*
" *degree of probability that the hypothesis is false?*

To answer this question, the author now plunges into the depths of Condorcets, *Calcul des probabilites*; and, after several pages of the most abstracted and general equations that can be imagined, and in which all physical ideas have quitted the mind, he makes it as clear as sunshine that the probability is = O!!! Q. E. *d.* The phenomena are now held as explained.

The inevitable conclusion which must, I think, follow from these ingenious labours is, that the universe cannot be any thing but what we see it to be. We have the good fortune to exist during one of those possible forms into which matter, after acting and reacting, during ages of ages, has worked itself into a shape which may endure for ever.

I have studied, with the utmost care and attention, this author's demonstration of this his fundamental proposition. I cannot say that it is not demonstrated; because I confess that I cannot form any distinct notion, either of his reasoning, or of the principles from which he reasons. I am willing to think that I misunderstand him; but I have given the proposition in his own words. I shall grant that it flows legitimately from his principles.

302. But it is surely very singular that this primordial law of nature, essential to matter, and the foundation of *all* its properties, *should not be observed in one single phenomenon of nature.* The author assumes and employs it under the name of the law of gravitation, the discovery of Sir Isaac Newton. But I apprehend that he is off his guard in this instance. Although an expert and elegant analogist, he does not seem to be aware that the proposition which he has demonstrated differs *toto cœlo* from the law of gravitation. This is remarkable; for I think that the mathematical reader will perceive that the proposition

which Mr Bue' considers as demonstrated is, that the velocity of the gravitating atom is inversely as the square of its distance from the atom toward which it gravitates and approaches. For the *little space, through which it approaches in every instant*, is adopted by every mechanician as the measure of the velocity. But in gravitation, it is not the velocity, but the acceleration, that is inversely as the square of the distance. I do not know one spontaneous phenomenon of nature, in which Mr Bue''s proposition is observed.

Thus will it always fare with us when we attempt to reason about the foundation of the powers of nature. Mr Bue' has heard of this my objection; but it is in vain to make a *refonte de l'ouvrage*, as I hear he is attempting. He must not shift his ground. If he has demonstrated this proposition, he can demonstrate no other from the same principles; and he must search for some other proof of the law of gravitation, *which he has shewn, by this proposition, to be incompatible with the primordial property of matter.*

I may here observe that Mr Bue''s ingenious explanation of many phenomena is not in the least hurt by this blundering proposition. He might have simply stated as a universal fact, that all matter gravitates, in the manner described by Newton, and he might then have applied this to the explanation of other phenomena, precisely as he has done; and the explanations would still have all that merit of great ingenuity, although I must consider them as altogether romantic and fanciful.

303. In the farther consideration of the mechanical phenomena of nature, as they are modified by the addition of the force of cohesion, it is necessary to divide the subjects of our examination into two classes, depending on the manner in which the particles are combined, so as to form masses of tangible matter. Those are either solid or fluid. Solid bodies are such as retain a certain shape

and resist (to speak metaphorically) in a certain degree, the forces which are employed to change this shape. Fluids exhibit no such firmness, and must be kept in vessels, otherwise their gravity would cause them to spread abroad on the ground. We shall first consider

The Mechanism of Solid Bodies.

The experiments with the object glasses, to which reference has been so often made, shew us that any piece of solid matter acts on another, at a very minute distance, with a repulsive force so great that we have no evidence that it can be overcome by any force that we can employ. This force was perceived in those experiments at a distance not less than the 5000th part of an inch, and it probably extended considerably farther, but with a diminished intensity.

We are fully warranted to say that the repulsion manifest in those experiments is the immediate cause of the physical contact of bodies, and is the force which we call pressure. By this repulsion, bodies act on our external fibres and nerves, and occasion in us the sensation of touch, or what we call *feeling*. This is perhaps our clearest notion of force of any kind. Gravity is perceived as an agent by the pressure of a heavy body on our hand; yet this pressure is not the agency of gravity, but of the corpuscular repulsion we are now considering. Gravity, by forcing the heavy body near enough to our hand, gives occasion to this repulsion, which most persons imagine to be the heaviness or weight of the body.

Being the cause of our sensation of touch, this repulsion gives us our first and most familiar notion of solidity, firmness, of corporeality. It is even the immediate cause of the philosophical notion of solidity, materiality, impenetrability, that is, the maintenance of the place occupied, exclusive of all other matter. One piece of matter cannot occupy a space already occupied by other matter, except by thrusting that other matter out of the space filled by it.

So many functions entitle this peculiar modification of corpuscular force to a very careful attention and study.

304. It appeared that the distance which produces physical contact is vastly greater than that in which adhesion and capillary attraction are exerted, and is therefore vastly greater than the distance at which the particles of the solid bodies cohere. Consequently, when one of those bodies is so near to another as to act on it, the actions of many particles of the one on many particles of the other are combined. Hence we derive a fundamental proposition.

305. The mutual action of two solid bodies in physical contact is always in the direction perpendicular to the common plane of contact.

Let AB Plate III. fig. 8. be the surface of a solid body, and let us suppose that the corpuscular force acts on a particle at all distances that do not exceed a certain line X. Let P be an external particle, whose distance from the surface AB is less than X. Let OCDE be a sphere whose centre is in the particle P, and whose radius is equal to X. It is plain that a part CDE of this sphere is occupied by the matter of the body. Every particle, therefore, contained in the spherical segment CDE acts on the particle P, and it is not acted on by any particle that is beyond the surface of this sphere. Now, suppose that the particle F repels P, in the direction, and with the force PH. There is another particle G, similarly situated on the other side of the perpendicular PD. It must repel P in the direction PI, and the measure PI of this force must be equal to PH, because the distance PG is equal to PF The forces PH and PI compose a force PK, the direction of which must evidently be perpendicular to the surface AB.

What has now been demonstrated concerning the action

on one particle P, is true of the action on every particle acted upon; and it is easily demonstrated to be true with respect to the action compounded of all those actions, and therefore the proposition is manifest.

The attentive reader will remark that this demonstration proceeds on the perfect equality of action of the particles F and G. This supposes that no particle such as F will be found within this spherical segment CDE which has not a companion G, similarly situated on the other side of the perpendicular PD; or, in other words, it supposes that all within this spherical segment is homogeneous. This, it must be owned, cannot be affirmed universally. But the sphere of this particular corpuscular force, which is the cause of physical and sensible contact, and impulsion, is so great in comparison with the natural distance of the particles, that we may reckon upon the homogeneity of the segment without the risk of any sensible error. There will be some irregularity, but in the small space of the 4000th or 5000th of an inch, it cannot be considerable; and the average motion will be, to all sense, perpendicular to the surface, just as gravity is to all sense perpendicular to the surface of the ocean, notwithstanding the mountains and other inequalities which occasion deviations and inequalities in natural gravitation. We cannot discover the irregularities in the case now before us as we can those of gravity. Yet we shall, after this, meet with some instances of them which cannot be disputed.

306. The question now is, whether this theoretical deduction from our notions of corpuscular action is agreeable to what we really observe. In answer to this, it must be observed that we cannot make the examination with such precision as we have been accustomed to in our astronomical discussions. When we recollect that the sphere of action cannot greatly exceed the 4000th part of an inch, it is evident that we may be mistaken in our notion of the exact position of the common plane of contact.

A scratch, or inequality, that we cannot perceive by the eye, may change the position in the point of contact unnoticed by us. But still, the experience which we have of this matter is abundantly agreeable to our proposition, and indeed may be pronounced accurately so. I know nothing that shews this more clearly than the motion of the balls on the billiard table. Thus, suppose a ball A fig. 9. impelled in the direction A *a*, and that another ball B lies partly in its way, so that it cannot pass without striking it. Suppose that in the instant of contact, the plane C *c* is the common tangent to both balls. The ball B will not move in a direction parallel to A *a*, but in the direction B *b*, perpendicular to C *c*. A person who has never seen the action of elastic balls in collision will be much surprised at seeing the ball take a direction so unexpected.

In like manner, if a series of balls, such as are represented in fig. 10. be struck by a ball moving in the direction A *a*, the last ball C of the series will be drawn off in a direction C *c*, perpendicular to the plane D *d*, which is the common tangent to C and the next ball. There is generally, however, a small deviation from the perpendicular toward the direction of the striking ball. Thus in fig. 9. the ball B does not move exactly in the perpendicular B *b*, but in a direction B β, a little nearer to the direction A *a*. But this is owing partly to imperfect elasticity, and partly to some friction by which the ball A *drags* a little into its given path the ball which it strikes. But in very elastic ivory balls this deviation is scarcely sensible.

307. There is another instance, in which the greatest precision in the direction and intensity of the force is necessary for insuring the intended effect, and where the complete success shews the truth of the principle on which the artist proceeds. This is in the construction of the pallets of a clock or watch. Fig. 11. represents the

swing wheel of a clock, urged round in the direction DEF by the rest of the wheel-work. It is prevented from running round by the point of the tooth A stopping on the pallet. This point of the tooth is urged forward in the direction of the tangent A α. A force in this direction would have little tendency to make the pallet turn round its centre G. But its action on the face of the pallet being in the direction A *a* perpendicular to that surface, pushes it outward, causing it to turn round G. In like manner, the tooth B, moving in the direction B β, would have no tendency to make the pallet turn round G. But the action being in the direction B *b*, has a proper efficacy for this purpose.

Many other examples may be had in the construction of machines. It may be remarked here, that there is nothing in which the genius and sagacity of an engineer is more certainly seen than in his attention to this maxim, and there is no article of practical mechanics that is more imperfectly understood by professional men. It is most evident that a right employment of this proposition is of the greatest importance in all machinery; for it is by such mutual contacts and pressures that all their motions and performances are effected. The proposition is by no means obvious, and the practice which it prescribes is frequently the most unlike to what our common notions would suggest to us, as in the two examples that have been given.

308. This proposition, when duly employed, makes a great change in the manner in which mechanical problems have been treated. Thus, in explaining the motion of a body which is impelled in the direction AB (fig. 12.) by a force AB, but which cannot move except along the solid path AC, the usual way is this. The force AB may be resolved into a force AC, in the direction which the body can really take, and a force AD perpendicular to that direction. This last force neither promotes nor op-

poses the motion in the direction AC, and therefore the body moves as if impelled by AC alone. All this is very just *in abstracto*, and the motion is truly ascertained. But it is not a narration of the procedure of nature. There is no such resolution of forces, but, on the contrary, a composition. We are not now engaged in an abstract proposition of dynamics, but in some natural question. Why cannot the body move in the direction AB? Because it is hindered by a solid body, on which it presses, and which reacts in a direction perpendicular to the touching surface. The motion which it takes is the diagonal of a parallelogram, one of whose sides is AB, and the other has the direction AE perpendicular to the path AC. Therefore draw BC perpendicular to AC, and complete the parallelogram ABCE, of which AC is the diagonal, and the expression of the force actually compounded of the motive force AB, and the reaction AE really exerted by the path in the point A, *but nowhere else*. The compound force causes the body to move along AC without any more pressure on the path, or any more reaction; whereas in the common way of considering the question, the resolution of the force is supposed continual, and therefore the pressure continual.

We shall meet with many cases where the investigation has been made extremely difficult by treating it in an abstracted manner, whereas it would have been extremely easy, had it been considered as nature presents it to our observation.

309. Having thus ascertained the direction in which an external atom, corpuscle, or larger mass of solid, that is firm matter, is acted on by another, we at the same time learn the direction of the reaction of that external body. It is also perpendicular to the touching surfaces. The natural consequence of this information seems to be, that when an external body is thus acted on (we shall

suppose it repelled,) it will be made to move off in that direction. Accordingly we see it so impelled, in the instance of the billiard ball. There seems nothing more necessary for explaining the whole effect. True, it has been perhaps only one sensible point of the body that has been so impelled! But as the body is solid, hard, and in a manner impenetrable, this impelled point cannot move off without pushing all the rest before it. We see no difficulty in this; and there seems nothing more wanting.

But there is a great deal yet unexplained. One particle only has been impelled; a small force suffices for moving one particle. How comes a reaction to be perceived as great as if each particle of the body had been separately acted on. The inertia of the most remote particle must be overcome, and its particular reaction seems to be exerted by the single particle that has really been impelled. It would seem that a part of the whole force employed has been somehow propagated through the substance of the ball to each of its particles, and also that the reaction of each of those particles has been propagated backwards to the particle really acted on, and that all these reactions have been exerted by this single particle.

But I do not think that we can conceive force to be a *thing* which can be transferred in this manner from one particle to another, or shared among the different particles of a body, as salt or sugar may be diffused through water. Our notion of the nature of force will never be very clear. But, besides this obscurity, we see that what has been hitherto said will not explain many circumstances of the motion produced by acting on one part only of a solid body. In some cases we see the impelled body move straight forward in the direction that we expect; but in other cases it not only goes forward, but also turns briskly round; and a proper attention will shew us that this combination of progressive and rotative motion is regulated by laws as steady as any in mechanics. There are principles, therefore, by which this propagation of moving force is regularly modified.

310. There is no such *propagation* of force. The united motion of the whole impelled body is an example of the *excitement* of those corpuscular forces which connect the particles of tangible matter. If the action or pressure has been great, and the body large, and of very tender friable texture, we often see that it is not all moved, although its parts are as universally in contact as those of another body which moves off completely by the same action. A blow given to a metal candlestick will knock it over. The same blow given to a china candlestick will probably leave the foot of it where it was, and the stalk only is knocked off. Why this difference? Why is the foot left behind? There has been a deficiency of connecting force. A hempen cord will twitch into rapid motion a large ball. But if we give the same sudden tug by means of a slender thread, we shall leave the ball behind, for the thread will be broken. Such examples shew pretty clearly how the motion of the remote part of the body is effected, either by pushing or pulling. It is the pressure of the *last particle* of the rod or rope that moves the ball; that pressure was brought into action by stretching the rope, or compressing the rod. Something drew the last particle but one of the rope a little way from the very last, and therefore brought into action, or excited the attractive force which is inherent in, or always accompanies the particles when at that particular distance from each other. What was this something? It was this: the last particle but two was drawn away from the last but one, and therefore attracted it, and drew it away from the last.

311. As I apprehend it to be of great importance that we have very clear and precise notions of the manner in

which the sensible forces which connect the particles of matter are brought into action by the action of external forces, I shall employ a little time in endeavouring to render it plain.

Let *a z* (fig 13.) be a single row of cohering atoms or particles, without weight, in a vertical position, and at their natural unconstrained distances from each other. Also let M be a material horizontal plane in physical contact with the row *a z*.

Now let the uppermost particle *a* recover its weight. It cannot remain in its present position, because it has no support. For we supposed the particles to be in their natural inactive distances. It is not acted on, that is, it is not supported by *b*. It must therefore descend by its gravity, and come so much nearer to *b* that *b* repels it with a force equal to its weight. *a* is now supported. But as *b* is supposed to act on *a*, *a* acts on *b*, and pushes it from its place toward *c*, therefore *b* cannot rest, or be in equilibrio, till it is as much repelled by *c* as by *a*. It therefore comes nearer to *c*, and is followed by *a*. But when *c* in this manner repels *b*, enabling it to support *a*, *b* repels *c*, pushing it nearer to *d*, and *c* cannot rest till it is as near to *d* as it is to *b*. But *d* is acted on by *c* as *c* was by *b*, and is made to approach to *e* and repel it. Without repeating the same thing of every particle, it is evident that all the particles, from *b* to M inclusive, will be brought so much nearer to one another that their mutual repulsion is precisely equal to the weight of *a*, and the last particle *z* acts on the plane M with the same force. Thus the plane M is said to carry the weight of *a*. But this is incorrect. The plane sustains a pressure equal to that weight; but it is the exerted elasticity or repulsion inherent in *z*, but dormant, till excited by the diminution of its distances.

Now let *b* become heavy. It will descend and come still nearer to *c*, till the increase of *c*'s repulsion is equal to the weight of *b*, and the whole repulsion exerted by *c* is equal to the sum of the weights of *a* and *b*. *b* is now repelling *c* with an equal force. Therefore *c* must approach *d* till their mutual repulsion is equal to the sum of the weights of *a* and *b*. It is evident that an equilibrium will not be established over the whole row of particles till all are as near to one another as *b* is to *c*. *z* will be at this distance from M, and M will sustain a pressure equal to the sum of the weights of *a* and *b*; a pressure which we may call 2.

Now let *c* also have weight. Without repeating all that has been said, it is very plain that the equilibrium will be re-established when *c* and all below it are so near one another as to exert a repulsion which we may call 3, and M sustains a pressure 3. *b* still presses with a force 2, and *a* with a force 1.

Prosecuting the examination in this way, it is obvious that the particles will be compressed so much the more as they are farther down in the row. *z* will press the plane with a force 24, *y* will press on *z* with a force 23, *x* will press *y* with a force 22, &c. &c., and *a* will still press with the force 1, the action of its elasticity, occasioned by its heaviness. We are warranted to say so, because we know that all tangible matter is elastic, and is actually compressed by its weight, and when it is so compressed, it is its springyness, and not its weight, that is the immediate agent in this pressure. We may even doubt (however difficult it may be to rid ourselves of the influence of habits) whether *a* would press at all, merely in consequence of its weight. For it is not, surely, from any previous knowledge of the nature of matter that we believe that a force of any kind can be transmitted through a piece of continuous matter, any more than through empty space. We have abundant evidence that this is the way in which the weight of bodies is felt and supported. We see that when substances which are very compressible are collected in a great heap, the lower portions are more compacted than what lies above them. Thus cotton wool or sheep's wool, when heaped very high, is much compressed at the bottom. We know common air is in a very compressed state, and presses on every square inch of the containing vessel with a force of nearly 15 pounds. We call this the pressure of the atmosphere. But it is really the elasticity of the air (occasioned indeed by its heaviness) that is pressing on the vessel. For, by shutting the cock in the upper part of the vessel, all communication with the atmosphere is cut off. Yet the air presses as much as before, and presses the lid of the vessel upward as much as it presses the bottom downwards. Therefore it is the inherent elasticity of the air that presses, and not the weight of the atmosphere. In like manner, it is the inherent elasticity or repulsive force of the particle *z* that thus presses on the plane M.

312. Having thus explained what is usually called the propagation of a dead pressure, I apprehend that it will be easy to conceive how a moving force is propagated through a coherent body. We have a similar excitement of corpuscular forces produced by the compression occasioned in one part of the solid mass, by the external force applied to that part. In short, this excitement must obtain over the whole cord, before the ball can be moved. The model (so to call it) which I described in § 310, will give the most distinct notion of this general excitement of force, whether they be pressures or tractions. Let the rod or cord be represented by a single row of little balls *a*, *b*, *c*,—*z*, (fig. 14.) connected by screw wires. At one end *z* it touches the body M which I would push away. I press on the other end *a*. This forces the ball *a* a little nearer to *b*, and *a* and *b* now repel one another, and *b* is pushed a little nearer to *c*, while I continue to press *a* forward. Thus *c* is repelled by *b*, and moves a little

nearer to *d*, and repels it. This goes on all the way to *z*, which is in contact with the body M. This will not yield till pressed with a certain force. This force can only be had in the elasticity of the spring which connects *y* with *z*. This spring must be compressed: the only means for doing this is to compress the spring which unites *x* and *y*. This must obtain the whole way between *z* and *a*. All the springs must be exerting protrusive force; therefore, all must be compressed, and in this simple case, all must be *equally* compressed. If any of them break, the body M is not moved.

In this model (let me still call it so) there is no propagation of a force originally in *a* to the other end of the series, nor any propagation backwards of the reaction of the body to be moved. All is effected by the excitement of the elasticity of the springs, and the compression caused by my pressure is the occasion for this excitement. The body M is not moved by the pressure of my hand propagated through the model, but solely by the elasticity of the spring *y z*. My hand is not resisted by the inertia of the body M propagated backwards to *a*, but solely by the elasticity of the compressed spring *a b*.

This model is a more perfect representation of nature than yet appears. Each little ball of which it consists may represent a particle of matter. We may remove M entirely. In this case, the spring *y z* will be no further compressed than suffices for exciting an elasticity able to remove the particle *z*. A very little will suffice for this. But the spring *x y* must be twice as much compressed; it must be compressed as much as *y z*, in order to move *y*, and as much more, in order to compress *y z*. In like manner, the spring *w x* must be thrice as much compressed as *y x*, and the spring *v w* must be four times as much compressed; and the spring *a b* must be twenty-four times as much. I shall feel therefore a reaction equal to what I should have felt had I applied my hand to the whole 24 balls, and had urged them into the same motion. But this is not the reaction of each ball propagated backward, but solely the great elasticity of the spring *a b* occasioned by its great compression.

313. When we urge a soft clay ball into quick motion by a stroke, or by the continued pressure of a rod, or by a spring, without striking it, we always observe that it is flattened or dimpled. Thus, if a strong spring be coiled up and held by a catch, and we lay a clay ball on it, and disengage the catch, the spring unbends and draws away the ball. If it fall into water and be picked up, we shall find it very much flattened and dimpled by the spring; so much the more as the unbending of the spring was more rapid. How was this done? The ball got no blow; and mere lying on the spring did not dimple it.

This will be still more sensibly seen, if the ball be impelled like a boy's trap-ball. The trapstick is struck at A, (Plate IV. fig.1.) and motion round the fulcrum C follows, and B is drawn off, but without any stroke; yet B is just as much dimpled as if the blow had been given it without any intervention. This impulse then is effected entirely by the same repulsive force which made B lie on the end of the trapstick without touching it. We have no evidence that it touched it when drawn off; nay though it may have been shattered to pieces.

Remark also that in the example mentioned above of the bent spring, we know that the motion was impressed on the ball by insensible degrees. We can tell the velocity acquired in every part of the motion till it quitted the spring.

314. Thus the model has given us a very clear notion of the procedure of nature; and it is a very accurate notion, and nowise misleads us. Nay, I must now say that it is not a model, but a real example, perfectly similar to the putting any piece of solid matter in motion. For all solid bodies that we are acquainted with have some elasti-

city, and mutual pressures are exerted all over them, whether by compression or dilatation. If, therefore, we limit our inquiries to the phenomena of the existing creation of God, our information is correct and adequate.

The question now is, whether our dynamical knowledge enables us to establish general principles which may suffice for explaining the phenomena, and for rendering this knowledge useful by increasing our power. I apprehend that it is, and that in this investigation we shall obtain much help from Father Boscovich's manner of considering these corpuscular forces. We shall thus discover some very general and simple laws by which this excitement of corpuscular force is regulated.

A very evident way of obtaining this knowledge will be to inquire, what analogy or relation there is between the force applied to one point of a solid body and the force that is thereby excited in another point of the body. We shall learn this again by discovering what external force will exactly balance the force thus excited. For the force excited must be equal to this external force, and must be in the opposite direction.

The elementary doctrines on this subject may be comprehended in the following propositions.

315. 1*st*. When any number of forces are in equilibrio by the intervention of a solid body, they are such as would be in equilibrio if they were all acting on a single atom.

A mass of solid matter may be the means of setting two or more forces, such as pressures, impulses, or mere solicitations, to motion, in *seeming* opposition to one another. That is, each may be prevented from producing that change of motion in the body which it would have produced had the other forces been away. This is called setting the forces in equilibrio with each other, and the forces are then said to balance each other. But it would be more proper to say that the body is in equilibrio.

This however is but a *seeming* opposition of the forces. A force cannot be in equilibrio with another unless both are applied to the same point. If another equal force be applied to the same point, in the opposite direction, all change of motion is prevented. Therefore when two forces are applied, and no change of motion ensues, we must conclude that these two forces are equal and opposite.

In applying these observations to the present question, therefore, we must say that any one of the external forces, such as A, Plate IV. fig. 2. is in equilibrio, not with another of the external forces B, but with the corpuscular forces which connect the rest of the body with the point or with the atom to which A is applied. In like manner, each of the other external forces is in immediate equilibrio, only with the force exerted by the point to which it is applied, which force is the corpuscular force connecting that particle with the rest of the body. Also we may say, in general, that the force exerted by any one of those particles is not a simple force, but the combined action of all the particles with which the particle acted on is *immediately* connected. This must be the case ninety-nine times in the hundred. The force exerted by the particle acted on by the external force, is the equivalent of all those immediately connecting forces. Now we have seen that when forces are thus excited in distant parts, the excitement must take place over all the intervening particles, perhaps differently in each. We have supposed all the forces to be in equilibrio. Therefore the equilibrium must obtain over all. For if any particle be not in equilibrio, it will not remain in its present situation, contrary to our supposition of perfect rest over all. The equilibrium therefore is general. Now consider how this equilibrium is produced. Every particle is attracting or repelling its adjoining particles, and is *equally* attracted or repelled by them. Therefore the whole corpuscular forces are made up of pairs, and in each pair the two forces are equal and opposite.

It is evident that if all these corpuscular forces were applied to one point they would be in equilibrio, because each is opposed by its equal. But it has been demonstrated that when a number of forces, acting on one point, are in equilibrio, if they be divided into parcels, the equivalents of those parcels would be in equilibrio if applied to one point. Now, in the present case, the forces exerted by the different points to which the external forces are applied, are each the equivalents of parcels of the corpuscular forces exerted all over the body. Therefore these equivalents would be in equilibrio if all were applied to one point. But each of those equivalents is equal and opposite to the external force with which it is in immediate equilibrium. Therefore these external forces are such as would be in equilibrio if all were applied to one point.

Such is the general law. But, in order that this may obtain, in order that the precise forces may be exerted which will produce this equilibrium, the body will assume certain positions or attitudes depending on the direction and intensity of the external forces. Those conditions will come into view by considering some of the most simple cases of this general law.

316. *2d.* If the body be in equilibrio between two external forces, they are equal and opposite, and the line joining the points of their application coincides with the line of direction in which the forces are exerted.

Suppose the point of a body Plate IV. fig. 3. to be urged by a force whose intensity and direction is represented by *a* A, while the point *b* is urged by a force whose intensity is represented by *b* B, equal to *a* A, and whose direction is opposite to *a* A; so that *b* B either coincides with *a* A or is parallel to it. It is plain that those two forces would be in equilibrio if applied to one point, because they are equal and opposite. But the body will not be in equilibrio, unless the line $a\,b$ between the points of application coincide with the production of A a or B b. For, when these two forces act, the point a is urged in the direction a A. But it is withheld by the corpuscular forces exerted between a and b, acting in the direction $a\,b$. Therefore the point a, being urged both in the direction a A and in the direction $a\,b$, tends to move in some intermediate direction a C. For the same reason, the point b will move in some intermediate direction b E. The body will therefore change its position. That is, it is not in equilibrio.

But if, as in fig. 2. the lines A a, $a\,b$, b B, form one straight line, it is evident that there will be no such motion of the points a and b. For a, being urged in the direction a A, and withheld in the opposite direction $a\,b$, and the force urging it in the direction $a\,b$ being equal to that urging it in the direction a A, the point a is in equilibrio. In the same manner, b is in equilibrio. The body will therefore be in perfect equilibrio between these two forces.

It is easy to see that in the first case, where the body was not in equilibrio, the motions of a and b tend to diminish the inclination of the line $a\,b$ to the direction of the forces, and would soon bring it into that direction, and then the body would be in equilibrio between them.

Thus then it is one condition for the equilibrium of two forces, that their line of direction shall coincide with the line joining the points to which they are applied. When this obtains, it is indifferent to what point of this line they are applied. The force b B may be applied at b, or e, or f, &c. This is an important circumstance.

3*d*. When a body is in equilibrio between three forces, then

317. (a) The directions of these three forces lie in one

plane. For this condition is necessary for their equilibrium if applied to one point. See DYNAMICS.

318. (*b*) The intensities of these forces are proportional to the sides of a triangle or parallelogram which have the same direction. For this also is necessary for their equilibrium in one point.

319. (*c*) If three forces are set in equilibrio by being applied to three different points of a rigid body, all their directions meet in one point, or they are all parallel; and any one of the forces is to any other of them reciprocally as the perpendiculars drawn to their directions from the point to which the remaining third external force is applied.

Let the three forces, acting in the directions AD, BE, CF, fig. 4. be applied to the points A, B, and C, of a rigid body. Let their intensities be represented by the length of these lines. Then it is to be proved that these directions either meet in one point T, or that they are parallel. Also, if CG and CH be perpendicular to AD and BE, we shall have AD : BE = CH : CG.

If the directions AD and BE are not parallel, let them meet in T. Join CT, and draw C α parallel to AD, and C β to BE.

Since we suppose these forces in equilibrio by the intervention of the rigid body, they are such as would balance if applied to one point with the same directions and intensities. Now the directions C α and C β are the same with AD and BE, and the figure C α T β is a parallelogram, and the forces AD and BE are in the same proportion as the sides C α and C β, and the third CF is as CT, and has that direction, because a force equal and opposite to CT would, if applied at C, balance the forces C α and C β. Therefore the direction of the force applied at C passes through the intersection T of the other two directions. Hence the first part of the proposition is demonstrated.

Secondly, by reason of the parallels $C\alpha$ and AT, and the parallels $C\beta$ and BT, the angles $C\beta G$ and $C\alpha H$ are equal. So are the right angles at G and H. Therefore the triangles $C\beta G$ and $C\alpha H$ are similar, and $C\alpha : C\beta =$ CH : CG. But $C\alpha : C\beta =$ AD : BE. Therefore we have AD : BE = CH : CG.

Such is the proof that may be deduced from a general proposition in dynamics. But it will greatly conduce to a better conception of the equilibrium in this fundamental case, if we trace the operation of the corpuscular connecting forces by which this equilibrium is *immediately* effected.

320. It is evident that by the joint action of the two forces acting in the directions AD and BE, and that of CF acting in the opposite direction to their equivalent, the point C is drawn away from the line AB. The body is on the stretch between C and the line AB, but between A and B there is a degree of compression. This state of things very much resembles what it would be if CA and CB were two threads, connecting C with A and with B, and having a slender rod or wire AB to keep those two points asunder. It is quite plain that if the body were pulled at C by a thread CF, and at A by a thread AT, and at B by the thread BT, those threads would have the directions of the forces. Now this could not be unless there were some solid matter between A and B, keeping the threads asunder. The threads CA, CB, are stretched, and the rod AB is compressed. We may perhaps conceive the internal mechanism still better if we suppose each of the lines CA, CB, and AB, to be spiral wires, twisted like cork-screws. When the forces pull at this elastic triangle, it is evident that the spirals CA and CB will be stretched a little beyond their natural dimensions, and are therefore exerting contractile forces. But AB is compressed into smaller bulk, and is therefore exerting expansive or protrusive forces.

Instead of thus pulling A, B, and C, in the directions AD, BE, and CF, we may suppose them pushed in the opposite directions DA, EB, and FC. In this case we should find CA and CB in a state of compression, and AB on the stretch. It may even happen that all the three lines are stretched, or all three compressed. This happens when the direction DA or EB divides the interior angle of the triangle. We always have one direction, such as CF, which does so. The other two directions AD and BE either (as in this figure) both divide the external angles CAK and CBN, or both divide the internal angles. A little attentive consideration will make this very evident, but it requires many words.

A force cannot be immediately balanced except by an equal and opposite force applied to the same point. The force CF does not immediately balance the forces AD and BE, but an equal and opposite force CR. This force CR arises from the combined action of the contractile forces of the threads CA and CB. It is their equivalent or resulting force. CR is therefore the diagonal of some parallelogram CPRQ, of which the side CP is to the side CQ as the contractile force excited along CB. In like manner, the force AD is in immediate equilibrium with the contractile force of CA, and the protrusive or expansive force of AB. The force AL must be the diagonal of some parallelogram, whose sides have the directions AC and BA, that is, of the parallelogram AKLI; and, lastly, BE is immediately balanced by its equal and opposite BO, the diagonal of the parallelogram BNOM, of which one side BM is the contractile force of CB, and the other BN is the protrusive force of AB.

Now, since every action is accompanied by an equal and contrary reaction, it follows that $AI = CP$, and $BM = CQ$, and $AK = BN$.

321. Draw CS perpendicular to AB. Then, because in any triangle, the sides are as the sines of the opposite

angles, we have AL : LI = sin. AIL : sin. LAI, = sin. CAB : sin. CAG. But, making CA the radius, it is plain that CS is the sine of the angle CAB, and CG is the sine of the angle CAG. Therefore

we have	AL : LI = CS : CG
and also	MO, or LI : BO = CH : CS
therefore	AL : BO = CH : CG
consequently	AD : BE = CH : CG

Q. E. *d.*

Thus have we demonstrated the proportion between the forces applied at A and B. We should have the same determination for the ratio AD : CF, or of BE : CF, by the same mode of procedure.

It remains to demonstrate, by tracing the operation of the excited corpuscular forces, that the three lines FC, AD, and BE, intersect in one point T. Therefore draw P a and Q b parallel to AB, meeting C α in a, and C β in b, and join a R and b R. The figure C a R b is a parallelogram. For, since CP and CQ are respectively equal to AI and BM, the triangles LKA and CP a are similar and equal. So are the triangles P a R and MOB. So are CQ b and ONB; and so are Q b R and ILA. Therefore C a and b R being, each of them, equal and parallel to LA, are equal and parallel, and therefore C b and a R are also equal and parallel, and the figure C a R b is a parallelogram.

Thus we see that CR is not only the diagonal of the parallelogram CQRP, and the equivalent of the two contractile forces CP and CQ, excited in the solid matter of the body, but that it is also the diagonal of the parallelogram C a R b, and the equivalent of the forces C a and C b, which are parallel and equal to LA and OB, or to AD and BE, the external forces. We see that although it is not in immediate equilibrium with those two forces, it would be in equilibrio with them, were they applied at C. Moreover, since we know that AD : BE = C α : C β, we have $C a : C b = C \alpha : C \beta$, and therefore the two similar parallelograms $C a R b$ and $C \alpha T \beta$ are about the same diameter; that is, the direction CR, or FC, coincides with CT, that is, it passes through the intersection of the directions AD and BE.

The force CF *remotely* balances the forces AD and BE, by exciting the contractile forces AI and BM, or CP and CQ, the equivalent of which, viz. CR, it balances immediately, while the expansive forces AK and BN balance one another.

The illustration by means of the threads CA and CB and the rod AB, by which the threads are kept apart, is quite accurate, and gives us very just and adequate notions of the unperceived corpuscular action by which the ostensible balance is effected.

It must also be noted, that the demonstration would have been precisely the same, to whatever points of the lines AT and BT the same forces had been applied in the same directions. Even at the point T or beyond it, the three forces would still have balanced each other. We shall always have the same equality and reciprocity of forces between the connected points; and this will always give the same form to the parallelogram $C \alpha T \beta$. The form of the triangle ABC is altogether indifferent.

This confirms, in an unquestionable manner, what was affirmed in dynamics, and is assumed by Sir Isaac Newton in his second Corollary from the Laws of Motion. Some mathematicians say that he has done this without sufficient authority, and that it should not have been assumed, but demonstrated. This is done here, on the authority of the third law of motion. Nothing else has been assumed.

322. It is scarcely necessary to repeat this discussion for the case where the two forces AD and BE do not intersect, being parallel. For it is evident that, in this

case, the third force CF *must* be parallel to the other two, and must be equal to this sum, or to their difference, according as the forces AD and BE act in the same or in opposite directions. For this must be the case when all act on one point. Then, as to the proportion of AD to BE, the investigation is precisely the same with the foregoing,. there will be the same excitement of contractile forces AI and BM, and of expansive forces AK and BN, and the same ultimate analogy AL : BO = CH : CG.

There is however a peculiar simplicity in this case of parallel forces, which is very convenient on numberless occasions; for which reason it is generally advisable to reduce all other cases to this, by resolving each force into others, parallel to certain lines properly chosen. The simplicity arises from this circumstance, that the lines CH and CG make one straight line, as in fig. 5. And, because CF is also parallel, it follows that the forces at A and B are inversely as the portions of any line PCQ, or ARV, or BXT, intercepted between the parallels. That is, AD : BE = CH : CG, = CQ : CP, = RV : RA, = XB : XT. In like manner AD : CF = BX : BT, = QC : QP, &c. &c.

323. Also AD : BE = BX : AR, by drawing perpendiculars, not from C on the directions AD and BE, but from A and B on the direction FC. This is peculiarly convenient on many occasions.

324. We now see that it is not enough for insuring the equilibrium of three external forces, by the intervention of a solid body, that they have such directions and proportions as would make them balance when acting on one atom. If the body be at perfect liberty, this balance will be effected, but the body will generally move and assume a particular attitude, (allow me so to call it), such that the perpendiculars drawn from one of the points of application on the directions of the forces acting on the other

two points, are reciprocally proportional to the intensities of those forces. In any other attitude of the body, the forces are *partly* employed in changing (as will appear by and by) the position or situation of the body among other bodies. This remark, and this distinction between attitude and situation or position, will be found to be important. Our chief concern at present is to obtain the proportions of the balancing forces; and even this was required, in order to learn the law which regulates the excitement of the corpuscular forces. Accordingly, we have learned the magnitude of the force excited at B fig. 4. by the application of the force AD at A. The force excited at B is evidently BO, and we discover this practically, by finding that AD is balanced by BE. We know that BO is equal, and opposite to this force BE.

325. We cannot proceed much farther without taking some notice of the *change* of attitude occasioned by the action of external forces. When things are in this state of equilibrium; suppose that some addition is made to the force AD, fig. 4. it will no longer balance AL, and the point A must be urged and must move towards T. The point B also will shift its place. The force CF must also be supposed to receive some change, that it may again balance AD and BE. But we do not at present attend so particularly to the condition of that point; and we may suppose it retained in its place by a competent force. The body must take a new attitude, as we do not suppose the directions of AB and BE to change, the angle β C α, and the opposite angle T of the parallelogram retains its magnitude. But it will be differently divided by CT. For it may be remarked that CT being taken as radius, CH and CG are the sines of the angles CTB and CTA. Therefore, as we supposed the force AD to be augmented, BE remaining the same, the ratio of CH to CG must increase. Therefore the angle

CTA must diminish and CTB must increase; the line CT shifting towards that side on which B is. Hence we see that increasing AD tends to give the body a sort of rotative motion round C in one direction; and that augmenting BE will tend to produce a rotative motion round C in the contrary direction. In short, the body must again assume such an attitude, that the perpendiculars CH and CG are again proportional to the forces applied at A and B.

326. Since we have AD : BE = CH : CG, we have AD × CG = BE × CH; that is, when the equilibrium obtains which ascertains the ratio of the excited forces, we have the rectangles or products of the forces by their perpendicular distances from C in the ratio of equality. This equation may be conceived as indicating an equality of energy or momentum for turning the body round the point C, for such rotation happens immediately, if either of the forces AD and BE be increased or diminished This way of considering the subject, is not our chief object at present, but it will be useful to keep it in mind. We may express it more briefly by making a represent the perpendicular from c on the direction of the force applied at A, and b the perpendicular on the direction of the force applied at B. This gives the equivalent equation AD × a = BE × b. Till this be accomplished, we have no method of ascertaining the force exerted in the different parts. We know however, (and it is of importance to have it familiarly in our recollection) that while this body is changing its attitude, the force exerted in every particle, is precisely competent to the motion which is there produced.

We have seen, that when a mass of rigid matter, or as it is usually denominated, a solid body, is the intermedium by which three forces are made to balance one another, their directions meet in one point. Hence it follows that these directions are in one plane, namely, the plane of the triangle ABC. Three forces may have their directions meeting in one point, and yet may be in different planes, as when they meet in the vertex D of the tetrahedron ABCD of Plate III. fig. 7. But in this case there is not an equilibrium, without the action of a fourth force, as we shall see by and by. At present we proceed to investigate the exerted forces when more than three points are considered. It is necessary to begin with forces acting all in one plane.

327. Let the forces AD, BE, CF, S *s*, Plate IV. fig. 6. be applied to the four points A, B, C, S, all in one plane, in which are also the lines of direction of the forces; and let all be in equilibrio. It is required to determine the ratio of the forces.

Any one of the forces, BE for example, may be stated as in opposition to two other forces AD and S *s*, tending to turn the body round the point C to which the remaining force CF is applied. BE may therefore be considered as the aggregate of two forces, B *e* and *e* E, of which the first balances AD, and the second balances S *s*.

The process of investigation may be the same as before. AL, equal and opposite to AD, is the force exerted at A, and it arises from a contractile force AI combined with a protrusive force AK. Making BN = AK, and drawing NO parallel to BC, we determine BO, the force exerted at B by the action of AD at A. In like manner, making S *l* equal and opposite to S *s*, we determine the exerted contractile force S *i*, and protrusive force S *k*. Make B *n* = S *k*, and *no* parallel to BC, we get the force B *o* exerted at B by the action of S *s* at S. Thence we learn that a force BE equal to the sum of BO, and B *o* will balance the forces at A and S, and thus obtain the ratio of the exerted forces. We also learn the force at C by making CP and CQ equal to AI and BM respectively, and C p and C q equal to S i and B m. Then forming the parallelograms CPRQ and C $p\,r\,q$, and drawing the diagonals CR and C r, and making them the sides of another parallelogram CRV r, we draw the diagonal CV. CV is the force exerted at C by the action of the three forces at A, B, and S, and C v, equal and opposite to CV, is the external force. Thus is the ratio of all the forces determined. All this is evident from § 320, and the composition of forces.

328. Supposing perpendiculars from C on each of the lines AD, BE, and S s, and representing them by the symbols a, b, and s, it is plain that $\mathrm{AD} \times a = \mathrm{BO} \times b$, and $\mathrm{S}\,s \times \mathrm{S} = \mathrm{B}\,o \times b$, and therefore $\overline{\mathrm{BO} + \mathrm{B}\,o} \times b$, or $\mathrm{BE} \times b = \mathrm{AD} \times a + \mathrm{S}\,s \times \mathrm{S}$. This equation may be considered as expressing the equality of the rotative momentum round C of the force BE to the combined momenta of AD and S s.

If the two directions AD and S s intersect in any point G, and if the forces are applied at that point, they will have the same rotative momentum as when applied at A and S. For the perpendiculars from C on their directions are not changed by changing the points in the lines of direction to which they are applied. Therefore, if instead of applying the forces AD and S s at G, we apply this equivalent G g, the effect should be the same also, since it is indifferent to what point in the line of its direction the force be applied, this equivalent may be applied to any point of the line G g H. Nay, what is not so obvious, if the forces AD and S s are jointly applied to any point of this line, the equilibrium with $\mathrm{BE} \times b$ will still obtain.

For let ABDR fig. 7. be a parallelogram having BR for its diagonal. From any point C draw perpendiculars C d, C a, and C r, on the sides BD, BA, and the diagonal BR; draw CL parallel to DB, CEF parallel to

DR, and CIH, EG, and DK parallel to BR, meeting the parallels CL, BD, and AR, in K, E, I, F, H. We have

$$BR \times Cr = BRHI$$
$$BA \times Ca = BAFE, = BRGE$$
$$DB \times Cd = LBDK$$

but $\quad LB, \text{or } CI : IE = BR \text{ (or } IH) :: BD$

$$CI \times BD = IH \times IE$$

and $\quad LBDK = EGHI$

therefore

$$BA \times Ca + BD \times cd = BRGE, + EGHI, = BRHI, = BR \times Cr.$$

329. The reader will doubtless perceive, that if there be any number of points beyond C to which external forces are applied, the same process may be followed for ascertaining the portion of the whole force BE which is in equilibrio with each of them. Therefore, conversely, it ascertains the force exerted in those different points by the action of an external force at B.

We also may perceive that we can always reduce the forces in equilibrio with BE to one, by continuing their directions till they intersect, and supposing their equivalent force applied at the point of intersection. Having done this, we may suppose, instead of BE, any number of forces applied to points on the same side of C, and in equilibrio with the equivalent of all the forces applied on the other side.

To do this in detail would be a tedious process. But the reader must, I apprehend, see by this time, that when the equilibrium obtains, we shall have an equation expressing it, and containing, on one side, the sum of the products of each force tending to turn the body in one direction by the perpendicular from C on the line of its direction; and on the other side an equal sum of similar products for the forces which tend to turn the body in the contrary direction.

Lastly, Since we saw that when there were only two forces AD and BE, fig. 4. the direction of the force at

C passed through the intersection T of AD and BE, or all were parallel, and that, in either case CF was equal and opposite to the equivalent of AD and BE, it follows that we can always reduce the most complicated case to that of two equal and opposite forces.

Hitherto the directions of all the forces, and also the points to which they are applied, have been supposed to lie in one and the same plane. But a body may be at once impelled in numberless directions in different planes, and the points of application may have any situation on the surface of the body. We must endeavour to simplify the consideration of this complicated case. This will be done by the help of the following proposition.

330. Let any number of external forces be applied to a solid body, in any points, and in any directions, and let all be in equilibrio. Let a plane pass through the point to which one of them is applied, and let all the forces be reduced to that plane, in the manner described under dynamics. Then I say that these reduced forces are such as would be in equilibrio were they applied to their respective points in that plane.

Let A, B, C, (fig. 8.) be three points of a body to which external forces are applied. Let the point C be supposed to lie in the plane of the paper, but the points A and B to be above that plane, in the air. Suppose the forces AD and BE to have their directions in the plane of the triangle ABC, and construct a figure expressing the excited forces, in the same way as was done in § 327. Let this figure be orthographically projected on the plane of the paper, by drawing perpendiculars A *a*, I *i*, K *k*, *L l*, &c. from every angle to the paper. We shall then have its orthographic projection. The reader's imagination may be assisted by supposing the original figure to be made of wire, and to have its point C on the paper, but all the rest above the paper, except the line CF, which goes through to the under side. Now let the paper be so held that the sun shines perpendicularly on it. The wire figure will cast a shadow on the paper, which will be its accurate orthographic projection. This is represented by fainter lines. The triangle ACB has $a\,C\,b$ for its shadow. $l\,k\,a\,i$ is the shadow of the parallelogram LKAI, $C\,p\,r\,q$ is the shadow of CPRQ, and $C\,f$ is the projection of CF, being the continuation of r C, which is the shadow of RC.

Now since this projection is made by lines perpendicular to the paper, and therefore parallel to one another, the plane $A\,a\,k\,K$ is parallel to the plane $I\,i\,l\,L$, and the plane $A\,a\,i\,I$ is parallel to the plane $K\,k\,l\,L$ (Eucl. xi. 15.) Therefore the intersections $l\,i$ and $k\,a$ with the plane of the paper are parallel (Eucl. xi. 16.). So are the intersections $a\,i$ and $k\,l$. Therefore $l\,k\,a\,i$ is a parallelogram. So, in like manner, is $m\,b\,n\,o$. Also, because AK = BN, and the lines $K\,k$, $A\,a$, $B\,b$, $N\,n$, are parallel, we have $k\,a = b\,n$. Therefore it is demonstrated, in the same way as in § 326, that if perpendiculars $C\,g$ and $C\,h$ are drawn from C on the directions $a\,l$ and $b\,o$, we shall have $a\,l \times C\,g = b\,o \times C\,h$, and the two forces $l\,a$ and $o\,b$ (which are the forces LA and OB reduced to this plane, and applied at a and at b) are in equilibrio, or have equal rotative momenta round C.

It is evident that we may proceed in the same way with any number of reduced forces that we have done with forces originally acting in the plane of the paper, by which means the proposition is demonstrated.

There is no occasion to consider the forces which are perpendicular to this plane. For the body being supposed in equilibrio, the forces are such as would be in equilibrio in any direction, if all were applied to one point. Therefore they would be in equilibrio if all were estimated in a direction perpendicular to this plane; and therefore the sum of all the forces acting in one direction would be equal to that of the forces acting in the opposite direction, so that no motion will ensue from those perpendicular forces alone.

From this proposition may be deduced a corollary expressing a remarkable property of the centre of position.

331. If there be applied to every particle of a solid body an equal and parallel force, and if a force equal to their sum be applied in the opposite direction, in a line passing through the centre of position, all will be in equilibrio. The body will neither be moved out of its place, nor will it change its attitude.

Let P (fig. 9.) be the centre of position of a solid body, and EGFH a section of the body by a plane passing through the centre of position in a direction parallel to that of the forces. Suppose a force applied to a particle A in this plane, acting in the direction A α, and let A α also represent the intensity of this force. Let the line FE, parallel to A α, pass through P. Then this line may represent the section of the plane EGFH by another plane perpendicular to it. In like manner, let the other particles B, C, D, &c. in the plane EGFH be actuated by forces B β, C κ, D δ, &c. all equal and parallel to A α. Let the same be supposed of every other particle of the body, both those that are above the plane EGFH of the figure and that below it. From every particle let perpendiculars A a, B b, C c, D d, &c. be drawn to the plane which cuts the plane EGFH in the line EF perpendicularly. Now, let all the forces A α, B β, &c. with their perpendiculars A a, B b, &c. be orthographically projected on the plane EGFH, in the manner described in the last proposition. The slightest consideration will shew us that the projection or shadow of each line representing a force must be a line equal and parallel to it, and therefore equal and parallel to A α or B β, &c. Also the perpendicular drawn from the particle to the upstanding plane is precisely equal to its projection, which is also perpendicular to EF.

Therefore the sum of all, or of any parcel of the perpendiculars to the upstanding plane, is equal to the sum of all the corresponding projections, perpendicular to EF.

Now, because all the forces are parallel, and their sum is equal to the force applied to some point in the line FE, in the direction FE, they would balance if applied all to the same point. Therefore the body will not be moved out of its place. Moreover, since, by the property of the centre of position, the sum of all the perpendiculars on one side of FE is equal to the sum of all those on the other side, and the forces are all equal, it follows that the sum of all the products of the forces by their perpendicular distances from the line FE on one side, is equal to the sum of similar products on the other side. Their rotative momenta are therefore equal, or they balance each other, and the body has no tendency to change its attitude. All motion is therefore stopped, whether progressive or rotative.

Note, that the complete equilibrium requires that the line of direction of the single opposing force pass through the very centre of position, because it is then only that we shall have the equality of the sums of the perpendiculars on the two sides of *every* plane that can be drawn through that line of direction. Should the line of direction be fe, fig. 9. although still in a plane passing through P, the rotative momentum on the side eHf will exceed that on the side eGf, and the body will turn in the direction eHf.

It is an obvious inference from this proposition, that if a heavy body be supported by a single point in the vertical line passing through its centre of position, then, 1*mo*, The force which thus supports the body acts directly upwards, and is equal to the aggregate or sum of the weight of all the particles; and 2*do*, The body will have no tendency to lean towards any side.

For the body may be conceived as having every particle actuated by an equal force, acting in parallel lines. Therefore, by this proposition, all are balanced, any plane being made to pass through the vertical line of support, the rotative momenta on the opposite sides of this plane are equal, by this proposition, and therefore the body will incline to neither side, but will remain in the same attitude, just as if the weight of every particle were united in the centre of position.

For this reason the point, determined in the way already described, and which we have called the centre of position, has been called the CENTRE OF GRAVITY, meaning by the term the point in which the whole gravity of the body is usually supposed to be concentrated. This however is an inaccurate conception. The weight of all the particles is not applied to the centre of gravity, nor does it act there by itself. But such must be the distance of the particles of the whole system, that a certain force may be mutually exerted by the support and the physically touching particle of the system, equal to the sum of the parallel and equal actions of gravity on the particles. It is not the gravity of the body which acts on the support, but a corpuscular repelling force. And it is really the centre of gravity, if we suppose the gravity of all the particles perfectly equal, and all acting in parallel lines. But it would not be the centre of the effort of gravity of a body placed very near to the centre of the earth, because, in that situation, the gravities of the different particles will neither be equal nor parallel. The name is, however, proper enough in respect of any body of moderate bulk on the surface of this globe.

332. Such are the more simple and elementary laws by which the excitement of corpuscular force is regulated. They may also be called the laws of the propagation of pressure. This is a very usual but inaccurate manner of expressing the phenomenon. It is very true that when the conditions here enumerated are observed, a pressure

being exerted on one point of a solid body, it is instantly accompanied by a determinate pressure, exerted by some other part of the body, balancing some other external pressure. But we have full evidence, by what was said in § 320, that the pressure that is observed in the remote part is the excitement of a corpuscular force belonging to that part, by means of an unknown number of corpuscular forces which connect the intervening particles of the body. When these intermediate forces are wanting in any part, the remote pressure does not appear. We shall find, once we begin the consideration of the forces actually existing, that these laws are as accurate as any theorem in geometry, confirming, in the most unexceptionable manner, the justness of the three propositions which we called the laws of motion, and the propriety of the mark and measure which we adopted of every change.

It is now time to consider the motions producible in a mass of solid matter by the action of an external force. We shall find this to be a subject considerably different from the motions in physical astronomy. In the celestial motions, the force which produces the change of motion is conceived as acting alike, or nearly alike, on every particle of the body, and acting without any intermedium. Every particle was conceived as influenced by a force corresponding with its situation, without any dependence on the state of the other particles. Jupiter and Mercury, if in the same place, would be equally moved, although Jupiter contains 5000 times as much matter as Mercury. A certain variation of velocity is acquired in that place, by every body; and the magnitude or quantity of this variation, can be varied only by a change of situation. But here, the external force is commonly applied to some part of the body, and the parts at a distance are moved, not by the external force, but by the corpuscular forces excited between the particles. This excitement requires the expenditure of the force employed, in order to generate any particular velocity in the whole mass, and the expenditure must be so much the greater, as the number of particles excited to motion is greater. The quantity of motion produced depends therefore on the force employed, and the velocity generated depends also on the mass of the body moved.

Sir Isaac Newton saw the necessity of a distinction between the *vis motrix*, or moving force, and the *vis acceleratrix*, or accelerating force. The first is measured by the quantity of motion produced or changed by it, and the last by the velocity generated.

The first observation of moment on this way of producing motion is, that if an external force act on a solid body, in a line passing through its centre of position or gravity, it will produce a motion of the whole body, in its own direction, every particle advancing alike, and therefore without any rotation or change of attitude.

333. Let p express the magnitude of the moving force, and m the number of particles in the body to be moved by it. It is plain, by the last proposition, that if a force $= \frac{p}{m}$, or the mth. part of p, be applied to each particle, in the opposite direction, all will be in equilibrio. Now suppose the force p to be annihilated in an instant. It is plain that the body will immediately begin to move in the direction opposite to that of the force p, and that every particle will move alike, being each impelled by an equal force $\frac{p}{m}$, just as a heavy body falls without changing its attitude, all parts being equally acted on. By the continued action of the force $\frac{p}{m}$ on every particle during some given time, a second for example, every particle acquires the velocity v in the same direction. The whole quantity of motion, therefore, will be $m\ v$, the exact

measure of the force p which held the body in equilibrio.

334. Suppose, on the other hand, that all the forces $\frac{p}{m}$ are annihilated in an instant, while p is pressing the body in its proper direction. Every particle is pressed in this direction with the force $\frac{p}{m}$; for all was in equilibrio when the external forces $\frac{p}{m}$ were applied to each particle. Each was therefore opposed to an equal and opposite force $\frac{p}{m}$. These forces, excited by the force p, must now act alone, and will produce the same motion, but in the opposite direction, that was produced by the external forces $\frac{p}{m}$, when the force p was removed. That is, by the action of those equal *excited* forces, every particle will begin to move in the direction of the force p, and will acquire the velocity v, if p continue to act on that body during a second, and the whole quantity of motion will be $m\,v$, the proper measure of the force p. The forces actually exerted in the different particles are not, strictly speaking, all equal; for, on the contrary, the force exerted in the particle to which p is immediately applied, is equal to the whole force p. But by far the greatest part of the force exerted in it is employed in pressing the particles with which it is immediately coherent, sufficiently near to the particles beyond them for exerting the necessary forces in them, to be expended in like manner in pressing those still more remote; and there remains only a portion $\frac{p}{m}$, sufficient for giving this particle the velocity v. This will appear very distinctly, by considering backwards from the most remote series of particles. These receive only as much pressure

as suffices for giving them this velocity, and they react as much on the penult series. The particles of this series receive, each of them, a pressure sufficient for balancing this reaction, and a pressure sufficient for giving them the velocity v. Thus each series in succession sustains more pressure as it is nearer to the particle acted on by the force p. But in every particle there is an overplus of force sufficient for producing the velocity v. There is no mystery or intricacy in this. For it is this overplus alone that was the balance for the forces $\frac{p}{m}$ that we at first supposed to be applied to each particle, and to be all balanced by the force p. There was, in that case, the same serieses of forces, greater and greater as we approach the point to which p is applied; but these are all employed in connecting the particles, and keeping them in that state of strain that is necessary for the excitement of that force in each particle which *immediately* balances the external force $\frac{p}{m}$ applied to each. All these connecting forces vanish, or cease to act, the instant that the force p is annihilated, and the body moves with the velocity v. For, in that instant, the strain being removed, the particles all resume their natural distances, in which they are inactive, and the body continues to move with the velocity already acquired. This indeed is not strictly true, because when the particles spring back to their natural distances, they commonly overpass them a little, as a pendulum let go from an oblique position does not stop at its natural position in the vertical, but passes to the other side, and oscillates some time, till it is brought to rest by various obstructions. The particles of the impelled body do the same thing, and vibrate a little on each side of their natural places, till this ceases, by the imperfection of their elasticity. That this is really the case, may be easily perceived by striking away a small bell or other elastic body. It may be heard ringing all the while it flies through the air. It may sometimes be more distinctly perceived in a child's plaything which imitates a mouse jumping out of a box. If this be so placed on the floor as that the mouse springs out along the floor, the spiral wire that is attached to it will sometimes be seen to expand and contract several times as it runs along the floor.

The other circumstance to be observed in this motion is, that it is merely progressive, without any rotation. This must be the case, since every particle has the same velocity v, no one gains upon another, and therefore, whatever position or bearing they have in the beginning, they retain it during the whole motion. The same progressive motion, free from all rotation, will be produced if the equivalent of any number of forces be in a line which passes through the centre of position. For a force equal and opposite to this equivalent, and applied in a line also passing through the centre, will balance those forces. Let the intensity of this equivalent be p, and let m be the number of particles in the body. Then if a force $\frac{p}{m}$ be applied to each particle, they will balance this equivalent, and be balanced by it. The effect, therefore, of the primitive external forces being equal, and opposite to the force p, which balances their equivalent, and this being equivalent with a force equal and opposite to $\frac{p}{m}$ applied to each particle, the effect of the primitive external forces is to impress this force $\frac{p}{m}$ on each particle, and therefore it will cause the whole body to advance alike, without rotation. It is very obvious that this permanent attitude is peculiar to this case. For if the same force, acting in the same direction, be applied in a line which passes on one side of the centre of position of the body, it is evident that that side will advance, in the very beginning of the motion, faster than the other. The body will therefore both advance and turn round. It is also obvious that this will be the most general result. For an action in a line passing through the centre is but one of millions of cases. Accordingly, it is very rarely that progressive motion is not combined with a motion of rotation. In the great movements of Nature, the celestial motions, we see those two kinds of motion always conjoined. All the planets on which observations can be made, turn round their axis while they revolve round the sun or other central body. A stone thrown from the hand, a bomb or cannon ball from the muzzle of the piece, are always observed to whirl as they fly. It requires particular contrivances to prevent this when inconvenient. On the other hand, this motion of rotation being subject to fixed principles, we can modify it so as to suit any purpose we may have in view. This will come under consideration by and by; and in the mean time we proceed with the remaining circumstances of progressive motion, as connected with the force employed.

335. The quantity of motion produced by the action of a force was taken for the measure of that moving force, upon every ground of good argument. (See DYNAMICS.) If any force be necessary for giving a certain motion to a cubic foot of matter, as much is surely necessary for giving the same motion to another cubic foot. Therefore it is most reasonable to call that a double force which gives the same motion to a double quantity of matter. The observations which have been made on the way in which force applied to one part of a firm body produces motion in the whole, perfectly agrees with this suggestion of common sense, and even confirms it, by shewing how those sensible pressures, which we call moving forces, arise sometimes from the combined action of accelerating forces, such as gravity and magnetism, whose action on

each atom is insensible. The sensible pressure of a mass of lead, used as the moving force acting on a machine, is usually considered as the accumulated gravitation of every particle. But we have seen that, although it be *equal* to the sum of those gravitations, it is really the corpuscular repulsion, or something similar, of the touching parts of the machine and weight, exerted, or brought into action by the compression occasioned by the heaviness of each particle. And we have seen how this mutual action comes to be as precisely measured by the number of particles of another body which it puts into motion, and the velocity generated in each, as the action of gravity on the weight itself is so measured. In other cases of moving forces, such as that of a spring, it cannot be measured in any other way. The sensible pressure of a spring is not the accumulated elastic force exerted by every particle of the spring. These elastic forces are only exerted between the particles of the spring; and they are probably not much inferior to the sensible pressure of the spring on the body moved by it. The sum total of these is probably immense; but it is unknown to us. This moving force cannot be measured in any other way than by the motion produced by it. But although the force of a spring must not be considered as the sum of the elasticities of the particles, it is still their combined action. The spring must unbend through some sensible space, that it may follow the yielding body, continually accelerating it, till the velocity becomes sensible. This could not be effected by the elasticity of the touching parts alone; their utmost motion is not the millionth of the millionth part of an inch, and would be instantly at an end by the yielding of the body. But each particle moving a little, the end of the spring is urged over a sensible space, and produces its effect. We have a clear evidence of this internal motion of compression. If we strike off any sonorous body, such as a metal plate, or a bell, by a smart blow, we shall hear it ringing all the while it flies through the air. Its particles were compressed in succession, till the whole was put into motion, and quitted the body which impelled it, and then, springing back to their natural distances from each other, they overpassed them, as a pendulum let go from an oblique position passes the vertical line which is its natural quiescent position. Just so the particles vibrate for a while on each side of their natural positions, till this motion is extinguished by their imperfect elasticity.

336. Thus it appears that the force of a spring, by which it can put other bodies in sensible motion, is not the sum, nor equal to the sum of the elastic forces exerted by each of its particles, although the joint action of them all is necessary for the ostensible effect. It is only in such cases as gravity, magnetism, and others which we call attractions or repulsions that the ostensible effect is at least equal to the sum of the forces acting on each particle. Both kinds of moving forces are measurable by the motions which they produce. And it may be observed here, that the perfect agreement between the measure of gravitation by considering the weight or pressure of a heavy body as the sum, or at least as equal to it, of the gravitation of each particle, and by the motion which this weight will produce in another body, is a proof of the propriety of the Cartesian measure of the force of moving bodies. Were we to suppose this force proportional to the square of the velocity produced, we should be led into numberless paralogisms, as any one will perceive by the example already given of the body dragged along a horizontal plane by another hanging by a cord. If we measure the force by its action in equal times, and this action by the quantity of motion generated, we shall find it always the same, as we have every reason to think that it is. The weight of the hanging body is the sole moving power, and it is always the same.

When it is employed to give motion to twice, or thrice, or four times as much matter, including its own, it generates $\frac{1}{2}$, $\frac{1}{3}$, or $\frac{1}{4}$ of the velocity, and always produces the same quantity of motion. But if we measure the force by the square of the velocity, the action diminishes in the same proportion that we increase the quantity of matter.

The values of the generated velocities may also be considered as the accelerative powers of the forces, in the circumstances of their employment. The powers themselves are generally known to us as pressures, or as solicitations to motion, and must be previously known, otherwise we cannot use them as terms of any proportion. They must be estimated by some measure of the same kind. We can compare the moving force of a bow with the action of gravity, by observing what weight will draw the bow to a given height, and compare this weight with the weight of the arrow. In the same way we can compare the attraction of a magnet with the force of a stream of wind, by balancing both by weights, and comparing the weights. But we cannot compare the impulsion of a moving body with any thing but another impulsion. By attempts to compare impulsion with pressures of any kind, mechanicians have been led into many puzzling difficulties. The comparison mentioned above, of the attraction of a magnet with the force of wind, may be thought contrary to this assertion. But we shall see that what we experience as the force of a stream of wind or water, is not an impulse but a pressure, accurately comparable with other pressures.

337. It only remains farther to be noticed, that in the preceding observations on the action of forces, we always mean to speak of the absolute pressure, energy, or strain really acting on the body. It is frequently necessary to distinguish this from the whole force residing in, or somehow attending the substance employed as the vehicle of

the power. Thus, in the example already alluded to on several occasions, the whole gravitation of the hanging body seems to be employed in dragging the equal body along the table. But the thread is only stretched by half of its weight. For it still descends with half of its natural velocity, and it only generates this half velocity in the other body. If there be any difficulty in conceiving this, it will become clear by comparing it with a case precisely similar, but which we conceive differently. A magnet attracts a pennyweight of iron, and in a minute generates in it a certain velocity. Add another pennyweight of iron, and the piece of twice the weight and bulk will acquire the same velocity, just as a piece of two pounds weight falls as fast as a piece of one pound. But join to the pennyweight of iron a pennyweight of brass, and the magnet will generate but half of the velocity. Here the same tendency of the iron toward the magnet has to drag along with it a mass which is without any tendency toward the magnet, just as the heavy body has to drag along the table a body which has its tendency downward taken away by the table.

We must, in every employment of the natural moving forces, deduct from the whole force competent to the subject employed, the part which is expended in making that subject follow up the yielding body so as to continue its action on it. This deduction is frequently of difficult investigation, particularly when we employ the strength of animals. The proportion of their force which is thus expended, varies with the velocity with which they must move, in order to continue their action. A certain velocity disables the animal from accelerating the body any more, because it requires all its strength to continue this velocity of its own body. This subject will be particularly considered, when we are occupied with the performance of machines.

338. We will now inquire into the modifications of the

motion produced by the action of an external force, when the direction of the force does not pass the centre of gravity. Since, in this case, an equal accelerating force is not exerted in every particle; they will not move alike, some will move faster than others. As the body is supposed to retain its continuity and its force, this unequal motion of its different parts must constitute a sort of rotation by which the body will change its attitude. What in the beginning was the uppermost, or the foremost, or the nethermost, will not continue so. It remains therefore to be considered, whether this motion of rotation be regulated by steady principles, and what is the relation subsisting between it and the progressive motions which have hitherto engaged all our attention, of what modifications it is susceptible, and how we may be able to direct or modify it.

This is not a matter of mere curiosity. The noblest mechanical art practised by man, depends entirely on those modifications; I mean the art of working a ship. A ship must be considered as a body at perfect liberty to move in every direction, and is susceptible of an infinite variety of motions besides the motion of progression, by which she advances in her course. She heels to leeward by pressure of the wind on her sails. She rolls and pitches by reason of the continual shifting of her support by the waves. She is made to yaw from her course by various buffetings which the different parts of the hull receive from the seas. She receives various motions, round various axes, according as the sails are spread or trimmed, and according as her bows strike the water during her progressive motion. All these tendencies to conversion, are not only useless but hurtful, and must be opposed by other forces, exerted by the rudder, in which office a part of its directing power is expended, so that the remainder is frequently almost unable to produce the conversion absolutely necessary for the safety of the ship. Thus, in a hard gale of wind, when it is necessary to sail on the other tack, it is very often impossible to turn the ship round any other way than by going to leeward, which is, in all cases, a great loss; and in working off a lee shore, highly dangerous. In cases of this kind, it requires great sagacity and judgment what movements to make, and how to make them, so as not to be baffled in the attempt. The success must evidently depend on the knowledge of the effect which different positions of the sails and rudder will have on the ship, and also of the various beatings of the waves upon her hull, in the different positions which she must assume. It is therefore unquestionable that a scientific knowledge of the movements of conversion, and the circumstances of force and direction which modify those movements, must tend greatly to enable the seaman to profit by his experience, by bringing every thing under general rules and principles of easy recollection. Such an intelligent seaman will thus learn how to trim his sails in such a manner that all the tendencies to conversion shall balance each other, and then the action of the rudder is kept ready for operating in a moment that any occasional conversion may be necessary. The difference in the steerage of a ship, according as the sails are trimmed, is most remarkable; when this is injudiciously done, the helm is never a moment still, but continually shifting from side to side, to check the deviations which the ship makes without ceasing from her course. On the other hand, when an intelligent seaman takes charge of the ship, and has set and trimmed her sails to his mind, the helm may be observed almost without motion, the helmsman has scarcely any thing to do, and a child may steer the ship.

Were there no other reasons, these considerations should recommend this branch of mechanical philosophy to the study of a Briton. But it is very interesting to the philosopher on another account. This earth is pendulous in free space, and at liberty to turn and move in every direction. The oblate figure of our globe, and the unequal gravitation of its different parts to the sun and the moon, occasions an external force, whose direction does not pass through the centre of gravity. By the action of this force, which is continually changing both its direction and its intensity, the axis of diurnal revolution incessantly changes its position, sometimes pointing to one star and sometimes to another. This motion is indeed very minute, but as it is partly continued according to a certain rule, it occasions a continual accumulation of one effect, viz. a change in the intersection of the equatorial circle with the plane of the ecliptic. This has made a great and conspicuous change on the whole appearances of the starry heavens. Just now, the whole heavens seem to turn round the star in the extremity of the tail of the smaller Bear, *ursa minor*, whereas in the days of Hipparchus they turned round a star near the nose of the Camelopard, and all the stars have changed their paths of diurnal motion, in consequence of this precession of the equinoxes.

We must therefore bestow some attention on this subject; but in this course of elementary instruction it is plain that nothing can with propriety be noticed but what is very simple and elementary. The subject, in general, is one of the most delicate and difficult in the whole of mechanical philosophy, and it is only the first mathematicians of Europe that have been able to treat it with any success. It is enough to say that Newton made mistakes of primary importance in his manner of treating it.

339. It is first incumbent on us to learn what change is made on the progressive motion by this difference in the action of the moving force; and here it is somewhat remarkable, that notwithstanding that the progressive motion of the body is accompanied by a motion of rotation,

in some cases extremely rapid, the progressive motion is the same as when the force acts in a line passing through the centre of gravity and produces no rotation. It was demonstrated in dynamics, that if any body D, of a collection or system A, B, D, be moved uniformly in a straight line, the centre of gravity moves uniformly in the same direction, and that its velocity is to that of the body D as the matter in that body is to the whole matter of the system. Therefore, if this motion has been the effect of any force, the motion of the centre is affected in the same manner as if this force had acted in the same direction on all the matter of the system collected in that centre. And, since this is equally true with respect to each body of the system, it follows that if each body be moving in any manner, by the action of different forces, the motion of the centre is the same as if all those forces were applied to the whole matter of the system collected there. Hence it followed, that if equal and opposite forces act on any two bodies, or on every pair of bodies of the system, the motion of the centre is not affected by those equal and opposite forces, because the equal and opposite motions resulting from this action compensate one another. Therefore, if a system of bodies are moving in such a manner that they cannot continue their motions without disturbing one another, either by collision, or because they are joined to one another by rods or strings, this disturbance consists in mutual action, each of which is accompanied by an equal reaction in the opposite direction. Such a system must therefore be considered as acted on by equal and opposite forces; and the motion of the centre is the same as if they did not act on one another at all. If the centre was at rest while their motions were not disturbed by one another, it will continue at rest notwithstanding the disturbance or total change of the individual motions, and if it was in motion, it will continue moving in the same direction, and with the same velocity as before.

340. Now any solid or cohering body is such a system, consisting of parts, which are so connected by material interpositions, that one part cannot move in any direction whatever, without either dragging the adjoining points along with it, or pushing them before it, or turning them aside, in one way or another; and these actions of the parts or particles of the body are all mutual, equal, and opposite. In whatever manner the particle a affects any remote particle p, the particle p affects a in a way precisely equal and opposite. Therefore, whatever motions may be occasioned to the different parts of this body by their mutual connection, the motion of the body, produced by the action of any external force or forces, is not affected by them, it being estimated and measured by the motion of the centre. We have seen that when a force p acts on a body containing the quantity of matter, or number of particles, in a direction passing through the centre of the body, it causes the whole body to advance with the velocity v, generating the quantity of motion $m v = p$. Every particle is affected by the accelerating force $\frac{p}{m}$, and advances with the velocity v.

Now let the same force $p = m v$ act on the same body, but in a direction which does not pass through the centre. The body will both advance and turn round; some parts will be moving faster than others, and in other directions, and some may even be moving backwards. But the motion of the centre of gravity will be the same as in the last ease, it will move in the same direction, and its velocity will be v as before.

It is plain that the centre of gravity will still describe a straight line, but that the other parts of the body must be moving in curve lines, of various curvature, according to their position respecting the centre, just as we observe in a coach wheel rolling along a smooth road. The centre of the wheel describes a line parallel to the road; but the other parts are sometimes above the centre and sometimes below it, sometimes before it and sometimes behind.

We must now endeavour to ascertain this motion of rotation, shewing, 1*st*, The position of the axis round which this rotation is performed; and 2*dly*, The velocity of this rotation, or the relation subsisting between it and the motion of progression.

For this purpose, we shall begin with a case of such simplicity that we shall be assisted in our first notions of the subject by the ordinary suggestions of common sense.

341. Let A and B (Plate IV. fig. 10.) be two equal balls, connected by an inflexible line, of which G is the middle point, and consequently the centre of gravity of the system. Let AB be so divided in P that AP : PB = 2 : 1. Now suppose that a force acts on this system at the point G, at right angles to AB, and let its intensity be such that in a given moment of time it would cause the system to describe the small space GI, perpendicular to AB. From what has been already said, it follows that the two balls will sustain equal impressions, and will therefore move equally fast; and that at the end of the movement, the system will have the position *ab* parallel to AB.

But let us now suppose that the same force has acted on the system in the point P, and still in the same direction, perpendicular to AB. We have just now learned that the motion of the centre G will be the same as in the former case. It will still be at I at the end of the moment, and will continue to move uniformly forward in that direction GI*g*. But, in order to ascertain the place of the two balls at the end of the moment, we must know their motions. And we must learn this by means of our knowledge of the forces or pressures by which they are urged forward. We learn this by inquiring what forces, applied to A and B in the opposite direction, will prevent them from acquiring this motion, that is, will just balance the force applied at P. Now we have seen that these forces must be in the inverse proportion of the distances of P from the balls. Therefore the force or pressure which urges B into motion is double of that which acts on A, and therefore will produce a double motion. The incipient motion of B will be double of that of A, and, at the end of the moment, the balls will be in such positions that Bβ is double of Aα. Therefore the line $\alpha\beta$ is no longer parallel to AB, but cuts it in C, so that BC is double of AC. But the centre G having moved over the same space GI as in the former case, it is plain that the line $\beta\alpha$ must cut the position $b\,a$ of the former case in the point I.

Thus we see that the system AB has changed its attitude as well as its situation, and that this change of attitude is the beginning of a rotation round the point G, which point alone moves uniformly in a straight line in the direction of the moving force. This rotation commences round an axis which passes through the centre of gravity G, perpendicular to the plane of the figure. For the motions of the two balls being parallel to the direction FP of the moving force, and to the line GI described by the centre, are necessarily in the same plane with those lines and the line AB. Since they are motions of rotation round an axis, each ball describing the circumference of a circle round it, this axis must be perpendicular to the planes of those circles.

342. Let us now consider the momentary rotation, and its relation to the motion of progression. Suppose the line BC to be connected with the balls, and to be turning along with them. Since the motion of the balls has been such that the line BC is not, at the end of the moment, in the position b I c, parallel to BC, but is inclined backwards so as to cut its former position in the point C, it appears that the point c has been moving backwards while G was advancing along GI. If about the centre I, with

the radius I *c*, we describe a small arch backwards, it will sensibly coincide with the line *c* C which is parallel to IG. The situation of the extremity of the revolving line, instead of being in *c* will be (as to all sense) in C, so that the line itself, instead of seeming to have turned round G, seems rather to have turned round C. For the upper end of it is still in the same place. This is actually observed in many instances. If a flat ruler, such as a gunter's scale, lying on a table, be struck laterally with a small blow, at a point about $\frac{2}{3}$ds or $\frac{3}{4}$ths of its length, it will turn in this very way, one end remaining still. But, to return to the two balls, we see in this momentary motion the beginning of a rotation, such that while G advances a small space GI, C describes (backwards) an arch of a circle equal to GI. Suppose this arch to be one degree. Then it will go completely round while G moves over a line 360 times greater than GI. We may therefore state the relation of the rotative and progressive motions by saying, that the system makes one revolution while the centre describes a line equal to the circumference of a circle whose radius is GC.

Having examplified this combination of motions by this very simple and palpable case, we shall now consider the subject more generally. But, that we may avoid some part of that complication of action which renders the consideration of the subject very intricate, we must still limit our attention to certain forms. We have seen that by applying the force at P below the centre of gravity, the system of two balls gets an incipient rotation round the point C above or beyond the centre. In like manner, had it been applied to the right hand of the centre, it would have produced a rotation round some point on the left hand of it. We therefore confine our attention at present to such shapes of body as have a middle plane, dividing the body through the centre, and having the matter similarly disposed on both sides. Such are all the solids formed by the revolution of a plane figure round an axis. Such are all the regular solids, all prismatic bodies, and a great variety of others. We shall farther suppose that the impelling force acts in this dividing plane. Thus we avoid any chance of rotation round an axis on either side of this plane. Even with this limitation, we shall have the same internal mechanism as in the cases that are excepted, and shall see that those exceptions make no difference in the general results.

Since the effect of a force applied to a solid body in a line passing through its centre of gravity, is to impress on every particle an equal accelerating force in its own direction, and since it was also demonstrated that such equal and parallel forces did not affect the relative motions and actions of bodies, it follows that this force, so applied, will make no change in any motion of rotation that the body may already have.

Now, let a solid body, of the class which has been selected for this introduction, be acted on by a force in the direction FP (Plate IV. fig. 4.) in the plane which passes through the centre of gravity G, and divides the body in the manner already mentioned, but so that the direction FP does not pass through the centre.

Let p express the magnitude of the force, or the quantity of motion which it would generate by acting uniformly for some given moment of time, and let m express the number of equal particles, or the quantity of matter in the body. This force will communicate to the centre of gravity G of this body, the velocity $\frac{p}{m}$, causing it to describe some line GI parallel to FP. Suppose that, at the same instant, an equal force is applied to the centre, in the opposite direction. It is evident that all progressive motion of the centre is now prevented. But it is as evident that there must be a rotation; for the two forces, though equal, are not directly opposed to each other.

Now let a third force, also equal to p, and acting in the same or a parallel direction, act on the centre of the body. This will cause the body to move with the velocity $\frac{p}{m}$ and will make no change in the rotation already produced. The body must (as I have said) move with the velocity $\frac{p}{m}$, because now the body is in the same state as if neither force had been applied at the centre, and as when the first force p alone was applied at F. But we made use of those forces in order to shew the certainty of the rotation, without making it different from what the force p alone will produce.

Since the centre must advance in a straight line, it does not partake of the rotation. This must therefore be performed round the centre, or round an axis passing through the centre. In order to find the position of this axis, we must recollect that the rotation is the same as when p was applied at F, and an opposing force was applied at the centre. In this case, the motion of the point F results from the action of the force in the direction FP, and that of the other force at G, propagated in the direction GF. The incipient motion of F must therefore be in the plane of these two forces, that is, the plane of NFG, that is, in a plane passing through the centre of gravity, in the direction of the moving force, and the axis round which the body turns is perpendicular to this plane.

We shall see, in due time, why this is limited to the incipient motion, and what circumstances make any change in it afterwards.

343. The next object is to ascertain the relation between the progressive and rotative motions. This is most conveniently determined by the space described by the centre during one revolution of the system. It may be enounced in the following manner.

Let GP (Plate IV. fig. 11.) be drawn from the centre of gravity, perpendicular to FP, the direction of the moving force. Let m express any particle or portion of the whole matter m of the body, and let r express its radius vector, or distance from the axis of rotation passing through the centre of gravity. The body will make one revolution while its centre G describes a line equal to the circumference of a circle whose radius is CG, $= \frac{\int m\, r^2}{m\,\mathrm{GP}}$.

For it is very obvious that whén a body both advances and turns round an axis, a point may be taken in a line connected with that axis and turning with it, at such a distance from the axis, that its velocity of rotation shall be equal to the progressive velocity of the centre. Therefore if, in the line PG produced, we take the point C so situated, it follows that the motion of that point, by its rotation round the axis, is equal, and, in this instant, is opposite to the motion of the centre, and therefore compensates it, so that C is at rest in absolute space. With the centre G, and radius GC, describe the circle CVB. Draw the radius vector GA, producing it till it meet the circle in V. Let GI be the velocity or uniform motion of the centre in some small moment of time. The motion of the point V is compounded of the rotation V z in the tangent, equal to GI and the progressive motion V v, also equal and parallel to GI. It therefore moves in absolute space with the velocity and in the direction V y, the diagonal of the rhombus V $v\,y\,z$. Draw z G, and from A draw A d parallel to V z, meeting z G in d, and A b equal and parallel to GI, and produce it till it meet CP in E. Complete the parallelogram A $b\,e\,d$, and draw the diagonal A e, producing it till it meet CP in H. Join CV and CA. It is plain that A e is the absolute motion of the particle A, being compounded of the progressive motion A b equal to GI, and A d the rotative motion of A round the centre of gravity.

The line $V v$ being parallel to GI, is perpendicular to CG, and $v y$, being parallel to $V z$, is perpendicular to VG. Therefore the angles $V v y$ and CGV are equal, and the isosceles triangles $V v y$ and CVG are similar. Therefore the angle CVG is equal to the angle $v V y$, or its equal $z V y$. Add the angle $GV y$ to both, and the angle $CV y$ is equal to the right angle $GV z$, and CV is perpendicular to $V y$. Again, $A b$ is perpendicular to CG, and $b e$ is perpendicular to AG. Therefore the angle CGV is equal to the angle $A b e$. Also $A b : b e = A b : A d, = V z : A d, = VG : AG, = CG : AG$. Therefore the triangles $b e A$ and GAC are similar, and the angle CAG is equal to the angle $A e b$ or $d A e$. Take away the common angle $GA e$, and the remainder $CA e$ is equal to the remainder $d AG$, and is therefore a right angle, and $A e$ is at right angles to AC.

Since what has now been demonstrated in relation to the particle A is equally true of another particle, it follows that the incipient absolute motions of every particle, resulting from the composition of these progressive and rotative motions, are found to be at right angles with the lines joining the particles with the point C, or with an axis passing through C, perpendicular to the plane NFG. And since the point or axis C is, in that instant, without motion, and yet the body is turning round while moving forward, the particles are turning round the momentary axis in C. This is a curious example of the composition of motions, and we shall see presently that it is of very frequent occurrence, and that a clear conception of it will greatly assist us in the discussion of many difficult and important questions.

It is obvious that when the circle CVB turns along with the body, it rolls along the line CC′ parallel to GI, as a coach wheel rolls along a level road, and that point of the circumference which is in contact with the line CC′ is, in that instant, without motion. This point of contact is therefore the momentary centre of motion for every particle of matter in the body. They are all beginning, as it were, to describe concentric arches having this point, or the axis passing through this point, for the centre. We are here speaking of their real absolute motions in free space. For, as the body is not supposed to change its form, the particles are also turning round the centre of gravity G, while it is carried forward in a straight line. When we consider the motions more particularly, we find that the point C describes a succession of cycloids, having the line CC′ for their base, and that all the other points of the circumference CVB describe cycloids, the particles within and without this circumference describing epicycloids, contracted or elongated, according to their situation. The centre of gravity G alone describes a straight line. In every position of the body, every particle is moving at right angles to the line joining it with that point of the circumference CVB which is in contact with the line CC′, and the velocity of every particle is as its distance from that point. For as Vy and Ae are described in the same time, and the angular motion of all the body is the same, the angles VCy and ACe are equal, and therefore the triangles are similar, and $Vy : Ae = VC : AC$, that is, the absolute velocity of the particles are as their distances from C.

The facts have now been stated and described, and we must, in the next place, shew how all these motions are produced by the excentric impulsion of the moving force, and by that means we may determine the situation of the axis of conversion thus assumed by the body. It is plain, in the first place, that the moving force P is equal to the sum of all the forces which, when applied at F, in the direction FP, produce all these motions. There is no other source from which they can be derived. We have considered P as a force which generates the quantity of motion $m v$. Therefore v is the velocity of the centre, by which it describes GI in some moment of time. A portion of this force, which we may distinguish as elemental by the symbol $m \dot{v}$, is employed in producing in the particle A the motion A e, which it is observed to take, with a velocity which is to the velocity in C or in N, that is, to the velocity v, as CA to CG. We have therefore $CG : CA = v : \frac{v.\,CA}{CG}$ and the velocity of A is expressed by $\frac{v.\,CA}{CG}$. The same symbol will conveniently enough express the quantity of motion in A, if all the particles be supposed equal, and be expressed by unity.

The method which we adopted for discovering the force excited in any remote point, by the action of a known force at another point, was by finding the proportion of two forces which would balance the body between them, when applied to those points, in the same directions; and they were ascertained to be inversely as the perpendiculars on their directions, drawn from the point to which the third force requisite for completing the balance was applied. We make use of C as the third point on this occasion, because it is really a fixed point, remaining at rest, and contributing to the excitation of the force at A, which gives it the motion A e. C is at rest, in consequence of the mutual compensation and balancing of all the forces which are acting on it in the instant of our examination. We are not, at present, interested in knowing the magnitude of any of those forces. It is enough that C is supported. Therefore to find the force $m v$ at F, which produces the velocity $\frac{v.\,CA}{CG}$ in A, institute the following analogy, $CP : CA = \frac{v.\,CA}{CG} : m \dot{v}$. This gives $m \dot{v} = \frac{v.\,CA^2}{CP.\,CG.}$: therefore taking in the motions of all the particles, we have $m\ v$, or $m\ v$, $= \frac{v. \int CA^2}{CP. CG.}$ Now (Elem. II. 12 & 13.) we have $CA^2 = GA^2 + GC^2 \pm 2\,GC.\,EG$, according as CE is greater or less than CG.

Therefore $\int CA^2 = \int GA^2 + \int CG^2 \pm 2\,CG.\int EG$.

But $\int CG^2 = m.\,CG^2$, because CG is an invariable line.

And $\pm 2\,CG.\int EG$ is $= v$, because the sum of all the $+$ EG is equal to the sum of all the $-$ EG, by the nature of the centre of gravity.

Therefore we have $m\ v = \frac{v. \int \overline{GA^2 + m.\,CG^2}}{CP.\ CG.}$.

Therefore $m\ v.$ CP. CG $= v. \int GA^2 + m\ v.\ CG^2$, or

$$m.\ CP.\ CG = \int GA^2 + m.\ CG^2, \text{ and}$$

$$\int GA^2 = m.\ CP.\ CG - m.\ CG^2, = m.\ GP.\ CG.$$

Therefore, finally, $CG = \frac{\int GA^2}{m.\ GP}, = \frac{\int m\ v^2}{m.\ GP}$.

Thus have we obtained the radius of the circle, whose circumference, unfolded, is the space described by the centre while the body makes one revolution.

There is another way in which we may proceed in the same investigation, which has the advantage of pointing out something of the internal procedure in producing those motions. We may examine how it happens that all the forces which act on C come to compensate each other. We may ascertain CG by means of this circumstance, that they are all balanced in that point.

A force NF is applied at F, in the direction FP, and we wish to know what effect is produced on C, in consequence of the particle A being dragged into the motion A e by the material connection between the three points

F, A, and C. Now the forces which are mutually exerted at these three points are such as would balance, if applied there in the same directions. And the three directions are either parallel, or they meet in one point. In the present case, NF and A e, which are two of the forces, meet in O. Therefore draw OC r. Then C r is the direction of the excitement in C by such an action at F as produces the motion A e in A. Draw PK and PL perpendicular to AO and CO. Also make the parallelogram A g e f, expressing by A f and A g the strains on the connections with F and C. Make F h = A f, and complete the parallelogram h F i N. Make C q = A g, and complete the parallelogram q C p r. Make k O = A e, and complete the parallelogram k m n O. Lastly, draw k t perpendicular to PO. We shall find from the process that F i = C p, and C r = O n, and NF = m O.

It is plain, from the investigations in a preceding article, that k O, O n (or k n) and O m are as the balancing forces at A, C, and F. Also k O was taken equal to A e, the force at A.

$$\begin{array}{lllll}
\text{Now} & k\,\mathrm{O} : k\,m = \sin.\ \mathrm{COP} & : \sin.\ k\,\mathrm{OP}, & = \mathrm{PL} : \mathrm{PK} \\
& k\,m : t\,m = \mathrm{CO} & : \mathrm{PO}, & = \mathrm{CP} : \mathrm{PL} \\
\text{therefore} & k\,\mathrm{O} : t\,m = & & \mathrm{CP} : \mathrm{PK}.
\end{array}$$

Now $t\,m$ expresses the force C r, reduced to the direction parallel to PO, that is, to the direction of the motion of C in its rotation round G. When all these forces $t\,m$ balance, C is at rest. Therefore, putting for k O the value which we found for A e, we have $\mathrm{CP} : \mathrm{PK} = \frac{v\,.\,\mathrm{CA}}{\mathrm{CG}}$:

$$\frac{v\,.\,\mathrm{CA}\,.\,\mathrm{PK}}{\mathrm{CP}\,.\,\mathrm{CG}}, = (\text{because } \mathrm{CA} : \mathrm{CE} = \mathrm{HP} : \mathrm{PK})\ \frac{v\ \mathrm{CE}\,.\,\mathrm{HP}}{\mathrm{CP}\,.\,\mathrm{CG}},$$

$$= \frac{v\,.\,\overline{\mathrm{CE}\,.\,\mathrm{CP} - \mathrm{CE}\,.\,\mathrm{CH}}}{\mathrm{CP}\,.\,\mathrm{CG}}, = \frac{v\,.\,\overline{\mathrm{CE}\,.\,\mathrm{CP} - \mathrm{CA}^2}}{\mathrm{CP}\,.\,\mathrm{CG}}, =$$

the force at C, arising from the reaction of A. Therefore the whole forces at C, arising from the reaction of all the particles, is $= \frac{v \,.\, CP \,.\int CE - v\int CA^2}{CP \,.\, CG}$. But the sum of all the lines such as CE is $m \,.\, CG$, by the nature of the centre of gravity, therefore the whole forces acting at C are $= \frac{v \,.\, CP \,.\, m \,.\, CG - v \int CA^2}{CP \,.\, CG}$, or $\frac{v \,.\, CP \,.\, m \,.\, CG}{CP \,.\, CG} - \frac{v\int CA^2}{CP.CG}$, that is, $m\,v - \frac{v\int CA^2}{CP.CG}$. Now, we have already seen that $\frac{v\int CA^2}{CP.CG}$ is equal to $m\,v$. Therefore the forces acting at C are $m\,v - m\,v$; that is, there is a complete compensation of opposing forces, in the very instant of impulsion, and C remains at rest.

On the other hand, inferring from the fact that C remains at rest, this mutual compensation of forces, we make $\frac{v \,.\, CP \,.\, m \,.\, CG - v\int CA^2}{CP \,.\, CG} = v$, and consequently $mv \,.\, CP \,.\, CG = \int v \,.\, CA^2$, which is $= . v\int GA^2 + mv\, CG^2$. Therefore, $m \,.\, CP \,.\, CG - m \,.\, CG^2 = \int GA^2$, that is, $m \,.\, GP \,.\, GC = \int GA^2$, and $CG = \frac{\int GA^2}{m \,.\, GP}$ as before.

The manner in which this theorem is usually demonstrated by the mechanicians, is more familiar, but it is not so immediately deduced from the actual state of things, viz. a body in free space, and unobstructed by forces of any kind. They begin as we did, by supposing the body impelled at F, in the direction FP, and resisted by an equal and opposite force at the centre of gravity. The rotation is then strictly performed round that centre, yet does not differ from what it would be without this opposing force. This supposition makes A move in the direction and with the velocity A d, while G has the velocity

GI. We have $GA : GV = A\,d : V\,z\ (= GI, = v)$, and then $GV : GA = v : \frac{v \cdot GA}{GV}, = \frac{v \cdot GA}{CG}$, from which we obtain, as by the other methods, $CG = \frac{\int GA^2}{m \cdot GP}$.

But this method is not so unexceptionable as those which we have followed, the state of things being so very different, nor does it so immediately suggest the composition of the two motions.

The intelligent reader has perhaps remarked, that the investigation of this theorem now before him proceeds on the supposition that all the particles of the body are in the plane of the figure, whereas some are above and some below or on the other side of that plane, and it may be doubted whether this simplification of the question be admissible. But a little reflection will show, that our conclusions are legitimate, at least with respect to the limited class of forms that we are considering. We may suppose all such bodies resolved into little prismatic elements, all of them perpendicular to the plane of the figure, and therefore parallel to the axis of rotation in C, or in G, which is also perpendicular to that plane. The whole of one of those prisms being equidistant from the axis, every atom of it has the same angular motion, and requires the same expenditure $m\,v$ of the force to excite it into motion as if it were in the plane of the figure. It is attached to this plane by its middle, or its centre of gravity, and therefore both halves of it will be equally urged forward. We shall see by and by what other circumstances must be attended to, when these elementary prisms are not bisected by the plane passing through the centre in the direction of the impelling force. In the mean time, we learn all the chief properties of this motion without the embarrassment arising from a farther complication. We may now consider a number of consequences of this fundamental proposition.

Cor. 1. The angular velocity of this motion, measured as usual by the number of revolutions made in a given time, or by the velocity of some point whose distance from the axis is unity, is proportional to the impelling force. For the velocity acquired by the centre of gravity being the same as if the body had no rotation, and this being proportional to the force, the time of describing the line equal to the circumference whose radius is CG is diminished in the same proportion, or the number of such circumferences described in some given time is augmented in the same proportion; therefore the number of revolutions made in a given time is increased in that proportion.

Cor. 2. The angular velocity is also proportional to the distance GP from the axis of gravity at which the body is impressed by the moving force. For CG, and consequently the circumference to be described during one revolution, is inversely as GP. Because, when the body is given, the quantity $\int GA^2$ is invariable, in whatever point P the power shall act. Since $m \cdot GC \cdot GP$ is equal to this constant quantity $\int GA^2$, GC must be inversely as GP. Therefore the space described in the time of one revolution, and consequently the time of describing it, is diminished in the same proportion.

Cor. 3. If GC be taken $=\frac{\int GA^2}{m \cdot GP}$, in the line perpendicular to FP, the point C will remain at rest in the first moment of the motion. For this reason C, determined in this manner, is called the *spontaneous centre of conversion.*

Cor. 4. The distance CP, between the point of impulsion and the spontaneous axis of conversion is equal to $\frac{\int PA^2}{m \cdot GP}$, or to $\frac{\int CA^2}{m \cdot CG}$. For it is easy to see, as before, that $\int PA^2 = \int GA^2 + m \cdot GP^2$, and therefore, $\frac{\int PA^2}{m \cdot GP}$

$= \frac{\int, GA^2}{m . GP} + \frac{m . GP^2}{m . GP}, = CG + GP, = CP$. This may be shown above

$$CA^2 = PA^2 + CP \pm 2 CP . PE$$

$\int CA^2 = \int PA^2 + m . CP . CP - CP . 2 GP$. And $m . CG . CP = \int CA^2$.

therefore $m . CP . CG = \int PA^2 + m . CP . CP - m . CP . 2GP$

and $\int PA^2 = m . CP . CG + m . CP . 2GP - m . CP . CP$

or $\int PA^2 = m . CP (CG + 2GP - CP) = m . CP . GP$

therefore $\frac{\int PA^2}{m . GP} = CP$.

It is also plain that $\frac{\int PA^2}{m . GP} = \frac{\int CA^2}{m . CG}$.

Cor. 5. We also have $GP = \frac{\int GA^2}{m . GC}$, so that if C be made the point of impulsion, P becomes the spontaneous centre of conversion corresponding to it.*

---

* When the author had arrived at this part of the article ROTATION, he was unable to proceed, in consequence of a severe illness, from which he never recovered. The reader is therefore referred to the article ROTATION in the Encyclopædia Britannica, for any farther information on this subject that he might have expected under the present article.—ED.

STRENGTH OF MATERIALS.

344. STRENGTH of Materials, in *Mechanics*, is a subject of so much importance, that in a nation so eminent as this for invention and ingenuity in all species of manufactures, and in particular so distinguished for its improvements in machinery of every kind, it is somewhat singular that no writer has treated it in the detail which its importance and difficulty demands. The man of science who visits our great manufactories is delighted with the ingenuity which he observes in every part, the innumerable inventions which come even from individual artisans, and the determined purpose of improvement and refinement which he sees in every workshop. Every cotton-mill appears an academy of mechanical science; and mechanical invention is spreading from these fountains over the whole kingdom: but the philosopher is mortified to see this ardent spirit so cramped by ignorance of principle, and many of these original and brilliant thoughts obscured and clogged with needless and even hurtful additions, and a complication of machinery which checks improvement even by its appearance of ingenuity. There is nothing in which this want of scientific education, this ignorance of principle, is so frequently observed as in the injudicious proportion of the parts of machines and other mechanical structures; proportions and forms of parts in which the strength and position are nowise regulated by the strains to which they are

exposed, and where repeated failures have been the only lessons.

It cannot be otherwise. We have no means of instruction, except two very short and abstracted treatises of the late Mr Emerson on the strength of materials.* We do not recollect a performance in our language from which our artists can get information. Treatises written expressly on different branches of the mechanical arts are totally silent on this, which is the basis and *only principle* of their performances. Who would imagine that PRICE'S BRITISH CARPENTER, the work of the first reputation in this country, and of which the sole aim is to teach the carpenter to erect solid and durable structures, does not contain one proposition or one reason by which one form of a thing can be shown to be stronger or weaker than another? We doubt very much if one carpenter in an hundred can give a reason to convince his own mind that a joist is stronger when laid on its edge than when laid on its broad side. We speak in this strong manner in hopes of exciting some man of science to publish a system of instruction on this subject. The limits of our work will not admit of a detail: but we think it necessary to point out the leading principles, and to give the traces of that systematic connexion by which all the knowledge already possessed of this subject may be brought together and properly arranged. This we shall now attempt in as brief a manner as we are able.

The strength of materials arises immediately or ultimately from the cohesion of the parts of bodies. Our examina-

---

* Since this article was published, several sets of experiments of very considerable importance have been made on the strength of materials. The reader is particularly referred to the article CARPENTRY, in the EDINBURGH ENCYCLOPÆDIA, vol. V. p. 494, Mr Barlow's ingenious *Essay on the Strength of Timber*, and Mr Tredgold's *Elementary Principles of Carpentry* (Lond. 1810),—a work of great practical utility.—ED.

tion of this property of tangible matter has as yet been very partial and imperfect, and by no means enables us to apply mathematical calculations with precision and success. The various modifications of cohesion, in its different appearances of perfect softness, plasticity, ductility, elasticity, hardness, have a mighty influence on the strength of bodies, but are hardly susceptible of measurement. Their texture also, whether uniform like glass and ductile metals, crystallized or granulated like other metals and freestone, or fibrous like timber, is a circumstance no less important; yet even here, although we derive some advantage from remarking to which of these forms of aggregation a substance belongs, the aid is but small. All we can do in this want of general principles is to make experiments on every class of bodies. Accordingly philosophers have endeavoured to instruct the public in this particular. The Royal Society of London at its very first institution made many experiments at their meetings, as may be seen in the first registers of the Society. Several individuals have added their experiments. The most numerous collection in detail is by Muschenbroek, professor of natural philosophy at Leyden. Part of it was published by himself in his *Essais de Physique*, in two vols 4to; but the full collection is to be found in his System of Natural Philosophy, published after his death by Lulofs, in three vols 4to. This was translated from the Low Dutch into French by Sigaud de la Fond, and published at Paris in 1760, and is a prodigious collection of physical knowledge of all kinds, and may almost suffice for a library of natural philosophy. But this collection of experiments on the cohesion of bodiës is not of that value which one expects. We presume that they were carefully made and faithfully narrated; but they were made on such small specimens, that the unavoidable natural inequalities of growth or texture produced irregularities in the results which bore too great a proportion to the whole quantities observed. We may make the same remark on the experiments of Couplet, Pitot, De la Hire, Du Hamel, and others of the French academy. In short, if we except the experiments of Buffon on the strength of timber, made at the public expense on a large scale, there is nothing to be met with from which we can obtain absolute measures which may be employed with confidence; and there is nothing in the English language, except a simple list by Emerson, which is merely a set of affirmations, without any narration of circumstances, to enable us to judge of the validity of his conclusions: but the character of Mr Emerson, as a man of knowledge and integrity, gives even to these assertions a considerable value.

But to make use of any experiments, there must be employed some general principle by which we can generalize their results. They will otherwise be only narrations of detached facts. We must have some notion of that intermedium, by the intervention of which an external force applied to one part of a lever, joist, or pillar, occasions a strain on a distant part. This can be nothing but the cohesion between the parts. It is this connecting force which is brought into action, or, as we more shortly express it, excited. This action is modified in every part by the laws of mechanics. It is this action which is what we call the *strength* of that part, and its effect is the strain on the adjoining parts; and thus it is the same force, differently viewed, that constitutes both the strain and the strength. When we consider it in the light of a resistance to fracture, we call it *strength*.

We call every thing a *force* which we observe to be ever accompanied by a change of motion; or, more strictly speaking, we infer the presence and agency of a force wherever we observe the state of things in respect of motion different from what we know to be the result of the action of all the forces which we know to act on the body. Thus when we observe a rope prevent a body from falling, we infer a moving force inherent in the rope with as much confidence as when we observe it drag the body along the ground. The *immediate action* of this force is undoubtedly exerted between the immediately adjoining parts of the rope. The immediate effect is the keeping the particles of the rope together. They ought to separate by any external force drawing the ends of the rope contrariwise; and we ascribe their not doing so to a mechanical force really opposing this external force. When desired to give it a name, we name it from what we conceive to be its effect, and therefore its characteristic, and we call it COHESION. This is merely a name for the fact; but it is the same thing in all our denominations. We know nothing of the causes but in the effects; and our name for the cause is in fact the name of the effect, which is COHESION. We mean nothing else by gravitation or magnetism. What do we mean when we say that Newton understood thoroughly the nature of gravitation, of the force of gravitation? or that Franklin understood the nature of the electric force? Nothing but this: Newton considered with patient sagacity the general facts of gravitation, and has described and classed them with the utmost precision. In like manner, we shall understand the nature of cohesion when we have discovered with equal generality the laws of cohesion, or general facts which are observed in the appearances, and when we have described and classed them with equal accuracy.

Let us therefore attend to the more simple and obvious phenomena of cohesion, and mark with care every circumstance of resemblance by which they may be classed. Let us receive these as the laws of cohesion, characteristic of its supposed cause, the force of cohesion. We cannot pretend to enter on this vast research. The modifications are innumerable: and it would require the penetration of more than Newton to detect the circumstance of similarity amidst millions of discriminating circumstances. Yet this is the only way of discovering which are the primary facts characteristic of the force, and which are the modifications. The study is immense, but by no means desperate; and we entertain great hopes that it will ere long be successfully prosecuted: but, in our particular predicament, we must content ourselves with selecting such general laws as seem to give us the most immediate information of the circumstances that must be attended to by the mechanician in his constructions, that he may unite strength with simplicity, economy, and energy.

145. 1. Then, it is a matter of fact, that all bodies are in a certain degree perfectly elastic; that is, when their form or bulk is changed by certain moderate compressions or distractions, it requires the continuance of the changing force to continue the body in this new state; and when the force is removed, the body recovers its original form. We limit the assertion to *certain moderate* changes: for instance, take a lead wire of one-fifteenth of an inch in diameter, and ten feet long; fix one end firmly to the ceiling, and let the wire hang perpendicular; affix to the lower end an index like the hand of a watch; on some stand immediately below let there be a circle divided into degrees, with its centre corresponding to the lower point of the wire: now turn this index twice round, and thus twist the wire. When the index is let go, it will turn backwards again, by the wire's untwisting itself, and make almost four revolutions before it stops; after which it twists and untwists many times, the index going backwards and forwards round the circle, diminishing however its arch of twist each time, till at last it settles precisely in its original position. This may be repeated for ever. Now, in this motion every part of the wire partakes equally of the twist. The particles are stretched, require force to keep them in their state of extension, and recover completely their relative positions. These are all the characters of what the mechanician calls *perfect* elasticity. This is a quality quite familiar in many

cases; as in glass, tempered steel, &c. but was thought incompetent to lead, which is generally considered as having little or no elasticity. But we make the assertion in the most general terms, with the limitation to moderate derangement of form. We have made the same experiment on a thread of pipe-clay, made by forcing soft clay through the small hole of a syringe by means of a screw; and we found it more elastic than the lead wire: for a thread of one-twentieth of an inch diameter and seven feet long allowed the index to make two turns, and yet completely recovered its first position.

346. 2. But if we turn the index of the lead wire four times round, and let it go again, it untwists again in the same manner, but it makes little more than four turns back again; and after many oscillations, it finally stops in a position almost two revolutions removed from its original position. It has now acquired a new arrangement of parts, and this new arrangement is permanent like the former; and, what is of particular moment, it is perfectly elastic. This change is familiarly known by the denomination of a SET. The wire is said to have TAKEN A SET. When we attend minutely to the procedure of nature in this phenomenon, we find that the particles have as it were slid on each other, still cohering, and have taken a new position, in which their connecting forces are in equilibrio: and in this change of relative situation, it appears that the connecting forces which maintained the particles in their first situation were not in equilibrio in some position intermediate between that of the first and that of the last form. The force required for changing this first form augmented with the change, but only to a certain degree; and during this process the connecting forces always tended to the recovery of this first form. But after the change of mutual position has passed a certain magnitude, the union has been partly destroyed, and the particles have been brought into new situations; such, that the forces which now connect

each with its neighbour tend, not to the recovery of the first arrangement, but to push them farther from it, into a new situation, to which they now verge, and require force to prevent them from acquiring. The wire is now in fact again perfectly elastic; that is, the forces which now connect the particles with their neighbours augment to a certain degree as the derangement from this new position augments. This is not reasoning from any theory. It is narrating facts, on which a theory is to be founded. What we have been just now saying is evidently a description of that sensible form of tangible matter which we call *ductility*. It has every gradation of variety, from the softness of butter to the firmness of gold. All these bodies have some elasticity; but we say they are not perfectly elastic, because they do not completely recover their original form when it has been greatly damaged. The whole gradation may be most distinctly observed in a piece of glass or hard sealing-wax. In the ordinary form glass is perhaps the most completely elastic body that we know, and may be bent till just ready to snap, and yet completely recovers its first form, and takes no set whatever; but when heated to such a degree as just to be visible in the dark, it loses its brittleness, and becomes so tough that it cannot be broken by any blow; but it is no longer elastic, takes any set, and keeps it. When more heated, it becomes as plastic as clay; but in this state is remarkably distinguished from clay by a quality which we call VISCIDITY, which is something like elasticity, of which clay and other bodies purely plastic exhibit no appearance. This is the joint operation of strong adhesion and softness. When a rod of perfectly soft glass is suddenly stretched a little, it does not at once take the shape which it acquires after some little time. It is owing to this, that in taking the impression of a seal, if we take off the seal while the wax is yet very hot, the sharpness of the impression is destroyed immediately Each part drawing its neighbour, and each part yielding, the prominent parts are pulled down and blunted, and the sharp hollows are pulled upwards and also blunted. The seal must be kept on till all has become not only stiff but hard.

This viscidity is to be observed in all plastic bodies which are homogeneous. It is not observed in clay, because it is not homogeneous, but consists of hard particles of argillaceous earth sticking together by their attraction for water. Something like it might be made of finely powdered glass and a clammy fluid such as turpentine. Viscidity has all degrees of softness till it degenerates to ropy fluidity like that of olive oil. Perhaps something of it may be found even in the most perfect fluid that we are acquainted with, as we observed in the experiments for ascertaining specific gravity.

There is in a late volume of the Philosophical Transactions a narrative of experiments, by which it appears that the thread of the spider is an exception to our first general law, and that it is perfectly ductile. It is there asserted, that a long thread of gossamer, furnished with an index, takes any position whatever; and that though the index be turned round any number of times (even many hundreds), it has no tendency to recover its first form. The thread takes completely any set whatever. We have not had an opportunity of repeating this experiment, but we have distinctly observed a phenomenon totally inconsistent with it. If a fibre of gossamer about an inch long be held by the end horizontally, it bends downward in a curve, like a slender slip of whalebone or a hair. If totally devoid of elasticity, and perfectly indifferent to any set, it would hang down perpendicularly without any curvature.

When ductility and elasticity are combined in different proportions, an immense variety of sensible modes of aggregation may be produced Some degree of both are probably to be observed in all bodies of complex constitution; that is, which consist of particles made up of many different kinds of atoms. Such a constitution of a body must afford many situations permanent, but easily deranged.

In all these changes of disposition which take place among the particles of a ductile body, the particles are at such a distance that they still cohere. The body may be stretched a little; and on removing the extending force, the body shrinks into its first form. It also resists moderate compressions; and when the compressing force is removed, the body swells out again. Now the corpuscular *fact* here is, that the particles are acted on by attractions and repulsions, which balance each other when no external force is acting on the body, and which augment as the particles are made, by any external cause, to recede from this situation of mutual inactivity; for since force is requisite to produce either the dilatation or the compression, and to maintain it, we are obliged, by the constitution of our minds, to infer that it is opposed by a force accompanying or inherent in every particle of dilatable or compressible matter; and as this necessity of employing force to produce a change indicates the agency of these corpuscular forces, and marks their kind, according as the tendencies of the particles appear to be toward each other in dilatation, or from each other in compression; so it also measures the degrees of their intensity. Should it require three times the force to produce a double compression, we must reckon the mutual repulsions triple when the compression is doubled; and so in other instances. We see from all this that the phenomena of cohesion indicate some relation between the centres of the particles. To discover this relation is the great problem in corpuscular mechanism, as it was in the Newtonian investigation of the force of gravitation. Could we discover this law of action between the corpuscles with the same certainty and distinctness, we might with equal confidence say what will be the result of any position which we give to the particles of bodies; but this is beyond our hopes. The law of gravitation is so simple, that the discovery or detection of it amid the variety of celestial phenomena required but one step; and in its own nature its possible combinations still do not greatly exceed the powers of human research. One is almost disposed to say that the Supreme Being has exhibited it to our reasoning powers as sufficient to employ with success our utmost efforts, but not so abstruse as to discourage us from the noble attempt. It seems to be otherwise with respect to cohesion. Mathematics informs us, that if it deviates sensibly from the law of gravitation, the simplest combinations will make the joint action of several particles an almost impenetrable mystery. We must therefore content ourselves, for a long time to come, with a careful observation of the simplest cases that we can propose, and with the discovery of secondary laws of action, in which many particles combine their influence. In pursuance of this plan, we observe,

347. 3. That whatever is the situation of the particles of a body with respect to each other, when in a quiescent state, they are kept in these situations by the balance of opposite forces. This cannot be refused, nor can we form to ourselves any other notion of the state of the particles of a body. Whether we suppose the ultimate particles to be of certain magnitudes and shapes, touching each other in single points of cohesion; or whether we (with Boscovich) consider them as at a distance from each other, and acting on each other by attractions and repulsions—we must acknowledge, in the first place, that the centres of the particles (by whose mutual distances we must estimate the distance of the particles) may and do vary their distances from each other. What else can we say when we observe a body increase in length, in breadth, and in thickness, by heating it, or when we see it diminish in all these dimensions by an external compression? A particle, therefore, situated in the midst of many others, and remaining in that situation, must be conceived as maintained in it by

the mutual balancing of all the forces which connect it with its neighbours. It is like a ball kept in its place by the opposite action of two springs. This illustration merits a more particular application. Suppose a number of balls ranged on the table in the angles of equilateral triangles, and that each ball is connected with the six which lie around it by means of an elastic wire curled like a cork-screw; suppose such another stratum of balls above this, and parallel to it, and so placed that each ball of the upper stratum is perpendicularly over the centre of the equilateral triangle below, and let these be connected with the balls of the under stratum by similar spiral wires. Let there be a third and a fourth, and any number of such strata, all connected in the same manner. It is plain that this may extend to any size and fill any space.—Now let this assemblage of balls be firmly contemplated by the imagination, and be supposed to shrink continually in all its dimensions, till the balls, and their distances from each other, and the connecting wires, all vanish from the sight as discrete individual objects. All this is very conceivable. It will now appear like a solid body, having length, breadth, and thickness; it may be compressed, and will again resume its dimensions; it may be stretched, and will again shrink; it will move away when struck; in short, it will not differ in its sensible appearance from a solid elastic body. Now when this body is in a state of compression, for instance, it is evident that any one of the balls is at rest, in consequence of the mutual balancing of the actions of all the spiral wires which connect it with those around it. It will greatly conduce to the full understanding of all that follows to recur to this illustration. The analogy or resemblance between the effects of this constitution of things and the effects of the corpuscular forces is very great; and wherever it obtains, we may safely draw conclusions from what we know would be the condition of a body of common tangible matter. We shall just give one instructive example, and then have done with this hypothetical body. We can suppose it of a long shape, resting on one point; we can suppose two weights A, B, suspended at the extremities, and the whole in equilibrio. We commonly express this state of things by saying that A and B are in equilibrio. This is very inaccurate. A is in fact in equilibrio with the united action of all the springs which connect the ball to which it is applied with the adjoining balls. These springs are brought into action, and each is in equilibrio with the joint action of all the rest. Thus through the whole extent of the hypothetical body, the springs are brought into action in a way and in a degree which mathematics can easily investigate. We need not do this: it is enough for our purpose that our imagination readily discovers that some springs are stretched, others are compressed, and that a pressure is excited on the middle point of support, and the support exerts a reaction which precisely balances it; and the other weight is, in like manner, in immediate equilibrio with the equivalent of the actions of all the springs which connect the last ball with its neighbours. Now take the analogical or resembling case, an oblong piece of solid matter, resting on a fulcrum, and loaded with two weights in equilibrio. For the actions of the connecting springs substitute the corpuscular forces, and the result will resemble that of the hypothesis.

Newton had said, that, as the great movements of the solar system were regulated by forces operating at a distance, and varying with the distance, so he strongly suspected (*valde suspicor*) that all the phenomena of cohesion, with all its modifications in the different sensible forms of aggregation, and in the phenomena of chemistry and physiology, resulted from the similar agency of forces varying with the distance of the particles. The learned Boscovich, in his celebrated Theory of Natural Philosophy, pursued this thought; and has shown, that if we suppose an ultimate atom of matter endowed with powers of attraction and repulsion, varying, both in kind and degree, with the distance, and if this force be the same in every atom, it may be regulated by such a relation to the distance from the neighbouring atom, that a collection of such may have all the sensible appearance of bodies in their different forms of solids, liquids, and vapours, elastic or unelastic, and endowed with all the properties which we perceive, by whose immediate operation the phenomena of motion by impulse, and all the phenomena of chemistry, and of animal and vegetable economy, may be produced. He shows, that notwithstanding a perfect sameness, and even a great simplicity in this atomical constitution, there will result from this union all that unspeakable variety of form and property which diversify and embellish the face of nature. Having already given an account of this celebrated work, we mention it only, by the bye, as far as a general notion of it will be of some service on the present occasion. For this purpose, we just observe that Boscovich conceives a particle of any individual species of matter to consist of an unknown number of particles of simpler constitution; each of which particles, in their turn, is compounded of particles still more simply constituted, and so on through an unknown number of orders, till we arrive at the simplest possible constitution of a particle of tangible matter, susceptible of length, breadth, and thickness, and necessarily consisting of four atoms of matter. And he shows that the more complex we suppose the constitution of a particle, the more must the sensible qualities of the aggregate resemble the observed qualities of tangible bodies. In particular, he shows how a particle may be so constituted, that although it act on one other particle of the same kind through a considerable interval, the interposition of a third particle of the same kind may render it totally, or almost totally, inactive; and therefore an assemblage of such particles would form such a fluid as air. All these curious inferences are made with incontrovertible evidence; and the greatest encouragement is thus given to the mathematical philosopher to hope, that, by cautious and patient proceeding in this way, we may gradually approach to a knowledge of the laws of cohesion, that will not shun a comparison even with the *Principia* of Newton. No step can be made in this investigation, but by observing with care, and generalizing with judgment, the phenomena, which are abundantly numerous, and much more at our command than those of the great and sensible motions of bodies. Following this plan, we observe,

348. 4. It is matter of fact, that every body has some degree of compressibility and dilatability; and when the changes of dimension are so moderate that the body completely recovers its original dimensions on the cessation of the changing force, the extensions or compressions are sensibly proportional to the extending or compressing forces; and therefore *the connecting forces are proportional to the distances of the particles from their quiescent, neutral, or inactive positions.* This seems to have been first viewed as a law of nature by the penetrating eye of Dr Robert Hooke, one of the most eminent philosophers of the last century. He published a cipher, which he said contained the theory of springiness, and of the motions of bodies by the action of springs. It was this, *c e i i i n o s s s t t u u.*—When explained in his dissertation, published some years after, it was *ut tensio sic vis.* This is precisely the proposition just now asserted as a general fact, a law of nature. This dissertation is full of curious observations of facts in support of his assertion. In his application to the motion of bodies, he gives his noble discovery of the balance-spring of a watch, which is founded on this law. The spring, as it is more and more coiled up, or unwound, by the motion of the balance, acts on it with a force proportional to the distance of the balance from its quiescent position. The balance therefore is acted on by an accelerating force, which varies in the same manner as the force of gravity acting on a pendulum swinging in a cycloid. Its vibrations

therefore must be performed in equal time, whether they are wide or narrow. In the same dissertation, Hooke mentions all the facts which John Bernoulli afterwards adduced in support of Leibnitz's whimsical doctrine of the force of bodies in motion, or the doctrine of the *vires vivæ;* a doctrine which Hooke might justly have claimed as his own, had he not seen its futility.

Experiments made since the time of Hooke show that this law is strictly true in the extent to which we have limited it, viz. in all the changes of form which will be completely undone by the elasticity of the body. It is nearly true to a much greater extent. James Bernoulli, in his dissertation on the elastic curve, relates some experiments of his own, which seem to deviate considerably from it; but on close examination they do not. The finest experiments are those of Coulomb, published in some late volumes of the memoirs of the Academy of Paris. He suspended balls by wires, and observed their motions of oscillation, which he found accurately corresponding with this law.

349. 5. It is universally observed, that when the dilatations have proceeded a certain length, a less addition of force is sufficient to increase the dilatation in the same degree. This is always observed when the body has been so far stretched that it takes a set, and does not completely recover its form. The like may be generally observed in compressions. Most persons will recollect, that in violently stretching an elastic cord, it becomes suddenly weaker, or more easily stretched. But these phenomena do not positively prove a diminution of the corpuscular force acting on one particle: it more probably arises from the disunion of some particles, whose action contributed to the whole or sensible effect. And in compressions we may suppose something of the same kind; for when we compress a body in one direction, it commonly bulges out in another; and in cases of very violent action some particles

may be disunited, whose transverse action had formerly balanced *part* of the compressing force. For the reader will see on reflection, that since the compression in one direction causes the body to bulge out in the transverse direction, and since this bulging out is in opposition to the transverse forces of attraction, it must employ some part of the compressing force. And the common appearances are in perfect uniformity with this conception of things. When we press a bit of dryish clay, it swells out and cracks transversely. When a pillar of wood is overloaded, it swells out, and small crevices appear in the direction of the fibres. After this it will not bear half of the load. This the carpenters call CRIPPLING; and a knowledge of the circumstances which modify it is of great importance, and enables us to understand some very paradoxical appearances, as will be shown by and bye.

This partial disuniting of particles formerly cohering is, we imagine, the chief reason why the totality of the forces which really oppose an external strain does not increase in the proportion of the extensions and compressions. But sufficient evidence will also be given that the forces which would connect one particle with one other particle do not augment in the accurate proportion of the change of distance; that in extensions they increase more slowly, and in compressions more rapidly.

But there is another cause of this deviation perhaps equally effectual with the former. Most bodies manifest some degree of ductility. Now what is this? The fact is, that the parts have taken a new arrangement, in which they again cohere. Therefore, in the passage to this new arrangement, the sensible forces, which are the joint result of many corpuscular forces, begin to respect this new arrangement instead of the former. This must change the simple law of corpuscular force, characteristic of the particular species of matter under examination. It does not require much reflection to convince us that the possible arrangements which the particles of a body may acquire, without appearing to change their nature, must be more numerous according as the particles are of a more complex constitution; and it is reasonable to suppose that the constitution of even the most simple kind of matter that we are acquainted with is exceedingly complex. Our microscopes show us animals so minute, that a heap of them must appear to the naked eye an uniform mass with a grain finer than that of the finest marble or razor hone; and yet each of these has not only limbs, but bones, muscular fibres, blood-vessels, fibres, and a blood consisting, in all probability, of globules organized and complex like our own. The imagination is here lost in wonder; and nothing is left us but to adore inconceivable art and wisdom, and to exult in the thought that we are the only spectators of this beautiful scene who can derive pleasure from the view. What is trodden under foot with indifference, even by the half-reasoning elephant, may be made by us the source of the purest and most unmixed pleasure. But let us proceed to observe,

350. 6. That the forces which connect the particles of tangible bodies change by a change of distance, not only in degree, but also in kind. A particle B, Fig. 1, is attracted by A when in the situation C or E. It is repelled by it when at D or F. It is not affected by it when in the situation B. The reader is requested carefully to remark, that this is not an inference founded on the authority of our mathematical figure. The figure is an expression (to assist the imagination) of facts in nature. It requires no force to keep the particles of a body in their quiescent situations; but if they are separated by stretching the body, they endeavour (pardon the figurative expression) to come to-

gether again. If they are brought nearer by compression, they endeavour to recede. This endeavour is manifested by the necessity of employing force to maintain the extension or condensation; and we represent this by the different position of our lines. But this is not all: the particle B, which is repelled by A when in the situation F or D, is neutral when at B, and is attracted when at C or E, may be placed at such a distance AG from A greater than AB that it shall be again repelled, or at such a distance AH that it shall again be attracted; and these alterations may be repeated again and again. This is curious and important, and requires something more than a bare assertion for its proof.

We have already mentioned the most curious and valuable observations of Sir Isaac Newton, by which it appears that light is thus alternately attracted and repelled by bodies. The rings of colour which appear between the object-glasses of long telescopes showed, that in the small interval of $\frac{1}{1000}$th of an inch, there are at least an hundred such changes observable, and that it is highly probable that these alternations extend to a much greater distance. At one of these distances the light actually converges towards the solid matter of the glass, which we express shortly, by saying that it is attracted by it, and that at the next distance it declines from the glass, or is repelled by it. The same thing is more simply inferred from the phenomena of light passing by the edges of knives and other opaque bodies. We refer the reader to the experiments themselves, the detail being too long for this place; and we request him to consider them minutely and attentively, and to form distinct notions of the inferences drawn from them. And we desire it to be remarked, that although Sir Isaac, in his discussion, always considers light as a set of corpuscles moving in free space, and obeying the actions of external forces like any other matter, the particular conclusion in which we are just now interested does not at all depend on this notion of the nature of light. Should we, with Des Cartes or Huygens, suppose light to be the undulation of an elastic medium, the conclusion will be the same. The undulations at certain distances are disturbed by forces directed towards the body, and at a greater distance, the disturbing forces tend *from* the body.

351. These and other facts already mentioned, are a few of many thousand, by which it is unquestionably proved that the particles of tangible matter are connected by forces acting at a distance, varying with the distance, and alternately attractive and repulsive. If we represent these forces by the ordinates of a curve, it is evident that this curve must cross the axis at all those distances where the forces change from attractive to repulsive, and the curve must have branches alternately above and below the axis.

All these alternations of attraction and repulsion take place at small and insensible distances. At all sensible distances the particles are influenced by the attraction of gravitation; and therefore this part of the curve must be a hyperbola whose equation is $y=\frac{a^3}{x^2}$. What is the form of the curve corresponding to the smallest distance of the particles? that is, what is the mutual action between the particles just before their coming into absolute contact? Analogy should lead us to suppose it to be repulsion: for solidity is the last and simplest form of bodies with which we are acquainted.—Fluids are more compounded, containing fire as an essential ingredient. We should conclude that this ultimate repulsion is insuperable, for the hardest bodies are the most elastic. We are fully entitled to say, that this repelling force exceeds all that we have ever yet applied to overcome it; nay, there are good reasons for saying that this ultimate repulsion, by which the particles are kept from mathematical contact, is really insuperable in its own nature, and that it is impossible to produce mathematical contact.

We shall just mention one of these, which we consider as unanswerable. Suppose two atoms, or ultimate particles of matter A and B. Let A be at rest, and B move up to it with the velocity 2; and let us suppose that it comes into mathematical contact, and impels it (according to the common acceptation of the word). Both move with the velocity 1. This is granted by all to be the final result of the collision. Now the instant of time in which this communication happens is no part either of the duration of the solitary motion of A, nor of the joint motion of A and B: it is the separation or boundary between them. It is at once the end of the first, and the beginning of the second, belonging equally to both. A was moving with the velocity 2. The distinguishing circumstance therefore of its mechanical state is, that it has a determination (however incomprehensible) by which it would move for ever with the velocity 2, if nothing changed it. This it has during the whole of its solitary motion, and therefore in the last instant of this motion. In like manner, during the whole of the joint motion, and therefore in the first instant of this motion, the atom A has a determination by which it would move for ever with the velocity 1. In one and the same instant, therefore, the atom A has two incompatible determinations. Whatever notion we can form of this state, which we call velocity, as a distinction of condition, the same impossibility of conception or the same absurdity occurs. Nor can it be avoided in any other way than by saying, that this change of A's motion is brought about by insensible gradations; that is, that A and B influence each other precisely as they would do if a slender spring were interposed.

The two magnets there spoken of are good representatives of two atoms endowed with mutual powers of repulsion; and the communication of motion is accomplished in both cases in precisely the same manner.

252. The simplest particle which can be a constituent of a body having length, breadth, and thickness, must consist of four such atoms, all of which combine their influence on each atom of another such particle. It is evident that the curve which expresses the forces that connect two such particles must be totally different from this original curve, this hylarchic principle. Supposing the last known, our mathematical knowledge is quite able to discover the first; but when we proceed to compose a body of particles, each of which consists of four such particles, we may venture to say, that the compound force which connects them is almost beyond our search, and that the discovery of the primary force from an *accurate* knowledge of the corpuscular forces of *this* particular matter is absolutely out of our power.

All that we can learn is, the possibility, nay the certainty, of an innumerable variety of external sensible forms and qualities, by which different kinds of matter will be distinguished, arising from the number, the order of composition, and the arrangement of the subordinate particles of which a particle of this or that kind of matter is composed. All these varieties will take place at those small and insensible distances which are between A and H, and may produce all that variety which we observe in the tangible or mechanical forms of bodies, such as elasticity, ductility, hardness, softness, fluidity, vapour, and all those unseen motions or actions which we observe in fusion and congelation, evaporation and condensation, solution and precipitation, crystallization, vegetable and animal assimilation and secretion, &c. &c. &c. while all bodies must be, in a certain degree, elastic, all must gravitate, and all must be incompenetrable.

This general and satisfactory resemblance between the

appearance of tangible matter and the legitimate consequence of this general hypothetical property of an atom of matter, affords a considerable probability that such is the origin of all the phenomena. We earnestly recommend to our readers a *careful* perusal of Boscovich's celebrated treatise. A careful perusal is necessary for seeing its value; and nothing will be got by a hasty look at it. The reader will be particularly pleased with the facility and evidence with which the ingenious author has deduced all the ordinary principles of mechanics, and with the explanation which he has given of fluidity, and his deduction from thence of the laws of hydrostatics. No part of the treatise is more valuable than the doctrine of the propagation of pressure through solid bodies. This, however, is but just touched on in the course of the investigation of the principles of mechanics. We shall borrow as much as will suffice for our present inquiry into the strength of materials; and we trust that our readers are not displeased with this general sketch of the doctrine (if it may be so called) of the cohesion of bodies. It is curious and important in itself, and is the foundation of all the knowledge we can acquire of the present article. We are sorry to say that it is as yet a new subject of study; but it is a very promising one, and we by no means despair of seeing the whole of chemistry brought by its means within the pale of mechanical science. The great and distinguishing agency in chemistry is heat, or fire the cause of heat; and one of its most singular effects is the conversion of bodies into elastic vapour. We have the clearest evidence that this is brought about by mechanical forces: for it can be opposed or prevented by external pressure, a very familiar mechanical force. We may perhaps find another mechanical force which will prevent fusion.

HAVING now made our readers familiar with the mode

of action in which cohesion operates in giving strength to solid bodies, we proceed to consider the strains to which this strength is opposed.

A piece of solid matter is exposed to four kinds of strains, pretty different in the manner of their operation.

1. It may be torn asunder, as in the case of ropes, stretchers, king-posts, tye-beams, &c.

2. It may be crushed, as in the case of pillars, posts, and truss-beams.

3. It may be broken across, as happens to a joist or lever of any kind.

4. It may be wrenched or twisted, as in the case of the axle of a wheel, the nail of a press, &c.

I. It may be pulled asunder.

353. This is the simplest of all strains, and the others are indeed modifications of it. To this the force of cohesion is *directly* opposed, with very little modification of its action by any particular circumstances.

When a long cylindrical or prismatic body, such as a rod of wood or metal, or a rope, is drawn by one end, it must be resisted at the other, in order to bring its cohesion into action. When it is fastened at one end, we cannot conceive it any other way than as equally stretched in all its parts; for all our observations and experiments on natural bodies concur in showing us that the forces which connect their particles, in any way whatever, are equal and opposite. This is called the *third law of motion;* and we admit its universality, while we affirm that it is purely experimental (see Physics). Yet we have met with dissertations by persons of eminent knowledge, where propositions are maintained inconsistent with this. During the dispute about the communication of motion, some of the ablest writers have said, that a spring compressed or stretched at the two ends was gradually less and less compressed or stretched from the extremities towards the middle: but the same writers acknowledged the universal equality of action and reaction, which is quite incompatible with this state of the spring. No such inequality of compression or dilatation has ever been observed; and a little reflection will show it to be impossible, in consistency with the equality of action and reaction.

Since all parts are thus equally stretched, it follows, that the strain in any transverse section is the same, as also in every point of that section. If therefore the body be supposed of a homogeneous texture, the cohesion of the parts is equable; and since every part is equally stretched, the particles are drawn to equal distances from their quiescent positions, and the forces which are thus excited, and now exerted in opposition to the straining force, are equal. This external force may be increased by degrees, which will gradually separate the parts of the body more and more from each other, and the connecting forces increase with this increase of distance, till at last the cohesion of some particles is overcome. This must be immediately followed by a rupture, because the remaining forces are now weaker than before.

It is the united force of cohesion, immediately before the disunion of the first particles, that we call the STRENGTH of the section. It may also be properly called its ABSOLUTE STRENGTH, being exerted in the simplest form, and not modified by any relation to other circumstances.

354. If the external force has not produced any permanent change on the body, and it therefore recovers its former dimensions when the force is withdrawn, it is plain that this strain may be repeated as often as we please, and the body which withstands it once will always withstand it. It is evident that this should be attended to in all constructions, and that in all our investigations on this subject this should be kept strictly in view. When we treat a

piece of soft clay in this manner, and with this precaution, the force employed must be very small. If we exceed this, we produce a permanent change. The rod of clay is not indeed torn asunder; but it has become somewhat more slender: the number of particles in a cross section is now smaller; and therefore, although it will again, in this new form, suffer, or allow an endless repetition of a *certain* strain without any farther permanent change, this strain is smaller than the former.

Something of the same kind happens in all bodies which receive a SETT by the strain to which they are exposed. All ductile bodies are of this kind. But there are many bodies which are not ductile. Such bodies break completely whenever they are stretched beyond the limit of their perfect elasticity. Bodies of a fibrous structure exhibit very great varieties in their cohesion. In some the fibres have no lateral cohesion, as in the case of a rope. The only way in which all the fibres can be made to unite their strength is, to twist them together. This causes them to bind each other so fast, that any one of them will break before it can be drawn out of the bundle. In other fibrous bodies, such as timber, the fibres are held together by some cement or gluten. This is seldom as strong as the fibre. Accordingly timber is much easier pulled asunder in a direction transverse to the fibres. There is, however, every possible variety in this particular.

In stretching and breaking fibrous bodies, the visible extension is frequently very considerable. This is not solely the increasing of the distance of the particles of the cohering fibre: the greatest part chiefly arises from drawing the crooked fibre straight. In this, too, there is great diversity; and it is accompanied with important differences in their power of withstanding a strain. In some woods, such as fir, the fibres on which the strength most depends are very straight. Such woods are commonly very elastic, do not take a sett, and break abruptly when overstrained: others, such as oak and birch, have their resisting fibres very undulating and crooked, and stretch very sensibly by a strain. They are very liable to take a sett, and they do not break so suddenly, but give warning by *complaining*, as the carpenters call it; that is, by giving visible signs of a derangement of texture. Hard bodies of an uniform glassy structure, or granulated like stones, are elastic through the whole extent of their cohesion, and take no sett, but break at once when overloaded.

Notwithstanding the immense variety which nature exhibits in the structure and cohesion of bodies, there are certain general facts of which we may now avail ourselves with advantage. In particular,

355. The absolute cohesion is proportional to the area of the section. This must be the case where the texture is perfectly uniform, as we have reason to think it is in glass and the ductile metals. The cohesion of each particle being alike, the whole cohesion must be proportional to their number, that is, to the area of the section. The same must be admitted with respect to bodies of a granulated texture, where the granulation is regular and uniform. The same must be admitted of fibrous bodies, if we suppose their fibres equally strong, equally dense, and similarly disposed through the whole section; and this we must either suppose, or must state the diversity, and measure the cohesion accordingly.

We may therefore assert, as a general proposition on this subject, that the absolute strength in any part of a body by which it resists being pulled asunder, or the force which must be employed to tear it asunder *in that part*, is proportional to the area of the section perpendicular to the extending force.

Therefore all cylindrical or prismatical rods are equally strong in every part, and will break alike in any part;

and bodies which have unequal sections will always break in the slenderest part. The length of the cylinder or prism has no effect on the strength; and the vulgar notion, that it is easier to break a very long rope than a short one, is a very great mistake. Also the absolute strengths of bodies which have similar sections are proportional to the squares of their diameters or homologous sides of the section.

The weight of the body itself may be employed to strain it and to break it. It is evident, that a rope may be so long as to break by its own weight. When the rope is hanging perpendicularly, although it is equally strong in every part, it will break towards the upper end, because the strain on any part is the weight of all that is below it. Its RELATIVE STRENGTH in any part, or power of withstanding the strain which is actually laid on it, is inversely as the quantity below that part.

356. When the rope is stretched horizontally, as in towing ship, the strain arising from its weight often bears a very sensible proportion to its whole strength.

Let AEB (Fig. 3.) be any portion of such a rope, and AC, BC be tangents to the curve into which its gravity bends it. Complete the parallelogram ACBD. It is well known that the curve is a catenaria, and that DC is perpendicular to the horizon; and that DC is to AC as the weight of the rope AEB to the strain to A.

In order that a suspended heavy body may be equally able in every part to carry its own weight, the section in that part must be proportional to the solid contents of all that is below it. Suppose it a conoidal spindle, formed by the revolution of the curve $A\,a\,e$ (Fig. 4.) round the axis CE. We must have $AC^2 : a\,c^2 = AEB$ sol. : $a\,E\,b$ sol. This condition requires the logarithmic curve for $A\,a\,e$, of which $C\,c$ is the axis.

These are the chief general rules which can be safely deduced from our clearest notions of the cohesion of bodies.

In order to make any practical use of them, it is proper to have some measures of the cohesion of such bodies as are commonly employed in our mechanics, and other structures where they are exposed to this kind of strain. These must be deduced solely from experiment. Therefore they must be considered as no more than general values, or as the averages of many particular trials. The irregularities are very great, because none of the substances are constant in their texture and firmness. Metals differ by a thousand circumstances unknown to us, according to their purity, to the heat with which they were melted, to the moulds in which they were cast, and the treatment they have afterwards received, by forging, wire-drawing, tempering, &c.

It is a very curious and inexplicable fact, that by forging a metal, or by frequently drawing it through a smooth hole in a steel plate, its cohesion is greatly increased. This operation undoubtedly deranges the natural situation of the particles. They are squeezed closer together in one direction; but it is not in the direction in which they resist the fracture. In this direction they are rather separated to a greater distance. The general density, however, is augmented in all of them except lead, which grows rather rarer by wire-drawing: but its cohesion may be more than tripled by this operation. Gold, silver, and brass, have their cohesion nearly tripled; copper and iron have it more than doubled. In this operation they also grow much harder. It is proper to heat them to redness after drawing a little. This is called *nealing* or *annealing*. It softens the metal again, and renders it susceptible of another drawing without the risk of cracking in the operation.

We do not pretend to give any explanation of this remarkable and very important fact, which has something resembling it in woods and other fibrous bodies, as will be mentioned afterwards.

The varieties in the cohesion of stones and other minerals, and of vegetable and animal substances, are hardly susceptible of any description or classification.

357. We shall take for the measure of cohesion the number of pounds avoirdupois which are just sufficient to tear asunder a rod or bundle of one inch square. From this it will be easy to compute the strength corresponding to any other dimension.

1*st*, METALS.

		lb.
Gold, cast		20,000 / 24,000
Silver, cast		40,000 / 43,000
Copper, cast	Japan	19,500
	Barbary	22,000
	Hungary	31,000
	Anglesea	34,000
	Sweden	37,000
Iron, cast		42,000 / 59,000
Iron, bar	Ordinary	68,000
	Stirian	75,000
	Best Swedish and Russian	84,000
	Horse-nails	71,006 *
Steel, bar,	Soft	120,000
	Razor temper	150,000

* This was an experiment by Muschenbroek, to examine the vulgar notion that iron forged from old horse-nails was stronger than all others, and shows its falsity.

		lb.
Tin, cast	Malacca	3,100
	Banca	3,600
	Block	3,800
	English block	5,200
	——— grain	6,500
Lead, cast		860
Regulus of antimony		1,000
Zinc		2,600
Bismuth		2,900

358. It is very remarkable that almost all the mixtures of metals are more tenacious than the metals themselves. The change of tenacity depends much on the proportion of the ingredients, and the proportion which produces the most tenacious mixture is different in the different metals. We have selected the following from the experiments of Muschenbroek. The proportion of ingredients here selected is that which produces the greatest strength.

Two parts of gold with one of silver	28,000
Five parts of gold with one of copper	50,000
Five parts of silver with one of copper	48,500
Four parts of silver with one of tin	41,000
Six parts of copper with one of tin	41,000
Five parts of Japan copper with one of Banca tin	57,000
Six parts of Chili copper with one of Malacca tin	60,000
Six parts of Swedish copper with one of Malacca tin	64,000
Brass consists of copper and zinc in an unknown proportion; its strength is	51,000
Three parts of block tin with one part of lead	10,200
Eight parts of block tin with one part of zinc	10,000
Four parts of Malacca tin with one part of regulus of antimony	12,000
Eight parts of lead with one of zinc	4,500
Four parts of tin with one of lead and one of zinc	13,000

These numbers are of considerable use in the arts. The mixtures of copper and tin are particularly interesting in the fabric of great guns. We see that, by mixing copper whose greatest strength does not exceed 37,000, with tin which does not exceed 6000, we produce a metal whose tenacity is almost double, at the same time that it is harder and more easily wrought. It is, however, more fusible, which is a great inconvenience. We also see that a very small addition of zinc almost doubles the tenacity of tin, and increases the tenacity of lead five times; and a small addition of lead doubles the tenacity of tin. These are economical mixtures. This is a very valuable information to the plumbers for augmenting the strength of water-pipes.

By having recourse to these tables, the engineer can proportion the thickness of his pipes (of whatever metal) to the pressures to which they are exposed.

2d, Woods.

359. We may premise to this part of the table the following general observations:

1. The wood immediately surrounding the pith or heart of the tree is the weakest, and its inferiority is so much more remarkable as the tree is older. In this assertion, however, we speak with some hesitation. Muschenbroek's *detail* of experiments is decidedly in the affirmative. Mr Buffon, on the other hand, says, that his experience has taught him that the heart of a sound tree is the strongest; but he gives no instances. We are certain, from many observations of our own, on very *large* oaks and firs, that the heart is much weaker than the exterior parts.

2. The wood next the bark, commonly called the *white* or *blea*, is also weaker than the rest; and the wood gradually increases in strength as we recede from the centre to the blea.

3. The wood is stronger in the middle of the trunk than at the springing of the branches or at the root; and the wood of the branches is weaker than that of the trunk.

4. The wood of the north side of all trees which grow in our European climates is the weakest, and that of the south east side is the strongest; and the difference is most remarkable in hedge-row trees, and such as grow singly. The heart of a tree is never in its centre, but always nearer to the north side, and the annual coats of wood are thinner on that side. In conformity with this, it is a general opinion of carpenters that timber is stronger whose annual plates are thicker. The trachea or air-vessels are weaker than the simple ligneous fibres. These air-vessels are the same in diameter and number of rows in trees of the same species, and they make the visible separation between the annual plates. Therefore when these are thicker, they contain a greater proportion of the simple ligneous fibres.

5. All woods are more tenacious while green, and lose very considerably by drying after the trees are felled.

The only author who has put it in our power to judge of the propriety of his experiments is Muschenbroek. He has described his method of trial minutely, and it seems unexceptionable. The woods were all formed into slips fit for his apparatus, and part of the slip was cut away to a parallelopiped of $\frac{1}{5}$th of an inch square, and therefore $\frac{1}{25}$th of a square inch in section. The absolute strengths of a square inch were as follows:

	lb.		lb.
361. Locust tree	20,100	Mulberry .	12,500
Jujeb . .	18,500	Willow . .	12,500
Beech, oak .	17,300	Ash . . .	12,000
Orange . .	15,500	Plum . .	11,800
Alder . .	13,900	Elder . .	10,000
Elm . .	13,200	Pomegranate	9,750

	lb.		lb.
Lemon . .	9,250	Quince . .	6,750
Tamarind .	8,750	Cypress . .	6,000
Fir	8,330	Poplar . .	5,500
Walnut . .	8,130	Cedar . . .	4,880
Pitch pine .	7,650		

Mr Muschenbroek has given a very minute detail of the experiments on the ash and the walnut, stating the weights which were required to tear asunder slips taken from the four sides of the tree, and on each side in a regular progression from the centre to the circumference. The numbers of this table corresponding to these two timbers may therefore be considered as the average of more than 50 trials made of each; and he says that all the others were made with the same care. We cannot therefore see any reason for not confiding in the results; yet they are considerably higher than those given by some other writers. Mr Pitot says, on the authority of his own experiments, and of those of Mr Parent, that 60 pounds will just tear asunder a square line of sound oak, and that it will bear 50 with safety. This gives 8640 for the utmost strength of a square inch, which is much inferior to Muschenbroek's valuation.

We may add to these,

362.	Ivory	16,270
	Bone	5,250
	Horn	8,750
	Whalebone	7,500
	Tooth of sea-calf	4,075

363. The reader will surely observe, that these numbers express something more than the utmost cohesion; for the weights are such as will very quickly, that is, in a minute or two, tear the rods asunder. It may be said in general, that two-thirds of these weights will sensibly impair the strength after a considerable while, and that one-half is the utmost that can remain suspended at them

without risk for ever; and it is this last allotment that the engineer should reckon upon in his constructions. There is, however, considerable difference in this respect. Woods of a very straight fibre, such as fir, will be less impaired by any load which is not sufficient to break them immediately.

According to Mr Emerson, the load which may be safely suspended to an inch square is as follows:

Iron	76,400
Brass	35,600
Hempen rope	19,600
Ivory	15,700
Oak, box, yew, plum-tree	7,850
Elm, ash, beech	6,070
Walnut, plum	5,360
Red fir, holly, elder, plane, crab	5,000
Cherry, hazle	4,760
Alder, asp, birch, willow	4,290
Lead	430
Freestone	914

He gives us a practical rule, that a cylinder whose diameter is d inches, loaded to one-fourth of its absolute strength, will carry as follows:

Iron	135	Cwt.
Good rope	22	
Oak	14	
Fir	9	

The rank which the different woods hold in this list of Mr Emerson's is very different from what we find in Muschenbroek's. But precise measures must not be expected in this matter. It is wonderful that in a matter of such unquestionable importance the public has not enabled some persons of judgment to make proper trials. They are beyond the abilities of private persons.

II. BODIES MAY BE CRUSHED.

364. It is of equal, perhaps greater, importance to know the strain which may be laid on solid bodies without danger of crushing them. Pillars and posts of all kinds are exposed to this strain in its simplest form; and there are cases where the strain is enormous, viz. where it arises from the oblique position of the parts; as in the struts, braces, and trusses, which occur very frequently in our great works.

It is therefore most desirable to have some general knowledge of the principle which determines the strength of bodies in opposition to this kind of strain. But unfortunately we are much more at a loss in this than in the last case. The mechanism of nature is much more complicated in the present case. It must be in some circuitous way that compression can have any tendency to tear asunder the parts of a solid body, and it is very difficult to trace the steps.

If we suppose the particles insuperably hard and in contact, and disposed in lines which are in the direction of the external pressures, it does not appear how any pressure can disunite the particles; but this is a gratuitous supposition. There are infinite odds against this precise arrangement of the lines of particles; and the compressibility of all kinds of matter in some degree shows that the particles are in a situation equivalent to distance. This being the case, and the particles, with their intervals, or what is equivalent to intervals, being in situations that are oblique with respect to the pressures, it must follow, that by squeezing them together in one direction, they are made to bilge out or separate in other directions. This may proceed so far that some may be thus pushed laterally beyond their limits of cohesion. The moment that this happens the resistance to compression is diminished, and the body will now be crushed together. We may form some notion of this by supposing a number of spherules, like small shot, sticking together by means of a cement. Compressing this in some particular direction causes the spherules to act among each other like so many wedges, each tending to penetrate through between the three which lie below it: and this is the simplest, and perhaps the only distinct notion we can have of the matter. We have reason to think that the constitution of very homegeneous bodies, such as glass, is not very different from this. The particles are certainly arranged symmetrically in the angles of some regular solids. It is only such an arrangement that is consistent with transparency, and with the free passage of light in *every* direction.

365. If this be the constitution of bodies, it appears probable that the strength, or the resistance which they are capable of making to an attempt to crush them to pieces, is proportional to the area of the section whose plane is perpendicular to the external force; for each particle being similarly and equally acted on and resisted, the whole resistance must be as their number; that is, as the extent of the section.

Accordingly this principle is assumed by the few writers who have considered this subject; but we confess that it appears to us very doubtful. Suppose a number of brittle or friable balls lying on a table uniformly arranged, but not cohering nor in contact, and that a board is laid over them and loaded with a weight; we have no hesitation in saying, that the weight necessary to crush the whole collection is proportional to their number or to the area of the section. But when they are in contact (and still more if they cohere), we imagine that the case is materially altered. Any individual ball is crushed only in consequence of its being bulged outwards in the direction perpendicular to the pressure employed. If this could be prevented by a hoop put round the ball like an equator, we cannot see how any force can crush it. Any thing therefore which makes this bulging outwards more difficult, makes a greater force necessary. Now this effect will be produced by the mere contact of the balls before the pressure is applied; for the central ball cannot swell outward laterally without pushing away the balls on all sides of it. This is prevented by the friction on the table and upper board, which is at least equal to one third of the pressure. Thus any interior ball becomes stronger by the mere vicinity of the others; and if we farther suppose them to cohere laterally, we think that its strength will be still more increased.

The analogy between these balls and the cohering particles of a friable body is very perfect. We should therefore expect that the strength by which it resists being crushed will increase in a greater ratio than that of the section, or the square of the diameter of similar sections; and that a square inch of any matter will bear a greater weight in proportion as it makes a part of a greater section. Accordingly, this appears in many experiments, as will be noticed afterwards. Muschenbroek, Euler, and some others, have supposed the strength of columns to be as the biquadrates of their diameters. But Euler deduced this from formulæ which occurred to him in the course of his algebraic analysis; and he boldly adopts it as a principle, without looking for its foundation in the physical assumptions which he had made in the beginning of his investigation. But some of his original assumptions were as paradoxical, or at least as gratuitous, as these results: and those, in particular, from which this proportion of the strength of columns was deduced, were almost foreign to the case; and therefore the inference was of no value. Yet it was received as a principle by Muschenbroek, and by the academicians of St Petersburgh. We make these very few observations, because the subject is of great practical importance; and it is a great obstacle to improvements when deference to a great name, joined to incapacity or indolence, causes authors to adopt his careless reveries as principles from which they are afterwards to draw important consequences. It must be acknowledged that we have not as yet established the relation between the dimensions and the strength of a pillar on solid mechanical principles. Experience plainly contradicts the general opinion, that the strength is proportional to the area of the section; but it is still more inconsistent with the opinion, that it is in the quadruplicate ratio of the diameters of similar sections. It would seem that the ratio depends much on the internal structure of the body; and experiment seems the only method for ascertaining its general laws.

366. If we suppose the body to be of a fibrous texture, having the fibres situated in the direction of the pressure, and slightly adhering to each other by some kind of cement, such a body will fail only by the bending of the fibres, by which they will break the cement and be detached from each other. Something like this may be supposed in wooden pillars. In such cases, too, it would appear that the resistance must be as the number of equally resisting fibres, and as their mutual support, jointly, and therefore as some function of the area of the section. The same thing must happen if the fibres are naturally crooked or undulated, as is observed in many woods, &c. provided we suppose some similarity in their form. Similarity of some kind must always be supposed, otherwise we need never aim at any general inferences.

In all cases therefore we can hardly refuse admitting, that the strength in opposition to compression is proportional to a function of the area of the section.

As the whole length of a cylinder or prism is equally pressed, it does not appear that the strength of a pillar is at all affected by its length. If indeed it be supposed to bend under the pressure, the case is greatly changed, because it is then exposed to a transverse strain; and this increases with the length of the pillar. But this will be considered with due attention under the next class of strains.

Few experiments have been made on this species of strength and strain. Mr Pitot says, that his experiments, and those of Mr Parent, show that the force necessary for crushing a body is nearly equal to that which will tear it asunder. He says that it requires something more than 60 pounds on every square line to crush a piece of sound oak. But the rule is by no means general: Glass, for instance, will carry a hundred times as much as oak in this way, that is, resting on it; but will not *suspend* above four or five times as much. Oak will suspend a great deal more than fir; but fir will carry twice as much as a pillar. Woods of a soft texture, although consisting of very tenacious fibres, are more easily crushed by their load. This softness of texture is chiefly owing to their fibres not being straight but undulated, and there being considerable vacuities between them, so that they are easily bent laterally and crushed. When a post is overstrained by its load, it is observed to swell sensibly in diameter. Increasing the load causes longitudinal cracks or shivers to appear, and it presently after gives way. This is called *crippling*.

In all cases where the fibres lie oblique to the strain the strength is greatly diminished, because the parts can then be made to slide on each other, when the cohesion of the cementing matter is overcome.

Muschenbroek has given some experiments on this subject; but they are cases of long pillars, and therefore do not belong to this place. They will be considered afterwards.

The only experiments of which we have seen any detail (and it is useless to insert mere assertions) are those of Mr Gauthey, in the 4th volume of Rozier's *Journal de Physique*. This engineer exposed to great pressures small rectangular parallelopipeds, cut from a great variety of stones, and noted the weights which crushed them. The following table exhibits the medium results of many trials on two very uniform kinds of freestone, one of them among the hardest and the other among the softest used in building.

367. Column 1st expresses the length of the section in French lines or 12ths of an inch; column 2d expresses the breadth; column 3d is the area of the section in square lines; column 4th is the number of ounces required to crush the piece; column 5th is the weight which was then borne by each square line of the section; and column 6th is the round numbers to which Mr Gauthey imagines that those in column 5th approximate.

Hard Stone.

No. of Exp.	Length of the Section	Breadth of the Section.	Area of the Section.	Ounces necessary to crush it.	Weight borne by each square inch.	
1	8	8	64	736	11,5	12
2	8	12	96	2625	27,3	24
3	8	16	128	4496	35,1	36
			Soft Stone.			
4	9	16	144	560	3,9	4
5	9	18	162	848	5,3	4,5
6	18	18	324	2928	9	9
7	18	24	432	5296	12,2	12

Little can be deduced from these experiments: The 1st and 3d, compared with the 5th and 6th, should furnish similar results; for the 1st and 5th are respectively half of the 3d and 6th: but the 3d is three times stronger (that is, a line of the 3d) than the first, whereas the 6th is only twice as strong as the 5th.

It is evident, however, that the strength increases much faster than the area of the section, and that a square line can carry more and more weight, according as it makes a part of a larger and larger section. In the series of experiments on the soft stone, the individual strength of a square line seems to increase nearly in the proportion of the section of which it makes a part.

Mr Gauthey deduces, from the whole of his numerous experiments, that a pillar of hard stone of Givry, whose section is a square foot, will bear with perfect safety 664,000 pounds, and that its extreme strength is 871,000, and the smallest strength observed in any of his experiments was 460,000. The soft bed of Givry stone had for its smallest strength 187,000, for its greatest 311,000, and for its safe load 249,000. Good brick will carry with safety 320,000; chalk will carry only 9000. The boldest piece of architecture in this respect which he has seen is a pillar in the church of All-Saints at Angers. It is 24 feet long and 11 inches square, and is loaded with 60,000, which is not $\frac{1}{7}$th of what is necessary for crushing it.

368. We may observe here by the way, that Mr Gauthey's measure of the suspending strength of stone is vastly small in proportion to its power of supporting a load laid above it. He finds that a prism of the hard bed of Givry, of a foot section, is torn asunder by 4600 pounds; and if it be firmly fixed horizontally in a wall, it will be broken by a weight of 56,000 suspended a foot from the wall. If it rest on two props at a foot distance, it will be broken by 206,000 laid on its middle. These experiments agree so ill with each other, that little use can be made of them. The subject is of great importance, and well deserves the attention of the patriotic philosopher.

369. A set of good experiments would be very valuable, because it is against this kind of strain that we must guard by judicious construction in the most delicate and difficult problems which come through the hands of the civil and military engineer. The construction of stone arches, and great wooden bridges, and

particularly the construction of the frames of carpentry called *centres* in the erection of stone bridges, are the most difficult jobs that occur. In the centres on which the arches of the bridge of Orleans were built, some of the pieces of oak were carrying upwards of two tons on every square inch of their scantling. All who saw it said that it was not able to carry the fourth part of the intended load. But the engineer understood the principles of his art, and ran the risk: and the result completely justified his confidence; for the centre did not complain in any part, only it was found too supple; so that it went out of shape while the haunches only of the arch were laid on it. The engineer corrected this by loading it at the crown, and thus kept it completely in shape during the progress of the work.

In the old Memoirs of the Academy of Petersburgh for 1778, there is a dissertation by Euler on this subject, but particularly limited to the strain on columns, in which the bending is taken into the account. Mr Fuss has treated the same subject with relation to carpentry in a subsequent volume. But there is little in these papers besides a dry mathematical disquisition, proceeding on assumptions which (to speak favourably) are extremely gratuitous. The most important consequence of the compression is wholly overlooked, as we shall presently see. Our knowledge of the mechanism of cohesion is as yet far too imperfect to entitle us to a confident application of mathematics. Experiments should be multiplied.

370. The only way we can hope to make these experiments useful is to pay a careful attention to the *manner* in which the fracture is produced. By discovering the general resemblances in this particular, we advance a step in our power of introducing mathematical measurement. Thus, when a cubical piece of chalk is slowly crushed between the chaps of a vice, we see it uniformly split in a

surface oblique to the pressure, and the two parts then slide along the surface of fracture. This should lead us to examine mathematically what relation there is between this surface of fracture and the necessary force; then we should endeavour to determine experimentally the position of this surface. Having discovered some general law or resemblance in this circumstance, we should try what mathematical hypothesis will agree with this. Having found one, we may then apply our simplest notions of cohesion, and compare the result of our computations with experiment. We are authorised to say, that a series of experiments have been made in this way, and that their results have been very uniform, and therefore satisfactory, and that they will soon be laid before the public as the foundation of successful practice in the construction of arches.

III. A BODY MAY BE BROKEN ACROSS.

371. The most usual, and the greatest strain, to which materials are exposed, is that which tends to break them transversely. It is seldom, however, that this is done in a manner perfectly simple; for when a beam projects horizontally from a wall, and a weight is suspended from its extremity, the beam is commonly broken near the wall, and the intermediate part has performed the functions of a lever. It sometimes, though rarely, happens that the pin in the joint of a pair of pincers or scissars is cut through by the strain; and this is almost the only case of a simple transverse fracture. Being so rare, we may content ourselves with saying, that in this case the strength of the piece is proportional to the area of the section.

372. Experiments were made for discovering the resistances made by two bodies to this kind of strain in the

following manner: Two iron bars were disposed horizontally at an inch distance; a third hung perpendicularly between them, being supported by a pin made of the substance to be examined. This pin was made of a prismatic form, so as to fit exactly the holes in the three bars, which were made very exact, and of the same size and shape. A scale was suspended at the lower end of the perpendicular bar, and loaded till it tore out that part of the pin which filled the middle hole. This weight was evidently the measure of the lateral cohesion of two sections. The side-bars were made to grasp the middle bar pretty strongly between them, that there might be no distance imposed between the opposite pressures. This would have combined the energy of a lever with the purely transverse pressure. For the same reason it was necessary that the internal parts of the holes should be no smaller than the edges. Great irregularities occurred in our first experiments from this cause, because the pins were somewhat tighter within than at the edges; but when this was corrected they were extremely regular. We employed three sets of holes, viz. a circle, a square (which was occasionally made a rectangle whose length was twice its breadth), and an equilateral triangle. We found in all our experiments the strength exactly proportional to the area of the section, and quite independent of its figure or position, and we found it considerably above the direct cohesion; that is, it took considerably more than twice the force to tear out this middle piece than to tear the pin asunder by a direct pull. A piece of fine freestone required 205 pounds to pull it directly asunder, and 575 to break it in this way.

373. The difference was very constant in any one substance, but varied from $\frac{4}{3}$ds to $\frac{6}{3}$ds in different kinds of matter, being smallest in bodies of a fibrous texture. But indeed we could not make the trial on any bodies of considerable cohesion, because they required such forces as our appa-

ratus could not support. Chalk, clay baked in the sun, baked sugar, brick, and freestone, were the strongest that we could examine.

But the more common case, where the energy of a lever intervenes, demands a minute examination.

374. Let DABC (Plate V. fig. 1.) be a vertical section of a prismatic solid (that is, of equal size throughout), projecting horizontally from a wall in which it is firmly fixed; and let a weight P be hung on it at B, or let any power P act at B in a direction perpendicular to AB. Suppose the body of insuperable strength in every part except in the vertical section DA, perpendicular to its length. It must break in this section only. Let the cohesion be uniform over the whole of this section; that is, let each of the adjoining particles of the two parts cohere with an equal force f.

There are two ways in which it may break. The part ABCD may simply slide down along the surface of fracture, provided that the power acting at B is equal to the accumulated force which is exerted by every particle of the section in the direction AD.

But suppose this effectually prevented by something that supports the point A. The action at P tends to make the body turn round A (or round a horizontal line passing through A at right angles to AB) as round a joint. This it cannot do without separating at the line DA. In this case the adjoining particles at D or at E will be separated horizontally. But their cohesion resists this separation. In order, therefore, that the fracture may happen, the energy or momentum of the power P, acting by means of the lever AB, must be superior to the accumulated energies of the particles. The energy of each depends not only on its cohesive force, but also on its situation; for the supposed insuperable firmness of the rest of the body makes it a lever turning round the fulcrum A, and the cohesion of each particle, such as D or

E, acts by means of the arm DA or EA. The energy of each particle will therefore be had by multiplying the force exerted by it in the instant of fracture by the arm of the lever by which it acts.

Let us therefore first suppose, that in the instant of fracture every particle is exerting an equal force f. The energy of D will be $f \times$ DA, and that of E will be $f \times$ EA, and that of the whole will be the sum of all these products. Let the depth DA of the section be called d, and let any undetermined part of it EA be called x, and then the space occupied by any particle will be $\dot{x}$. The cohesion of this space may be represented by $f\dot{x}$, and that of the whole by $f d$. The energy by which each element $\dot{x}$ of the line DA, or d, resists the fracture, will be $f x \dot{x}$, and the whole accumulated energies will be $f \times \int x \dot{x}$. This we know to be $f \times \frac{1}{2} d^2$, or $f d \times \frac{1}{2} d$. It is the same therefore as if the cohesion $f d$ of the whole section had been acting at the point G, which is in the middle of DA.

The reader who is not familiarly acquainted with the fluxionary calculus may arrive at the same conclusion in another way. Suppose the beam, instead of projecting horizontally from a wall, to be hanging from the ceiling, in which it is firmly fixed. Let us consider how the equal cohesion of every part operates in hindering the lower part from separating from the upper by opening round the joint A. The equal cohesion operates just as equal gravity would do, but in the opposite direction. Now we know, by the most elementary mechanics, that the effect of this will be the same as if the whole weight were concentrated in the centre of gravity G of the line DA, and that this point G is in the middle of DA. Now the number of fibres being as the length d of the line, and the cohesion of each fibre being $=f$, the cohesion of the whole line is $f \times d$ or $f d$.

The accumulated energy therefore of the cohesion in the instant of fracture is $f\,d \times \frac{1}{2}\,d$. Now this must be equal or just inferior to the energy of the power employed to break it. Let the length AB be called l; then $P \times l$ is the corresponding energy of the power. This gives us $f\,d\,\frac{1}{2}\,d = p\,l$ for the equation of equilibrium corresponding to the vertical section ADCB.

Suppose now that the fracture is not permitted at DA, but at another section $\delta\,\alpha$ more remote from B. The body being prismatic, all the vertical sections are equal; and therefore $f\,d\,\frac{1}{2}\,d$ is the same as before. But the energy of the power is by this means increased, being now $= P \times B\,\alpha$, instead of $P \times BA$: Hence we see that when the prismatic body is not insuperably strong in all its parts, but equally strong throughout, it must break close at the wall, where the strain or energy of the power is greatest. We see, too, that a power which is just able to break it at the wall is unable to break it anywhere else; also an absolute cohesion $f\,d$, which can withstand the power p in the section DA, will not withstand it in the section $\delta\,\alpha$, and will withstand more in the section $d'\,a'$.

This teaches us to distinguish between absolute and relative strength. The relative strength of a section has a reference to the strain actually exerted on that section. This relative strength is properly measured by the power which is just able to balance or overcome it, when applied at its proper place. Now since we had $f\,d\,\frac{1}{2}\,d = p\,l$, we have $p = \frac{f\,d\,\frac{1}{2}\,d}{l}$ for the measure of the strength of the section DA, in relation to the power applied at B.

If the solid is a rectangular beam, whose breadth is b, it is plain that all the vertical sections are equal, and that AG or $\frac{1}{2}\,d$ is the same in all. Therefore the equation expressing the equilibrium between the momentum of the external force and the accumulated momenta of cohesion will be $p\,l = f\,d\,b \times \frac{1}{2}\,d$.

The product $d\,b$ evidently expresses the area of the section of fracture, which we may call s, and we may express the equilibrium thus, $p\,l = f\,s\,\frac{1}{2}\,d$, and $2\,l : d = f\,s : p$.

Now $f\,s$ is a proper expression of the absolute cohesion of the section of fracture, and p is a proper measure of its strength in relation to a power applied at B. We may therefore say, that *twice the length of a rectangular beam is to the depth as the absolute cohesion to the relative strength.*

Since the action of equable cohesion is similar to the action of equal gravity, it follows, that whatever is the figure of the section, the relative strength will be the same as if the absolute cohesion of all the fibres were acting at the centre of gravity of the section. Let g be the distance between the centre of gravity of the section and the axis of fracture, we shall have $p\,l = f\,s\,g$, and $l : g = f\,s : p$. It will be very useful to recollect this analogy in words: "*The length of a prismatic beam of any shape is to the height of the centre of gravity above the lower side, as the absolute cohesion to the strength relative to this length.*"

Because the relative strength of a rectangular beam is $\frac{f\,b\,d\,\frac{1}{2}\,d}{l}$ or $\frac{f b d^2}{2\,l}$, it follows, that the relative strengths of different beams are proportional to the absolute cohesion of the particles, to the breadth, and to the square of the depth directly, and to the length inversely; also in prisms whose sections are similar, the strengths are as the cubes of the diameters.

375. Such are the more general results of the mechanism of this transverse strain, in the hypothesis that all the particles are exerting equal forces in the instant of fracture. We are indebted for this doctrine to the celebrated Galileo; and it was one of the first specimens of the application of mathematics to the science of nature.

We have not included in the preceding investigation that action of the external force by which the solid is drawn sidewise, or tends to slide along the surface of fracture. We have supposed a particle E (fig. 1.) to be pulled only in the direction E e, perpendicular to the section of fracture, by the action of the crooked lever BAE. But it is also pulled in the direction EA; and its reaction is in some direction ϵ E, compounded of ϵf, by which it resists being pulled outwards; and ϵ e, by which it resists being pulled downwards. We are but imperfectly acquainted with the force ϵe, and only know that their accumulated sum is equal to the force p: but in all important cases which occur in practice, it is unnecessary to attend to this force; because it is so small in comparsion of the forces in the direction E e, as we easily conclude from the usual smallness of AD in comparison of AB.

376. The hypothesis of equal cohesion, exerted by all the particles in the instant of fracture, is not conformable to nature: for we know, that when a force is applied transversely at B, the beam is bent downwards, becoming convex on the upper side; that side is therefore on the stretch. The particles at D are farther removed from each other than those at E, and are therefore *actually exerting* greater cohesive forces. We cannot say with certainty and precision in what proportion each fibre is extended. It seems most probable that the extensions are proportional to the distances from A. We shall suppose this to be really the case. Now recollect the general law which we formerly said was *observed* in all moderate extensions, *viz.* that the attractive forces exerted by the dilated particles were proportional to their dilatations. Suppose now that the beam is so much bent that the particles at D are exerting their utmost force, and that this fibre is just ready to break or actually breaks. It is plain that a total fracture must immediately ensue; because the force which was superior to the full cohesion of the

particle at D, and a certain portion of the cohesion of all the rest, will be more than superior to the full cohesion of the particle next within D, and a smaller portion of the cohesion of the remainder.

Now let F represent, as before, the full force of the exterior fibre D, which is exerted by it in the instant of its breaking, and then the force exerted at the same instant by the fibre E will be had by this analogy AD : AE, or $d : x = f : \frac{f x}{d}$, and the force really exerted by the fibre E is $f \times \frac{x}{d}$.

The force exerted by a fibre whose thickness is $\dot{x}$ is therefore $\frac{f x \dot{x}}{d}$; but this force resists the strain by acting by means of the lever EA or x. Its energy or momentum is therefore $\frac{f x^2 \dot{x}}{d}$, and the accumulated momenta of all the fibres in the line AE will be $f \times$ sum of $\frac{x^2 \dot{x}}{d}$. This, when x is taken equal to d, will express the momentum of the whole fibres in the line AD. This, therefore, is $f \frac{\frac{1}{3} d^3}{d}$, or $f \frac{1}{3} d^2$, or $f d \times \frac{1}{3} d$. Now $f d$ expresses the absolute cohesion of the whole line AD. The accumulated momentum is therefore the same as if the absolute cohesion of the whole line were exerted at $\frac{1}{3}$d of AD from A.

377. From these premises it follows that the equation expressing the equilibrium of the strain and cohesion is $p\,l = f d \times \frac{1}{3} d$; and hence we deduce the analogy, "*As thrice the length is to the depth, so is the absolute cohesion to the relative strength.*"

This equation and this proportion will equally apply to

rectangular beams whose breadth is b; for we shall then have $p\,l = f\,b\,d \times \frac{1}{3}\,d$.

We also see that the relative strength is proportional to the absolute cohesion of the particles, to the breadth, and to the square of the depth directly, and to the length inversely: for p is the measure of the force with which it is resisted, and $p = \frac{f\,b\,d\,\frac{1}{3}\,d}{l}, = \frac{f\,b\,d^2}{3\,l}$. In this respect, therefore, this hypothesis agrees with the Galilean; but it assigns to every beam a smaller proportion of the absolute cohesion of the section of fracture, in the proportion of 3 to 2. In the Galilean hypothesis this section has a momentum equal to $\frac{1}{2}$ of its absolute strength, but in the other hypothesis it is only $\frac{1}{3}$d. In beams of a different form the proportion may be different.

As this is a most important proposition, and the foundation of many practical maxims, we are anxious to have it clearly comprehended, and its evidence perceived by all. Our better informed readers will therefore indulge us while we endeavour to present it in another point of view, where it will be better seen by those who are not familiarly acquainted with the fluxionary calculus.

378. Fig. 2. of Plate V. is a perspective view of a three sided beam, projecting horizontally from a wall, and loaded with a weight at B just sufficient to break it. DABC is a vertical plane through its highest point D, in the direction of its length. $a\,D\,a$ is another vertical section perpendicular to AB. The piece being supposed of insuperable strength everywhere, except in the section $a\,D\,a$, and the cohesion being also supposed insuperable along the line $a\,A\,a$, it can break nowhere but in this section, and by turning round $a\,A\,a$ as round a hinge. Make $D\,d$ equal to AD, and let $D\,d$ represent the absolute cohesion of the fibre at D, which absolute cohesion we expressed by the symbol f. Let a plane $a\,d\,a$ be made to pass through $a\,a$ and d, and let $d\,a'\,a'$ be another cross sec-

tion. It is plain that the prismatic solid contained between the two sections *a D a* and *a′ d a′* will represent the full cohesion of the whole section of fracture; for we may conceive this prism as made up of lines such as F *f*, equal and parallel to D *d* representing the absolute cohesion of each particle such as F. The pyramidal solid *d* D *a a*, cut off by the plane *d a a*, will represent the cohesions *actually exerted* by the different fibres in the instant of fracture. For take any point E in the surface of fracture, and draw E *e* parallel to AB, meeting the plane *a d a* in *e*, and let *e* AE be a vertical plane. It is evident that D *d* is to E *e* as AD to AE; and therefore (since the forces exerted by the different fibres are as their extension, and their extension as their distances from the axis of fracture) E *e* will represent the force actually exerted by the fibre in E while D is exerting its full force D *d*. In like manner, the plane F F *f f* expresses the cohesion exerted by all the fibres in the line F F, and so on through the whole surface. Therefore the pyramid *d a a* D expresses the accumulated exertion of the whole surface of fracture.

Farther, suppose the beam to be held perpendicular to the horizon, with the end B uppermost, and that the weight of the prism contained between the two sections *a* D *a* and *a′ d a′* (now horizontal) is just able to overcome the full cohesion of the section of fracture. The weight of the pyramid *d* D *a a* will also be just able to overcome the cohesions *actually exerted* by the different fibres in the instant of fracture, because the weight of each fibre, such as E *e*, is just superior to the cohesion actually exerted at E.

Let *o* be the centre of gravity of the pyramidal solid, and draw *o* O perpendicular to the plane *a* D *a*. The whole weight of the solid *d* D *a a* may be conceived as accumulated in the point *o*, and as acting on the point O, and it will have the same tendency to separate the two cohering surfaces as when each fibre is hanging by its respective point. For this reason the point O may be called the *centre of actual effort* of the unequal forces of cohesion. The momentum, therefore, or energy by which the cohering surfaces are separated, will be properly measured by, the weight of the solid $d\,\mathrm{D}\,a\,a$ multiplied by OA; and this product is equal to the product of the weight p multiplied by BA, or by l. Thus, suppose that the cohesion along the line AD only is considered. The whole cohesion will be represented by a triangle $\mathrm{A\,D}\,d$. $\mathrm{D}\,d$ represents f, and AD is d, and $\mathrm{A}\,d$ is x. Therefore $\mathrm{A\,D}\,d$ is $\frac{1}{2}$ $f\,d$. The centre of gravity o of the triangle $\mathrm{A\,D}\,d$ is in the intersection of a line drawn from A to the middle of $\mathrm{D}\,d$ with a line drawn from d to the middle of AD; and therefore the line $o\,O$ will make $\mathrm{AO} = \frac{2}{3}$ of $\mathrm{A}\,d$. Therefore the actual momentum of cohesion is $f \times \frac{1}{2}\,d \times \frac{2}{3}\,d, =$ $f \times d \times \frac{1}{3}\,d = f\,d \times \frac{1}{3}\,d$, or equal to the absolute cohesion acting by means of the lever $\frac{d}{3}$. If the section of fracture is a rectangle, as in a common joist, whose breadth $a\,a$ is $= b$, it is plain that all the vertical lines will be equal to AD, and their cohesions will be represented by triangles like $\mathrm{A\,D}\,d$; and the whole actual cohesion will be represented by a wedge whose bases are vertical planes, and which is equal to half of the parallelopiped $\mathrm{AD} \times \mathrm{D}\,d$ $\times\, a\,a$, and will therefore be $= \frac{1}{2}\,f\,b\,d$; and the distance AO of its centre of gravity from the horizontal line AA will be $\frac{2}{3}$ of AD. The momentum of cohesion of a joist will therefore be $\frac{1}{2}\,f\,b\,d \times \frac{2}{3}\,d$, or $f\,b\,d\,\frac{1}{3}\,d$, as we have determined in the other way.

The beam represented in the figure is a triangular prism. The pyramid $\mathrm{D}\,a\,a\,d$ is $\frac{1}{3}$ of the prism $a\,a\,\mathrm{D}\,d\,a'a'$. If we make s represent the surface of the triangle $a\,\mathrm{D}\,a$, the pyramid is $\frac{1}{3}$ of $f\,s$. The distance AO of its centre of gravity from the horizontal line AA′ is $\frac{1}{2}$ of AD, or $\frac{1}{2}\,d$. Therefore the momentum of actual cohesion is $\frac{1}{3}\,f\,s \times \frac{1}{2}\,d, =$ $f\,s\,\frac{1}{6}\,d$; that is, it is the same as if the full cohesion of all the fibres were accumulated at a point I whose distance from A is $\frac{1}{6}$ of AD or d; or (that we may see its value in every point of view) it is $\frac{1}{6}$ of the momentum of the full cohesion of all the fibres when accumulated at the point D, or acting at the distance $d=$ AD.

This is a very convenient way of conceiving the momentum of actual cohesion, by comparing it with the momentum of absolute cohesion applied at the distance AD from the axis of fracture. The momentum of the absolute cohesion applied at D is to the momentum of actual cohesion in the instant of fracture as AD to AI. Therefore the length of AI, or its proportion to AD, is a sort of index of the strength of the beam. We shall call it the INDEX, and express it by the symbol i.

Its value is easily obtained. The product of the absolute cohesion by AI must be equal to that of the actual cohesion by AO. Therefore say, " as the prismatic solid $a\,a\,\mathrm{D}\,d\,a'\,a'$ is to the pyramidal solid $a\,a\,\mathrm{D}\,d$, so is AO to AI." We are assisted in this determination by a very convenient circumstance. In this hypothesis of the actual cohesions being as the distances of the fibres from A, the point O is the centre of oscillation or percussion of the surface D $a\,a$ turning round the axis $a\,a$; for the momentum of cohesion of the line FF is $\mathrm{FF} \times \mathrm{F}f \times \mathrm{EA} = \mathrm{FF} \times \mathrm{EA}^2$, because F$f$ is equal to EA. Now AO, by the nature of the centre of gravity, is equal to the sum of all these momenta divided by the pyramid $a\,a\,\mathrm{D}\,d$; that is, by the sum of all the $\mathrm{FF} \times \mathrm{F}f$; that is, by the sum of all the $\mathrm{FF} \times \mathrm{EA}$. Therefore $\mathrm{AO} = \frac{\text{sum of } \mathrm{FF} \times \mathrm{EA}^2}{\text{sum of } \mathrm{FF} \times \mathrm{EA}}$, which is just the value of the distance of the centre of percussion of the triangle $a\,a$ D from A. Moreover, if G be the centre of gravity of the triangle a D a, we shall have DA to GA as the absolute cohesion to the sum of the cohesions actually exerted in the instant of fracture; for, by the nature of this centre of gravity, AG is equal to

$\frac{\text{sum of FF} \times \text{EA}}{\text{sum of FF}}$, and the sum of FF $\times$ AG is equal to the sum of FF $\times$ EA. But the sum of all the lines FF is the triangle a D a, and the sum of all the FF $\times$ EA is the sum of all the rectangles FF ff; that is, the pyramid d D a a. Therefore a prism whose base is the triangle a D a, and whose height is AG, is equal to the pyramid, or will express the sum of the actual cohesions; and a prism, whose base is the same triangle, and whose height is D d or D a, expresses the absolute cohesion. Therefore DA is to GA as the absolute cohesion to the sum of the actual cohesions.

Consequently we have DA : GA = OA : IA.

Therefore, whatever be the form of the beam, that is, whatever be the figure of its section, find the centre of oscillation O, and the centre of gravity G of this section. Call their distances from the axis of fracture o and g. Then AI or $i = \frac{o\,g}{d}$, and the momentum of cohesion is $f\,s \times \frac{o\,g}{d}$, where s is the area of fracture.

This index is easily determined in all the cases which generally occur in practice. In a rectangular beam AI is $\frac{1}{3}$ d of AD; in a cylinder (circular or elliptic) AI is $\frac{5}{16}$ths of AD, &c.

In this hypothesis, that the cohesion actually exerted by each fibre is as its extension, and that the extensions of the fibres are as their distances from A (Plate V. fig. 1.), it is plain that the forces exerted by the fibres D, E, &c. will be represented by the ordinates D d, E e, &c. to a straight line A d. And we learn from the principles of ROTATION that the centre of percussion O is in the ordinate which passes through the centre of gravity of the triangle AD d, or (if we consider the whole section having breadth as well as depth) through the centre of gravity of the solid bounded by the planes DA, d A; and we found that this point O was the centre of effort of the cohesions *actually* exerted in the instant of fracture, and that I was the centre of an *equal* momentum, which would be produced if all the fibres were accumulated there and exerted their *full* cohesion.

This consideration enables us to determine, with equal facility and neatness, the strength of a beam in any hypothesis of forces. The above hypothesis was introduced with a cautious limitation to moderate strains, which produced no permanent change of form, or no sett as the artists call it: and this suffices for all purposes of practice, seeing that it would be imprudent to expose materials to more violent strains. But when we compare this theory with experiments in which the pieces are really broken, considerable deviations may be expected, because it is very probable that in the vicinity of rupture the forces are no longer proportional to the extensions.

379. That no doubt may remain as to the justness and completeness of the theory, we must shew how the relative strength may be determined in any other hypothesis. Therefore suppose that it has been established by experiment on any kind of solid matter, that the forces actually exerted in the instant of fracture by the fibres at D, E, &c. are as the ordinates D d', E e', &c. of any curve line A e' d'. We are supposed to know the form of this curve, and that of the solid which is bounded by the vertical plane through AD, and by the surface which passes through this curve A e' d' perpendicularly to the length of the beam. We know the place of the centre of gravity of this curve surface or solid, and can draw a line through it parallel to AB, and cutting the surface of fracture in some point O. This point is also the centre of effort of all the cohesions actually exerted; and the product of AO, and of the solid which expresses the actual cohesions, will give the momentum of cohesion equivalent to the former

$fs\frac{og}{d}$. Or we may find an index AI, by making AI a fourth proportional to the full cohesion of the surface of fracture, to the accumulated actual cohesions, and to AO; and then $fs \times i$ (= AI) will be the momentum of cohesion; and we shall still have I for the point in which all the fibres may be supposed to exert their full cohesion f, and to produce a momentum of cohesion equal to the real momentum of the cohesions actually exerted, and the relative strength of the beam will still be $p = \frac{fsi}{l}$ or $\frac{fsgo}{dl}$ Thus, if the forces be as the squares of the extensions (still supposed to be as the distances from A), the curve A c' a' will be the common parabola, having AB for its axis and AD for the tangent at its vertex. The area AD d' will be $\frac{1}{3}$d AD $\times$ D d; and in the case of a rectangular beam, AO will be $\frac{3}{4}$ths AD, and AI will be $\frac{1}{4}$th of AD.

We may observe here in general, that if the forces actually exerted in the instant of fracture be as any power q of the distance from A, the index AI will be $= \frac{\text{AD}}{q+2}$ for a rectangular beam, and the momentum of cohesion will always be *(cæteris paribus)* as the breadth and as the square of the depth; nay, this will be the case whenever the action of the fibres D and E is expressed by any *similar functions* of d and x. This is evident to every reader acquainted with the fluxionary calculus.

As far as we can judge from experience, no simple algebraic power of the distance will express the actual cohesions of the fibres. No curve which has either AD or AB for its tangent will suit. The observations which we made in the beginning show, that although the curve of Pl. III. fig. 1. must be sensibly straight in the vicinity of the points of intersection with the axis, in order to agree with our observations which show the moderate extensions to be as the extending forces, the curve *must* be concave towards the axis in all its attractive branches, because it cuts it again. Therefore the curve A e' d' of Plate V. fig. 1. must make a finite angle with AD or AB, and it must, in all probability, be also concave towards AD in the neighbourhood of d'. It may however be convex in some part of the intermediate arch. We have made experiments on the extensions of different bodies, and find great diversities in this respect: But in all, the moderate extensions were as the forces, and this with great accuracy till the body took a sett, and remained longer than formerly when the extending force was removed.

We must now remark, that this correction of the Galilean hypothesis of equal forces was suggested by the bending which is observed in all bodies which are strained transversely. Because they are bent, the fibres on the convex side have been extended. We cannot say in what proportion this obtains in the different fibres. Our most distinct notions of the internal equilibrium between the particles render it highly probable that their extension is proportional to their distance from that fibre which retains its former dimensions. But by whatever law this is regulated, we see plainly that the actions of the stretched fibres must follow the proportions of some function of this distance, and that therefore the relative strength of a beam is in all cases susceptible of mathematical determination.

380. We also see an intimate connection between the strain and the curvature. This suggested to the celebrated James Bernoulli the problem of the ELASTIC CURVE, *i. e.* the curve into which an extensible rigid body will be bent by a transverse strain. His solution in the *Acta Lipsiæ* 1694 and 1695 is a very beautiful specimen of mathematical discussion; and we recommend it to the

perusal of the curious reader. He will find it very perspicuously treated in the first volume of his works, published after his death, where the wide steps which he had taken in his investigation are explained so as to be easily comprehended. His nephew Dan. Bernoulli has given an elegant abridgment in the Petersburgh Memoirs for 1729. The problem is too intricate to be fully discussed in a work like this; but it is also too intimately connected with our present subject to be entirely omitted. We must content ourselves with showing the leading mechanical property of this curve, from which the mathematician may deduce all its geometrical properties.

381. When a bar of uniform depth and breadth, and of a given length, is bent into an arch of a circle, the extension of the outer fibres is proportional to the curvature; for, because the curves formed by the inner and outer sides of the beam are similar, the circumferences are as the radii, and the radius of the inner circle is to the difference of the radii, as the length of the inner circumference is to the difference of the circumferences. The difference of the radii is the depth of the beam, the difference of the circumferences is the extension of the outer fibres, and the inner circumference is supposed to be the primitive length of the beam. Now the second and third quantities of the above analogy, viz. the depth and length of the beam, are constant quantities, as is also their product. Therefore the product of the inner radius and the extension of the outer fibre is also a constant quantity, and the whole extension of the outer fibre is inversely as the radius of curvature, or is directly as the curvature of the beam.

The mathematical reader will readily see, that into whatever curve the elastic bar is bent, the whole extension of the outer fibre is equal to the length of a similar curve, having the same proportion to the thickness of the beam that the length of the beam has to the radius of curvature.

Now let ADCB (Pl. V. fig. 3.) be such a rod, of uniform breadth and thickness, firmly fixed in a vertical position, and bent into a curve AEFB by a weight W suspended at B, and of such a magnitude that the extremity B has its tangent perpendicular to the action of the weight, or parallel to the horizon. Suppose too that the extensions are proportional to the extending forces. From any two points E and F draw the horizontal ordinates EG, FH. It is evident that the exterior fibres of the sections E *e* and F *f* are stretched by forces which are in the proportion of EG to FH (these being the long arms of the levers, and the equal thicknesses E *e*, F *f* being the short arms). Therefore (by the hypothesis) their extensions are in the same proportion. But because the extensions are proportional to some similar functions of the distance from the axes of fracture E and F, the extension of any fibre in the section E *e* is to the contemporaneous extension of the similarly situated fibre in the section F *f*, as the extension of the exterior fibre in the section E *e* is to the extension of the exterior fibre in the section F *f*: therefore the whole extension of E *e* is to the whole extension of F *f* as EG to FH, and EG is to FH as the curvature in E to the curvature in F.

Here let it be remarked, that this proportionality of the curvature to the extension of the fibres is not limited to the hypothesis of the proportionality of the extensions to the extending forces. It follows from the extension in the different sections being as some similar function of the distance from the axis of fracture; an assumption which cannot be refused.

This then is the fundamental property of the elastic curve, from which its equation, or relation between the abscissa and ordinate, may be deduced in the usual forms, and all its other geometrical properties. These are foreign to our purpose; and we shall notice only such properties as have an immediate relation to the strain and strength of the different parts of a flexible body, and which in particular serve to explain some difficulties in the valuable experiments of Mr Buffon on the Strength of Beams.

382. We observe, in the first place, that the elastic curve cannot be a circle, but is gradually more incurvated as it recedes from the point of application B of the straining forces. At B it has no curvature; and if the bar were extended beyond B there would be no curvature there. In like manner, when a beam is supported at the ends and loaded in the middle, the curvature is greatest in the middle; but at the props, or beyond them, if the beam extend farther, there is no curvature. Therefore when a beam projecting 20 feet from a wall is bent to a certain curvature at the wall by a weight suspended at the end, and a beam of the same size projecting 20 feet is bent to the very same curvature at the wall by a greater weight at 10 feet distance, the figure and the mechanical state of the beam in the vicinity of the wall is different in these two cases, though the curvature at the very wall is the same in both. In the first case every part of the beam is incurvated; in the second, all beyond the 10 feet is without curvature. In the first experiment the curvature at the distance of five feet from the wall is $\frac{3}{4}$ths of the curvature at the wall; in the second, the curvature at the same place is but $\frac{1}{2}$ of that at the wall. This must weaken the long beam in this whole interval of five feet, because the greater curvature is the result of a greater extension of the fibres.

383. In the next place, we may remark, that there is a certain determinate curvature for every beam which cannot be exceeded without breaking it; for there is a certain separation of two adjoining particles that puts an end to their cohesion. A fibre can therefore be extended

only a certain proportion of its length. The ultimate extension of the outer fibres must bear a certain determinate proportion to its length, and this proportion is the same with that of the thickness (or what we have hitherto called the depth) to the radius of ultimate curvature, which is therefore determinate.

384. A beam of uniform breadth and depth is therefore most incurvated where the strain is greatest, and will break in the most incurvated part. But by changing its form, so as to make the strength of its different sections in the ratio of the strain, it is evident that the curvature may be the same throughout, or may be made to vary according to any law. This is a remark worthy of the attention of the watchmaker. The most delicate problem in practical mechanics is so to taper the balance-spring of a watch that its wide and narrow vibrations may be isochronous. Hooke's principle *ut tensio sic vis* is not sufficient when we take the *inertia* and motion of the spring itself into the account. The figure into which it bends and unbends has also an influence. Our readers will take notice that the artist aims at an accuracy which will not admit an error of $\frac{1}{86400}$th, and that Harrison and Arnold have actually attained it in several instances. The taper of a spring is at present a nostrum in the hands of each artist, and he is careful not to impart his secret.

Again, since the depth of the beam is thus proportional to the radius of ultimate curvature, this ultimate or breaking curvature is inversely as the depth. It may be expressed by $\frac{1}{d}$.

385. When a weight is hung on the end of a prismatic beam, the curvature is nearly as the weight and the length directly, and as the breadth and the cube of the depth inversely; for the strength is $= f\,\frac{b\,d^2}{3\,l}$. Let us suppose

that this produces the ultimate curvature $\frac{1}{d}$. Now let the beam be loaded with a smaller weight w, and let the curvature produced be C, we have this analogy $f\frac{b\,d^2}{3\,l} : w = \frac{1}{d} : \text{C}$, and $\text{C} = \frac{3\,l\,w}{f b d^3}$. It is evident that this is also true of a beam supported at the ends and loaded between the props; and we see how to determine the curvature in its different parts, whether arising from the load, or from its own weight, or from both.

386. When a beam is thus loaded at the end or middle, the loaded point is pulled down, and the space through which it is drawn may be called the DEFLECTION. This may be considered as the sub-tense of the angle of contact, or as the versed sine of the arch into which the beam is bent, and is therefore as the curvature when the length of the arches is given (the flexure being moderate), and as the square of the length of the arch when the curvature is given. The deflection therefore is as the curvature and as the square of the length of the arch jointly; that is, as $\frac{3\,l\,w}{f\,b\,d^3} \times l^2$, or as $\frac{3\,l^3\,w}{f\,b\,d^3}$. The deflection from the primitive shape is therefore as the bending weight and the cube of the length directly, and as the breadth and cube of the depth inversely.

387. In beams just ready to break, the curvature is as the depth inversely, and the deflection is as the square of the length divided by the depth; for the ultimate curvature at the breaking part is the same whatever is the length; and in this case the deflection is as the square of the length.

We have been the more particular in our consideration of this subject, because the resulting theorems afford us the finest methods of examining the laws of corpuscular action, that is, for discovering the variation of the force of cohesion by a change of distance. It is true it is not the atomical law, or HYLARCHIC PRINCIPLE, as it may justly be called, which is thus made accessible, but the specific law of the particles of the substance or kind of matter under examination. But even this is a very great point; and coincidences in this respect among the different kinds of matter are of great moment. We may thus learn the nature of the corpuscular action of different substances, and perhaps approach to a discovery of the *mechanism* of chemical affinities. For that chemical actions are insensible cases of local motion is undeniable, and local motion is the province of mechanical discussion; nay, we see that these hidden changes are produced by mechanical forces in many important cases, for we see them promoted or prevented by means purely mechanical. The conversion of bodies into elastic vapour by heat can at all times be prevented by a *sufficient* external pressure. A strong solution of Glauber's salt will congeal in an instant by agitation, giving out its latent heat; and it will remain fluid for ever, and return its latent heat in a close vessel which it completely fills. Even water will by such treatment freeze in an instant by agitation, or remain fluid for ever by confinement. We know that heat is produced or extricated by friction, that certain compounds of gold or silver with saline matters explode with irresistible violence by the smallest pressure or agitation. Such facts should rouse the mathematical philosopher, and excite him to follow out the conjectures of the illustrious Newton, encouraged by the ingenious attempts of Boscovich; and the proper beginning of this study is to attend to the laws of attraction and repulsion exerted by the particles of cohering bodies, discoverable by experiments made on their actual extensions and compressions. The experiments of simple extensions and compressions are quite insufficient, because the total stretching of a wire is so small a quantity, that the mistake of the 1000th part of an inch occasions an irregularity which deranges any progression so as to make it useless. But by the bending of bodies, a distension of $\frac{1}{100}$th of an inch may be easily magnified in the deflection of the spring ten thousand times. We know that the investigation is intricate and difficult, but not beyond the reach of our present mathematical attainments; and it will give very fine opportunities of employing all the address of analysis. In the last century and the beginning of the present this was a sufficient excitement to the first geniuses of Europe. The cycloid, the catenaria, the elastic curve, the velaria, the caustics, were reckoned an abundant recompense for much study; and James Bernoulli requested, as an honourable monument, that the logarithmic spiral might be inscribed on his tombstone. The reward for the study to which we now presume to incite the mathematicians is the almost unlimited extension of natural science, important in every particular branch. To go no further than our present subject, a great deal of important practical knowledge respecting the strength of bodies is derived from the single observation, that in the moderate extensions which happen before the parts are overstrained, the forces are nearly in the proportion of the extensions or separations of the particles. To return to our subject.

388. James Bernoulli, in his second dissertation on the elastic curve, calls in question this law, and accommodates his investigation to any hypothesis concerning the relation of the forces and extensions. He relates some experiments of lute strings where the relation was considerably different. Strings of three feet long,

Stretched by	2,	4,	6,	8,	10 pds.
Were lengthened	9,	17,	23,	27,	30 lines.

But this is a most exceptionable form of the experiment. The strings were twisted, and the mechanism of the extensions is here exceedingly complicated, combined with compressions and with transverse twists, &c. We made experiments on fine slips of the gum caoutchouc, and on the juice of the berries of the white bryony, of which a single grain will draw to a thread of two feet long, and again return into a perfectly round sphere. We measured the diameter of the thread by a microscope with a micrometer, and thus could tell in every state of extension the proportional number of particles in the sections. We found, that though the whole range in which the distance of the particles was changed in the proportion of 13 to 1, the extensions did not *sensibly* deviate from the proportion of the forces. The same thing was observed in the caoutchouc as long as it perfectly recovered its first dimensions. And it is on the authority of these experiments that we presume to announce this as a law of nature.

389. Dr Robert Hooke was undoubtedly the first who attended to this subject, and assumed this as a law of nature. Mariotte indeed was the first who expressly used it for determining the strength of beams: this he did about the 1679, correcting the simple theory of Galileo. Leibnitz indeed, in his dissertation *de Resistentia Solidorum*, in the *Acta Eruditorum* 1684, introduces this consideration, and wishes to be regarded as the discoverer; and he is always acknowledged as such by the Bernoullis and others who adhered to his peculiar doctrines. But Mariotte had published the doctrine in the most express terms long before; and Bulfinger, in the *Comment. Petropol.* 1729, completely vindicates his claim. But Hooke was unquestionably the discoverer of this law. It made the foundation of his theory of springs, announced to the Royal Society about the year 1661, and read in 1666. On this occasion he mentions many things on the strength of bodies as quite familiar to his thoughts, which are immediate deductions from this principle; and among these *all* the facts which John Bernoulli adduces

in support of Leibnitz's notions about the force of bodies in motion.

390. But even with this first correction of Mariotte, the mechanism of transverse strain is not fully nor justly explained. The force acting in the direction BP (Plate V. fig. 1.), and bending the body ABCD, not only stretches the fibres on the side opposite to the axis of fracture, but compresses the side AB, which becomes concave by the strain. Indeed it cannot do the one without doing the other: For in order to stretch the fibres at D, there must be some fulcrum, some support, on which the virtual lever BAD may press, that it may tear asunder the stretched fibres. This fulcrum must sustain both the pressure arising from the cohesion of the distended fibres, and also the action of the external force, which immediately tends to cause the prominent part of the beam to slide along the section DA. Let BAD therefore be considered as a crooked lever, of which A is the fulcrum. Let an external force be applied at B, in the direction BP, and let a force equal to the accumulated cohesion of AD be applied at O in the direction opposite to AB, that is, perpendicular to AO; and let these two forces be supposed to balance each other by the intervention of the lever. In the first place, the force at O must be to the force at B as AB to AO: Therefore, if we make AK equal and opposite to AO, and AL equal and opposite to AB, the common principles of mechanics inform us that the fulcrum A is affected in the same manner as if the two forces AK and AL were immediately applied to it, the force AK being equal to the weight P, and AL equal to the accumulated cohesion actually exerted in the instant of fracture. The fulcrum is therefore really pressed in the direction AM, the diagonal of the parallelogram, and it must resist in the direction and with the force MA; and this power of resistance, this support, must be furnished by the repulsive forces exerted

by *those particles only* which are in a state of actual compression. The force AK, which is equal to the external force P, must be resisted in the direction KA by the lateral cohesion of the whole particles between D and A (the particle D is not only drawn forward but downward). This prevents the part CDAB from sliding down along the section DA.

391. This is fully verified by experiment. If we attempt to break a long slip of cork, or any such very compressible body, we always observe it to bulge out on the concave side before it cracks on the other side. If it is a body of fibrous or foliated texture, it seldom fails splintering off on the concave side; and in many cases this splintering is very deep, even reaching half way through the piece. In hard and granulated bodies, such as a piece of freestone, chalk, dry clay, sugar, and the like, we generally see a considerable splinter or shiver fly off from the hollow side. If the fracture be slowly made by a force at B gradually augmented, the formation of the splinter is very distinctly seen. It forms a triangular piece like *a* I *b*, which generally breaks in the middle. We doubt not but that attentive observation would show that the direction of the crack on each side of I is not very different from the direction AM and its correspondent on the other side. This is by no means a circumstance of idle curiosity, but intimately connected with the mechanism of cohesion.

392. Let us see what consequences result from this state of the case respecting the strength of bodies. Let D Δ KC (Plate V. fig. 4.) represent a vertical section of a prism of compressible materials, such as a piece of timber. Suppose it loaded with a weight P hung at its extremity. Suppose it also of such a constitution that all the fibres in AD are in a state of dilatation, while those in A Δ are in a state of compression. In the instant of fracture the particles at D and E are withheld by forces D *d*, E *e*, and

the particles at Δ and E repel, resist, or support, with forces Δ δ, E ε.

Some line, such as *d e* A ε δ, will limit all these ordinates, which represent the forces actually exerted in the instant of fracture. If the forces are as the extensions and compressions, as we have great reason to believe, *d e* A and A ε δ will be two straight lines. They will form *one* straight line *d* A δ, if the forces which resist a certain dilatation are equal to the forces which resist an equal compression. But this is quite accidental, and is not strictly true in any body. In most bodies which have any considerable firmness, the compressions made by any external force are not so great as the dilatations which the same force would produce; that is, the repulsions which are excited by any supposed degree of compression are greater than the attractions excited by the same degree of dilatation. Hence it will generally follow, that the angle *d* A D is less than the angle δ A Δ, and the ordinates D *d*, E *e*, &c. are less than the corresponding ordinates Δ δ, E ε, &c.

But whatever be the nature of the line *d* A δ, we are certain of this, that the whole area A D *d* is equal to the whole area A Δ δ: for as the force at B is gradually increased, and the parts between A and D are more extended, and greater cohesive forces are excited, there is always such a degree of repulsive forces excited in the particles between A and Δ that the one set precisely balances the other. The force at B, acting perpendicularly to AB, has no tendency to push the whole piece closer on the part next the wall or to pull it away. The sum of the attractive and repulsive forces actually excited must therefore be equal. These sums are represented by the two triangular areas, which are therefore equal.

The greater we suppose the repulsive forces corresponding to any degree of compression, in comparison with the attractive forces corresponding to the same degree of extension, the smaller will A Δ be in comparison of AD. In a piece of cork or sponge, A Δ may chance to be equal to AD, or even to exceed it; but in a piece of marble, A Δ will perhaps be very small in comparison of AD.

393. Now it is evident that the repulsive forces excited between A and Δ have no share in preventing the fracture. They rather contribute to it, by furnishing a fulcrum to the lever, by whose energy the cohesion of the particles in AD is overcome. Hence we see an important consequence of the compressibility of the body. Its power of resisting this transverse strain is diminished by it, and so much the more diminished as the stuff is more compressible.

This is fully verified by some very curious experiments made by Du Hamel. He took 16 bars of willow 2 feet long and ½ an inch square, and supporting them by props under the ends, he broke them by weights hung on the middle. He broke 4 of them by weights of 40, 41, 47 and 52 pounds: the mean is 45. He then cut 4 of them ⅓d through on the upper side, and filled up the cut with a thin piece of harder wood stuck in pretty tight. These were broken by 48, 54, 50, and 52 pounds; the mean of which is 51. He cut other four ½ through, and they were broken by 47, 49, 50, 46; the mean of which is 48. The remaining four were cut ⅔ds; and their mean strength was 42.

Another set of his experiments is still more remarkable.

Six battens of willow, 36 inches long and 1½ square, were broken by 525 pounds at a medium.

Six bars were cut ⅓d through, and the cut filled with a wedge of hard wood stuck in with a little force: these broke with 551.

Six bars were cut half through, and the cut was filled in the same manner: they broke with 542.

Six bars were cut ¾ths through; these broke with 530.

A batten cut $\frac{3}{4}$ths through, and loaded till nearly broken, was unloaded, and the wedge taken out of the cut. A thicker wedge was put in tight, so as to make the batten straight again by filling up the space left by the compression of the wood: this batten broke with 577 pounds.

From this it is plain that more than $\frac{2}{3}$ds of the thickness (perhaps nearly $\frac{3}{4}$ths) contributed nothing to the strength.

The point A is the centre of fracture in this case; and in order to estimate the strength of the piece, we may suppose that the crooked lever virtually concerned in the strain is DAB. We must find the point I, which is the centre of effort of all the attractive forces, or that point where the full cohesion of AD must be applied, so as to have a momentum equal to the accumulated momenta of all the variable forces. We must in like manner find the centre of effort i of the repulsive or supporting forces exerted by the fibres lying between A and Δ.

It is plain, and the remark is important, that this last centre of effort is the real fulcrum of the lever, although A is the point where there is neither extension nor contraction; for the lever is supported in the same manner as if the repulsions of the whole line A Δ were exerted at that point. Therefore let S represent the surface of fracture from A to D, and f represent the absolute cohesion of a fibre at D in the instant of fracture. We shall have $f\text{S} \times \overline{\text{I} + i} = p\,l$, or $l : \text{I} + i = f\text{S} : p$; that is, the length AB is to the distance between the two centres of effort I and i, as the absolute cohesion of the section between A and D is to the relative strength of the section.

It would be perhaps more accurate, to make AI and Ai equal to the distances of A from the horizontal lines passing through the centres of gravity of the triangles dAD and δ A Δ. It is only in this construction that the points I and i are the centres of real effort of the accumulated attractions and repulsions. But I and i, determined as we have done, are the points where the full, equal, actions may be all applied, so as to produce the same moments. The final results are the same in both cases. The attentive and duly informed reader will see that Mr Bulfinger, in a very elaborate dissertation on the strength of beams in the *Comment. Petropolitan.* 1729, has committed several mistakes in his estimation of the actions of the fibres. We mention this because his reasonings are quoted and appealed to as authorities by Muschenbroek and other authors of note. The subject has been considered by many authors on the continent. We recommend to the reader's perusal the very minute discussions in the Memoirs of the Academy of Paris for 1702 by Varignon, the Memoirs for 1708 by Parent, and particularly that of Coulomb in the *Mem. par les Scavans Etrangers*, tom. vii.

It is evident, from what has been said above, that if S and s represent the surfaces of the sections above and below A, and if G and g are the distances of their centres of gravity from A, and O and o the distances of their centres of oscillation, and D and d their whole depths, the momentum of cohesion wil be $\frac{f\,\text{S}\cdot\text{G}\cdot\text{O}}{\text{D}} + \frac{f\,s.g.o.}{d} = p\,l.$

If (as is most likely) the forces are proportional to the extensions and compressions, the distances AI and A i, which are respectively $= \frac{\text{G}\cdot\text{O}}{\text{D}}$ and $\frac{g\,.\,o}{d}$ are respectively $= \frac{1}{3}$ DA, and $\frac{1}{3}$ Δ A; and when taken together are $= \frac{1}{3}$ D Δ. If, moreover, the extensions are equal to the compressions in the instant of fracture, and the body is a rectangular prism like a common joist or beam, then DA and Δ A are also equal; and therefore the momentum of cohesion is $f\,b \times \frac{1}{2}\,d \times \frac{1}{3}\,d = \frac{f\,b\,d^2}{6}$, $= f\,b\,d \times \frac{1}{6}\,d = p\,l.$

Hence we obtain this analogy, "Six times the length is to

the depth as the absolute cohesion of the section is to its relative strength."

394. Thus we see that the compressibility of bodies has a very great influence on their power of withstanding a transverse strain. We see that in the most favourable supposition of equal dilatations and compressions, the strength is reduced to one half of the value of what it would have been had the body been incompressible. This is by no means obvious; for it does not readily appear how compressibility, which does not diminish the cohesion of a single fibre, should impair the strength of the whole. The reason, however, is sufficiently convincing when pointed out. In the instant of fracture a smaller portion of the section is actually exerting cohesive forces, while a part of it is only serving as a fulcrum to the lever, by whose means the strain on the section is produced. We see too that this diminution of strength does not so much depend on the sensible compressibility, as on its proportion to the dilatability by equal forces. When this proportion is small, A Δ is small in comparison of AD, and a greater portion of the whole fibre is exerting attractive forces. The experiments already mentioned of Du Hamel de Monceau on battens of willow, shew that its compressibility is nearly equal to its dilatability. But the case is not very different in tempered steel. The famous Harrison, in the delicate experiments which he made while occupied in making his longitude watch, discovered that a rod of tempered steel was nearly as much diminished in its length, as it was augmented by the same external force. But it is not by any means certain that this is the proportion of dilatation and compression which obtains in the very instant of fracture. We rather imagine that it is not. The forces are nearly as the dilatations till very near breaking; but we think that they diminish when the body is just going to break. But it

seems certain that the forces which resist compression increase faster than the compressions, even before fracture. We know incontestably that the ultimate resistances to compression are insuperable by any force which we can employ. The repulsive forces therefore (in their whole extent) increase faster than the compressions, and are expressed by an asymptotic branch of the Boscovichian curve formerly explained. It is therefore probable, especially in the more simple substances, that they increase faster, even in such compressions as frequently obtain in the breaking of hard bodies. We are disposed to think that this is always the case in such bodies as do not fly off in splinters on the concave side; but this must be understood with the exception of the permanent changes which may be made by compression, when the bodies are crippled by it. This always increases the compression itself, and causes the neutral point to shift still more towards D. The effect of this is sometimes very great and fatal.

Experiment alone can help us to discover the proportion between the dilatability and compressibility of bodies. The strain now under consideration seems the best calculated for this research. Thus, if we find that a piece of wood an inch square requires 12,000 pounds to tear it asunder by a direct pull, and that 200 pounds will break it transversely by acting 10 inches from the section of fracture, we must conclude that the neutral point A is in the middle of the depth, and that the attractive and repulsive forces are equal. Any notions that we can form of the constitution of such fibrous bodies as timber, make us imagine that the *sensible* compressions, including what arises from the bending up of the compressed fibres, is much greater than the real corpuscular extensions. One may get a general conviction of this unexpected proposition by reflecting on what must happen during the fracture. An undulated fibre can only be drawn straight, and then the corpuscular extension begins; but it may be bent up by compression to any degree, the corpuscular compression being little affected all the while. This observation is very important; and though the forces of corpuscular repulsion may be almost insuperable by any compression that we can employ, a *sensible* compression may be produced by forces not enormous, sufficient to cripple the beam. Of this we shall see very important instances afterwards.

395. It deserves to be noticed, that although the relative strength of a prismatic solid is extremely different in the three hypotheses now considered, yet the proportional strengths of different pieces follow the same ratio; namely, the direct ratio of the breadth, the direct ratio of the square of the depth, and the inverse ratio of the length. In the first hypothesis (of equal forces) the strength of a rectangular beam was $\frac{f b d^2}{2 l}$; in the second (of attractive forces proportional to the extensions) it was $\frac{f b d^2}{3 l}$; and in the third (equal attractions and repulsions proportional to the extensions and compressions) it was $\frac{f b d^2}{6 l}$, or more generally $\frac{f b d^2}{m l}$, where m expresses the unknown proportion between the attractions and repulsions corresponding to an equal extension and compression.

396. Hence we derive a piece of useful information, which is confirmed by unexcepted experience, that the strength of a piece depends chiefly on its depth, that is, on that dimension which is in the direction of the strain. A bar of timber of one inch in breadth and two inches in depth is four times as strong as a bar of only one inch deep, and it is twice as strong as a bar two inches broad and one deep; that is, a joist or lever is always strongest when laid on its edge.

397. There is therefore a choice in the manner in which

the cohesion is opposed to the strain. The general aim must be to put the centre of effort I as far from the fulcrum or the neutral point A as possible, so as to give the greatest energy or momentum to the cohesion. Thus, if a triangular bar projecting from a wall is loaded with a weight at its extremity, it will bear thrice as much when one of the sides is uppermost as when it is undermost. The bar of Fig. 2. would be three times as strong if the side AB were uppermost and the edge DC undermost.

398. Hence it follows that the strongest joist that can be cut out of a round tree is not the one which has the greatest quantity of timber in it, but such that the product of its breadth by the square of its depth shall be the greatest possible. Let ABCD (fig. 5.) be the section of this joist inscribed in the circle, AB being the breadth and AD the depth. Since it is a rectangular section, the diagonal BD is a diameter of the circle, and BAD is a right angled triangle. Let BD be called a and BA be called x; then AD is $= \sqrt{a^2 - x^2}$. Now we must have $AB \times AD^2$, or $x \times \overline{a^2 - x^2}$, or $a^2 x - x^3$, a maximum. Its fluxion $a^2 \dot{x} - 3 x^2 \dot{x}$ must be made $= o$, or $a^2 = 3 x^2$, or $x^2 = \frac{a^2}{3}$. If therefore we make DE $= \frac{1}{3}$ DB, and draw EC perpendicular to BD, it will cut the circumference in the point C, which determines the depth BC and the breadth CD.

Because BD : BC = CD : CE, we have the area of the section BC·CD = BD·CE. Therefore the different sections having the same diagonal BD are proportional to their heights CE. Therefore the section BCDA is less than the section B c D a, whose four sides are equal. The joist so shaped, therefore, is both stronger, lighter, and cheaper.

399. The strength of ABCD is to that of a B c D as 10,000 to 9186, and the weight and expence as

10,000 to 10,607; so that ABCD is preferable to aBcD in the proportion of 10,607 to 9186, or nearly 115 to 100.

From the same principles it follows that a hollow tube is stronger than a solid rod containing the same quantity of matter. Let Fig. 6. represent the section of a cylindric tube, of which AF and BE are the exterior and interior diameters and C the centre. Draw BD perpendicular to BC, and join DC. Then, because $BD^2 = CD - CB$, BD is the radius of a circle containing the same quantity of matter with the ring. If we estimate the strength by the first hypothesis, it is evident that the strength of the tube will be to that of the solid cylinder, whose radius is BD, as $BD^2 \times AC$ to $BD^2 \times BD$; that is, as AC to BD: for BD^2 expresses the cohesion of the ring or the circle, and AC and BD are equal to the distances of the centres of effort (the same with the centres of gravity) of the ring and circle from the axis of fracture.

The proportion of these strengths will be different in the other hypotheses, and is not easily expressed by a general formula; but in both it is still more in favour of the ring or hollow tube.

The following very simple solution will be readily understood by the intelligent reader. Let O be the centre of oscillation of the exterior circle, o the centre of oscillation of the inner circle, and w the centre of oscillation of the ring included between them. Let M be the quantity of surface of the exterior circle, m that of the inner circle, and μ that of the ring.

We have $\mathrm{F}w = \frac{\mathrm{M \cdot FO} - m \cdot \mathrm{F}o}{\mu}, = \frac{5\,\mathrm{FC}^2 + \mathrm{EC}^2}{4\,\mathrm{FC}}$, and the strength of the ring $= \frac{f\mu \times \mathrm{F}w}{2}$, and the strength of the same quantity of matter in the form of a solid cylinder is $f\mu \times \frac{5}{8}\,\mathrm{BD}$; so that the strength of the ring is to that of the solid rod of equal weight as F w to $\frac{5}{4}$ BD, or nearly as FC to BD. This will easily appear by recollecting that FO is $= \frac{\text{sum of } p \, . \, r^2}{m \, . \, \text{FC}}$, and that the momentum of cohesion is $\frac{fm \cdot \text{FC} \cdot \text{F}a}{2\,\text{FC}} = \frac{fm \cdot \text{F}o}{2}$ for the inner circle, &c.

Emerson has given a very inaccurate approximation to this value in his *Mechanics*, 4to.

400. This property of hollow tubes is accompanied also with greater stiffness; and the superiority in strength and stiffness is so much the greater as the surrounding shell is thinner in proportion to its diameter.

401. Here we see the admirable wisdom of the author of nature in forming the bones of animal limbs hollow. The bones of the arms and legs have to perform the office of levers, and are thus opposed to very great transverse strains. By this form they become incomparably stronger and stiffer, and give more room for the insertion of muscles, while they are lighter and therefore more agile; and the same wisdom has made use of this hollow for other valuable purposes of the animal economy. In like manner, the quills in the wings of birds acquire by their thinness the very great strength which is necessary, while they are so light as to give sufficient buoyancy to the animal in the rare medium in which it must live and fly about. The stalks of many plants, such as all the grasses, and many reeds, are in like manner hollow, and thus possess an extraordinary strength. Our best engineers now begin to imitate nature by making many parts of their machines hollow, such as their axles of cast iron, &c.; and modern philosophical instrument makers now form the axes and framings of their great astronomical instruments in the same manner.

In the supposition of homogeneous texture, it is plain that the fracture happens as soon as the particles at D are separated beyond their utmost limit of cohesion. This is a determined quantity, and the piece bends till this degree of extension is produced in the outermost fibre. It follows that the smaller we suppose the distance between A and D, the greater will be the curvature which the beam will acquire before it breaks. Greater depth therefore makes a beam not only stronger but also stiffer. But if the parallel fibres can slide on each other, both the strength and the stiffness will be diminished. Therefore if, instead of one beam D Δ KC, we suppose two, DABC and A Δ KB, not cohering, each of them will bend, and the extension of the fibres AB of the under beam will not hinder the compression of the adjoining fibres AB of the upper beam.

402. The two together therefore will not be more than twice as strong as one of them (supposing DA = A Δ) instead of being four times as strong; and they will bend as much as either of them alone would bend by half the load. This may be prevented, if it were possible to unite the two beams all along the seam AB, so that the one shall not slide on the other. This may be done in small works, by gluing them together with a cement as strong as the natural lateral cohesion of the fibres. If this cannot be done (as it cannot in large works), the sliding is prevented by JOGGLING the beams together; that is, by cutting down several rectangular notches in the upper side of the lower beam, and making similar notches in the under side of the upper beam, and filling up the square spaces with pieces of very hard wood firmly driven in, as represented in Fig. 7. Some employ iron bolts by way of joggles. But when the joggle is much harder than the wood into which it is driven, it is very apt to work loose, by widening the hole into which it is lodged. The same thing is sometimes done by scarfing the one upon the other, as represented in Fig. 8.; but this wastes more timber, and is not so strong, because the mutual

hooks which this method forms on each beam are very apt to tear each other up. By one or other of these methods, or something similar, may a compound beam be formed, of any depth, which will be almost as stiff and strong as an entire piece.

403. On the other hand, we may combine strength with pliableness, by composing our beam of several thin planks laid on each other, till they make a proper depth, and leaving them at full liberty to slide on each other. It is in this manner that coach-springs are formed, as is represented in Fig. 9. In this assemblage there must be no joggles nor bolts of any kind put through the planks or plates; for this would hinder their mutual sliding. They must be kept together by straps which surround them, or by something equivalent.

404. The preceding observations show the propriety of some maxims of construction, which the artists have derived from long experience.

Thus, if a mortice is to be cut out of a piece which is exposed to a cross strain, it should be cut out from that side which becomes concave by the strain, as in Fig. 10. but by no means as in Fig. 11.

If a piece is to be strengthened by the addition of another, the added piece must be joined to the side which grows convex by the strain, as in Fig. 12. and 13.

Before we go any farther, it will be convenient to recal the reader's attention to the analogy between the strain on a beam projecting from a wall, and loaded at the extremity, and a beam supported at both ends and loaded in some intermediate point. It is sufficient on this occasion to read attentively what is delivered in our article on Roofs. We learn there that the strain on the middle point C Fig. 13. of a rectangular beam AB, supported on props at A and B, is the same as if the part CA projected from a wall, and were loaded with the half of the weight W suspended at A. The momentum of the strain is therefore $\frac{1}{2}$ W $\times$ $\frac{1}{2}$ AB, $=$ W $\times$ $\frac{1}{4}$ AB $= p \frac{1}{4} l$, or $\frac{p l}{4}$. The momentum of cohesion must be equal to this in every hypothesis.

Having now considered in sufficient detail the circumstances which affect the strength of any section of a solid body that is strained transversely, it is necessary to take notice of some of the chief modifications of the strain itself. We shall consider only those that occur most frequently in our constructions.

The strain depends on the external force, and also on the lever by which it acts.

405. It is evidently of importance, that since the strain is exerted in any section by means of the cohesion of the parts intervening between the section under consideration and the point of application of the external force, the body must be able in all these intervening parts to propagate or excite the strain in the remote section. In every part it must be able to resist the strain excited in that part. It should therefore be equally strong; and it is useless to have any part stronger, because the piece will nevertheless break where it is not stronger throughout; and it is useless to make it stronger (relatively to its strain) in any part, for it will nevertheless equally fail in the part that is too weak.

Suppose then, in the first place, that the strain arises from a weight suspended at one extremity, while the other end is firmly fixed in a wall. Supposing also the cross sections to be all rectangular, there are several ways of shaping the beam so that it shall be equally strong throughout. Thus it may be equally deep in every part, the upper and under surfaces being horizontal planes. The condition will be fulfilled by making all the horizontal sections triangles, as in Fig. 14. The two sides are vertical planes meeting in an edge at the extremity L. For the equation expressing the balance of strain and strength is $pl = fbd^2$. Therefore since d^2 is the same throughout, and also p, we must have $fb = l$, and b (the breadth AD of any section ABCD) must be proportional to l (or AL), which it evidently is.

Or, if the beam be of uniform breadth, we must have d^2 everywhere proportional to l. This will be obtained by making the depths the ordinates of a common parabola, of which L is the vertex and the length is the axis. The upper or under side may be a straight line, as in Fig. 15. or the middle line may be straight, and then both upper and under surfaces will be curved. It is almost indifferent what is the shape of the upper and under surfaces, provided the distances between them in every part be as the ordinates of a common parabola.

Or, if the sections are all similar, such as circles, squares, or any other similar polygons, we must have d^3 or b^3 proportional to l, and the depths or breadths must be as the ordinates of a cubical parabola.

406. It is evident that these are also the proper forms for a lever moveable round a fulcrum, and acted on by a force at the extremity. The force comes in the place of the weight suspended in the cases already considered; and as such levers always are connected with another arm, we readily see that both arms should be fashioned in the same manner. Thus in Fig. 14. the piece of timber may be supposed a kind of steelyard, moveable round a horizontal axis OP, in the front of the wall, and having the two weights P and π in equilibrio. The strain occasioned by each at the section in which the axis OP is placed must be the same, and each arm OL and O λ must be equally strong in all its parts. The longitudinal sections of each arm must be a triangle, a common parabola, or a cubic parabola, according to the conditions previously given.

And, moreover, all these forms are equally strong: For any one of them is equally strong in all its parts, and they are all supposed to have the same section at the front of the wall or at the fulcrum. They are not, however, equally stiff. The first, represented in Fig. 14. will bend least upon the whole, and the one formed by the cubic parabola will bend most. But their curvature at the very fulcrum will be the same in all.

It is also plain, that if the lever is of the second or third kind, that is, having the fulcrum at one extremity, it must still be of the same shape; for in abstract mechanics it is indifferent which of the three points is considered as the axis of motion. In every lever the two forces at the extremities act in one direction, and the force in the middle acts in the opposite direction, and the great strain is always at that point. Therefore a lever, such as Fig. 14. moveable round an axis passing horizontally through λ, and acting against an obstacle at OP, is equally able in all its parts to resist the strains excited in those parts.

The same principles and the same construction will apply to beams, such as joists, supported at the ends L and λ Fig. 14., and loaded at some intermediate part OP. This will appear evident by merely inverting the directions of the forces at these three points, or by referring to our article on Roofs.

407. Hitherto we have supposed the external straining force as acting only in one point of the beam. But it may be uniformly distributed all over the beam. To make a beam in such circumstances equally strong in all its parts, the shape must be considerably different from the former.

408. Thus suppose the beam to project from a wall.

If it be of equal breadth throughout, its sides being vertical planes, parallel to each other and to the length, the vertical section in the direction of its length must be a triangle instead of a common parabola; for the weight uniformly distributed over the part lying beyond any section, is as the length beyond that section: and since it may all be conceived as collected at its centre of gravity, which is the middle of that length, the lever by which this load acts or strains the section is also proportional to the same length. The strain on the section (or momentum of the load) is as the square of that length. The section must have strength in the same proportion. Its strength being as the breadth and the square of the depth, and the breadth being constant, the square of the depth of any section must be as the square of its distance from the end, and the depth must be as that distance; and therefore the longitudinal vertical section must be a triangle.

But if all the transverse sections are circles, squares, or any other similar figures, the strength of every section, or the cube of the diameter, must be as the square of the lengths beyond that section, or the square of its distance from the end; and the sides of the beam must be a semi-cubical parabola.

If the upper and under surfaces are horizontal planes, it is evident that the breadth must be as the square of the distance from the end, and the horizontal sections may be formed by arches of the common parabola, having the length for their tangent at the vertex.

By recurring to the analogy so often quoted between a projecting beam and a joist, we may determine the proper form of joists which are uniformly loaded through their whole length.

409. This is a frequent and important case, being the office of joists, rafters, &c. and there are some circumstances which must be particularly noticed, because they are not so obvious, and have been misunderstood. When a beam AB Fig. 16. is supported at the ends, and a weight is laid on any point P, a strain is excited in every part of the beam. The load on P causes the beam to press on A and B, and the props react with forces equal

and opposite to these pressures. The load at P is to the pressures at A and B as AB to PB and PA, and the pressures at A is to that at B as PB to PA; the beam therefore is in the same state, with respect to strain in every part of it, as if it were resting on a prop at P, and were loaded at the ends with weights equal to the two pressures on the props; and observe, these pressures are such as will balance each other, being inversely as their distances from P. Let P represent the weight or load at P. The pressure on the prop P must be $P \times \frac{PA}{AB}$ This is therefore the reaction of the prop B, and is the weight which we may suppose suspended at B, when we conceive the beam resting on a prop at P, and carrying the balancing weights at A and B.

The strain occasioned at any other point C, by the load P at P, is the same with the strain at C, by the weight $P \times \frac{PA}{AB}$ hanging at B, when the beam rests on P, in the manner now supposed; and it is the same if the beam, instead of being balanced on a prop at P, had its part AP fixed in a wall. This is evident. Now we have shewn at length that the strain at C, by the weight $P \times \frac{PA}{AB}$ hanging at B, is $P \times \frac{PA}{AB} \times BC$. We desire it to be particularly remarked that the pressure at A has no influence on the strain at C, arising from the action of any load between A and C; for it is indifferent how the part AP of the projecting beam PB is supported. The weight at A just performs the same office with the wall in which we suppose the beam to be fixed. We are thus particular, because we have seen even persons not unaccustomed to discussions of this kind puzzled in their conceptions of this strain.

Now let the load P be laid on some point p between C

and P. The same reasoning shows us that the point is (with respect to strain) in the same state as if the beam were fixed in a wall, embracing the part p B, and a weight $= \mathrm{P} \times \frac{p\,\mathrm{B}}{\mathrm{AB}}$ were hung on at A, and the strain at C is $\mathrm{P} \times \frac{p\,\mathrm{B}}{\mathrm{AB}} \times \mathrm{AC}$.

410. In general, therefore, the strain on any point C, arising from a load P laid on another point P, is proportional to the rectangle of the distances of P and C from the ends nearest to each. It is $\mathrm{P} \times \frac{\mathrm{PA} \times \mathrm{CB}}{\mathrm{AB}}$, or $\mathrm{P} \times \frac{p\,\mathrm{B} \times \mathrm{CA}}{\mathrm{AB}}$, according as the load lies between C and A or between C and B.

Cor. 1. The strains which a load on any point P occasions on the points C, c, lying on the same side of P, are as the distances of these points from the end B. In like manner the strains on E and e are as EA and e A.

Cor. 2. The strain which a load occasions in the part on which it rests is as the rectangle of the parts on each side. Thus the strain occasioned at C by a load is to that at D by the same load as AC × CB to AD × DB. It is therefore greatest in the middle.

411. Let us now consider the strain on any point C arising from a load uniformly distributed along the beam. Let AP be represented by x, and Pp by $\dot{x}$, and the whole weight on the beam by a. Then

The weight on Pp is - - - $= a\,\frac{\dot{x}}{\mathrm{AB}}$

Pressure on B by the weight on Pp $= a\,\frac{\dot{x}}{\mathrm{AB}} \times \frac{x}{\mathrm{AB}}$.

Or - - - - - - - - - - - $= a\,\frac{x\dot{x}}{\mathrm{AB}^2}$.

Pres. on B by the whole wt. on AC $= a\,\frac{\frac{1}{2}\mathrm{AC}^2}{\mathrm{AB}^2} = a\,\frac{\mathrm{AC}^2}{2\,\mathrm{AB}}$

Strain at C by the weight on AC $= a\,\frac{AC^2 \times BC}{2\,AB^2}$.

Strain at C by the weight on BC $= a\,\frac{BC^2 \times AC}{2\,AB^2}$.

Do. by the whole weight on AB $= a\,\frac{AC^2 \times BC + BC^2 \times AC}{2\,AB^2}$,

$= a\,\frac{AC \times BC \times \overline{AC + CB}}{2\,AB^2}, = a\,\frac{AC \times BC}{2\,AB}$.

Thus we see that the strain is proportional to the rectangle of the parts, in the same manner as if the load a had been laid directly on the point C, and is indeed equal to one half of the strain which would be produced at C by the load a laid on there.

412. It was necessary to be thus particular, because we see in some elementary treatises of mechanics, published by authors of reputation, mistakes which are very plausible, and mislead the learner. It is there said, that the pressure at B from a weight uniformly diffused along AB is the same as if it were collected at its centre of gravity, which would be the middle of AB; and then the strain at C is said to be this pressure at B multiplied by BC. But surely it is not difficult to see the difference of these strains. It is plain that the pressure of gravity downwards on any point between the end A and the point C has no tendency to diminish the strain at C, arising from the upward reaction of the prop B; whereas the pressure of gravity between C and B is almost in direct opposition to it, and must diminish it. We may however avoid the fluxionary calculus with safety by the consideration of the centre of gravity, by supposing the weights of AC and BC to be collected at their respective centres of gravity; and the result of this computation will be the same as above; and we may use either method, although the weight is not uniformly distributed, provided only that we know in what manner it is distributed.

This investigation is evidently of importance in the practice of the engineer and architect, informing them what support is necessary in the different parts of their constructions. We shall consider some cases of this kind in the article Roofs.

413. It is now easy to form a joist, so that it shall have the same relative strength in all its parts.

I. To make it equally able in all its parts to carry a given weight laid on any point C taken at random, or uniformly diffused over the whole length, the strength of the section at the point C must be as $AC \times CB$. Therefore,

1. If the sides are parallel vertical planes, the square of the depth (which is the only variable dimension) or CD^2, must be as $AC \times CB$, and the depths must be ordinates of an ellipse.

2. If the transverse sections are similar, we must make CD^3 as $AC \times CB$.

3. If the upper and under surfaces are parallel, the breadth must be as $AC \times CB$.

II. If the beam is necessarily loaded at some given point C, and we would have the beam equally able in all its parts to resist the strain arising from the weight at C, we must make the strength of every transverse section between C and either end as its distance from that end. Therefore,

1. If the sides are parallel vertical planes, we must make $CD^2 : EF^2 = AC : AE$.

2. If the sections are similar, then $CD^3 : EF^3 = AC : AE$.

3. If the upper and under surfaces are parallel, then breadth at C : breadth at E = AC : AE.

414. The same principles enable us to determine the strain and strength of square or circular plates, of different extent, but equal thickness. This may be comprehended in this general proposition.

Similar plates of equal thickness supported all round will carry the same absolute weight, uniformly distributed, or resting on similar points, whatever is their extent.

Suppose two similar oblong plates of equal thickness, and let their lengths and breadths be L, l, and B, b. Let their strength or momentum of cohesion be C, c, and the strains from the weights W, w, be S, s,

Suppose the plates supported at the ends only, and resisting fracture transversely. The strains, being as the weights and lengths, are as WL and $w\,l$, but their cohesion are as the breadths; and since they are of equal relative strength, we have $\text{WL} : w\,l = \text{B} : b$, and $\text{WL}\,b = w\,l\,\text{B}$ and $\text{L} : l = w\,\text{B} : \text{W}\,b$: but since they are of similar shapes $\text{L} : l = \text{B} : b$, and therefore $w = \text{W}$.

The same reasoning holds again when they are also supported along the sides, and therefore holds when they are supported all round (in which case the strength is doubled).

And if the plates are of any other figure, such as circles or ellipses, we need only conceive similar rectangles inscribed in them. These are supported all round by the continuity of the plates, and therefore will sustain equal weights; and the same may be said of the segments which lie without them, because the strengths of any similar segments are equal, their lengths being as their breadths.

Therefore the thickness of the bottoms of vessels holding heavy liquors or grains should be as their diameters, and as the square root of their depths jointly.

Also the weight which a square plate will bear is to that which a bar of the same matter and thickness will bear as twice the length of the bar to its breadth.

415. There is yet another modification of the strain which tends to break a body transversely, which is of very frequent occurrence, and in some cases must be very carefully attended to, viz. the strain arising from its own weight.

When a beam projects from a wall, every section is

strained by the weight of all that projects beyond it. This may be considered as all collected at its centre of gravity. Therefore the strain on any section is in the joint ratio of the weight of what projects beyond it, and the distance of its centre of gravity from the section.

416. The determination of this strain, and of the strength necessary for withstanding it, must be more complicated than the former, because the form of the piece which results from this adjustment of strain and strength influences the strain. The general principle must evidently be, that the strength or momentum of cohesion of every section must be as the product of the weight beyond it, multiplied by the distance of its centre of gravity. For example:

Suppose the beam DLA Fig. 17. to project from the wall, and that its sides are parallel vertical planes, so that the depth is the only variable dimension. Let $LB = x$ and $B\,b = y$. The element $B\,b\,c\,C$ is $= y\,\dot{x}$. Let G be the centre of gravity of the part lying without $B\,b$, and g be its distance from the extremity L. Then $x - g$ is the arm of the lever by which the strain is excited in the section $B\,b$. Let $B\,b$ or y be as some power m of LB; that is, let $y = x^m$. Then the contents of $L\,B\,b$ is $\frac{x^{m+1}}{m+1}$. The momentum of gravity round a horizontal axis at L is $y\,\dot{x}\,x = x^{m+1}\dot{x}$, and the whole momentum round the axis is $\frac{x^{m+2}}{m+2}$. The distance of the centre of gravity from L is had by dividing this momentum by the whole weight, which is $\frac{x^{m+1}}{m+1}$. The quotient or g is $\frac{x \times \overline{m+1}}{m+2}$. And the distance of the centre of gravity from the section $B\,b$ is $x - \frac{x \times m+1}{m+2}, = \frac{x \times \overline{m+2} - x \times \overline{m+1}}{m+2}, = \frac{x}{m+2}$. Therefore the strain on the section $B\,b$ is had by multiplying $\frac{x^{m+1}}{m+1}$

by $\frac{x}{m+2}$. The product is $\frac{x^{m+2}}{\overline{m+2} \times \overline{m+1}}$. This must be as the square of the depth, or as y^2. But y is as x^m, and y^2 as x^{2m}. Therefore we have $m+2=2m$, and $m=2$; that is, the depth must be as the square of the distance from the extremity, and the curve L b A is a parabola touching the horizontal line in L.

417. It is easy to see that a conoid formed by the rotation of this figure round DL will also be equally able in every section to bear its own weight.

We need not prosecute this farther. When the figure of the piece is given, there is no difficulty in finding the strain; and the circumstance of equal strength to resist this strain is chiefly a matter of curiosity.

418. It is evident, from what has been already said, that a projecting beam becomes less able to bear its own weight, as it projects farther. Whatever may be the strength of the section DA, the length may be such that it will break by its own weight. If we suppose two beams A and B of the same substance and similar shapes, that is, having their lengths and diameters in the same proportion; and farther suppose that the shorter can just bear its own weight; then the longer beam will not be able to do the same: For the strengths of the sections are as the cubes of the diameters, while the strains are as the biquadrates of the diameters; because the weights are as the cubes, and the levers by which these weights act in producing the strain are as the lengths or as the diameters.

419. These considerations show us, that in all cases where the strain is affected by the weight of the parts of the machine or structure of any kind, the smaller bodies are more able to withstand it than the greater; and there seems to be bounds set by nature to the size of machines constructed of any given materials. Even when the weight of the parts of the machine is not taken into the

account, we cannot enlarge them in the same proportion in all their parts. Thus a steam-engine cannot be doubled in all its parts, so as to be still efficient. The pressure on the piston is quadrupled. If the lift of the pump be also doubled in height while it is doubled in diameter, the load will be increased eight times, and will therefore exceed the power. The depth of lift, therefore, must remain unchanged; and in this case the machine will be of the same relative strength as before, independent of its own weight. For the beam being doubled in all its dimensions, its momentum of cohesion is eight times greater, which is again a balance for a quadruple load acting by a double lever.—But if we now consider the increase of the weight of the machine itself, which must be supported, and which must be put in motion by the intervention of its cohesion, we see that the large machine is weaker and less efficient than the small one.

There is a similar limit set by nature to the size of plants and animals formed of the same matter. The cohesion of an herb could not support it if it were increased to the size of a tree, nor could an oak support itself if 40 or 50 times bigger, nor could an animal of the make of a long-legged spider be increased to the size of a man; the articulations of its legs could not support it.

420. Hence may be understood the prodigious superiority of the small animals both in strength and agility. A man by falling twice his own height may break his firmest bones. A mouse may fall 20 times its height without risk; and even the tender mite or wood-louse may fall unhurt from the top of a steeple. But their greatest superiority is in respect of nimbleness and agility. A flea can leap above 500 times its own length, while the strength of the human muscles could not raise the trunk from the ground on limbs of the same construction.

The angular motions of small animals (in which consists their nimbleness or agility) must be greater than those of large animals, supposing the force of the muscular fibre to be the same in both. For supposing them similar, the number of equal fibres will be as the square of their linear dimensions; and the levers by which they act are as their linear dimensions. The energy therefore of the moving force is as the cube of these dimensions. But the momentum of inertia, or $\int p\, r^2$, is as the 4th power: Therefore the angular velocity of the greater animals is smaller. The number of strokes which a fly makes with its wings in a second is astonishingly great; yet, being voluntary, they are the effects of its agility.

We have hitherto confined our attention to the simplest form in which this transverse strain can be produced. This was quite sufficient for showing us the mechanism of nature by which the strain is resisted; and a very slight attention is sufficient for enabling us to reduce to this every other way in which the strain can be produced. We shall not take up the reader's time with the application of the same principles to other cases of this strain, but refer him to the article Roofs. In that article we have shown the analogy between the strain on the section of a beam projecting from a wall and loaded at the extremity, and the strain on the same section of a beam simply resting on supports at the ends, and loaded at some intermediate point or points. The strain on the middle C of a beam AB Fig. 18. so supported, arising from a weight laid on there, is the same with the strain which half that weight hanging at B would produce on the same section C if the other end of the beam were fixed in a wall. If therefore 1000 pounds hung on the end of a beam projecting 10 feet from a wall will just break it at the wall, it will require 4000 pounds on its middle to break the same beam resting on two props 10 feet asunder. We have also shown in that article the additional strength which will be given to this beam by extending both ends beyond the props, and there framing it firmly into other pillars or supports.

421. We can hardly add any thing to what has been said in that article, except a few observations on the effects of the obliquity of the external force. We have hitherto supposed it to act in the direction BP Fig. 4. perpendicular to the length of the beam. Suppose it to act in the direction BB, oblique to BA. In the article Roof we supposed the strain to be the same as if the force p acted at the distance AB′, but still perpendicular to AB: so it is. But the strength of the section A Δ is not the same in both cases; for by the obliquity of the action the piece DCKΔ is pressed to the other. We are not sufficiently acquainted with the corpuscular forces to say precisely what will be the effect of the pressure arising from this obliquity; but we can clearly see, in general, that the point A, which in the instant of fracture is neither stretched nor compressed, must now be farther up, or nearer to D; and therefore the number of particles which are exerting cohesive forces is smaller, and therefore the strength is diminished. Therefore, when we endeavour to proportion the strength of a beam to the strain arising from an external force acting obliquely, we make too liberal allowance by increasing this external force in the ratio of AB to AB′. We acknowledge our inability to assign the proper correction. But this circumstance is of very great influence. In many machines, and many framings of carpentry, this oblique action of the straining force is unavoidable; and the most enormous strains to which materials are exposed are generally of this kind. In the frames set up for carrying the ring-stones of arches, it is hardly possible to avoid them: for although the judicious engineer disposes his beams so as to sustain only pressures in the direction of their lengths, tending either to crush them or to tear them asunder, it

frequently happens that, by the settling of the work, the pieces come to cheek and bear on each other transversely, tending to break each other across. This we have remarked upon in the article Roofs, with respect to a truss by Mr Price. Now when a cross strain is thus combined with an enormous pressure in the direction of the length of the beam, it is in the utmost danger of snapping suddenly across. This is one great cause of the carrying away of masts. They are compressed in the direction of their length by the united force of the shrouds, and in this state the transverse action of the wind soon completes the fracture.

422. When considering the compressing strains to which materials are exposed, we deferred the discussion of the strain on columns, observing that it was not, in the cases which usually occur, a simple compression, but was combined with a transverse strain, arising from the bending of the column. When the column ACB Fig. 19. resting on the ground at B, and loaded at top with a weight A, acting in the vertical direction AB, is bent into a curve ACB, so that the tangent at C is perpendicular to the horizon, its condition somewhat resembles that of a beam firmly fixed between B and C, and strongly pulled by the end A, so as to bend it between C and A. Although we cannot conceive how a force acting on a straight column AB in the direction AB can bend it, we may suppose that the force acted first in the horizontal direction A *b*, till it was bent to this degree, and that the rope was then gradually removed from the direction A *b* to the direction AB, increasing the force as much as is necessary for preserving the same quantity of flexure.

423. The first author (we believe) who considered this important subject with scrupulous attention was the celebrated Euler, who published, in the Berlin Memoirs for 1757, his Theory of the Strength of Columns. The

general proposition established by this theory is, that the strength of prismatical columns is in the direct quadruplicate ratio of their diameters, and the inverse duplicate ratio of their lengths. He prosecuted this subject in the Petersburgh Commentaries for 1778, confirming his former theory. We do not find that any other author has bestowed much attention on it, all seeming to acquiesce in the determinations of Euler, and to consider the subject as of very great difficulty, requiring the application of the most refined mathematics. Muschenbroek has compared the theory with experiment; but the comparison has been very unsatisfactory, the difference from the theory being so enormous as to afford no argument for its justness. But the experiments do not contradict it, for they are so anomalous as to afford no conclusion or general rule whatever.

To say the truth, the theory can be considered in no other light than as a specimen of ingenious and very artful algebraic analysis. Euler was unquestionably the first analyst in Europe for resource and address. He knew this, and enjoyed his superiority, and without scruple admitted any physical assumptions which gave him an opportunity of displaying his skill. The inconsistency of his assumptions with the known laws of mechanism gave him no concern; and when his algebraic processes led him to any conclusion which would make his readers stare, being contrary to all our usual notions, he frankly owned the paradox, but went on in his analysis, saying, "*Sed analysi magis fidendum.*" Mr Robins has given some very risible instances of this confidence in his analysis. Nay, so fond was he of this kind of amusement, that, after having published an untenable Theory of Light and Colours, he published several memoirs, explaining the aberration of the heavenly bodies, and deducing some very wonderful consequences, fully confirmed by experience, from the Newtonian principles, which were opposite and totally inconsistent with his own theory, merely because the Newtonian theory gave him "*occasionem analyseos promovendæ.*" We are thus severe in our observations, because his theory of the strength of columns is one of the strongest instances of this wanton kind of proceeding, and because his followers in the Academy of St Petersburgh, such as Mr Fuss, Lexell, and others, adopt his conclusions, and merely echo his words. We are not a little surprised to see Mr Emerson, a considerable mathematician, and a man of very independent spirit, hastily adopting the same theory, of which we doubt not but our readers will easily see the falsity.

Euler considers the column ACB Fig. 19. as in a condition precisely similar to that of an elastic rod bent into the curve by a cord AB connecting its extremities. In this he is not mistaken. But he then draws CD perpendicular to AB, and considers the strain on the section C as equal to the momentum or mechanical energy of the weight A acting in the direction DB upon the lever *x c* D, moveable round the fulcrum *c*, and tending to tear asunder the particles which cohere along the section *c* C *x*. This is the same principle (as Euler admits) employed by James Bernoulli in his investigation of the elastic curve ACB. Euler considers the strain on the section *c x* as the same with what it would sustain if the same power acted in the horizontal direction EF on a point E as far removed from C as the point D is. We have reasoned in the same manner (as has been observed) in the article Roofs, where the obliquity of action was inconsiderable. But in the present case, this substitution leads to the greatest mistakes, and has rendered the whole of this theory false and useless. It would be just if the column were of materials which are incompressible. But it is evident, from what has been said above, that by the compression of the parts the real fulcrum of the lever shifts away from the point *c*, so much the more as the compression is greater.

In the great compressions of loaded columns, and the almost unmeasurable compressions of the truss beams in the centres of bridges, and other cases of chief importance, the fulcrum is shifted far over towards x, so that very few fibres resist the fracture by their cohesion; and these few have a very feeble energy or momentum, on account of the short arm of the lever by which they act. This is a most important consideration in carpentry, yet it makes no element of Euler's theory. The consequence of this is, that a very small degree of curvature is sufficient to cause the column or strutt to snap in an instant, as is well known to every experienced carpenter. The experiment by Muschenbroek, which Euler makes use of in order to obtain a measure of strength in a particular instance, from which he might deduce all others by his theorem, is an incontestible proof of this. The force which broke the column is not the twentieth part of what is necessary for breaking it by acting at E in the direction EF. Euler takes no notice of this immense discrepancy, because it must have caused him to abandon the speculation with which he was then amusing himself.

424. We cannot find room at present to enter minutely upon the refutation of this theory; but we can easily show its uselessness, by its total inconsistency with common observation. It results legitimately from this theory, that if CD have no magnitude, the weight A can have no momentum, and the column cannot be broken. True, it cannot be broken in this way, snapped by a transverse fracture, if it do not bend; but we know very well that it can be crushed or crippled, and we see this frequently happen. This circumstance or event does not enter into Euler's investigation, and therefore the theory is imperfect at least, and useless. Had this crippling been introduced in the form of a physical assumption, every topic of reasoning employed in the process must have been laid aside, as the intelligent reader will easily see

But the theory is not only imperfect, but false. The ordinary reader will be convinced of this by another legitimate consequence of it. Fig. 20. is the same with Fig. 106. of *Emerson's Mechanics*, where this subject is treated on Euler's principles, and represents a crooked piece of matter resting on the ground at F, and loaded at A with a weight acting in the vertical direction AF. It results from Euler's theory that the strains at *b*, B, D, E, &c. are as *b c*, BC, DI, EK, &c. Therefore the strains at G and H are nothing; and this is asserted by Emerson and Euler as a serious truth; and the piece may be thinned *ad infinitum* in these two places, or even cut through, without any diminution of its strength. The absurdity of this assertion strikes at first hearing. Euler asserts the same thing with respect to a point of contrary flexure. Farther discussion is, we apprehend, needless.

425. This theory must therefore be given up. Yet these dissertations of Euler in the Petersburgh Commentaries deserve a perusal, both as very ingenious specimens of analysis, and because they contain maxims of practice which are important. Although they give an erroneous measure of the comparative strength of columns, they show the immense importance of preventing all bendings, and point out with accuracy where the tendencies to bend are greatest, and how this may be prevented by very small forces, and what a prodigious accession of force this gives the column. There is a valuable paper in the same volume by Fuss *on the Strains on framed Carpentry*, which may also be read with advantage.

426. It will now be asked, what shall be substituted in place of this erroneous theory? What is the true proportion of the strength of columns? We acknowledge our inability to give a satisfactory answer. Such can be obtained only by a previous knowledge of the proportion between the extensions and compressions produced by equal forces, by the knowledge of the absolute compres-

sions producible by a given force, and by a knowledge of the degree of that derangement of parts which is termed crippling. These circumstances are but imperfectly known to us, and there lies before us a wide field of experimental inquiry. Fortunately the force requisite for crippling a beam is prodigious, and a very small lateral support is sufficient to prevent that bending which puts the beam in imminent danger. A judicious engineer will always employ transverse bridles, as they are called, to stay the middle of long beams, which are employed as pillars, strutts, or truss beams, and are exposed, by their position, to enormous pressures in the direction of their lengths. Such stays may be observed, disposed with great judgment and economy, in the centres employed by Mr Perronet in the erection of his great stone arches. He was obliged to correct this omission made by his ingenious predecessor in the beautiful centres of the bridge of Orleans, which we have no hesitation in affirming to be the finest piece of carpentry in the world.

It only remains on this head to compare these theoretical deductions with experiment.

427. Experiments on the transverse strength of bodies are easily made, and accordingly are very numerous, especially those made on timber, which is the case most common and most interesting. But in this great number of experiments there are very few from which we can draw much practical information. The experiments have in general been made on such small scantlings, that the unavoidable natural inequalities bear too great a proportion to the strength of the whole piece. Accordingly, when we compare the experiments of different authors, we find them differ enormously, and even the experiments by the same author are very anomalous. The completest series that we have yet seen is that detailed by Belidor in his *Science des Ingenieurs*. They are contained in the following table. The pieces were sound,

even-grained oak. The column b contains the breadths of the pieces in inches; the column d contains their depths; the column l contains their lengths; column p contains the weights (in pounds) which broke them when hung on their middles; and m is the column of averages or mediums.

N	b	d	l	p	m	
1	1	1	18	400 415 405	406	The ends lying loose.
2	1	1	18	600 600 624	608	The ends firmly fixed.
3	2	1	18	810 795 812	805	Loose.
4	1	2	18	1570 1580 1590	1580	Loose.
5	1	1	36	185 195 180	187	Loose.
6	1	1	36	285 280 285	283	Fixed.
7	2	2	36	1550 1620 1585	1585	Loose.
8	$1\frac{2}{3}$	$2\frac{1}{3}$	36	1665 1675 1640	1660	Loose.

428. By comparing Experiments 1st and 3d, the strength appears proportional to the breadth.

Experiments 3d and 4th shew the strength proportional to the square of the depth.

Experiments 1st and 5th shew the strength nearly in the inverse proportion of the lengths, but with a sensible deficiency in the longer pieces.

Experiments 5th and 7th shew the strengths proportional to the breadths, and the square of the depth.

Experiments 1st and 7th shew the same thing, compounded with the inverse proportion of the length: the deficiency relative to the length is not so remarkable here.

Experiments 1st and 2d, and experiments 5th and 6th, shew the increase of strength, by fastening the ends, to be in the proportion of 2 to 3. The theory gives the proportion of 2 to 4. But a difference in the manner of fixing may produce this deviation from the theory, which only supposed them to be held down at places beyond the props, as when a joist is held in the walls, and also rests on two pillars between the walls. (See what is said on this subject under the article Roof.)

The chief source of irregularity in such experiments is the fibrous, or rather plated texture of timber. It consists of annual additions, whose cohesion with each other is vastly weaker than that of their own fibres. Let Fig. 21. represent the section of a tree, and ABCD, *a b c d* the section of two battens that are to be cut out of it for experiment, and let AD and *a d* be the depths, and DC, *d c* the breadths. The batten ABCD will be the strongest, for the same reason that an assemblage of planks set edgewise will form a stronger joist than planks laid above each other like the plates of a coach-spring. Mr Buffon found by many trials that the strength of ABCD was to that of *a b c d* (in oak) nearly as 8 to 7. The authors of the different experiments were not careful that their battens had their plates all disposed similarly with respect to the strain. But even with this precaution they would not have afforded sure grounds of computation for large works; for great beams occupy much, if not the whole, of the section of the tree; and from this it has happened that their strength is less than in proportion to that of a small lath or batten. In short, we can trust no experiments but such as have been made on large beams. These must be very rare, for they are most expensive and laborious, and exceed the abilities of most of those who are disposed to study this matter.

But we are not wholly without such authority. Mr Buffon and Mr Du Hamel, two of the first philosophers and mechanicians of the age, were directed by government to make experiments on this subject, and were supplied with ample funds and apparatus. The relation of their experiments is to be found in the Memoirs of the French Academy for 1740, 1741, 1742, 1768; as also in Du Hamel's valuable performances *sur l'Exploitation des Arbres, et sur la Conservation et le Transport de Bois.* We earnestly recommend these dissertations to the perusal of our readers, as containing much useful information relative to the strength of timber, and the best methods of employing it. We shall here give an abstract of Mr Buffon's experiments.

429. He relates a great number which he had prosecuted during two years on small battens. He found that the odds of a single layer, or part of a layer, more or less, or even a different disposition of them, had such influence that he was obliged to abandon this method, and to have recourse to the largest beams that he was able to break. The following table exhibits one series of experiments on bars of sound oak, clear of knots, and four inches square. This is a specimen of all the rest.

Column 1st is the length of the bar in feet clear between the supports.

Column 2d is the weight of the bar (the second day after it was felled) in pounds. Two bars were tried of each length. Each of the first three pairs consisted of two cuts of the same tree. The one next the root was always found the heaviest, stiffest, and strongest. Indeed Mr Buffon says that this was invariably true, that the heaviest was always the strongest; and he recommends it as a certain or sure rule for the choice of timber. He finds that this is always the case when the timber has grown vigorously, forming very thick annual layers. But he also observes that this is only during the advances of the tree to maturity; for the strength of the different circles approaches gradually to equality during the tree's healthy growth, and then it decays in these parts in a contrary order. Our toolmakers assert the same thing with respect to beech: yet a contrary opinion is very prevalent; and wood with a fine, that is, a small grain, is frequently preferred. Perhaps no person has ever made the trial with such minuteness as Mr Buffon, and we think that much deference is due to his opinion.

Column 3d is the number of pounds necessary for breaking the tree in the course of a few minutes.

Column 4th is the inches which it bent down before breaking.

Column 5th is the time at which it broke.

1	2	3	4	5
7	60 56	5350 5275	3,5 4,5	29 22
8	68 63	4600 4500	3,75 4,7	15 13
9	77 71	4100 3950	4,85 5,5	14 12
10	84 82	3625 3600	5,83 6,5	15 15
12	100 98	3050 2925	7, 8,	

The experiments on other sizes were made in the same way. A pair at least of each length and size was taken. The mean results are contained in the following table. The beams were all square, and their sizes in inches are placed at the head of the columns, and their lengths in feet are in the first column.

	4	5	6	7	8	A
7	5312	11525	18950	32200	47649	11525
8	4550	9787	15525	26050	39750	10085
9	4025	8308	13150	22350	32800	8964
10	3612	7125	11250	19475	27750	8068
12	2987	6075	9100	16175	23450	6723
14		5300	7475	13225	19775	5763
16		4350	6362	11000	16375	5042
18		3700	5562	9245	13200	4482
20		3225	4950	8375	11487	4034
22		2975				3667
24		2162				3362
28		1775				2881

Mr Buffon had found by numerous trials that oak-timber lost much of its strength in the course of drying or seasoning; and therefore, in order to secure uniformity, his trees were all felled in the same season of the year, were squared the day after, and tried the third day. Trying them in this green state gave him an opportunity of observing a very curious and unaccountable phenomenon. When the weights were laid briskly on, nearly sufficient to break the log, a very sensible smoke was observed to issue from the two ends with a sharp hissing noise. This continued all the while the tree was bending and cracking. This shows that the log is affected or strained through its whole length; indeed this must be inferred from its bending through its whole length. It also shows us the great effects of the compression. It is a pity Mr Buffon did not take notice whether this smoke issued from the upper or compressed half of the section only, or whether it came from the whole.

430. We must now make some observations on these experiments, in order to compare them with the theory which we have endeavoured to establish.

Mr Buffon considers the experiments with the 5 inch bars as the standard of comparison, having both extended these to greater lengths, and having tried more pieces of each length.

Our theory determines the relative strength of bars of the same section to be inversely as their lengths. But (if we except the five experiments in the first column) we find a very great deviation from this rule. Thus the 5-inch bar of 28 feet long should have half the strength of that of 14 feet, or 2650; whereas it is but 1775. The bar of 14 feet should have half the strength of that of 7 feet, or 5762; whereas it is but 5300. In like manner, the fourth of 11525 is 2881; but the real strength of the 28 feet bar is 1775. We have added a column A, which exhibits the strength which each of the 5-inch bars ought to have by the theory. This deviation is most distinctly seen in Fig. 22. where BK is the scale of lengths, B being at the point 7 of the scale and K at 28. The ordinate CB is = 11525, and the other ordinates DE, GK, &c. are respectively $= \frac{7\ \text{CB}}{\text{Length}}$. The lines DF, GH, &c. are made = 4350, 1775, &c. expressing the strengths given by experiment. The 10 feet bar and the 24 feet bar are remarkably anomalous. But all are deficient, and the defect has an evident progression from the first to the last. The same thing may be shown of the other columns, and even of the first, though it is very small in that column. It may also be observed in the experiments of Belidor, and in all that we have seen. We cannot doubt therefore of its being a law of nature, depending on the true principles of cohesion, and the laws of mechanics.

But it is very puzzling, and we cannot pretend to give a satisfactory explanation of the difficulty. The only effect which we can conceive the length of a beam to have, is to increase the strain at the section of fracture by employing the intervening beam as a lever. But we do not distinctly see what change this can produce in the mode of action of the fibres in this section, so as either to change their cohesion or the place of its centre of effort: yet something of this kind must happen.

We see indeed some circumstances which must contribute to make a smaller weight sufficient, in Mr Buffon's experiments, to break a long beam than in the exact inverse proportion of its length.

In the first place, the weight of the beam itself augments the strain as much as if half of it were added in the form of a weight. Mr Buffon has given the weights of every beam on which he made experiments, which is very nearly 74 pounds per cubic foot. But they are much too small to account for the deviation from the theory. The half weights of the 5-inch beams of 7, 14, and 28 feet length are only 45, 92, and 182 pounds; which makes the real strains in the experiments 11560, 5390, and 1956; which are far from having the proportions of 4, 2, and 1.

Buffon says that healthy trees are universally strongest at the root end; therefore when we use a longer beam, its middle point, where it is broken in the experiment, is in a weaker part of the tree. But the trials of the 4-inch beams show that the difference from this cause is almost insensible.

The length must have some mechanical influence which the theory we have adopted has not yet explained. It may not however be inadequate to the task. The very ingenious investigation of the elastic curve, by James Bernoulli and other celebrated mathematicians, is perhaps as refined an application of mathematical analysis as we know. Yet in this investigation it was necessary, in order to avoid almost insuperable difficulties, to take the simplest possible case, viz. where the thickness is exceedingly small in comparison with the length. If the thickness be considerable, the quantities neglected in the calculus are too great to permit the conclusion to be accurate, or very nearly so. Without being able to define the form into which an elastic body of considerable thickness will be bent, we can say with confidence, that in an extreme case, where the compression in the concave side is very great, the curvature differs considerably from the Bernoullian curve. But, as our investigation is incomplete and very long, we do not offer it to the reader.

431. The following more familiar considerations will, we apprehend, render it highly probable that the relative strength of beams decreases faster than in the inverse ratio of their length. The curious observation by Mr Buffon of the vapour which issued with a hissing noise from the ends of a beam of green oak, while it was breaking by the load on its middle, shows that the whole length of the piece was affected: indeed it must be, since it is

bent throughout. We have shown above, that a certain definite curvature of a beam of a given form is always accompanied by rupture. Now suppose the beam A of 10 feet long, and the beam B of 20 feet long, bent to the same degree, at the place of their fixture in the wall; the weight which hangs on A is nearly double of that which must hang on B. The form of any portion, suppose 5 feet, of these two beams, immediately adjoining to the wall, is considerably different. At the distance of 5 feet the curvature of A is $\frac{1}{2}$ of its curvature at the wall. The curvature of B in the corresponding point is $\frac{3}{4}$ths of the same curvature at the wall. Through the whole of the intermediate 5 feet, therefore, the curvature of B is greater than that of A. This must make it weaker throughout. It must occasion the fibres to slide more on each other (that it may acquire *this* greater curvature), and thus affect their lateral union; and therefore those which are stronger will not assist their weaker neighbours. To this we must add, that in the shorter beams the force with which the fibres are pressed laterally on each other is double. This must impede the mutual sliding of the fibres which we mentioned a little ago; nay, this lateral compression may change the law of longitudinal cohesion (as will readily appear to the reader who is acquainted with Boscovich's doctrines), and increase the strength of the very surface of fracture, in the same way, however inexplicable, as it does in metals when they are hammered or drawn into wire.

The reader must judge how far these remarks are worthy of his attention. The engineer will carefully keep in mind the important fact, that a beam of quadruple length, instead of having $\frac{1}{4}$th of the strength, has only about $\frac{1}{6}$th; and the philosopher should endeavour to discover the cause of this diminution, that he may give the artist a more accurate rule of computation.

432. Our ignorance of the law by which the cohesion

of the particles changes by a change of distance, hinders us from discovering the precise relation between the curvature and the momentum of cohesion; and all we can do is to multiply experiments, upon which we may establish some *empirical* rules for calculating the strength of solids. Those from which we must reason at present are too few and too anomalous to be the foundation of such an empirical formula. We may, however, observe, that Mr Buffon's experiments give us considerable assistance in this particular: For if to each of the numbers of the column for the 5-inch beams, corrected by adding half the weight of the beam, we add the constant number 1245, we shall have a set of numbers which are very nearly reciprocals of the lengths. Let 1245 be called c, and let the weight which is known by experiment to be necessary for breaking the 5-inch beam of the length a be called P. We shall have $\frac{\overline{P+c} \times a}{l} - c = p$. Thus the weight necessary for breaking the 7-foot bar is 11560. This added to 1245, and the sum multiplied by 7, gives $\overline{P+c} \times a =$ 89635. Let l be 18; then $\frac{89635}{18} - 1245 = 3725, = p$, which differs not more than $\frac{1}{40}$th from what experiment gives us. This rule holds equally well in all the other lengths except the 10 and 24 foot beams, which are very anomalous. Such a formula is abundantly exact for practice, and will answer through a much greater variety of length, though it cannot be admitted as a true one; because, in a certain very great length, the strength will be nothing. For other sizes the constant number must change in the proportion of d^3, or perhaps of p.

433. The next comparison which we have to make with the theory is the relation between the strength and the square of the depth of the section. This is made by comparing with each other the numbers in any horizontal

line of the table. In making this comparison we find the numbers of the 5-inch bars uniformly greater than the rest. We imagine that there is something peculiar to these bars: They are in general heavier than in the proportion of their section, but not so much as to account for all their superiority. We imagine that this set of experiments, intended as a standard for the rest, has been made at one time, and that the season has had a considerable influence. The fact however is, that if this column be kept out, or the numbers which represent the strengths be uniformly diminished about $\frac{1}{16}$th, the different sizes will deviate very little from the ratio of the square of the depth, as determined by theory. There is, however, a small deficiency in the larger beams.

We have been thus anxious in the examination of these experiments, because they are the only ones which have been related in sufficient detail, and made on a proper scale for giving us data from which we can deduce confidential maxims for practice. They are so troublesome and expensive that we have little hopes of seeing their number greatly increased; yet surely our navy board would do an unspeakable service to the public by appropriating a fund for such experiments under the management of some man of science.

434. There remains another comparison, which is of chief importance, namely, the proportion between the ABSOLUTE COHESION and the RELATIVE STRENGTH. It may be guessed, from the very nature of the thing, that this must be very uncertain. Experiments on the absolute strength must be confined to very small pieces, by reason of the very great forces which are required for tearing them asunder. The values therefore deduced from them must be subject to great inequalities. Unfortunately we have got no detail of any experiments; all that we have to depend on is two passages of Muschenbroek's *Essais de Physique*; in one of which he says, that a piece of sound oak

$\frac{27}{100}$ths of an inch square is torn asunder by 1150 pounds; and in the other, that an oak plank 12 inches broad and 1 thick will just suspend 189,163 pounds. These give for the cohesion of an inch square 15,755 and 15,763 pounds. Bouguer, in his *Traite' du Navire*, says, that it is very well known that a rod of sound oak $\frac{1}{4}$th of an inch square will be torn asunder by 1000 pounds. This gives 16,000 for the cohesion of a square inch. We shall take this as a round number, easily used in our computations. Let us compare this with Mr Buffon's trials of beams four inches square.

The absolute cohesion of this section is $16,000 \times 16 = 256,000$. Did every fibre exert its whole force in the instant of fracture, the momentum of cohesion would be the same as if it had all acted at the centre of gravity of the section at 2 inches from the axis of fracture, and is therefore 512,000. The 4-inch beam, 7 feet long, was broken by 5312 pounds hung on its middle. The half of this, or 2656 pounds, would have broken it, if suspended at its extremity, projecting $3\frac{1}{2}$ feet or 42 inches from a wall. The momentum of this strain is therefore $2656 \times 42, = 111552$. Now this is in equilibrio with the actual momentum of cohesion, which is therefore 111552 instead of 512000. The strength is therefore diminished in the proportion of 512000 to 111552, or very nearly of 459 to 1.

As we are quite uncertain as to the place of the centre of effort, it is needless to consider the full cohesion as acting at the centre of gravity, and producing the momentum 512,000; and we may convert the whole into a simple multiplier m of the length, and say, *as* m *times the length is to the depth, so is the absolute cohesion of the section to the relative strength.* Therefore let the absolute cohesion of a square inch be called f, the breadth b, the depth d, and the length l (all in inches), the relative strength, or the external force p, which balances it, is $\frac{f b d^2}{9,18 l}$, or in round numbers $\frac{f b d^2}{9 l}$; for $m = 2 \times 4,59$.

This great diminution of strength cannot be wholly accounted for by the inequality of the cohesive forces exerted in the instant of fracture; for in this case we know that the centre of effort is at one third of the height in a rectangular section (because the forces really exerted are as the extensions of the fibres). The relative strength would be $\frac{f b d^2}{3 l}$, and p would have been 8127 instead of 2656.

We must ascribe this diminution (which is three times greater than that produced by the inequality of the cohesive forces) to the compression of the under part of the beam; and we must endeavour to explain in what manner this compression produces an effect which seems so little explicable by such means.

As we have repeatedly observed, it is a matter of nearly universal experience that the forces *actually* exerted by the particles of bodies, when stretched or compressed, are very nearly in the proportion of the distances to which the particles are drawn from their natural positions. Now, although we are certain that, in enormous compressions, the forces increase faster than in this proportion, this makes no sensible change in the present question, because the body is broken before the compressions have gone so far; nay, we imagine that the compressed parts are crippled in most cases even before the extended parts are torn asunder. Muschenbroek asserts this with great confidence with respect to oak, on the authority of his own experiments. He says, that although oak will suspend half as much again as fir, it will not support, as a pillar, two-thirds of the load which fir will support in that form.

We imagine therefore that the mechanism in the *present* case is nearly as follows:

Let the beam DCK Δ (fig. 23.) be loaded at its extremity with the weight P, acting in the direction KP, perpendicular to DC. Let D Δ be the section of fracture. Let DA be about $\frac{1}{3}$d of D Δ. A will be the particle or fibre which is neither extended or compressed. Make $\Delta\delta : D\,d = DA : A\,\Delta$. The triangles DA$d$, Δ A δ, will represent the accumulated attracting and repelling forces. Make AI and A $i = \frac{1}{3}$d DA and $\frac{1}{3}$d Δ A. The point I will be that to which the full cohesion D d or f of the particles in AD must be applied, so as to produce the same momentum which the variable forces at I, D, &c. really produce at their several points of application. In like manner, i is the centre of similar effort of the repulsive forces excited by the compression between A and Δ, and it is the real fulcrum of a bended lever I $i\,k$, by which the whole effect is produced. The effect is the same as if the full cohesion of the stretched fibres in AD were accumulated in I, and the full repulsion of all the compressed fibres in A Δ were accumulated in i. The forces which are balanced in the operation are the weight P, acting by the arm $k\,i$, and the full cohesion of AD acting by the arm I i The forces exerted by the compressed fibres between A and Δ only serve to give support to the lever, that it may exert its strain.

We imagine that this does not differ much from the real procedure of nature. The position of the point A may be different from what we have deduced from Mr Buffon's experiments, compared with Muschenbroek's value of the absolute cohesion of a square inch. If this last should be only 12000, DA must be greater than we have here made it, in the proportion of 12000 to 16000. For I i must still be made $= \frac{2}{3}$d A Δ, supposing the forces to be proportional to the extensions and compressions. There can be no doubt that a part only of the cohesion of D Δ operates in resisting the fracture in all substances which have any compressibility; and it is confirmed by the experiments of Mr Du-Hamel on willow, and the inferences are by no means confined to that species of timber. We say, therefore, that when the beam is broken, the cohesion of AD alone is exerted, and that each fibre exerts a force proportional to its extension; and the accumulated momentum is the same as if the full cohesion of AD were acting by the lever I $i = \frac{1}{3}$d of D Δ.

It may be said, that if only $\frac{1}{3}$d of the cohesion of oak be exerted, it may be cut $\frac{2}{3}$ds through without weakening it. But this cannot be, because the cohesion of the whole is employed in preventing the lateral slide so often mentioned. We have no experiments to determine that it *may not* be cut through $\frac{1}{3}$d without loss of its strength.

This must not be considered as a subject of mere speculative curiosity: It is intimately connected with all the practical uses which we can make of this knowledge; for it is almost the only way that we can learn the compressibility of timber. Experiments on the direct cohesion are indeed difficult, and exceedingly expensive if we attempt them in large pieces. But experiments on compression are almost impracticable. The most instructive experiments would be, first to establish, by a great number of trials, the transverse force of a modern batten; and then to make a great number of trials of the diminution of its strength, by cutting it through on the concave side. This would very nearly give us the proportion of the cohesion which really operates in resisting fractures. Thus, if it be found that one half of the beam may be cut on the under side without diminution of its strength (taking care to drive in a slice of harder wood) we may conclude that the point A is at the middle, or somewhat above it.

Much lies before the curious mechanician, and we are

as yet very far from a scientific knowledge of the strength of timber.

481. In the mean time, we may derive from these experiments of Buffon a very useful practical rule, without relying on any value of the absolute cohesion of oak. We see that the strength is nearly as the breadth, as the square of the depth, and as the inverse of the length. It is most convenient to measure the breadth and depth of the beam in inches, and its length in feet. Since, then, a beam four inches square and seven feet between the supports is broken by 5312 pounds, we must conclude that a batten one inch square, and one foot between the supports, will be broken by 581 pounds. Then the strength of any other beam of oak, or the weight which will just break it when hung on its middle, is $581\,\frac{b\,d^2}{l}$.

But we have seen that there is a very considerable deviation from the inverse proportion of the lengths, and we must endeavour to accommodate our rule to this deviation. We found, that by adding 1245 to each of the ordinates or numbers in the column of the 5-inch bars, we had a set of numbers very nearly reciprocal of the lengths; and if we make a similar addition to the other columns in the proportion of the cubes of the sizes, we have nearly the same result. The greatest error (except in the case of experiments which are very irregular) does not exceed $\frac{1}{13}$th of the whole. Therefore, for a radical number, add to the 5312 the number 640, which is to 1245 very nearly as 4^3 to 5^3. This gives 5952. The 64th of this is 93, which corresponds to a bar of one inch square and seven feet long. Therefore 93×7 will be the reciprocal corresponding to a bar of one foot. This is 651. Take from this the present empirical correction, which is $\frac{b\,40}{b\,4}$, or 10, and there remains 641 for the strength of the bar. This gives us for a general rule $p = 651 \frac{b\,d^2}{l} - 10\,b\,d^2$.

Example. Required the weight necessary to break an oak beam eight inches square and 20 feet between the props, $p = 651 \times \frac{8 \times 8^2}{20} - 10 \times 8 \times 8^2$. This is 11545, whereas the experiment gives 11487. The error is very small indeed. The rule is most deficient in comparison with the five-inch bars, which we have already said appear stronger than the rest.

The following process is easily remembered by such as are not algebraists.

Multiply the breadth in inches twice by the depth, and call this product f Multiply f by 651, and divide by the length in feet. From the quotient take 10 times f. The remainder is the number of pounds which will break the beam.

We are not sufficiently sensible of our principles to be confident that the correction 10 f should be in the proportion of the section, although we think it most probable. It is quite empirical, founded on Buffon's experiments. Therefore the safe way of using this rule is to suppose the beam square, by increasing or diminishing its breadth till equal to the depth. Then find the strength by this rule, and diminish or increase it for the change which has been made in its breadth. Thus, there can be no doubt that the strength of the beam given as an example is double of that of a beam of the same depth and half the breadth.

The reader cannot but observe that all this calculation relates to the very greatest weight which a beam will bear for a very few minutes. Mr Buffon uniformly found that two-thirds of this weight sensibly impaired its strength, and frequently broke it at the end of two or three months. One half of this weight brought the beam to a certain bend, which did not increase after the first minute or two, and may be borne by the beam for any length of time. But the beam contracted a bend, of which it did not recover any considerable portion. One-third seemed to have no permanent effect on the beam; but it recovered its rectilineal shape completely, even after having been loaded several months, provided that the timber was seasoned when first loaded; that is to say, one-third of the weight which would quickly break a seasoned beam, or one-fourth of what would break one just felled, may lie on it for ever without giving the beam a sett.

We have no detail of experiments on the strength of other kinds of timber; only Mr Buffon says, that fir has about $\frac{6}{10}$ths of the strength of oak; Mr Parent makes it $\frac{10}{12}$ths; Emerson, $\frac{2}{3}$ds, &c.

We have been thus minute in our examination of the mechanism of this transverse strain, because it is the greatest to which the parts of our machines are exposed. We wish to impress on the minds of artists the necessity of avoiding this as much as possible. They are improving in this respect, as may be seen by comparing the centres on which stone arches of great span are now turned with those of former times. They were formerly a load of mere joists resting on a multitude of posts, which obstructed the navigation, and were frequently losing their shape by some of the posts sinking into the ground. Now they are more generally trusses, where the beams abutt on each other, and are relieved from transverse strains. But many performances of eminent artists are still very injudiciously exposed to cross strains. We may instance one which is considered as a fine work, viz. the bridge at Walton on Thames. Here every beam of the great arch is a joist, and it hangs together by framing. The finest piece of carpentry that we have seen is the centre employed in turning the arches of the bridge at

Orleans, described by Perronet. In the whole there is not one cross strain. The beam, too, of Hornblower's steam-engine is very scientifically constructed.

482. IV. The last species of strain which we are to examine is that produced by twisting. This takes place in all axles which connect the working parts of machines.

Although we cannot pretend to have a very distinct conception of that modification of the cohesion of a body by which it resists this kind of strain, we can have no doubt that, when all the particles act alike, the resistance must be proportional to the number. Therefore if we suppose the two parts ABCD, ABFE Fig. 24. of the body EFCD to be of insuperable strength, but cohering more weakly in the common surface AB, and that one part ABCD is pushed laterally in the direction AB, there can be no doubt that it will yield only there, and that the resistance will be proportional to the surface.

483. In like manner, we can conceive a thin cylindrical tube, of which KAH Fig. 25. is the section, as cohering more weakly in that section than anywhere else. Suppose it to be grasped in both hands, and the two parts twisted round the axis in opposite directions, as we would twist the two joints of a flute, it is plain that it will first fail in this section, which is the circumference of a circle, and the particles of the two parts which are contiguous to this circumference will be drawn from each other laterally. The total resistance will be as the number of equally resisting particles, that is, as the circumference (for the tube being supposed very thin, there can be no sensible difference between the dilatation of the external and internal particles). We can now suppose another tube within this, and a third within the second, and so on till we reach the centre. If the particles of each ring exerted the same force (by suffering the same

dilatation in the direction of the circumference), the resistance of each ring of the section would be as its circumference and its breadth (supposed indefinitely small), and the whole resistance would be as the surface; and this would represent the resistance of a solid cylinder. But when a cylinder is twisted in this manner by an external force applied to its circumference, the external parts will suffer a greater circular extension than the internal; and it appears that this extension (like the extension of a beam strained transversely) will be proportional to the distance of the particles from the axis. We cannot say that this is demonstrable, but we can assign no proportion that is more probable. This being the case, the forces simultaneously exerted by each particle will be as its distance from the axis. Therefore the whole force exerted by each ring will be as the square of its radius, and the accumulated force actually exerted will be as the cube of the radius; that is, the accumulated force exerted by the whole cylinder, whose radius is CA, is to the accumulated force exerted *at the same time* by the part whose radius is CE, as CA^3 to CE^3.

The whole cohesion now exerted is just two-thirds of what it would be if all the particles were exerting the same attractive forces which are just now exerted by the particles in the external circumference. This is plain to any person in the least familiar with the fluxionary calculus. But such as are not may easily see it in this way.

Let the rectangle AC *c a* Fig. 25. be set upright on the surface of the circle along the line CA, and revolve round the axis C *c*. It will generate a cylinder whose height is C *c* or A *a*, and having the circle KAH for its base. If the diagonal C *a* be supposed also to revolve, it is plain that the triangle *c* C *a* will generate a cone of the same height, and having for its base the circle described by the revolution of *c a*, and the point C for its apex. The cylindrical surface generated by A a will express the whole cohesion exerted by the circumference AHK, and the cylindrical surface generated by E e will represent the cohesion exerted by the circumference ELM, and the solid generated by the triangle CA a will represent the cohesion exerted by the whole circle AHK, and the cylinder generated by the rectangle AC c a will represent the cohesion exerted by the same surface if each particle had suffered the extension A a.

Now it is plain, in the first place, that the solid generated by the triangle e EC is to that generated by a AC as EC^3 to AC^3. In the next place, the solid generated by a AC is two thirds of the cylinder, because the cone generated by c C a is one-third of it.

484. We may now suppose the cylinder twisted till the particles in the external circumference lose their cohesion. There can be no doubt that it will now be wrenched asunder, all the inner circles yielding in succession. Thus we obtain one useful piece of information, viz. that a body of homogeneous texture resists a *simple twist* with two-thirds of the force with which it resists an attempt to force one part laterally from the other, or with one-third part of the force which will cut it asunder by a square-edged tool. For to drive a square-edged tool through a piece of lead, for instance, is the same as forcing a piece of the lead as thick as the tool laterally away from the two pieces on each side of the tool. Experiments of this kind do not seem difficult, and they would give us very useful information.

485. When two cylinders AHK and BNO Fig. 25. are wrenched asunder, we must conclude that the external particles of each are just put beyond their limits of cohesion, are equally extended, and are exerting equal forces. Hence it follows, that in the instant of fracture the sum total of the forces actually exerted are as the squares of the diameters.

For drawing the diagonal C e, it is plain that E e, $=$ A a, expresses the distension of the circumference ELM, and that the solid generated by the triangle CE e, expresses the cohesion exerted by the surface of the circle ELM, when the particles in the circumference suffer the extension E e equal to A a. Now the solids generated by CA a and CE e being respectively two-thirds of the corresponding cylinders, are as the squares of the diameters.

486. Having thus ascertained the real strength of the section, and its relation to its absolute lateral strength, let us examine its strength relative to the external force employed to break it. This examination is very simple in the case under consideration. The straining force must act by some lever, and the cohesion must oppose it by acting on some other lever. The centre of the section may be the neutral point whose position is not disturbed.

Let F be the force exerted laterally by an exterior particle. Let a be the radius of the cylinder, x the indeterminate distance of any circumference, and $\dot{x}$ the indefinitely small interval between the concentric arches; that is, let $\dot{x}$ be the breadth of a ring and x its radius. The forces being as the extensions, and the extensions as the distances from the axis, the cohesion actually exerted at any part of any ring will be $f\frac{x\,\dot{x}}{a}$. The force exerted by the whole ring (being as the circumference or as the radius) will be $f\frac{x^2\,\dot{x}}{a}$. The momentum of cohesion of a ring, being as the force multiplied by its lever, will be $f\frac{x^3\,\dot{x}}{a}$. The accumulated momentum will be the sum or fluent of $f\frac{x^3\,\dot{x}}{a}$; that is, when $x = a$, it will be $\frac{1}{4}f\frac{a^4}{a}$, $= \frac{1}{4}f\,a^3$.

487. Hence we learn that the strength of an axle, by which it resists being wrenched asunder by a force acting at a given distance from the axis, is as the cube of its diameter.

But farther, $\frac{1}{4} f a^3$ is $= f a^2 \times \frac{1}{4} a$. Now $f a^2$ represents the full lateral cohesion of the section. The momentum therefore is the same as if the full lateral cohesion were accumulated at a point distant from the axis by $\frac{1}{4}$th of the radius, or $\frac{1}{8}$th of the diameter of the cylinder.

Therefore let F be the number of pounds which measures the lateral cohesion of a circular inch, d the diameter of the cylinder in inches, and l the length of the lever by which the straining force p is supposed to act, we shall have $F \times \frac{1}{8} d^3 = p\, l$, and $F \frac{d^3}{8l} = p$.

We see in general that the strength of an axle, by which it resists being wrenched asunder by twisting, is as the cube of its diameter.

488. We see also that the internal parts are not acting so powerfully as the external. If a hole be bored out of the axle of half its diameter, the strength is diminished only $\frac{1}{8}$th, while the quantity of matter is diminished $\frac{1}{4}$th. Therefore hollow axles are stronger than solid ones containing the same quantity of matter. Thus let the diameter be 5 and that of the hollow 4: then the diameter of another solid cylinder having the same quantity of matter with the tube is 3. The strength of the solid cylinder of the diameter 5 may be expressed by 5^3 or 125. Of this the internal part (of the diameter 4) exerts 64; therefore the strength of the tube is 125—64, = 61. But the strength of the solid axle of the same quantity of matter and diameter 3 is 3^3, or 27, which is not half of that of the tube.

489. Engineers, therefore, have of late introduced this improvement in their machines, and the axles of cast iron

are all made hollow when their size will admit it. They have the additional advantage of being much stiffer, and of affording much better fixure for the flaunches, which are used for connecting them with the wheels or levers by which they are turned and strained. The superiority of strength of hollow tubes over solid cylinders is much greater in this kind of strain than in the former or transverse. In this last case the strength of this tube would be to that of the solid cylinder of equal weight as 61 to 32½ nearly.

490. The apparatus which we mentioned on a former occasion for trying the lateral strength of a square inch of solid matter, enabled us to try this theory of twist with all desirable accuracy. The bar which hung down from the pin in the former trials was now placed in a horizontal position, and loaded with a weight at the extremity. Thus it acted as a powerful lever, and enabled us to wrench asunder specimens of the strongest materials. We found the results perfectly conformable to the theory, in as far as it determined the proportional strength of different sizes and forms: but we found the ratio of the resistance to twisting to the simple lateral resistance considerably different; and it was some time before we discovered the cause.

491. We had here taken the simplest view that is possible of the action of cohesion in resisting a twist. It is frequently exerted in a very different way. When, for instance, an iron axle is joined to a wooden one by being driven into one end of it, the extensions of the different circles of particles are in a very different proportion. A little consideration will show that the particles in immediate contact with the iron axle are in a state of violent extension; so are the particles of the exterior surface of the wooden part, and the intermediate parts are less strained. It is almost impossible to assign the exact proportion of the cohesive forces exerted in the different

parts. Numberless cases can be pointed out where parts of the axle are in a state of compression, and where it is still more difficult to determine the state of the other particles. We must content ourselves with the deductions made from this simple case, which is fortunately the most common. In the experiments just now mentioned, the centre of the circle is by no means the neutral point, and it is very difficult to ascertain its place: but when this consideration occurred to us, we easily freed the experiments from this uncertainty, by extending the lever to both sides, and by means of a pulley applied equal force to each arm, acting in opposite directions. Thus the centre became the neutral point, and the resistance to twist was found to be $\frac{2}{3}$ds of the simple lateral strength.

492. We beg leave to mention here that our success in these experiments encouraged us to extend them much farther. We hoped by these means to discover the absolute cohesion of many substances, which would have required an enormous apparatus and a most unmanageable force to tear them asunder directly. But we could reason with confidence from the resistance to twist (which we could easily measure), provided that we could ascertain the proportion of the direct and the lateral strengths. Our experiments on chalk, finely prepared clay, and white bees wax, (of one melting and one temperature), were very consistent and satisfactory. But we have hitherto found great irregularities in this proportion in bodies of a fibrous texture like timber. These are the most important cases, and we still hope to be able to accomplish our project, and to give the public some valuable information. This being our sole object, it was our duty to mention the method which promises success, and thus excite others to the task; and it will be no mortification to us to be deprived of the honour of being the first who thus adds to the stock of experimental knowledge.

When the matter of the axle is of the most simple texture, such as that of metals, we do not conceive that the length of the axle has any influence on the fracture. It is otherwise if it be of a fibrous texture like timber: the fibres are bent before breaking, being twisted into spirals like a cork-screw. The length of the axle has somewhat of the influence of a lever in this case, and it is easier wrenched asunder if long. Accordingly we have found it so; but we have not been able to reduce this influence to calculation.

493. The reader is requested to accept of these endeavours to communicate information on this important and difficult subject. We are duly sensible of their imperfection, but flatter ourselves that we have in many instances pointed out the method which must be pursued for improving our knowledge on this subject; and we have given the English reader a more copious list of experiments on the strength of materials than he will meet with in our language. Many useful deductions might be made from these premises respecting the manner of disposing and combining the strength of materials in our structures. The best form of joists, mortises, tenons, scarphs; the rules for joggling, tabling, faying, fishing, &c. practised in the delicate art of mast-making, are all founded on this doctrine: but the discussion of these would be equivalent to writing a complete treatise of carpentry. We hope that this will be executed by some intelligent mechanician, for there is no mechanic art that is more susceptible of scientific treatment. Such a treatise, if well executed, could not fail of being well received by the public in this age of mechanical improvement *

---

* For farther information upon this subject, the reader is referred to the article CARPENTRY, and to the other articles where this subject is treated of, in the *Edinburgh Encyclopædia*, conducted by Dr BREWSTER, ED.

CARPENTRY.

494. CARPENTRY is the art of framing timber for the purposes of architecture, machinery, and, in general, for all considerable structures.

It is not intended in this article to give a full account of carpentry as a *mechanical* art, or to describe the various ways of executing its different works, suited to the variety of materials employed, the processes which must be followed for fashioning and framing them for our purposes, and the tools which must be used, and the manner in which they must be handled: This would be an occupation for volumes; and though of great importance, must be entirely omitted here. Our only aim at present will be to deduce, from the principles and laws of mechanics, and the knowledge which experience and judicious inferences from it have given us concerning the strength of timber, in relation to the strain laid on it, such maxims of construction as will unite economy with strength and efficacy.

This object is to be attained by a knowledge, 1st, of the strength of our materials, and of the absolute strain that is to be laid on them; 2dly, of the modifications of this strain, by the place and direction in which it is exerted, and the changes that can be made by a proper disposition of the parts of our structure; and, 3dly, having disposed every piece in such a manner as to derive the utmost advantage from its relative strength, we must know how to form the joints and other connections, in such a manner as to secure the advantages derived from this disposition.

495. This is evidently a branch of mechanical science which makes carpentry a *liberal* art, constitutes part of the learning of the ENGINEER, and distinguishes him from the workman. Its importance in all times and states of civil society is manifest and great. In the present condition of these kingdoms, raised, by the active ingenuity and energy of our countrymen, to a pitch of prosperity and influence unequalled in the history of the world, a condition which consists chiefly in the superiority of our manufactures, attained by prodigious multiplication of engines of every description, and for every species of labour, the SCIENCE (so to term it) of carpentry is of immense consequence. We regret therefore exceedingly, that none of our celebrated artists have done honour to themselves and their country, by digesting into a body of consecutive doctrines the results of their great experience, so as to form a system from which their pupils might derive the first principles of their education. The many volumes called COMPLETE INSTRUCTORS, MANUALS, JEWELS, &c. take a much humbler flight, and content themselves with instructing the mere workman, or sometimes give the master builder a few approved forms of roofs and other framings, with the rules for drawing them on paper; and from thence forming the working draughts which must guide the saw and the chissel of the workman. Hardly any of them offer any thing that can be called a principle, applicable to many particular cases, with the rules for this adaptation. We are indebted for the greatest part of our knowledge of this subject to the labours of literary men, chiefly foreigners, who have

published in the memoirs of the learned academies dissertations on different parts of what may be termed the *science of carpentry*.

496. The theory of carpentry is founded on two distinct portions of mechanical science, namely, a knowledge of the strains to which framings of timber are exposed, and a knowledge of their *relative* strength.

We shall therefore attempt to bring into one point of view the propositions of mechanical science that are more immediately applicable to the art of carpentry, and are to be found in various parts of this work, particularly under *Roof* and *Strength of Materials*. From these propositions we hope to deduce such principles as shall enable an attentive reader to comprehend distinctly what is to be aimed at in framing timber, and how to attain this object with certainty: and we shall illustrate and confirm our principles by examples of pieces of carpentry which are acknowledged to be excellent in their kind.

497. The most important proposition of general mechanics to the carpenter is that which exhibits the composition and resolution of forces; and we beg our practical readers to endeavour to form very distinct conceptions of it, and to make it very familiar to their mind. When accommodated to their chief purposes, it may be thus expressed:

1. If a body, or any part of a body, be at once pressed in the two directions AB, AC (Plate VI. fig. 1), and if the intensity or force of those pressures be in the proportion of these two lines, the body is affected in the same manner as if it were pressed by a single force acting in the direction AD, which is the diagonal of the parallelogram ABDC formed by the two lines, and whose intensity has the same proportion to the intensity of each of the other two that AD has to AB or AC.

Such of our readers as have *studied* the laws of motion, know that this is fully demonstrated. We refer them to the article Dynamics, where it is treated at some length. The practitioner in carpentry will get more useful confidence in the doctrine, if he will shut his book, and verify the theoretical demonstrations by actual experiments. They are remarkably easy and convincing. Therefore it is our request that the artist, who is not so habitually acquainted with the subject, do not proceed further till he has made it quite familiar to his thoughts. Nothing is so conducive to this as the actual experiment; and since this only requires the trifling expence of two small pulleys and a few yards of whipcord, we hope that none of our practical readers will omit it.

2. Let the threads A *d*, AF *b*, and AE *c* (fig. 2.), have the weights *d*, *b*, and *c*, appended to them, and let two of the threads be laid over the pulleys F and E. By this apparatus the knot A will be drawn in the directions AB, AC, and AK. If the sum of the weights *b* and *c* be greater than the single weight *d*, the assemblage will of itself settle in a certain determined form; if you pull the knot A out of its place, it will always return to it again, and will rest in no other position. For example, if the three weights are equal, the threads will always make equal angles, of 120 degrees each, round the knot. If one of the weights be three pounds, another four, and the third five, the angle opposite to the thread stretched by five pounds will always be square, &c. When the knot A is thus in equilibrio, we must infer, that the action of the weight *d*, in the direction A *d*, is in direct opposition to the combined action of *b*, in the direction AB, and of *c*, in the direction AC. Therefore, if we produce *d* A to any point D, and take AD to represent the magnitude of the force, or pressure exerted by the weight *d*, the pres-

sures exerted on A by the weights *b* and *c*, in the directions AB, AC, are in fact equivalent to a pressure acting in the direction AD, whose intensity we have represented by AD. If we now measure off by a scale on AF and AE the lines AB and AC, having the same proportions to AD that the weights *b* and *c* have to the weight *d*, and if we draw DB and DC, we shall find DC to be equal and parallel to AB, and DB equal and parallel to AC; so that AD is the diagonal of a parallelogram ABDC. We shall find this always to be the case, whatever are the weights made use of; only we must take care that the weight which we cause to act without the intervention of a pulley be less than the sum of the other two: if any one of the weights exceeds the sum of the other two, it will prevail, and drag them along with it.

Now, since we know that the weight *d* would just balance an equal weight *g*, pulling directly upwards by the intervention of the pulley G; and since we see that it just balances the weights *b* and *c*, acting in the directions AB, AC, we must infer, that the knot A is affected in the same manner by those two weights, or by the single weight *g*; and therefore, that *two pressures, acting in the directions, and with the intensities,* AB, AC, *are equivalent to a single pressure having the direction and proportion of* AD. In like manner, the pressures AB, AK, are equivalent to AH, which is equal and opposite to AC. Also AK and AC are equivalent to AI, which is equal and opposite to AB.

498. We shall consider this combination of pressures a little more particularly.

Suppose an upright beam BA Fig. 3. pushed in the direction of its length by a load B, and abutting on the ends of two beams AC, AD, which are firmly resisted at their extreme points C and D, which rest on two blocks, but are nowise joined to them: these two beams can resist no way but in the directions CA, DA; and therefore the pressures which they sustain from the beam BA are in the directions AC, AD. We wish to know how much each sustains? Produce BA to E, taking AE from a scale of equal parts, to represent the number of tons or pounds by which BA is pressed. Draw EF and EG parallel to AD and AC; then AF, measured on the same scale, will give us the number of pounds by which AC is strained or crushed, and AG will give the strain on AD.

It deserves particular remark here, that the length of AC or AD has no influence on the strain, arising from the thrust of BA, while the directions remain the same. The effects, however, of this strain are modified by the length of the piece on which it is exerted. This strain compresses the beam, and will therefore compress a beam of double length twice as much. This may change the form of the assemblage. If AC, for example, be very much shorter than AD, it will be much less compressed: The line CA will turn about the centre C, while DA will hardly change its position; and the angle CAD will grow more open, the point A sinking down. The artist will find it of great consequence to pay a very minute attention to this circumstance, and to be able to see clearly the change of shape which necessarily results from these mutual strains. He will see in this the cause of failure in many very great works. By thus changing shape, strains are often produced in places where there were none before, and frequently of the very worst kind, tending to break the beams across.

The dotted lines of this figure shew another position of the beam AD′. This makes a prodigious change, not only in the strain on AD′, but also in that on AC. Both of them are much increased; AG is almost doubled, and AF is four times greater than before. This addition was made to the figure, to shew what enormous strains may be produced by a very moderate force AE, when it is exerted on a very obtuse angle.

The 4th and 5th figures will assist the most uninstructed reader in conceiving how the very same strains AF, AG, are laid on these beams, by a weight simply hanging from a billet resting on A, pressing hard on AD, and also leaning a little on AC; or by an upright piece AE, joggled on the two beams AC, AD, and performing the office of an ordinary king-post. The reader will thus learn to call off his attention from the means by which the strains are produced, and learn to consider them abstractedly, merely as strains, in whatever situation he finds them, and from whatever cause they arise.

We presume that every reader will perceive, that the proportions of these strains will be precisely the same if every thing be inverted, and each beam be drawn or pulled in the opposite direction. In the same way that we have substituted a rope and weight in Fig. 4. or a king-post in Fig. 5. for the loaded beam BA of Fig. 3. we might have substituted the framing of Fig. 6. which is a very usual practice. In this framing, the batten DA is stretched by a force AG, and the piece AC is compressed by a force AF. It is evident, that we may employ a rope, or an iron rod hooked on at D, in place of the batten DA, and the strains will be the same as before.

This seemingly simple matter is still full of instruction; and we hope that the well informed reader will pardon us, though we dwell a little longer on it for the sake of the young artist.

By changing the form of this framing, as in Fig. 7. we produce the same strains as in the disposition represented by the dotted lines in Fig. 3. The strains on both the battens AD, AC, are now greatly increased.

The same consequences result from an improper change of the position of AC. If it is placed as in Fig. 8. the strains on both are vastly increased. In short, the rule is general; that the more open we make the angle against which the push is exerted, the greater are the strains which are brought on the strutts or ties which form the sides of the angle.

The reader may not readily conceive the piece AC of Fig. 8. as sustaining a compression; for the weight B appears to hang from AC as much as from AD. But his doubts will be removed by considering whether he could employ a rope in place of AC. He cannot: But AD may be exchanged for a rope. AC is therefore a *strutt*, and not a *tie.*

In Fig. 9. AD is again a strutt, butting on the block D, and AC is a tie: and the batten AC may be replaced by a rope. While AD is compressed by the force AG, AC is stretched by the force AF.

If we give AC the position represented by the dotted lines, the compression of AD is now AG′, and the force stretching AC′ is now AF′; both much greater than they were before. This disposition is analogous to Fig. 8. and to the dotted lines in Fig. 3. Nor will the young artist have any doubts of AC′ being on the stretch, if he consider whether AD can be replaced by a rope. It cannot, but AC′ may; and it is therefore not compressed, but stretched.

In Fig. 10. all the three pieces, AC, AD, and AB, are ties, on the stretch. This is the complete inversion of Fig. 3.; and the dotted position of AC induces the same changes in the forces AF′, AG′, as in Fig. 3.

Thus have we gone over all the varieties which can happen in the bearings of three pieces on one point. All calculations about the strength of carpentry are reduced to this case: for when more ties or braces meet in a point (a thing that rarely happens), we reduce them to three, by substituting for any two the force which results from their combination, and then combining this with another; and so on.

The young artist must be particularly careful not to mistake the kind of strain that is exerted on any piece of the framing, and suppose a piece to be a brace which is really a tie. It is very easy to avoid all mistakes in this matter by the following rule, which has no exception.

499. Take notice of the direction in which the piece acts from which the strain proceeds. Draw a line in that direction *from* the point on which the strain is exerted; and let its length (measured on some scale of equal parts) express the magnitude of this action in pounds, hundreds, or tons. From its *remote* extremity draw lines parallel to the pieces on which the strain is exerted. The line parallel to one piece will necessarily cut the other, or its direction produced: If it cut the piece itself, that piece is compressed by the strain, and it is performing the office of a strutt or brace: if it cut its direction produced, the piece is stretched, and it is a tie. In short, the strains on the pieces AC, AD, are to be estimated in the direction of the points F and G *from* the strained point A. Thus, in Fig. 3. the upright piece BA, loaded with the weight B, presses the point A in the direction AE: so does the rope AB in the other figures, or the batten AB in Fig. 5.

In general, if the straining piece is within the angle formed by the pieces which are strained, the strains which they sustain are of the opposite kind to that which it exerts. If it be pushing, they are drawing; but if it be within the angle formed by their directions produced, the strains which they sustain are of the same kind. All the three are either drawing or pressing. If the straining piece lie within the angle formed by one piece and the produced direction of the other, its own strain, whether compression or extension, is of the same kind with that of the most remote of the other two, and opposite to that of the nearest. Thus, in Fig. 9. where AB is drawing, the remote piece AC is also drawing, while AD is pushing or resisting compression.

In all that has been said on this subject, we have not spoken of any joints. In the calculations with which we are occupied at present, the resistance of joints has no share; and we must not suppose that they exert any force which tends to prevent the angles from changing. The joints are supposed perfectly flexible, or to be like compass joints; the pin of which only keeps the pieces together when one or more of the pieces draws or pulls. The carpenter must always suppose them all compass joints, when he calculates the thrusts and draughts of the different pieces of his frames. The strains on joints, and their power to produce or balance them, are of a different kind, and require a very different examination.

500. Seeing that the angles which the pieces make with each other are of such importance to the magnitude and the proportion of the excited strains, it is proper to find out some way of readily and compendiously conceiving and expressing this analogy.

In general, the strain on any piece is proportional to the straining force. This is evident.

Secondly, the strain on any piece AC is proportional to the sine of the angle which the straining force makes with the other piece directly, and to the sine of the angle which the pieces make with each other inversely.

For it is plain, that the three pressures AE, AF, and AG, which are exerted at the point A, are in the proportion of the lines AE, AF, and FE, (because FE is equal to AG). But because the sides of a triangle are proportional to the sines of the opposite angles, the strains are proportional to the sines of the angles AFE, AEF, and FAE. But the sine of AFE is the same with the sine of the angle CAD, which the two pieces AC and

AD make with each other; and the sine of AEF is the same with the sine of EAD, which the straining piece BA makes with the piece AC. Therefore we have this analogy, Sin. CAD : Sin. EAD = AE : AF, and $AF = AE \times \frac{\text{Sin. EAD}}{\text{Sin. CAD}}$. Now the sine of angles are most conveniently conceived as decimal fractions of the radius, which is considered as unity. Thus, *Sin.* 30° is the same thing with 0.5, or ½; and so of others. Therefore, to have the strain on AC, arising from any load AE acting in the direction AE, multiply AE by the sine of EAD, and divide the product by the sine of CAD.

This rule shews how great the strains must be when the angle CAD becomes very open, approaching to 180 degrees. But when the angle CAD becomes very small, its sine (which is our divisor) is also very small; and we should expect a very great quotient in this case also. But we must observe, that in this case the sine of EAD is also very small; and this is our multiplier. In such a case, the quotient cannot exceed unity.

But it is unnecessary to consider the calculation by the tables of sines more particularly. The angles are seldom known any otherwise but by drawing the figure of the frame of carpentry. In this case, we can always obtain the measures of the strains from the same scale, with equal accuracy, by drawing the parallelogram AFCG.

501. Hitherto we have considered the strains excited at A only as they affect the pieces on which they are exerted. But the pieces, in order to sustain, or be subject to any strain, must be supported at their ends C and D; and we may consider them as mere intermediums, by which these strains are made to act on those points of support: Therefore AF and AG are also measures of the forces which press or pull at C and D. Thus we learn the supports which must be found for these points.

These may be infinitely various. We shall attend only to such as somehow depend on the framing itself.

502. Such a structure as Fig. 11. very frequently occurs, where a beam BA is strongly pressed to the end of another beam AD, which is prevented from yielding, both because it lies on another beam HD, and because its end D is hindered from sliding backwards. It is indifferent from what this pressure arises: we have represented it as owing to a weight hung on at B, while B is withheld from yielding by a rod or rope hooked to the wall. The beam AD may be supposed at full liberty to exert all its pressure on D, as if it were supported on rollers lodged in the beam HD; but the loaded beam BA presses both on the beam AD and on HD. We wish only to know what strain is borne by AD?

All bodies act on each other in the direction perpendicular to their touching surfaces; therefore the support given by HD is in a direction perpendicular to it. We may therefore supply its place at A by a beam AC, perpendicular to HD, and firmly supported at C. In this case, therefore, we may take AE as before, to represent the pressure exerted by the loaded beam, and draw EG perpendicular to AD, and EF parallel to it, meeting the perpendicular AC in F. Then AG is the strain compressing AD, and AF is the pressure on the beam HD.

503. It may be thought, that, since we assume as a principle that the mutual pressures of solid bodies are exerted perpendicular to their touching surfaces, this balance of pressures, in framings of timbers, depends on the directions of their butting joints: but it does not, as will readily appear by considering the present case. Let the joint or abutment of the two pieces BA, AD, be mitred, in the usual manner, in the direction f A f'. Therefore, if A e be drawn perpendicular to A f, it will be the direction of the actual pressure exerted by the loaded beam BA on the beam AD. But the reaction of AD, in the

opposite direction A t, will not balance the pressure of BA; because it is not in the direction precisely opposite. BA will therefore slide along the joint, and press on the beam HD. AE represents the load on the mitre joint A. Draw E e perpendicular to A e, and E f parallel to it. The pressure AE will be balanced by the reactions e A and f A: or, the pressure AE produces the pressures A e and A f; of which A f must be resisted by the beam HD, and A e by the beam AD. The pressure A f not being perpendicular to HD, cannot be fully resisted by it; because (by our assumed principle) it reacts only in a direction perpendicular to its surface. Therefore draw $f\,p$, $f\,i$ parallel to HD, and perpendicular to it. The pressure A f will be resisted by HD with the force p A; but there is required another force i A, to prevent the beam BA from slipping outwards. This must be furnished by the reaction of the beam DA. In like manner, the other force A e cannot be fully resisted by the beam AD, or rather by the prop D, acting by the intervention of the beam; for the action of that prop is exerted through the beam in the direction DA. The beam AD, therefore, is pressed to the beam HD by the force A e, as well as by A f. To find what this pressure on HD is, draw $e\,g$ perpendicular to HD, and $e\,o$ parallel to it, cutting EG in r. The forces g A and o A will resist, and balance A e.

Thus we see, that the two forces A e and A f, which are equivalent to AE, are equivalent also to A p, A i, A o, and A g. But because A f and e E are equal and parallel, and E r and $f\,i$ are also parallel, as also $e\,r$ and $f\,p$, it is evident, that $i\,f$ is equal to r E, or to o F, and i A is equal to r e, or to G g. Therefore the four forces A g, A o, A p, A i, are equal to AG and AF. Consequently AG is the compression of the beam AD, or the force pressing it on D, and AF is the force pressing it on the beam HD. The proportion of these pressures, therefore, is not affected by the form of the joint.

This remark is important; for many carpenters think the form and direction of the butting joint of great importance; and even the theorist, by not prosecuting the general principle through *all* its consequences, may be led into an error. The form of the joint is of no importance, in as far as it affects the strains in the direction of the beams; but it is often of great consequence, in respect to its own firmness, and the effect it may have in bruising the piece on which it acts, or being crippled by it.

504. The same compression of AB, and the same thrust on the point D by the intervention of AD, will obtain, in whatever way the original pressure on the end A is produced. Thus supposing that a chord is made fast at A, and pulled in the direction AE, and with the same force, the beam AD will be equally compressed, and the prop D must react with the same force.

But it often happens that the obliquity of the pressure on AD, instead of compressing it, stretches it; and we desire to know what tension it sustains. Of this we have a familiar example in a common roof. Let the two rafters AC, AD Fig. 12., press on the tie-beam DC. We may suppose the whole weight to press vertically on the ridge A, as if a weight B were hung on there. We may represent this weight by the portion A *b* of the vertical or plumb line, intercepted between the ridge and the beam. Then drawing *b f* and *b g* parallel to AD and AC, A *g* and A *f* will represent the pressures on AC and AD. Produce AC till CH be equal to A *f*. The point C is forced out in this direction, and with a force represented by this line. As this force is not perpendicularly across the beam, it evidently stretches it; and this extending force must be withstood by an equal force pulling it in the opposite direction. This must arise from a similar oblique thrust of the opposite rafter on the other end D. We concern ourselves only with this extension at present; but we see that the cohesion of the beam does nothing but supply the balance to the extending forces. It must still be supported externally, that it *may resist*, and, by resisting obliquely, be stretched. The points C and D are supported on the walls, which they press in the directions CK and DO, parallel to A *b*. If we draw HK parallel to DC, and HI parallel to CK (that is, to A *b*), meeting DC produced in I, it follows from the composition of forces, that the point C would be supported by the two forces KC and IC. In like manner, making DN = A *g*, and completing the parallelogram DMNO, the point D would be supported by the forces OD and MD. If we draw *g o* and *f k* parallel to DC, it is plain that they are equal to NO and CK, while A *o* and A *k* are equal to DO and CK, and A *b* is equal to the sum of DO and CK (because it is equal to A *o* + A *k*.) The weight of the roof is equal to its vertical pressure on the walls.

Thus we see, that while a pressure on A, in the direction A *b*, produces the strains A *f* and A *g*, on the pieces AC and AD, it also excites a strain CI or DM in the piece DC. And this completes the mechanism of a frame; for all derive their efficacy from the triangles of which they are composed, as will appear more clearly as we proceed.

505. But there is more to be learned from this. The consideration of the strains on the two pieces AD and AC, by the action of a force at A, only shewed them as the means of propagating the same strains in their own direction to the points of support. But, by adding the strains exerted in DC, we see that the frame becomes an intermedium, by which exertions may be made on other bodies, in certain directions and proportions; so that this frame may become part of a more complicated one, and,

as it were, an element of its constitution. It is worth while to ascertain the proportion of the pressures CK and DO, which are thus exerted on the walls. The similarity of triangles gives the following analogies:

$$\mathrm{DO} : \mathrm{DM} = \mathrm{A}\,b : b\,\mathrm{D}$$

$$\mathrm{CI}, \text{ or } \mathrm{DM} : \mathrm{CK} = \mathrm{C}\,b : \mathrm{A}\,b$$

$$\text{Therefore } \mathrm{DO} : \mathrm{CK} = \mathrm{C}\,b : b\,\mathrm{D}.$$

Or, *the pressures on the points* C *and* D, *in the direction of the straining force* A *b*, *are reciprocally proportional to the portions of* DC *intercepted by* A *b*.

Also, since A *b* is = DO + CK, we have

$\mathrm{A}\,b : \mathrm{CK} = \mathrm{C}\,b + b\,\mathrm{D}$ (or CD) $: b\,\mathrm{D}$, and

$\mathrm{A}\,b : \mathrm{DO} = \mathrm{CD} : b\,\mathrm{C}.$

In general, any two of the three parallel forces A *b*, DO, CK, are to each other in the reciprocal proportion of the parts of CD, intercepted between their directions and the direction of the third.

And this explains a still more important office of the frame ADC. If one of the points, such as D, be supported, an external power acting at A, in the direction A *b*, and with an intensity which may be measured by A *b*, may be set in equilibrio, with another acting at C, in the direction CL, opposite to CK, or A *b*, and with an intensity represented by CK: for since the pressure CH is partly withstood by the force IC, or the firmness of the beam DC supported at D, the force KC will complete the balance. When we do not attend to the support at D, we conceive the force A *b* to be balanced by KC, or KC to be balanced by A *b*. And, in like manner, we may neglect the support or force acting at A, and consider the force DO as balanced by CK.

506. Thus our frame becomes a lever, and we are able to trace the interior mechanical procedure which gives it its efficacy: it is by the intervention of the forces of cohesion, which connect the points to which the external

forces are applied with the supported point or fulcrum, and with each other.

These strains or pressures A *b*, DO, and CK, not being in the directions of the beams, may be called *transverse*. We see that by their means a frame of carpentry may be considered as a solid body: but the example which brought this to our view is too limited for explaining the efficacy which may be given to such constructions. We shall therefore give a general proposition, which will more distinctly explain the procedure of nature, and enable us to trace the strains as they are propagated through all the parts of the most complicated framing, finally producing the exertion of its most distant points.

507. We presume that the reader is now pretty well habituated to the conception of the strains as they are propagated along the lines joining the points of a frame, and we shall therefore employ a very simple figure.

Let the strong lines ACBD Fig. 13. represent a frame of carpentry. Suppose that it is pulled at the point A by a force acting in the direction AE, but that it rests on a fixed point C, and that the other extreme point B is held back by a power which resists in the direction BF: It is required to determine the proportion of the strains excited in its different parts, the proportion of the external pressures at A and B, and the pressure which is produced on the obstacle or fulcrum C?

It is evident that each of the external forces at A and B tend one way, or to one side of the frame, and that each would cause it to turn round C if the other did not prevent it; and that if, notwithstanding their action, it is turned neither way, the forces in actual exertion are in equilibrio by the intervention of the frame. It is no less evident that these forces concur in pressing the frame on the prop C. Therefore, if the piece CD were away, and if the joints C and D be perfectly flexible, the pieces

CA, CB would be turned round the prop C, and the pieces AD, DB would also turn with them, and the whole frame change its form. This shews, by the way, and we desire it to be carefully kept in mind, that the firmness or stiffness of framing depends entirely on the triangles bounded by beams which are contained in it. An open quadrilateral may always change its shape, the sides revolving round the angles. A quadrilateral may have an infinity of forms, without any change of its sides, by merely pushing two opposite angles towards each other, or drawing them asunder. But when the three sides of a triangle are determined, its shape is also invariably determined; and if two angles be held fast, the third cannot be moved. It is thus that, by inserting the bar CD, the figure becomes unchangeable; and any attempt to change it by applying a force to an angle A, immediately excites forces of attraction or repulsion between the particles of the stuff which forms its sides. Thus it happens, in the present instance, that a change of shape is prevented by the bar CD. The power at A presses its end against the prop; and in doing this it puts the bar AD on the stretch, and also the bar DB. Their places might therefore be supplied by cords or metal wires. Hence it is evident that DC is compressed, as is also AC; and, for the same reason, CB is also in a state of compression; for either A or B may be considered as the point that is impelled or withheld. Therefore DA and DB are stretched, and are resisting with attractive forces. DC and CB are compressed, and are resisting with repulsive forces. DB is also acting with repulsive forces, being compressed in like manner: and thus the support of the prop, combined with the firmness of DC, puts the frame ADBC into the condition of the two frames in Fig. 8. and Fig. 9. Therefore the external force at A is really in equilibrio with an attracting force acting in the direction AD, and a repulsive force acting in the direction AK. And since all the connecting forces are mutual and equal, the point D is pulled or drawn in the direction DA. The condition of the point B is similar to that of A, and D is also drawn in the direction DB. Thus the point D, being urged by the forces in the directions DA and DB, presses the beam DC on the prop, and the prop resists in the opposite direction. Therefore the line DC is the diagonal of the parallelogram, whose sides have the proportion of the forces which connect D with A and B. This is the principle on which the rest of our investigation proceeds. We may take DC as the representation and measure of their joint effect. Therefore draw CH, CG, parallel to DA, DB, cutting AE, BF in L and O, and cutting DA, DB in I and M. Complete the parallelograms ILKA, MONB. Then DG and AI are the equal and opposite forces which connect A and D; for GD = CH, = AI. In like manner DH and BM are the forces which connect D and B.

The external force at A is in immediate equilibrio with the combined forces, connecting A with D and with C. AI is one of them: Therefore AK is the other; and AL is the compound force with which the external force at A is in immediate equilibrium. This external force is therefore equal and opposite to BO; and AL is to BO as the external force at A to the external force at B. The prop C resists with forces equal to those which are propagated to it from the points D, A, and C. Therefore it resists with forces CH, CG, equal and opposite to DG, DH; and it resists the compressions KA, NB, with equal and opposite forces C *k*, C *n*. Draw *k l*, *n o* parallel to AD, BD, and draw C *l* Q, C *o* P: It is plain that *k* CH *l* is a parallelogram equal to KAIL, and that C *l* is equal to AL. In like manner C *o* is equal to BO. Now the forces C *k*, CH, exerted by the prop, compose the force C *l*; and C *n*, CG compose the force C *o*. These two forces C *l*, C *o* are equal and parallel to AL and BO; and therefore they are equal and opposite to the external forces acting at A and B. But they are (primitively) equal and opposite to the pressures (or at least the compounds of the pressures) exerted on the prop, by the forces propagated to C from A, D, and B. Therefore the pressures exerted on the prop are the same as if the external forces were applied there in the same directions as they are applied to A and B. Now if we make CV, CZ equal to C l and C o, and complete the parallelogram CVYZ; it is plain that the force YC is in equilibrio with l C and o C. Therefore the pressures at A, C, and B, are such as would balance if applied to one point.

Lastly, in order to determine their proportions, draw CS and CR perpendicular to DA and DB. Also draw A d, B f perpendicular to CQ and CP; and draw C g, C i perpendicular to AE, BF.

The triangles CPR and BP f are similar, having a common angle P, and a right angle at R and f.

In like manner the triangles CQS and AQ d are similar. Also the triangles CHR, CGS are similar, by reason of the equal angles at H and G, and the right angles at R and S. Hence we obtain the following analogies:

$$\begin{aligned} Co : CP &= On : PB, = CG : PB \\ CP : CR &= PB : fB \\ CR : CS &= CH : CG \\ CS : CQ &= Ad : AQ \\ CQ : Cl &= AQ : Kl, = AQ : CH. \end{aligned}$$

Therefore, by equality,

$$\begin{aligned} Co : Cl &= Ad : fB \\ \text{or } BO : AL &= Cg : Ci. \end{aligned}$$

That is, the external forces are reciprocally proportional to the perpendiculars drawn from the prop on the lines of their direction. *

The learned reader will perceive, that this analogy is precisely the same with that of forces which are in equilibrio by the intervention of a

This proposition (sufficiently general for our purpose) is fertile in consequences, and furnishes many useful instructions to the artist. The strains LA, OB, CY, that are excited, occur in many, we may say in all, framings of carpentry, whether for edifices or engines, and are the sources of their efficacy. It is also evident, that the doctrine of the transverse strength of timber is contained in this proposition ; for every piece of timber may be considered as an assemblage of parts, connected by forces which act in the direction of the lines which join the strained points on the matter which lies between those points, and also act on the rest of the matter, exciting those lateral forces which produce the inflexibility of the whole. See STRENGTH OF MATERIALS.

Thus it appears that this proposition contains the principles which direct the artist to frame the most powerful levers ; to secure uprights by shores or braces, or by ties and ropes ; to secure scaffoldings for the erection of spires,

---

lever. In fact, this whole frame of carpentry is nothing else than a *built* or *framed lever* in equilibrio. It is acting in the same manner as a solid, which occupies the whole figure compressed in the frame, or as a body of any size and shape whatever, that will admit the three points of application A, C, and B. It is always in equilibrio in the case first stated ; because the pressure *produced* at B by a force applied to A is always such as balances it. The reader may also perceive, in this proposition, the analysis or tracing of those internal mechanical forces which are indispensably requisite for the functions of a lever. The mechanicians have been extremely puzzled to find a legitimate demonstration of the equilibrium of a lever ever since the days of Archimedes. Mr Vince has the honour of first demonstrating, most ingeniously, the principle assumed by Archimedes, but without sufficient ground, for *his* demonstration : but Mr Vince's demonstration is only a putting the mind into that perplexed state which makes it acknowledge the proposition, but without a clear perception of its truth. The difficulty has proceeded from the abstract notion of a lever, conceiving it as a mathematical line—inflexible, without reflecting how it is inflexible —for the very source of this indispensable quality furnishes the mechanical connection between the remote pressures and the fulcrum ; and this supplies the demonstration (without the least difficulty) of the desperate case of a straight lever urged by parallel forces. See ROTATION.

and many other most delicate problems of his art. He also learns, from this proposition, how to ascertain the strains that are produced, without his intention, by pieces which he intended for other offices, and which, by their transverse action, put his work in hazard. In short, this proposition is the key to the science of his art.

We would now counsel the artist, after he has made the tracing of the strains and thrusts through the various parts of a frame familiar to his mind, and even amused himself with some complicated fancy framings, to read over with care the articles STRENGTH OF MATERIALS and ROOF. He will now conceive its doctrines much more clearly than when he was considering them as abstract theories. The mutual action of the woody fibres will now be easily comprehended, and his confidence in the results will be greatly increased.

508. There is a proposition (see the article ROOF) which has been called in question by several very intelligent persons; and they say that Belidor has demonstrated, in his SCIENCE DES INGENIEURS, that a beam firmly fixed at both ends is not twice as strong as when simply lying on the props, and that its strength is increased only in the proportion of 2 to 3; and they support this determination by a list of experiments recited by Belidor, which agree *precisely* with it. Belidor also says, that Pitot had the same result in his experiments. These are respectable authorities; but Belidor's reasoning is any thing but demonstration; and his experiments are described in such an imperfect manner, that we cannot build much on them. It is not said in what manner the battens were secured at the ends, any farther than that it was by *chevalets*. If by this word is meant a tressle, we cannot conceive how they were employed; but we see it sometimes used for a wedge or key. If the battens were wedged in the holes, their resistance to fracture may be made what we please: they may be loose, and therefore

resist little more than when simply laid on the props. They may be (and probably were) wedged very fast, and bruised or crippled.

Our proposition mentioned distinctly the security given to the ends of the beams. They were mortised into remote posts. Our *precise* meaning was, that they were simply kept from rising by these mortises, but at full liberty to bend up between E and I, and between G and K. Our assertion was not made from theory alone (although we think the reasoning incontrovertible), but was agreeable to numerous experiments made in those precise circumstances. Had we mortised the beams firmly into two very stout posts, which could not be drawn nearer to each other by bending, the beam would have borne a *much* greater weight, as we have verified by experiment. We hope that the following mode of conceiving this case will remove all doubts.

Let LM be a long beam, Plate VI. Fig. 14. divided into six equal parts, in the points D, B, A, C, E. Let it be firmly supported at L, B, C, M. Let it be cut through at A, and have compass joints at B and C. Let FB, GC be two equal uprights, resting on B and C, but without any connection. Let AH be a similar and equal piece, to be occasionally applied at the seam A. Now let a thread or wire AGE be extended over the piece GC, and made fast at A, G, and E. Let the same thing be done on the other side of A. If a weight be now laid on at A, the wires AFD, AGE will be strained, and may be broken. In the instant of fracture we may suppose their strains to be represented by A *f* and A *g*. Complete the parallelogram, and A *a* is the magnitude of the weight. It is plain that nothing is concerned here but the cohesion of the wires; for the beam is sawed through at A, and its parts are perfectly moveable round B and C.

Instead of this process apply the piece AH below A, and keep it there by straining the same wire BHC over it. Now lay on a weight. It must press down the ends of BA and CA, and cause the piece AH to strain the wire BHC. In the instant of fracture of *the same* wire, its resistances H b and H c must be equal to A f and A g, and the weight h H which breaks them must be equal to A a.

Lastly, employ all the three pieces FB, AH, GC, with the same wire attached as before. There can be no doubt but that the weight which breaks all the four wires must be $= a\,A + h\,H$, or twice A a.

The reader cannot but see that the wires perform the very same office with the fibres of an entire beam LM held fast in the four holes D, B, C, and E, of some upright posts.

In the experiments for verifying this, by breaking slender bars of fine deal, we get complete demonstration, by measuring the curvatures produced in the parts of the beam thus held down, and comparing them with the curvature of a beam simply laid on the props B and C: and there are many curious inferences to be made from these observations, but we have not room for them in this place.

509. We may observe, by the way, that we learn from this case, that purlins are able to carry twice the load when notched into the rafters that they carry when mortised into them, which is the most usual manner of framing them. So would the binding joists of floors; but this would double the thickness of the flooring. This method should be followed in every possible case, such as breast summers, lintels over several pillars, &c. These should never be cut off and mortised into the sides of every upright; numberless cases will occur which shew the importance of the maxim.

We must here remark, that the proportion of the spaces BC and CM, or BC and LB, has a very sensible effect on the strength of the beam BC; but we have not yet satisfied our minds as to the *rationale* of this effect. It is undoubtedly connected with the serpentine form of the curve of the beam before fracture. This should be attended to in the construction of the springs of carriages. These are frequently supported at a middle point (and it is an excellent practice), and there is a certain proportion which will give the easiest motion to the body of the carriage. We also think that it is connected with that deviation from the best theory observable in Buffon's experiments on various lengths of the same scantling. The force of the beams diminished much more than in the inverse proportion of their lengths.

510. We have seen that it depends entirely on the position of the pieces in respect of their points of ultimate support, and of the direction of the external force which produces the strains, whether any particular piece is in a state of extension or of compression. The knowledge of this circumstance may greatly influence us in the choice of the construction. In many cases we may substitute slender iron rods for massive beams, when the piece is to act the part of a tie. But we must not invert this disposition; for when a piece of timber acts as a strut, and is in a state of compression, it is next to certain that it is not equally compressible in its opposite sides through the whole length of the piece, and that the compressing force on the abutting joint is not acting in the most equable manner all over the joint. A very trifling inequality in either of these circumstances (especially in the first) will compress the beam more on one side than on the other. This cannot be without the beam's bending, and becoming concave on that side on which it is most compressed. When this happens, the frame is in danger of being crushed, and soon going to ruin. It is therefore indispensably necessary to make use of beams in all cases where struts are required of considerable length, rather than of metal rods of slender dimensions, unless in situa-

tions where we can effectually prevent their bending, as in trussing a girder internally, where a cast iron strut may be firmly cased in it, so as not to bend in the smallest degree. In cases where the pressures are enormous, as in the very oblique struts of a centre or arch frame, we must be particularly cautious to do nothing which can facilitate the compression of either side. No mortises should be cut near to one side; no lateral pressures, even the slightest, should be allowed to touch it. We have seen a pillar of fir 12 inches long and one inch in section, when loaded with three tons, snap in an instant when pressed on one side by 16 pounds, while another bore 4½ tons without hurt, because it was inclosed (loosely) in a stout pipe of iron.

In such cases of enormous compression, it is of great importance that the compressing force bear equally on the whole abutting surface. The German carpenters are accustomed to put a plate of lead over the joint. This prevents, in some measure, the penetration of the end fibres. Mr Perronet, the celebrated French architect, formed his abutments into arches of circles, the centre of which was the remote end of the strut. By this contrivance the unavoidable change of form of the triangle made no partial bearing of either angle of the abutment. This always has a tendency to splinter off the heel of the beam where it presses strongest. It is a very judicious practice.

When circumstances allow it, we should rather employ ties than struts for securing a beam against lateral strains. When an upright pillar, such as a flag-staff, a mast, or the uprights of a very tall scaffolding, are to be shoared up, the dependence is more certain on those braces that are stretched by the strain than on those which are compressed. The scaffolding of the iron bridge near Sunderland had some ties very judiciously disposed, and others with less judgment.

We should proceed to consider the transverse strains as

they affect the various parts of a frame of carpentry; but we have very little to add to what has been said already under the subject of STRENGTH OF MATERIALS, and under the subject of ROOF. What we shall add in this article will find a place in our occasional remarks on different works. It may, however, be of use to recal to the reader's memory the following propositions.

511. 1. When a beam AB, Fig. 15. is firmly fixed at the end A, and a straining force acts perpendicularly to its length at any point B, the strain occasioned at any section C between B and A is proportional to CB, and may therefore be represented by the product $W \times CB$; that is, by the product of the number of tons, pounds, &c. which measure the straining force, and the number of feet, inches, &c. contained in CB. As the loads on a beam are easily conceived, we shall substitute this for any other straining force.

2. If the strain or load is uniformly distributed along any part of the beam lying beyond C, (that is, further from A), the strain at C is the same as if the load were all collected at the middle point of that part; for that point is the centre of gravity of the load.

3. The strain on any section D of a beam AB, Fig. 16. resting freely on two props A and B, is $w \times \frac{AD \times DB}{AB}$ (See ROOF, and STRENGTH OF MATERIALS.) Therefore,

4. The strain on the middle point, by a force applied there, is one-fourth of the strain which the same force would produce, if applied to one end of a beam of the same length, having the other end fixed.

5. The strain on any section C of a beam, resting on two props A and B, occasioned by a force applied perpendicularly to another point D, is proportional to the rectangle of the exterior segments, or is equal to

$w \times \frac{AC \times DB}{AB}$ Therefore,

The strain at C occasioned by the pressure on D, is the same with the strain at D occasioned by the same pressure on C.

6. The strain on any section D, occasioned by a load uniformly diffused over any part EF, is the same as if the two parts ED, DF of the load were collected at their middle points *e* and *f*. Therefore,

The strain on any part D, occasioned by a load uniformly distributed over the whole beam, is one half of the strain that is produced when the same load is laid on at D; and,

The strain on the middle point C, occasioned by a load uniformly distributed over the whole beam, is the same which half that load would produce if laid on at C.

7. A beam supported at both ends on two props B and C, Fig. 14., will carry twice as much when the ends beyond the props are kept from rising, as it will carry when it rests loosely on the props,

8. Lastly, the transverse strain on any section, occasioned by a force applied obliquely, is diminished in the proportion of the sine of the angle which the direction of the force makes with the beam. Thus, if it be inclined to it in an angle of thirty degrees, the strain is one half of the strain occasioned by the same force acting perpendicularly.

On the other hand, the RELATIVE STRENGTH of a beam, or its power in any particular section to resist any transverse strain, is proportional to the absolute cohesion of the section directly, to the distance of its centre of effort from the axis of fracture directly, and to the distance from the strained point inversely.

Thus in a rectangular section of the beam, of which

b is the breadth, d the depth (that is, the dimension in the direction of the straining force), measured in inches, and f the number of pounds which one square inch will just support without being torn asunder, we must have $f \times b \times d^2$, proportional to $w \times$ CB, Fig. 15. Or, $f \times b \times d^2$, multiplied by some number m, depending on the nature of the timber, must be equal to $w \times$ CB. Or, in the case of the section C of Fig. 16. that is strained by the force of w applied at D, we must have $m \times f\,b\,d^2 = w \times \frac{AC \times DB}{AB}$. Thus if the beam is of sound oak, m is very nearly $= \frac{1}{9}$ (see STRENGTH OF MATERIALS.) Therefore we have $\frac{f\,b\,d^2}{9} = w \times \frac{AC \times CB}{AB}$.

Hence we can tell the precise force w which any section C can just resist when that force is applied in any way whatever. For the above mentioned formula gives $w = \frac{f\,b\,d^2}{9\,CB}$, for the case represented by Fig. 15. But the case represented in Fig. 16. having the straining force applied at D, gives the strain at C $(= w) = f \times \frac{b\,d^2 \times AB}{9\,AC \times CB}$.

Example. Let an oak beam, four inches square, rest freely on the props A and B, seven feet apart, or 84 inches. What weight will it just support at its middle point C, on the supposition that a square inch rod will just carry 16,000 pounds, pulling it asunder?

The formula becomes $w = \frac{16000 \times 4 \times 16 \times 84}{9 \times 42 \times 42}$, or $w = \frac{86016000}{15876}$, $= 5418$ pounds. This is very near what was employed in Buffon's experiment, which was 5312.

Had the straining force acted on a point D, half way

between C and B, the force sufficient to break the beam at C would be $= \frac{16000 \times 4 \times 16 \times 84}{9 \times 42 \times 21} = 10836$ lbs.

Had the beam been sound red fir, we must have taken $f = 10{,}000$ nearly, and m nearly 8; for although fir be less cohesive than oak in the proportion of 5 to 8 nearly, it is less compressible, and its axis of fracture is therefore nearer to the concave side.

512. Having considered at sufficient length the strains of different kinds which arise from the form of the parts of a frame of carpentry, and the direction of the external forces which act on it, whether considered as impelling or as supporting its different parts, we must now proceed to consider the means by which this form is to be secured, and the connections by which those strains are excited and communicated.

The joinings practised in carpentry are almost infinitely various, and each has advantages which make it preferable in some circumstances. Many varieties are employed merely to please the eye. We do not concern ourselves with these: Nor shall we consider those which are only employed in connecting small works, and can never appear on a great scale; yet even in some of these, the skill of the carpenter may be discovered by his choice; for in all cases, it is wise to make every, even the smallest, part of his work as strong as the materials will admit. He will be particularly attentive to the changes which will necessarily happen by the shrinking of timber as it dries, and will consider what dimensions of his framings will be affected by this, and what will not; and will then dispose the pieces which are less essential to the strength of the whole, in such a manner that their tendency to shrink shall be in the same direction with the shrinking of the whole framing. If he do otherwise, the seams will widen, and parts will be split asunder. He will dispose his boardings in such a manner as to contribute to the stiffness of the whole, avoiding at the same time the giving them positions which will produce lateral strains on truss beams which bear great pressures; recollecting, that although a single board has little force, yet many united have a great deal, and may frequently perform the office of very powerful struts.

Our limits confine us to the joinings which are most essential for connecting the parts of a single piece of a frame when it cannot be formed of one beam, either for want of the necessary thickness or length; and the joints for connecting the different sides of a trussed frame.

513. Much ingenuity and contrivance have been bestowed on the manner of building up a great beam of many thicknesses, and many singular methods are practised as great nostrums by different artists: but when we consider the manner in which the cohesion of the fibres performs its office, we shall clearly see that the simplest are equally effectual with the most refined, and that they are less apt to lead us into false notions of the strength of the assemblage.

514. Thus, were it required to build up a beam for a great lever or a girder, so that it may act nearly as a beam of the same size of one log—it may either be done by plain joggling, as in Plate VII. Fig. 1. A, or by scarfing, as in Fig. 1. B or C.

515. If it is to act as a lever, having the gudgeon on the lower side at C, we believe that most artists will prefer the form B and C; at least this has been the case with nine-tenths of those to whom we have proposed the question. The best informed only hesitated; but the ordinary artists were all confident in its superiority; and we found their views of the matter very coincident. They considered the upper piece as grasping the lower in its hooks; and several imagined that, by driving the one very tight on the other, the beam would be stronger than an entire log: but if we attend care-

fully to the internal procedure in the loaded lever, we shall find the upper one clearly the strongest. If they are formed of equal logs, the upper one is thicker than the other by the depth of the joggling or scarfing, which we suppose to be the same in both; consequently, if the cohesion of the fibres in the intervals is able to bring the uppermost filaments into full action, the form A is stronger than B, in the proportion of the greater distance of the upper filaments from the axis of the fracture: this may be greater than the difference of the thickness, if the wood is very compressible. If the gudgeon be in the middle, the effect, both of the joggles and the scarfings, is considerably diminished; and if it is on the upper side, the scarfings act in a very different way. In this situation, if the loads on the arms are also applied to the upper side, the joggled beam is still more superior to the scarfed one. This will be best understood by resolving it in imagination into a trussed frame. But when a gudgeon is thus put upon that side of the lever which grows convex by the strain, it is usual to connect it with the rest by a powerful strap, which embraces the beam, and causes the opposite point to become the resisting point. This greatly changes the internal actions of the filaments, and, in some measure, brings it into the same state as the first, with the gudgeon below. Were it possible to have the gudgeon on the upper side, and to bring the whole into action without a strap, it would be the strongest of all; because, in general, the resistance to compression is greater than to extension. In every situation the joggled beam has the advantage, and it is the easiest executed.

We may frequently gain a considerable accession of strength by this building up of a beam; especially if the part which is stretched by the strain be of oak, and the other part be fir. Fir being so much superior to oak as a pillar (if Muschenbroek's experiments may be confided in), and oak so much preferable as a tie, this construction seems to unite both advantages. But we shall see much better methods of making powerful levers, girders, &c. by trussing.

Observe, that the efficacy of both methods depends entirely on the difficulty of causing the piece between the cross joints to slide along the timber to which it adheres. Therefore, if this be moderate, it is wrong to make the notches deep; for as soon as they are so deep that their ends have a force sufficient to push the slice along the line of junction, nothing is gained by making them deeper; and this requires a greater expenditure of timber.

Scarfings are frequently made oblique, as in Fig. 2. but we imagine that this is a bad practice. It begins to yield at the point, where the wood is crippled and splintered off, or at least bruised out a little: as the pressure increases, this part, by squeezing broader, causes the solid parts to rise a little upwards, and gives them some tendency, not only to push their antagonists along the base, but even to tear them up a little. For similar reasons, we disapprove of the favourite practice of many artists, to make the angles of their scarfings acute, as in Fig. 3. This often causes the two pieces to tear each other up. The abutments should always be perpendicular to the directions of the pressures. Lest it should be forgotten in its proper place, we may extend this injunction also to the abutments of different pieces of a frame, and recommend it to the artist even to attend to the shrinking of the timbers by drying. When two timbers abut obliquely, the joint should be most full at the obtuse angle of the end; because, by drying, that angle grows more obtuse, and the beam would then be in danger of splintering off at the acute angle.

516. It is evident, that the nicest work is indispensably necessary in building up a beam. The parts must abut

on each other completely, and the smallest play or void takes away the whole efficacy. It is usual to give the butting joints a small taper to one side of the beam, so that they may require moderate blows of a maul to force them in, and the joints may be perfectly close when the external surfaces are even on each side of the beam. But we must not exceed in the least degree; for a very taper wedge has great force; and if we have driven the pieces together by very heavy blows, we leave the whole in a state of violent strain, and the abutments are perhaps ready to splinter off by a small addition of pressure. This is like too severe a proof for artillery; which, though not sufficient to burst the pieces, has weakened them to such a degree, that the strain of ordinary service is sufficient to complete the fracture. The *workman* is tempted to exceed in this, because it smooths off and conceals all uneven seams; but he must be watched. It is not unusual to leave some abutments open enough to admit a thin wedge reaching through the beam. Nor is this a bad practice, if the wedge is of materials which are not compressed by the driving or the strain of service. Iron would be preferable for this purpose, and for the joggles, were it not that by its too great hardness it cripples the fibres of timber to some distance. In consequence of this, it often happens that, in beams which are subjected to desultory and sudden strains (as in the levers of reciprocating engines,) the joggles or wedges widen the holes, and work themselves loose: Therefore skilful engineers never admit them, and indeed as few bolts as possible, for the same reason: but when resisting a steady or dead pull, they are not so improper, and are frequently used.

517. Beams are built up not only to increase their dimensions in the direction of the strain (which we have hitherto called their depth), but also to increase their breadth or the dimensions perpendicular to the strain.

We sometimes double the breadth of a girder which is thought too weak for its load, and where we must not increase the thickness of the flooring. The mast of a great ship of war must be made bigger athwartship, as well as fore and aft. This is one of the nicest problems of the art; and professional men are by no means agreed in their opinions about it. We do not presume to decide; and shall content ourselves with exhibiting the different methods.

518. The most obvious and natural method is that shewn in Fig. 4. It is plain that (independent of the connection of cross bolts, which are used in them all when the beams are square) the piece C cannot bend in the direction of the plane of the figure without bending the piece D along with it. This method is much used in the French navy; but it is undoubtedly imperfect. Hardly any two great trees are of equal quality, and swell and shrink alike. If C shrinks more than D, the feather of C becomes loose in the groove wrought in D to receive it; and when the beam bends, the parts can slide on each other like the plates of a coach spring; and if the bending is in the direction *e f*, there is nothing to hinder this sliding but the bolts, which soon work themselves loose in the bolt-holes.

519. Fig. 5. exhibits another method. The two halves of the beam are tabled into each other in the same manner as in Fig 1. It is plain that this will not be affected by the unequal swelling or shrinking, because this is insensible in the direction of the fibres; but when bent in the direction *a b*, the beam is weaker than Fig. 4. bent in the direction *e f*. Each half of Fig. 4. has, in every part of its length, a thickness greater than half the thickness of the beam. It is the contrary in the alternate portions of the halves of Fig. 5. When one of them is bent in the direction AB, it is plain that it drags the other with it by means of the cross butments of its tables, and

there can be no longitudinal sliding. But unless the work is accurately executed, and each hollow completely filled up by the table of the other piece, there will be a lateral slide along the cross joints sufficient to compensate for the curvature; and this will hinder the one from compressing or stretching the other in conformity to this curvature.

520. The imperfection of this method is so obvious, that it has seldom been practised: but it has been combined with the other, as is represented in Fig. 6. where the beams are divided along the middle, and the tables in each half are alternate, and alternate also with the tables of the other half. Thus 1, 3, 4, are prominent, and 5, 2, 6, are depressed. This construction evidently puts a stop to both sides, and obliges every part of both pieces to move together. *a b* and *c d* show sections of the built-up beam corresponding to AB and CD.

No more is intended in this practice by any intelligent artist, than the causing the two pieces to act together in all their parts, although the strains may be unequally distributed on them. Thus, in a built-up girder, the binding joints are frequently mortised into very different parts of the two sides. But many seem to aim at making the beam stronger than if it were of one piece; and this inconsiderate project has given rise to many whimsical modes of tabling and scarfing, which we need not regard.

521. The practice in the British dock-yards is somewhat different from any of these methods. The pieces are tabled as in Fig. 6., but the tables are not thin parallelopipeds, but thin prisms. The two outward joints or visible seams are straight lines, and the table No. 1. rises gradually to its greatest thickness in the axis. In like manner, the hollow 5 for receiving the opposite table, sinks gradually from the edge to its greatest depth in the axis. Fig. 7. represents a section of a round piece of

timber built up in this way, where the full line EFGH is the section corresponding to AB of Fig. 6. and the dotted line EGFH is the section corresponding to CD.

This construction, by making the external seam straight, leaves no lodgment for water, and looks much fairer to the eye: but it appears to us that it does not give such firm hold when the mast is bent in the direction EH. The exterior parts are most stretched and most compressed by this bending; but there is hardly any abutment in the exterior parts of these tables. In the very axis, where the abutment is the firmest, there is little or no difference of extension and compression.

But this construction has an advantage, which we imagine much more than compensates for these imperfections, at least in the particular case of a round mast: it will draw together by hooping incomparably better than any of the others. If the cavity be made somewhat too shallow for the prominence of the tables, and if this be done uniformly along the whole length, it will make a somewhat open seam; and this opening can be regulated with the utmost exactness from end to end by the plane. The heart of those vast trunks is very sensibly softer than the exterior circles: Therefore, when the whole is hooped, and the hoops hard driven, and at considerable intervals between each spell—we are confident that all may be compressed till the seam disappears; and then the whole makes one piece, *much* stronger than if it were an original log of that size; because the middle has become, by compression, as solid as the crust, which was naturally firmer, and resisted farther compression. We verified this beyond a doubt, by hooping a built stick of a timber which has this inequality of firmness in a remarkable degree, and it was nearly twice as strong as another of the same size.

Our mastmakers are not without their fancies and whims; and the manner in which our masts and yards are generally built up, is not near so simple as Fig. 6.: but it consists of the same essential parts, acting in the very same manner, and derives all its efficacy from the principles which are here employed.

522. This construction is particularly suited to the situation and office of a ship's mast. It has no bolts; or, at least, none of any magnitude, or that make very important parts of its construction. The most violent strains perhaps that it is exposed to, is that of twisting, when the lower yards are close braced up by the force of many men acting by a long lever. This form resists a twist with peculiar energy: it is therefore an excellent method for building up a great shaft for a mill. The way in which they are usually built up is by reducing a central log to a polygonal prism, and then filling it up to the intended size by *planting* pieces of timber along its sides, either spiking them down, or cocking them into it by a feather, or joggling them by slips of hard wood sunk into the central log and into the slips. *N. B.* Joggles of elm are sometimes used in the middle of the large tables of masts; and when sunk into the firm wood near the surface, they must contribute much to the strength. But it is very necessary to employ wood not much harder than the pine; otherwise it will soon enlarge its bed, and become loose; for the timber of these large trunks is very soft.

523. The most general reason for piecing a beam is to increase its length. This is frequently necessary, in order to procure tie-beams for very wide roofs. Two pieces must be scarfed together.—Numberless are the modes of doing this; and almost every master carpenter has his favourite nostrum. Some of them are very ingenious: But here, as in other cases, the most simple are commonly the strongest. We do not imagine that any, the most ingenious, is equally strong with a tie consisting of two pieces of the same scantling laid over each other for a

certain length, and firmly bolted together. We acknowledge that this will appear an artless and clumsy tie-beam; but we only say that it will be stronger than any that is more artificially made up of the same thickness of timber. This, we imagine, will appear sufficiently certain.

The simplest and most obvious scarfing, (after the one now mentioned) is that represented in Fig. 8. No. 1. & 2. If considered merely as two pieces of wood joined, it is plain that, as a tie, it has but half the strength of an entire piece, supposing that the bolts (which are the only connections) are fast in their holes. No. 2. requires a bolt in the middle of the scarf to give it that strength; and, in every other part, is weaker on one side or the other.

But the bolts are very apt to bend by the violent strain, and require to be strengthened by uniting their ends by iron plates; in which case it is no longer a wooden tie. The form of No. 1. is better adapted to the office of a pillar than No. 2.; especially if its ends be formed in the manner shewn in the elevation No. 3. By the sally given to the ends, the scarf resists an effort to bend it in that direction. Besides, the form of No 2. is unsuitable for a post; because the pieces, by sliding on each other by the pressure, are apt to splinter off the tongue which confines their extremity.

Fig. 9. and 10. exhibit the most approved form of a scarf, whether for a tie or for a post. The key represented in the middle is not essentially necessary; the two pieces might simply meet square there. This form, without a key, needs no bolts (although they strengthen it greatly); but, if worked very true and close, and with square abutments, will hold together, and will resist bending in any direction. But the key is an ingenious and a very great improvement, and will force the parts together with perfect tightness. The same precaution must be observed that we mentioned on another occasion, not to produce a constant internal strain on the parts by overdriving the key. The form of Fig. 9. is by far the best; because the triangle of Fig. 10. is much easier splintered off by the strain, or by the key, than the square wood of Fig. 9. It is far preferable for a post, for the reason given when speaking of Fig. 8. No. 1. & 2. Both may be formed with a sally at the ends equal to the breadth of the key. In this shape, Fig. 9. is vastly well suited for joining the parts of the long corner posts of spires and other wooden towers. Fig. 9. No. 2. differs from No. 1. only by having three keys. The principle and the longitudinal strength are the same. The long scarf of No. 2. tightened by the three keys, enables it to resist a bending much better.

None of these scarfed tie-beams can have more than one-third of the strength of an entire piece, unless with the assistance of iron plates; for if the key be made thinner than one-third, it has less than one-third of the fibres to pull by.

We are confident therefore, that when the beads of the bolts are connected by plates, the simple form of Fig. 8. No. 1. is stronger than those more ingenious scarfings. It may be strengthened against lateral bendings by a little tongue, or by a sally; but it cannot have both.

524. The strongest of all methods of piecing a tie-beam would be to set the parts end to end, and grasp them between other pieces on each side, as in Fig. 11. This is what the ship-carpenter calls *fishing* a beam; and is a frequent practice for occasional repairs. Mr Perronet used it for the tie-beams or stretchers, by which he connected the opposite feet of a centre, which was yielding to its load, and had pushed aside one of the piers above four inches. Six of these not only withstood a strain of 1800 tons, but, by wedging behind them, he brought the feet of the truss $2\frac{1}{2}$ inches nearer. The stretchers were 14 inches by 11 of sound oak, and could have with-

stood three times that strain. Mr Perronet, fearing that the great length of the bolts employed to connect the beams of these stretchers would expose them to the risk of bending, scarfed the two side pieces into the middle piece. The scarfing was of the triangular kind (*Trait de Jupiter*), and only an inch deep, each face being two feet long, and the bolt passed through close to the angle.

In piecing the pump rods, and other wooden stretchers of great engines, no dependence is had on scarfing; and the engineer connects every thing by iron straps. We doubt the propriety of this, at least in cases where the bulk of the wooden connection is not inconvenient. These observations must suffice for the methods employed for connecting the parts of a beam; and we now proceed to consider what are more usually called the joints of a piece of carpentry.

525. Where the beams stand square with each other, and the strains are also square with the beams, and in the plane of the frame, the common mortise and tenon is the most perfect junction. A pin is generally put through both, in order to keep the pieces united, in opposition to any force which tends to part them. Every carpenter knows how to bore the hole for this pin, so that it shall draw the tenon tight into the mortise, and cause the shoulder to butt close, and make neat work; and he knows the risk of tearing out the bit of the tenon beyond the pin, if he draw it too much. We may just observe, that square holes and pins are much preferable to round ones for this purpose, bringing more of the wood into action, with less tendency to split it. The ship carpenters have an ingenious method of making long wooden bolts, which do not pass completely through, take a very fast hold, though not nicely fitted to their holes, which they must not be, lest they should be crippled in driving. They call it *foxtail wedging*. They stick into the point of the bolt a very thin wedge of hard wood, so as to pro-

ject a proper distance; when this reaches the bottom of the hole by driving the bolt, it splits the end of it, and squeezes it hard to the side. This may be practised with advantage in carpentry. If the ends of the mortise are widened inwards, and a thin wedge be put into the end of the tenon, it will have the same effect, and make the joint equal to a dovetail. But this risks the splitting the piece beyond the shoulder of the tenon, which would be unsightly. This may be avoided as follows: Let the tenon T, Fig. 12 have two *very* thin wedges *a* and *c* stuck in near its angles, projecting equally; at a very small distance within these, put in two shorter ones *b*, *d*, and more within these if necessary. In driving this tenon, the wedges *a* and *c* will take first, and split off a thin slice, which will easily bend without breaking. The wedges *b*, *d*, will act next, and have a similar effect, and the others in succession. The thickness of all the wedges taken together must be equal to the enlargement of the mortise toward the bottom

When the strain is transverse to the plane of the two beams, the principles laid down in the article STRENGTH OF MATERIALS, will direct the artist in placing his mortise. Thus the mortise in a girder for receiving the tenon of a binding joist of a floor should be as near the upper side as possible, because the girder becomes concave on that side by the strain. But as this exposes the tenon of the binding joist to the risk of being torn off, we are obliged to mortise farther down. The form (Fig. 13.) generally given to this joint is extremely judicious. The sloping part *a b* gives a very firm support to the additional bearing *e d*, without much weakening of the girder. This form should be copied in every case where the strain has a similar direction.

526. The joint that most of all demands the careful attention of the artist, is that which connects the ends of beams, one of which pushes the other very obliquely,

putting it into a state of extension. The most familiar instance of this is the foot of a rafter pressing on the tie-beam, and thereby *drawing* it away from the other wall. When the direction is very oblique (in which case the extending strain is the greatest), it is difficult to give the foot of the rafter such a hold of the tie beam as to bring many of its fibres into the proper action. There would be little difficulty if we could allow the end of the tie-beam to project to a small distance beyond the foot of the rafter: but, indeed, the dimensions which are given to tie-beams, for other reasons, are always sufficient to give enough of abutment when judiciously employed. Unfortunately this joint is much exposed to failure by the effects of the weather. It is much exposed, and frequently perishes by rot, or becomes so soft and friable that a very small force is sufficient, either for pulling the filaments out of the tie-beam, or for crushing them together. We are therefore obliged to secure it with particular attention, and to avail ourselves of every circumstance of construction.

One is naturally disposed to give the rafter a deep hold by a long tenon; but it has been frequently observed in old roofs that such tenons break off. Frequently they are observed to tear up the wood that is above them, and push their way through the end of the tie-beam. This, in all probability, arises from the first fagging of the roof, by the compression of the rafters and of the head of the king-post. The head of the rafter descends, the angle with the tie-beam is diminished by the rafter revolving round its step in the tie-beam. By this motion the heel or inner angle of the rafter becomes a fulcrum to a very long and powerful lever much loaded. The tenon is the other arm, very short, and being still fresh, it is therefore very powerful. It therefore forces up the wood that is above it, tearing it out from between the cheeks of the mortise, and then pushes it along. Carpenters have therefore given up long tenons, and give to the toe of the tenon a shape which abuts firmly, in the direction of the thrust, on the solid bottom of the mortise, which is well supported on the under side by the wall-plate. This form has the farther advantage of having no tendency to tear up the end of the mortise. It is represented in Fig. 14. The tenon has a small portion *a b* cut perpendicular to the surface of the tie-beam, and the rest *b c* is perpendicular to the rafter.

But if the tenon is not sufficiently strong (and it is not so strong as the rafter, which is thought not to be stronger than is necessary), it will be crushed, and then the rafter will shade out along the surface of the beam. It is therefore necessary to call in the assistance of the whole rafter. It is in this distribution of the strain among the various abutting parts that the varieties of joints and their merits chiefly consist. It would be endless to describe every nostrum, and we shall only mention a few that are most generally approved of.

527. The aim in Fig. 15. is to make the abutments exactly perpendicular to the thrusts. It does this very precisely; and the share which the tenon and the shoulder have of the whole may be what we please, by the portion of the beam that we notch down. If the wall-plate lie duly before the heel of the rafter, there is no risk of straining the tie across or breaking it, because the thrust is made direct to that point where the beam is supported. The action is the same as against the joggle on the head or foot of a king-post. We have no doubt but that this is a very effectual joint. It is not, however, much practised. It is said that the sloping seam at the shoulder lodges water; but the great reason seems to be a secret notion that it weakens the tie-beam. If we consider the direction in which it acts as a tie, we must acknowledge that this form takes the best method for bringing the whole of it into action.

Fig. 16. exhibits a form that is more general, but certainly worse. What part of the thrust that is not borne by the tenon acts obliquely on the joint of the shoulder, and gives the whole a tendency to rise up and slide outward.

The shoulder joint is sometimes formed like the dotted line *a b c d e f g* of Fig. 16. This is much more agreeable to the true principle, and would be a very perfect method, were it not that the intervals *b d* and *d f* are so short that the little wooden triangles *b c d*, *d e f*, will be easily pushed off their bases *b d*, *d f*.

Fig. 17. seems to have the most general approbation. It is the joint recommended by Price, and copied into all books of carpentry as the *true joint* for a rafter foot. The visible shoulder-joint is flush with the upper surface of the tie-beam. The angle of the tenon at the tie nearly bisects the obtuse angle formed by the rafter and the beam, and is therefore somewhat oblique to the thrust. The inner shoulder *a c* is nearly perpendicular to *b d*. The lower angle of the tenon is cut off horizontally as at *c d*. Fig. 18. is a section of the beam and rafter foot, shewing the different shoulders.

We do not perceive the peculiar merit of this joint. The effect of the three oblique abutments *a b*, *a c*, *e d*, is undoubtedly to make the whole bear on the outer end of the mortise, and there is no other part of the tie-beam that makes immediate resistance. Its only advantage over a tenon extending in the direction of the thrust is, that it will not tear up the wood above it. Had the inner shoulder had the form *e c i*, having its face *i c* perpendicular, it would certainly have acted more powerfully in stretching many filaments of the tie-beam, and would have had much less tendency to force out the end of the mortise. The little bit *c i* would have prevented the sliding upwards along *e c*. At any rate, the joint *a b* being flush with the beam, prevents any sensible abutment on the shoulder *a c*.

Fig. 17. No. 2. is a simpler, and, in our opinion, a preferable joint. We observe it practised by the most eminent carpenters for all oblique thrusts; but it surely employs less of the cohesion of the tie-beam than might be used without weakening it, at least when it is supported on the other side by the wall-plate.

Fig. 17. No. 3. is also much practised by the first carpenters.

Fig. 19. is proposed by Mr Nicholson as preferable to Fig. 17. No. 3 because the abutment of the inner part is better supported. This is certainly the case; but it supposes the whole rafter to go to the bottom of the socket, and the beam to be thicker than the rafter. Some may think that this will weaken the beam too much, when it is no broader than the rafter is thick; in which case they think that it requires a deeper socket than Nicholson has given it. Perhaps the advantages of Nicholson's construction may be had by a joint like Fig. 19. No. 2.

528. Whatever is the form of these butting joints, great care should be taken that all parts bear alike, and the artist will attend to the magnitude of the different surfaces. In the general compression, the greater surfaces will be less compressed, and the smaller will therefore change most. When all has settled, every part should be *equally* close. Because great logs are moved with difficulty, it is very troublesome to try the joint frequently to see how the parts fit; therefore we must expect less accuracy in the interior parts. This should make us prefer those joints whose efficacy depends chiefly on the visible joint.

It appears from all that we have said on this subject, that a very small part of the cohesion of the tie-beam is sufficient for withstanding the horizontal thrust of a roof, even though very low pitched. If, therefore, no other use is made of the tie-beam, one much slenderer may be used, and blocks may be firmly fixed to the ends, on which the rafters might abut, as they do on the joggles on the head and foot of a king-post. Although a tie-beam has commonly floors or ceilings to carry, and sometimes the workshops and store-rooms of a theatre, and therefore requires a great scantling, yet there frequently occur in machines and engines very oblique stretchers, which have no other office, and are generally made of dimensions quite inadequate to their situation, often containing ten times the necessary quantity of timber. It is therefore of importance to ascertain the most perfect manner of executing such a joint. We have directed our attention to the principles that are really concerned in the effect. In all hazardous cases, the carpenter calls in the assistance of iron straps; and they are frequently necessary, even in roofs, notwithstanding this superabundant strength of the tie-beam. But this is generally owing to bad construction of the wooden joint, or to the failure of it by time. Straps will be considered in their place.

There needs but little to be said of the joints at a joggle worked out of solid timber; they are not near so difficult as the last. When the size of a log will allow the joggle to receive the whole breadth of the abutting brace, it ought certainly to be made with a square shoulder; or, which is still better, an arch of a circle, having the other end of the brace for its centre. Indeed this in general will not sensibly differ from a straight line perpendicular to the brace. By this circular form, the settling of the roof makes no change in the abutment; but when there is not sufficient stuff for this, we must avoid bevel joints at the shoulders, because these always tend to make the brace slide off. The brace in Fig. 20. must not be joined as at *a*, but as at *b*, or some equivalent manner. Observe the joints at the head of the main posts of Drury Lane Theatre, Fig. D.

529. When the very oblique action of one side of a frame of carpentry does not extend but compress the piece on which it abuts (as in Plate VI. Fig. 11.), there is no difficulty in the joint. Indeed a joining is unnecessary, and it is enough that the pieces abut on each other; and we have only to take care that the mutual pressure be equally borne by all the parts, and that it do not produce lateral pressures, which may cause one of the pieces to slide on the butting joint. A very slight mortise and tenon is sufficient at the joggle of a king-post with a rafter or straining beam. It is best, in general, to make the butting plain, bisecting the angle formed by the sides, or else perpendicular to one of the pieces. In Fig. 20. No. 2. where the straining beam *a b* cannot slip away from the pressure, the joint *a* is preferable to *b*, or indeed to any uneven joint, which never fails to produce very unequal pressures on the different parts, by which some are crippled, others are splintered off, &c.

530. When it is necessary to employ iron straps for strengthening a joint, a considerable attention is necessary, that we may place them properly. The first thing to be determined is the direction of the strain. This is learned by the observations in the beginning of this article. We must then resolve this strain into a strain parallel to each piece, and another perpendicular to it. Then the strap which is to be made fast to any of the pieces must be so fixed, that it shall resist in the direction parallel to the piece. Frequently this cannot be done; but we must come as near to it as we can. In such cases we must suppose that the assemblage yields a little to the pressures which act on it. We must examine what change of shape a small yielding will produce. We must now see how this will affect the iron strap which we have already supposed attached to the joint in some manner that we thought suitable. This settling will perhaps draw the pieces away from it, leaving it loose and unserviceable

(this frequently happens to the plates which are put to secure the obtuse angles of butting timbers, when their bolts are at some distance from the angles, especially when these plates are laid on the inside of the angles); or it may cause it to compress the pieces harder than before; in which case it is answering our intention. But it may be producing cross strains, which may break them; or it may be crippling them. We can hardly give any general rules; but the reader will do well to read what is said in the article Roof. He will there see the nature of the strap or stirrup, by which the king-post carries the tie-beam. The strap that we observe most generally ill placed is that which connects the foot of the rafter with the beam. It only binds down the rafter, but does not act against its horizontal thrust. It should be placed farther back on the beam, with a bolt through it, which will allow it to turn round. It should embrace the rafter almost horizontally near the foot, and should be notched square with the back of the rafter. Such a construction is represented in Fig. 21. By moving round the eye-bolt, it follows the rafter, and cannot pinch and cripple it, which it always does in its ordinary form. We are of opinion that straps which have eye-bolts in the very angles, and allow all motion round them, are of all the most perfect. A branched strap, such as may at once bind the king-post and the two braces which butt on its foot, will be more serviceable if it have a joint. When a roof warps, those branched straps frequently break the tenons, by affording a fulcrum in one of their bolts. An attentive and judicious artist will consider how the beams will act on such occasions, and will avoid giving rise to these great strains by levers. A skilful carpenter never employs many straps, considering them as auxiliaries foreign to his art, and subject to imperfections in workmanship which he cannot discern nor amend. We must refer the reader to Nicholson's Carpenter and Join-

ER'S ASSISTANT for a more particular account of the various forms of stirrups, screwed rods, and other iron work for carrying tie-beams, &c.

As for those that are necessary for the turning joints of great engines constructed of timber, they make no part of the art of carpentry.

531. After having attempted to give a systematic view of the principles of framing carpentry, we shall conclude, by giving some examples which will illustrate and confirm the foregoing principles.

532. Fig. 1. Plate VIII. is the roof of the chapel of the Royal Hospital at Greenwich, constructed by Mr S. Wyatt.

	Inch. Scantling.
AA, Is the tie-beam, 57 feet long, spanning 51 feet clear	14 by 12
CC, Queen-posts	9×12
D, Braces	9×7
E, Truss beam	10×7
F, Straining piece	6×7
G, Principal rafters	10×7
H, A cambered beam for the platform	9.7
B, An iron string, supporting the tie-beam	2×2

The trusses are 7 feet apart, and the whole is covered with lead, the boarding being supported by horizontal ledgers, *h*, *h*, of 6 by 4 inches.

This is a beautiful roof, and contains less timber than most others of the same dimensions. The parts are all disposed with great judgment. Perhaps the iron rod is unnecessary; but it adds great stiffness to the whole.

The iron straps at the rafter feet would have had more effect if not so oblique. Those at the head of the posts are very effective.

We may observe, however, that the joints between the straining beam and its braces are not of the best kind, and tend to bruise both the straining beam and the truss beam above it.

533. Fig. 2. is the roof of St Paul's, Covent Garden, constructed by Mr Wapshot in 1796.

AA, Tie-beam, spanning 50 feet 2 inches............16.12
B, Queen-post..9x8
C, Truss beam...10x8
D, King-post (14 at the joggle)..........................9x8
E, Brace...8x7½
FF, Principal brace (at bottom).........................10x8½
HH, Principal rafter (at bottom)........................10x8½
g g, Studs supporting the rafter..........................8x8

This roof far excels the original one put up by Inigo Jones. One of its trusses contains 198 feet of timber. One of the old roof had 273, but had many inactive timbers, and others ill disposed *. The internal truss FCF is admirably contrived for supporting the exterior rafters, without any pressure on the far projecting ends of the tie-beam. The former roof had bent them greatly, so as to appear ungraceful.

We think that the camber (six inches) of the tie-beam is rather hurtful; because, by settling, the beam lengthens; and this must be accompanied by a *considerable* sinking of the roof. This will appear by calculation.

534. Fig. 3. is the roof of Birmingham Theatre, constructed by Mr George Saunders. The span is 80 feet clear, and the trusses are 10 feet apart.

A, Is an oak corbel...9x5
B, Inner plate..9x9
C, Wall plate...8x5½
D, Pole plate,...7x5
E, Beam...15x15
F, Straining beam...12x9

* The figure of Inigo Jones's roof, given in the Encyclopædia Britannica, art. Roof, Fig. 22. is very erroneous. Ed.

G, Oak king-post (in the shaft)..........................9×9
H, Oak queen-post (in the shaft)........................7×9
I, Principal rafters..9×9
K, Common ditto....... ....4×2½
L, Principal braces.............. ....9 and 6×9
M, Common ditto...6×9
N, Purlins...7×5
Q, Straining sill...5½×9

This roof is a fine specimen of British carpentry, and is one of the boldest and lightest roofs in Europe. The straining sill Q gives a firm abutment to the principal braces, and the space between the posts is 19½ feet wide, affording roomy workshops for the carpenters and other workmen connected with a theatre. The contrivance for taking double hold of the wall, which is very thin, is excellent. There is also added a beam (marked R), bolted down to the tie-beams. The intention of this was to prevent the total failure of so bold a trussing, if any of the tie-beams should fail at the end by rot.

535. Akin to this is Fig. 4., the roof of Drury Lane theatre, 80 feet 3 inches in the clear, and the trusses 15 feet apart, constructed by Mr Edward Grey Saunders.

A, Beams..10 by 7
B, Rafters.......... ....7×7
C, King-posts...12×7
D, Struts....................... ....5×7
E, Purlins...9×5
G, Pole plates... .5×5
I, Common rafters..5×4
K, Tie-beam to the main truss... ....15×12
L, Posts to ditto..15×12
M, Principal braces to ditto......... ....14 and 12×12
N, Struts...8×12
P, Straining beams.......................................12×12

The main beams are trussed in the middle space with oak trusses 5 inches square. This was necessary for its width of 32 feet, occupied by the carpenters, painters, &c. The great space between the trusses affords good store-rooms, dressing-rooms, &c.

It is probable that this roof has not its equal in the world for lightness, stiffness, and strength. The main truss is so judiciously framed, that each of them will safely bear a load of near 300 tons; so it is not likely that they will ever be quarter loaded. The division of the whole into three parts makes the exterior roofings very light. The strains are admirably kept from the walls, and the walls are even firmly bound together by the roof. They also take off the dead weight from the main truss one-third.

536. The intelligent reader will perceive that all these roofs are on one principle, depending on a truss of three pieces and a straight tie-beam. This is indeed the great principle of a truss, and is a step beyond the roof with two rafters and a king-post. It admits of much greater variety of forms, and of greater extent. We may see that even the middle part may be carried to any space, and yet be flat at top; for the truss beam may be supported in the middle by an inverted king-post (of timber, not iron), carried by iron or wooden ties from its extremities: And the same ties may carry the horizontal tie-beam K; for till K be torn asunder, or M, M, and P be crippled, nothing can fail.

The roof of St Martin's church in the Fields, is constructed on good principles, and every piece properly disposed. But although its span does not exceed 40 feet from column to column, it contains more timber in a truss than there is in one of Drury Lane theatre. The roof of the chapel at Greenwich, that of St Paul's, Covent Garden, that of Birmingham, and that of Drury Lane theatres, form a series gradually more perfect.

Such specimens afford excellent lessons to the artists. We therefore account them a useful present to the public.

537. There is a very ingenious project offered to the public by Mr Nicholson (*Carpenter's Assistant*, p. 68.) He proposes iron rods for king-posts, queen-posts, and all other situations where beams perform the office of ties. This is in prosecution of the notions which we have given in the article Roof. He receives the feet of the braces and struts in a socket very well connected with the feet of his iron king-post; and he secures the foot of his queen-posts from being pushed inwards, by interposing a straining sill. He does not even mortise the foot of his principal rafter into the end of the tie-beam, but sets it in a socket like a shoe, at the end of an iron bar, which is bolted into the tie-beam a good way back. All the parts are formed and disposed with the precision of a person thoroughly acquainted with the subject; and we have not the smallest doubt of the success of the project, and the complete security and durability of his roofs, and we expect to see many of them executed. We abound in iron, but we must send abroad for building timber. This is therefore a valuable project; at the same time, however, let us not over-rate its value. Iron is but about 12 times stronger than red fir, and is more than 12 times heavier; nor is it cheaper, weight for weight, or strength for strength.

Our illustrations and examples have been chiefly taken from roofs, because they are the most familiar instances of the difficult problems of the art. We could have wished for more room even on this subject. The construction of dome roofs has been (we think) mistaken, and the difficulty is much less than is imagined. We mean in respect of strength; for we grant that the obliquity of the joints, and a general intricacy, increases the trouble of workmanship exceedingly.

538. Wooden bridges form another class equally difficult and important; but our limits are already overpassed, and will not admit them. The principle on which they should all be constructed, without exception, is that of a truss, avoiding all lateral bearings on any of the timbers. In the application of this principle, we must farther remark, that the angles of our truss should be as acute as possible; therefore we should make it of as few and as long pieces as we can, taking care to prevent the bending of the truss beams by bridles, which embrace them, but without pressing them to either side. When the truss consists of many pieces, the angles are very obtuse; and the thrusts increase nearly in the duplicate proportion of the number of angles. The proper maxims will readily occur to the artist who considers with attention the specimens of centres or coombs, which we shall give when treating the subject of CENTRES.

539. With respect to the frames of carpentry which occur in engines and great machines, the varieties are such that it would require a volume to treat of them properly. The principles are already laid down; and if the reader be really interested in the study, he will engage in it with seriousness, and cannot fail of being instructed. We recommend to his consideration, as a specimen of what may be done in this way, the working beam of Hornblower's steam-engine. When the beam must act by chains hung from the upper end of arch heads, the framing there given seems very scientifically constructed; at the same time, we think that a strap of wrought iron, reaching the whole length of the upper bar, would be vastly preferable to those partial plates which the engineer has put there, for the bolts will soon work loose.

But when arches are not necessary, the form employed by Mr Watt is vastly preferable, both for simplicity and for strength. It consists of a simple beam AB (Fig. 5.) having the gudgeon C on the upper side. The two piston rods are attached to wrought iron joints A and B. Two strong struts DC, EC, rest on the upper side of the gudgeon, and carry an iron string ADEB, consisting of three pieces, connected with the struts by proper joints of wrought iron. A more minute description is not needed for a clear conception of the principle. No part of this is exposed to a cross strain; even the beam AB might be sawed through at the middle. The iron string is the only part which is stretched; for AC, DC, EC, BC, are all in a state of compression. We have made the angles equal, that all may be as great as possible, and the pressure on the struts and strings a minimum. Mr Watt makes them much lower, as A *d e* B, or A δ ε B. But this is for economy, because the strength is almost insuperable. It might be made with wooden strings; but the workmanship of the joints would more than compensate the cheapness of the materials.

540. We offer this article to the public with deference, and we hope for an indulgent reception of our essay on a subject which is in a manner new, and would require much study. We have bestowed our chief attention on the strength of the construction, because it is here that persons of the profession have the most scanty information. We beg them not to consider our observations as too refined, and that they will study them with care. One principle runs through the whole; and when that is clearly conceived and familiar to the mind, we venture to say that the practitioner will find it of easy application, and that he will improve every performance by a continual reference to it.

ROOF.

541. The word Roof expresses the covering of a house or building, by which its inhabitants or contents are protected from the injuries of the weather. The Greeks, who have perhaps excelled all nations in taste, and who have given the most perfect model of architectonic ordonnance within a certain limit, never erected a building which did not exhibit the roof in the distinctest manner; and though they borrowed much of their model from the orientals, as will be evident to any one who compares their architecture with the ruins of Persepolis, and of the tombs in the mountains of Sheeraz, they added that form of roof which their own climate taught them was necessary for sheltering them from the rains. The roofs in Persia and Arabia are flat, but those of Greece are without exception sloping. It seems therefore a gross violation of the true principles of taste in architecture (at least in the regions of Europe), to take away or to hide the roof of a house; and it must be ascribed to that rage for novelty which is so powerful in the minds of the rich. Our ancestors seemed to be of a very different opinion, and turned their attention to the ornamenting of their roofs as much as any other part of a building. They showed them in the most conspicuous manner, running them up to a great height, broke them into a thousand fanciful shapes, and

stuck them full of highly dressed windows. We laugh at this, and call it Gothic and clumsy; and our great architects, not to offend any more in this way, conceal the roof altogether by parapets, balustrades, and other contrivances. Our forefathers certainly did offend against the maxims of true taste, when they enriched a part of a house with marks of elegant habitation, which every spectator must know to be a cumbersome garret: but their successors no less offend, who take off the cover of the house altogether, and make it impossible to know whether it is not a mere screen or colonnade we are looking at.

542. We cannot help thinking that Sir Christopher Wren erred when he so industriously concealed the roof of St Paul's church in London. The whole of the upper order is a mere screen. Such a quantity of wall would have been intolerably offensive, had he not given it some appearance of habitation by the mock windows or niches. Even in this state it is gloomy, and it is odd, and is a puzzle to every spectator.—There should be no puzzle in the design of a building any more than in a discourse. It has been said that the double roof of our great churches which have aisles is an incongruity, looking like a house standing on the top of another house. But there is not the least occasion for such a thought. We know that the aisle is a shed, a cloister. Suppose only that the lower roof or shed is hidden by a balustrade, it then becomes a portico, against which the connoisseur has no objection: yet there is no difference; for the portico must have a cover, otherwise it is neither a shed, cloister, nor portico, any more than a building without a roof is a house. A house without a visible roof is like a man abroad without his hat; and we may add, that the whim of concealing the chimnies, now so fashionable, changes a house to a barn or storehouse. A house should not be a copy of any thing. It has a title to be an original; and a screen-like

house and a pillar-like candlestick are similar solecisms in taste.

543. The arehitect is anxious to present a fine object, and a very simple outline discusses all his concerns with the roof. He leaves it to the carpenter, whom he frequently puzzles (by his arrangements) with coverings almost impossible to execute. Indeed it is seldom that the idea of a roof is admitted by him into his great compositions; or if he does introduce it, it is from mere affectation, and we may say pedantry. A pediment is frequently stuck up in the middle of a grand front, in a situation where a roof cannot perform its office; for the rain that is supposed to flow down its sides must be received on the top of the level buildings which flank it. This is a manifest incongruity. The tops of dressed windows, trifling porches, and sometimes a projecting portico, are the only situations in which we see the figure of a roof correspond with its office. Having thus lost sight of the principle, it is not surprising that the draughtsman (for he should not be called architect) runs into every whim: and we see pediment within pediment, a round pediment, a hollow pediment, and the greatest of all absurdities, a broken pediment. Nothing could ever reconcile us to the sight of a man with a hat without its crown, because we cannot overlook the use of a hat.

544. But when one builds a house, ornament alone will not do. We must have a cover; and the enormous expence and other great inconveniences which attend the concealment of this cover by parapets, balustrades, and screens, have obliged architects to consider the pent roof as admissible, and to regulate its form. Any man of sense, not under the influence of prejudice, would be determined in this by its fitness for answering its purpose. A high pitched roof will undoubtedly shoot off the rains and snows better than one of a lower pitch. The wind will not so easily blow the dropping rain in between the

slates, nor will it have so much power to strip them off. A high pitched roof will exert a smaller thrust on the walls, both because its strain is less horizontal, and because it will admit of lighter covering. But it is more expensive, because there is more of it. It requires a greater size of timbers to make it equally strong, and it exposes a greater surface to the wind.

545. There have been great changes in the pitch of roofs: our forefathers made them very high, and we make them very low. It does not, however, appear, that this change has been altogether the effect of principle. In the simple unadorned habitations of private persons, every thing comes to be adjusted by an experience of inconveniences which have resulted from too low pitched roofs; and their pitch will always be nearly such as suits the climate and covering. Our architects, however, go to work on different principles. Their professed aim is to make a beautiful object. The sources of the pleasures arising from what we call *taste* are so various, so complicated, and even so whimsical, that it is almost in vain to look for principle in the rules adopted by our professed architects. We cannot help thinking that much of their practice results from a *pedantic* veneration for the beautiful productions of Grecian architecture. Such architects as have written on the principles of the art in respect of proportions, or what they call the ORDONNANCE, are very much puzzled to make a chain of reasoning; and the most that they have made of the Greek architecture is, that it exhibits a nice adjustment of strength and strain. But when we consider the extent of this adjustment, we find that it is wonderfully limited. The whole of it consists of a basement, a column, and an entablature; and the entablature, it is true, exhibits something of a connection with the framework and roof of a wooden building; and we believe that it really originated from this in the hands of the orientals, from whom the Greeks cer-

tainly borrowed their forms and their combinations. We could easily show in the ruins of Persepolis, and among the tombs in the mountains (which were long prior to the Greek architecture), the fluted column, the base, the Ionic and Corinthian capital, and the Doric arrangement of lintels, beams, and rafters, all derived from unquestionable principle. The only addition made by the Greeks was the pent roof; and the changes made by them in the subordinate forms of things, are such as we should expect from their exquisite judgment of beauty.

But the whole of this is very limited; and the Greeks, after making the roof a chief feature of a house, went no farther, and contented themselves with giving it a slope suited to their climate. This we have followed, because in the milder parts of Europe we have no cogent reason for deviating from it; and if any architect should deviate greatly in a building where the outline is exhibited as beautiful, we should be disgusted; but the disgust, though felt by almost every spectator, has its origin in nothing but habit. In the professed architect or man of education, the disgust arises from pedantry: for there is not such a close connection between the form and uses of a roof as shall give precise determinations; and the mere form is a matter of indifference.

546. We should not therefore reprobate the high-pitched roofs of our ancestors, particularly on the continent It is there where we see them in all the extremity of the fashion, and the taste is by no means exploded as it is with us. A baronial castle in Germany and France is seldom rebuilt in the pure Greek style, or even like the modern houses in Britain; the high pitched roofs are retained. We should not call them Gothic, and ugly because Gothic, till we show their principle to be false or tasteless. Now we apprehend that it will be found quite the reverse; and that though we cannot bring ourselves to think them beautiful, we ought to think them so. The

construction of the Greek architecture is a transference of the practices that are necessary in a wooden building to a building of stone. To this the Greeks have adhered, in spite of innumerable difficulties. Their marble quarries, however, put it in their power to retain the proportions which habit had rendered agreeable. But it is next to impossible to adhere to these proportions with freestone or brick, when the order is of magnificent dimensions. Sir Christopher Wren saw this: for his mechanical knowledge was equal to his taste. He composed the front of St Paul's church in London of two orders, and he coupled his columns; and still the lintels which form the architrave are of such length that they could carry no additional weight, and he was obliged to truss them behind. Had he made but one order, the architrave could not have carried its own weight. It is impossible to execute a Doric entablature of this size in brick. It is attempted in a very noble front, the Academy of Arts in St Petersburgh. But the architect was obliged to make the mutules, and other projecting members of the corniche, of granite, and many of them broke down by their own weight.

547. Here is surely an error in principle. Since stone is the chief material of our buildings, ought not the members of ornamental architecture to be refinements on the essential and unaffected parts of a simple stone-building There is almost as much propriety in the architecture of India, where a dome is made in imitation of a lily or other flower inverted, as in the Greek imitation of a wooden building. The principles of masonry, and not of carpentry, should be seen in our architecture, if we would have it according to the rules of just taste. Now we affirm that this is the characteristic feature of what is called the Gothic architecture. In this no dependence is had on the transverse strength of stone. No lintels are to be seen; no extravagant projections. Every stone is pressed

to its neighbours, and none is exposed to a transverse strain. The Greeks were enabled to execute their colossal buildings only by using immense blocks of the hardest materials. The Norman mason could raise a building to the skies without using a stone which a labourer could not carry to the top on his back. Their architects studied the principles of equilibrium; and, having attained a wonderful knowledge of it, they indulged themselves in exhibiting remarkable instances. We call this false taste, and say that the appearance of insecurity is the greatest fault. But this is owing to our habits: our thoughts may be said to run in a wooden train, and certain simple maxims of carpentry are familiar to our imagination; and in the careful adherence to these consists the beauty and symmetry of the Greek architecture. Had we been as much habituated to the equilibrium of pressure, this apparent insecurity would not have met our eye; we would have perceived the strength, and we should have relished the ingenuity.

548. The Gothic architecture is perhaps entitled to the name of rational architecture; and its beauty is founded on the characteristic distinction of our species. It deserves cultivation; not the pitiful, servile, and unskilled copying of the monuments; this will produce incongruities and absurdities equal to any that have crept into the Greek architecture: but let us examine with attention the nice disposition of the groins and spaundrels; let us study the tracery and knots, not as ornaments, but as useful members; let us observe how they have made their walls like honey-combs, and admire their ingenuity as we pretend to admire the instinct infused by the great architect into the bee. All this cannot be understood without mechanical knowledge; a thing which few of our professional architects have any share of. Thus would architectonic taste be a mark of skill; and the person who presents the design of a building would know how to execute it,

without committing it entirely to the mason and carpenter.

These observations are not a digression from our subject. The same principles of mutual pressure and equilibrium have a place in roofs and many wooden edifices; and if they had been as much studied as the Normans and Saracens seem to have studied such of them as were applicable to their purposes, we might have produced wooden buildings as far superior to what we are familiarly acquainted with, as the bold and wonderful churches still remaining in Europe are superior to the timid productions of our stone architecture. The centres used in building the bridge of Orleans, is an instance of what may be done in this way.

549. The Norman architects frequently roofed with stone. Their wooden roofs were in general very simple, and their professed aim was to dispense with them altogether. Fond of their own science, they copied nothing from a wooden building, and ran into a similar fault with the ancient Greeks. The parts of their buildings which were necessarily of timber were made to imitate stone-buildings; and Gothic ornament consists in cramming every thing full of arches and spaundrels. Nothing else is to be seen in their timber works, nay even in their sculpture.

550. But there appears to have been a rivalship in old times between the masons and the carpenters. Many of the baronial halls are of prodigious width, and are roofed with timber: and the carpenters appeared to have borrowed much knowledge from the masons of those times, and their wide roofs are frequently constructed with great ingenuity. Their aim, like the masons, was to throw a roof over a very wide building without employing great logs of timber. We have seen roofs 60 feet wide, without having a piece of timber in them above 10 feet long and 4 inches square. The Parliament House and Tron

Church of Edinburgh, the great hall of Tarnaway castle near Elgin, are specimens of those roofs. They are very numerous on the continent. Indeed Britain retains few monuments of private magnificence. Aristocratic state never was so great with us; and the rancour of our civil wars gave most of the performances of the carpenter to the flames. Westminster Hall exhibits a specimen of the false taste of the Norman roofs. It contains the essential parts indeed, very properly disposed; but they are hidden, or intentionally covered, with what is conceived to be ornamental; and this is an imitation of stone arches, crammed in between slender pillars which hang down from the principal frames, trusses, or rafters. In a pure Norman roof, such as Tarnaway hall, the essential parts are exhibited as things understood, and therefore relished. They are refined and ornamented; and it is here that the inferior kind of taste or the want of it may appear. And here we do not mean to defend all the whims of our ancestors; but we assert that it is no more nécessary to consider the members of a roof as a thing to be concealed like a garret, than the members of a ceiling, which form the most beautiful part of the Greek architecture. Should it be said that a roof is only a thing to keep off the rain, it may be answered, that a ceiling is only to keep off the dust, or the floor to be trodden under foot, and that we should have neither compartments in the one nor inlaid work or carpets on the other. The structure of a roof may therefore be exhibited with propriety, and made an ornamental feature. This has been done even in Italy. The church of St Maria Maggiore in Rome and several others are specimens: but it must be acknowledged, that the forms of the principal frames of these roofs, which resemble those of our modern buildings, are very unfit for agreeable ornament. As we have already observed, our imaginations have not been made sufficiently familiar with the principles, and we are rather alarmed than pleased with the appearance of the immense logs of timber which form the couples of these roofs, and hang over our heads with every appearance of weight and danger. It is quite otherwise with the ingenious roofs of the German and Norman architects. Slender timbers, interlaced with great symmetry, and thrown by necessity into figures which are naturally pretty, form altogether an object which no carpenter can view without pleasure. And why should the gentleman refuse himself the same pleasure of beholding scientific ingenuity?

551. The roof is in fact the part of the building which requires the greatest degree of skill, and where science will be of more service than in any other part. The architect seldom knows much of the matter, and leaves the task to the carpenter. The carpenter considers the framing of a great roof as the touchstone of his art: and nothing indeed tends so much to show his judgment and his fertility of resource.

552. It must therefore be very acceptable to the artist to have a clear view of the principles by which this difficult problem may be solved in the best manner, so that the roof may have all the strength and security that can be wished for, without an extravagant expence of timber and iron. We have said that mechanical science can give great assistance in this matter. We may add that the framing of carpentry, whether for roofs, floors, or any other purpose, affords one of the most elegant and most satisfactory applications which can be made of mechanical science to the arts of common life. Unfortunately the practical artist is seldom possessed even of the small portion of science which would almost insure his practice from all risk of failure; and even our most experienced carpenters have seldom any more knowledge than what arises from their experience and natural sagacity. The most approved author in our language is Price in his British Carpenter. Mathurin Jousse is in like manner

the author most in repute in France; and the publications of both these authors are void of every appearance of principle. It is not uncommon to see the works of carpenters of the greatest reputation tumble down, in consequence of mistakes from which the most elementary knowledge would have saved them.

553. We shall attempt, in this article, to give an account of the leading principles of this art, in a manner so familiar and palpable, that any person who knows the common properties of the lever, and the composition of motion, shall so far understand them as to be able, on every occasion, so to dispose his materials, with respect to the strains to which they are to be exposed, that he shall always know the effective strain on every piece, and shall, in most cases, be able to make the disposition such as to derive the greatest possible advantage from the materials which he employs.

554. It is evident that the whole must depend on the principles which regulate the strength of the materials, relative to the manner in which this strength is exerted, and the manner in which the strain is laid on the piece of matter. With respect to the first, this is not the proper place for considering it, and we must refer the reader to the article STRENGTH OF MATERIALS. We shall just borrow from that article two or three propositions suited to our purpose.

The force with which the materials of our edifices, roofs, floors, machines, and framings of every kind, resist being broken or crushed, or pulled asunder, is, immediately or ultimately, the cohesion of their particles. When a weight hangs by a rope, it tends either immediately to break all the fibres, overcoming the cohesion among the particles of each, or it tends to pull one parcel of them from among the rest, with which they are joined. This union of the fibres is brought about by some kind of gluten, or by twisting, which causes them to bind each other so hard that any one will break rather than come out, so much is it withheld by friction. The ultimate resistance is therefore the cohesion of the fibre ; the force or strength of all fibrous materials, such as timber, is exerted in much the same manner. The fibres are either broken or pulled out from among the rest. Metals, stone, glass, and the like, resist being pulled asunder by the simple cohesion of their parts.

The force which is necessary for breaking a rope or wire is a proper measure of its strength. In like manner, the force necessary for tearing directly asunder any rod of wood or metal, breaking all its fibres, or tearing them from among each other, is a proper measure of the united strength of all these fibres. And it is the simplest strain to which they can be exposed, being just equal to the sum of the forces necessary for breaking or disengaging each fibre. And, if the body is not of a fibrous structure, which is the case with metals, stones, glass, and many other substances, this force is still equal to the simple sum of the cohesive forces of each particle which is separated by the fracture. Let us distinguish this mode of exertion of the cohesion of the body by the name of its ABSOLUTE STRENGTH.

When solid bodies are, on the contrary, exposed to great compression, they can resist only to a certain degree. A piece of clay or lead will be squeezed out; a piece of freestone will be crushed to powder; a beam of wood will be crippled, swelling out in the middle, and its fibres lose their mutual cohesion, after which it is easily crushed by the load. A notion may be formed of the manner in which these strains are resisted by conceiving a cylindrical pipe filled with small shot, well shaken together, so that each spherule is lying in the closest manner possible, that is, in contact with six others in the same vertical plane (this being the position in which the shot will take the least room). Thus each touches the rest in six points: Now suppose them all united, in these six points only, by some cement. This assemblage will stick together and form a cylindrical pillar, which may be taken out of its mould. Suppose this pillar standing upright, and loaded above. The supports arising from the cement act obliquely, and the load tends either to force them asunder laterally, or to make them slide on each other: either of these things happening, the whole is crushed to pieces. The resistance of fibrous materials to such a strain is a little more intricate, but may be explained in a way very similar.

A piece of matter of any kind may also be destroyed by wrenching or twisting it. We can easily form a notion of its resistance to this kind of strain, by considering what would happen to the cylinder of small shot if treated in this way.

And lastly, a beam, or a bar of metal, or a piece of stone or other matter, may be broken transversely. This will happen to a rafter or joist supported at the ends when overloaded, or to a beam having one end stuck fast in a wall and a load laid on its projecting part. This is the strain to which materials are most commonly exposed in roofs; and, unfortunately, it is the strain which they are the least able to bear; or rather it is the manner of application which causes an external force to excite the greatest possible immediate strain on the particles. It is against this that the carpenter must chiefly guard, avoiding it when in his power, and, in every case, diminishing it as much as possible. It is necessary to give the reader a clear notion of the great weakness of materials in relation to this transverse strain. But we shall do nothing more, referring him to the article STRENGTH OF MATERIALS.

555. Let ACDB Plate IX. Fig. 1. represent the side of a beam projecting horizontally from a wall in which it is firmly fixed, and let it be loaded with a weight W ap-

pended to its extremity. This tends to break it; and the least reflection will convince any person that if the beam is equally strong throughout, it will break in the line CD, even with the surface of the wall. It will open at D, while C will serve as a sort of joint, round which it will turn. The cross section through the line CD is, for this reason, called the *section of fracture*, and the horizontal line, drawn through C on its under surface, is called the *axis of fracture*. The fracture is made by tearing asunder the fibres, such as DE or FG. Let us suppose a real joint at C, and that the beam is really sawed through along CD. and that in place of its natural fibres threads are substituted all over the section of fracture. The weight now tends to break these threads; and it is our business to find the force necessary for this purpose.

It is evident that DCA may be considered as a bended lever, of which C is the fulcrum. If f be the force which will just balance the cohesion of a thread when hung on it so that the smallest addition will break it, we may find the weight which will be sufficient for this purpose when hung on at A, by saying, $AC : CD = f : \varphi$, and φ will be the weight which will just break the thread, by hanging φ by the point A. This gives us $\varphi = f \times \frac{CD}{CA}$. If the weight be hung on at a, the force just sufficieut for breaking the same thread will be $= f \times \frac{CD}{Ca}$. In like manner the force φ, which must be hung on at A in order to break an equally strong or an equally resisting fibre at F, must be $= f \times \frac{CF}{CA}$. And so on of all the rest.

If we suppose all the fibres to exert equal resistances at the instant of fracture, we know, from the simplest elements of mechanics, that the resistance of all the particles in the line CD, each acting equally in its own place, is the same as if all the individual resistances were united in the middle point g. Now this total resistance is the resistance or strength f of each particle, multiplied by the number of particles. This number may be expressed by the line CD, because we have no reason to suppose that they are at unequal distances. Therefore, in comparing different sections together, the number of particles in each are as the sections themselves. Therefore DC may represent the number of particles in the line DC. Let us call this line the depth of the beam, and express it by the symbol d. And since we are at present treating of roofs whose rafters and other parts are commonly of uniform breadth, let us call AH or BI the breadth of the beam, and express it by b, and let CA be called its length, l. We may now express the strength of the whole line CD by $f \times d$, and we may suppose it all concentrated in the middle point g. Its mechanical energy, therefore, by which it resists the energy of the weight w, applied at the distance l, is $f.$ CD. Cg, while the momentum of w is $w.$ CA. We must therefore have $f.$ CD. C$g. = w.$ CA, or $f\,d.\ \frac{1}{2}\,d = w.\ l$, and $f\,d : w = l : \frac{1}{2}\,d$, or $f\,d : w = 2\,l : d$. That is, twice the length of the beam is to its depth as the absolute strength of one of its vertical planes to its relative strength, or its power of resisting this transverse fracture.

It is evident, that what has been now demonstrated of the resistance exerted in the line CD, is equally true of every line parallel to CD in the thickness or breadth of the beam. The absolute strength of the whole section of fracture is properly represented by $f.\ d.\ b$, and we still have $2\,l : d = f\,d\,b : w$; or twice the length of the beam is to its depth as the absolute strength to the relative strength. Suppose the beam 12 feet long and one foot deep; then whatever is its absolute strength, the 24th part of this will break it if hung at its extremity.

But even this is too favourable a statement; all the fibres are supposed to meet alike in the instant of fracture. But this is not true. At the instant that the fibre at D breaks, it is stretched to the utmost, and is exerting its whole force. But at this instant the fibre at g is not so much stretched, and it is not then exerting its utmost force. If we suppose the extension of the fibres to be as their distance from C, and the actual exertion of each to be as their extensions, it may easily be shown (see STRENGTH OF MATERIALS), that the whole resistance is the same as if the full force of all the fibres were united at a point r distant from C by one-third of CD. In this case we must say, that the absolute strength is to the relative strength as three times the length to the depth; so that the beam is weaker than by the former statement in the proportion of two to three.

Even this is more strength than experiment justifies; and we can see an evident reason for it. When the beam is strained, not only are the upper fibres stretched, but the lower fibres are compressed. This is very distinctly seen, if we attempt to break a piece of cork cut into the shape of a beam: this being the case, C is not the centre of fracture. There is some point c which lies between the fibres which are stretched and those that are compressed. This fibre is neither stretched nor squeezed; and this point is the real centre of fracture: and the lever by which a fibre D resists, is not DC, but a shorter one Dc; and the energy of the whole resistances must be less than by the second statement. Till we know the proportion between the dilatability and compressibility of the parts, and the relation between the dilatations of the fibres and the resistances which they exert in this state of dilatation, we cannot positively say where the point c is situated, nor what is the sum of the actual resistances, or the point where their action may be supposed concentrated. The firmer woods, such as oak and chesnut, may be supposed to be but slightly compressible; we know that willow and other soft woods are very compressible. These last must therefore be weaker: for it is evident, that the fibres which are in a state of compression do not resist the fracture. It is well known, that a beam of willow may be cut through from C to *g* without weakening it in the least, if the cut be filled up by a wedge of hard wood stuck in.

We can only say, that very sound oak and red fir have the centre of effort so situated, that the absolute strength is to the relative strength in a proportion not less than that of three and a half times the length of the beam to its depth. A square inch of sound oak will carry about 8000 pounds. If this bar be firmly fixed in a wall, and project about 12 inches, and be loaded at the extremity with 200 pounds, it will be broken. It will just bear 190, its relative strength being $\frac{1}{42}$ of its absolute strength; and this is the case only with the finest pieces, so placed that their annual plates or layers are in a vertical position. A larger log is not so strong transversely, because its plates lie in various directions round the heart.

556. These observations are enough to give us a distinct notion of the vast diminution of the strength of timber when the strain is across it; and we see the justice of the maxim which we inculcated, that the carpenter, in framing roofs, should avoid as much as possible the exposing his timbers to transverse strains. But this cannot be avoided in all cases. Nay, the ultimate strain, arising from the very nature of a roof, is transverse. The rafters must carry their own weight, and this tends to break them across: an oak beam a foot deep will not carry its own weight if it project more than 60 feet. Besides this, the rafters must carry the lead, tyling, or slates. We must therefore consider this transverse strain a little more particularly, so far as to know what strain will be laid on any part by any unavoidable load, laid on either at that or at any other.

557. We have hitherto supposed, that the beam had one of its ends fixed in a wall, and that it was loaded at the other end. This is not an usual arrangement, and was taken merely as affording a simple application of the mechanical principles. It is much more usual to have the beam supported at the ends, and loaded in the middle. Let the beam FEGH (Fig. 2.) rest on the props E and G, and be loaded at its middle point C with a weight W It is required to determine the strain at the section CD? It is plain that the beam will receive the same support, and suffer the same strain, if, instead of the blocks E and G, we substitute the ropes E *f c*, G *h g*, going over the pulleys *f* and *g*, and loaded with proper weights *e* and *g*. The weight *e* is equal to the support given by the block E; and *g* is equal to the support given by G. The sum of *e* and *g* is equal to W; and, on whatever point W is hung, the weights *e* and *g* are to W in the proportion of DG and DE to GE. Now, in this state of things, it appears that the strain on the section CD arises immediately from the upward action of the ropes F *f* and H *h*, or the upward pressions of the blocks E and G; and that the office of the weight W is to oblige the beam to oppose this strain. Things are in the same state in respect of strain as if a block were substituted at D for the weight W, and the weights *e* and *g* were hung on at E and G; only the directions will be opposite. The beam tends to break in the section CD, because the ropes pull it upwards at E and G, while a weight W holds it down at C. It tends to open at D, and C becomes the centre of fracture. The strain therefore is the same as if the half ED were fixed in the wall, and a weight equal to *g*, that is, to the half of W, were hung on at G.

Hence we conclude, that a beam supported at both ends, but not fixed there, and loaded in the middle, will carry four times as much weight as it can carry at its extremity, when the other extremity is fast in a wall.

The strain occasioned at any point L by a weight W, hung on at any other point D, is $=W \times \frac{DE}{EG} \times LG$. For EG is to ED as W is to the pressure occasioned at G. This would be balanced by some weight *g* acting over the pulley *h*; and this tends to break the beam at L, by acting on the lever GL. The pressure at G is $W. \frac{DE}{EG}$, and therefore the strain at L is $W. \frac{DE}{EG}. LG$.

In like manner, the strain occasioned at the point D by the weight W hung on there, is $W \times \frac{DE}{EG} \times DG$; which is therefore equal to $\frac{1}{2}$ W, when D is the middle point.

Hence we see, that the general strain on the beam arising from one weight, is proportionable to the rectangle of the parts of the beam, (for $\frac{W.DE.DG}{EG}$ is as DE.DG), and is greatest when the load is laid on the middle of the beam.

We also see, that the strain at L, by a load at D, is equal to the strain at D by the same load at L. And the strain at L, from a load at D, is to the strain by the same load at L as DE to LE. These are all very obvious corollaries; and they sufficiently inform us concerning the strains which are produced on any part of the timber by a load laid on any other part.

If we now suppose the beam to be fixed at the two ends, that is, firmly framed, or held down by blocks at I and K, placed beyond E and G, or framed into posts, it will carry twice as much as when its ends were free. For suppose it sawn through at CD; the weight W hung on there will be just sufficient to break it at E and G. Now restore the connection of the section CD, it will require another weight W to break it there at the same time.

Therefore, when a rafter, or any piece of timber, is firmly connected with the three fixed points G, E, I, it will bear a greater load between any two of them than if its connection with the remote point were removed; and if it be fastened in four points, G, E, I, K, it will be twice as strong in the middle part as without the two remote connections.

One is apt to expect from this that the joist of a floor will be much strengthened by being firmly built in the wall. It is a little strengthened; but the hold which can thus be given it is much too short to be of any sensible service; and it tends greatly to shatter the wall, because, when it is bent down by a load, it forces up the wall with the momentum of a long lever. Judicious builders therefore take care not to bind the joists tight in the wall. But when the joists of adjoining rooms lie in the same direction, it is a great advantage to make them of one piece. They are then twice as strong as when made in two lengths.

558. It is easy to deduce from these premises the strain on any point which arises from the weight of the beam itself, or from any load which is uniformly diffused over the whole or any part. We may always consider the whole of the weight which is thus uniformly diffused over any part as united in the middle point of that part; and if the load is not uniformly diffused, we may still suppose it united at its centre of gravity. Thus, to know the strain at L arising from the weight of the whole beam, we may suppose the whole weight accumulated in its middle point D. Also the strain at L, arising from the weight of the part ED, is the same as if this weight were accumulated in the middle point *d* of ED; and it is the same as if half the weight of ED were hung on at D. For the real strain at L is the upward pressure at G, acting by the lever GL. Now calling *e* the

weight of the part DE: this upward pressure will be $\frac{e \times d\mathrm{E}}{\mathrm{EG}}$, or $\frac{\frac{1}{2}e \times \mathrm{DE}}{\mathrm{EG}}$.

Therefore the strain on the middle of a beam, arising from its own weight, or from any uniform load, is the weight of the beam or its load $\frac{\mathrm{ED}}{\mathrm{EG}} \times \mathrm{DG}$; that is, half the weight of the beam or load multiplied or acting by the lever DG; for $\frac{\mathrm{ED}}{\mathrm{EG}}$ is $\frac{1}{2}$.

Also the strain at L, arising from the weight of the beam, or the uniform load, is $\frac{1}{2}$ the weight of the beam or load acting by the lever LG. It is therefore proportional to LG, and is greatest of all at D. Therefore a beam of uniform strength throughout, uniformly loaded, will break in the middle.

559. It is of importance to know the relation between the strains arising from the weights of the beams, or from any uniformly diffused load, and the relative strength. We have already seen, that the relative strength is $f\frac{db.d}{m\,l}$, where m is a number to be discovered by experiment for every different species of materials. Leaving out every circumstance but what depends on the dimensions of the beam, viz. d, b, and l, we see that the relative strength is in the proportion of $\frac{d^2 b}{l}$, that is, as the breadth and the square of the depth directly, and the length inversely

Now, to consider first the strain arising from the weight of the beam itself, it is evident that this weight increases in the same proportion with the depth, the breadth, and the length of the beam. Therefore its power of resisting this strain must be as its depth directly, and the square of its length inversely. To consider this in a more popular manner, it is plain that the increase of breadth makes no change in the power of resisting the actual strain, because the load and the absolute strength increase in the same proportion with the breadth. But by increasing the depth, we increase the resisting section in the same proportion, and therefore the number of resisting fibres and the absolute strength: but we also increase the weight in the same proportion. This makes a compensation, and the relative strength is yet the same. But by increasing the depth, we have not only increased the absolute strength, but also its mechanical energy: For the resistance to fracture is the same as if the full strength of each fibre was exerted at the point which we called the centre of effort; and we showed, that the distance of this from the under side of the beam was a certain portion (a half, a third, a fourth, &c.) of the whole depth of the beam. This distance is the arm of the lever by which the cohesion of the wood may be supposed to act. Therefore this arm of the lever, and consequently the energy of the resistance, increases in the proportion of the depth of the beam, and this remains uncompensated by any increase of the strain. On the whole, therefore, the power of the beam to sustain its own weight increases in the proportion of its depth. But, on the other hand, the power of withstanding a given strain applied at its extremity, or to any aliquot part of its length, is diminished as the length increases, or is inversely as the length; and the strain arising from the weight of the beam also increases as the length. Therefore the power of resisting the strain actually exerted on it by the weight of the beam, is inversely as the square of the length. On the whole, therefore, the power of a beam to carry its own weight, varies in the proportion of its depth directly, and the square of its length inversely.

As this strain is frequently a considerable part of the whole, it is proper to consider it apart, and then to reckon only on what remains for the support of any extraneous load.

560. In the next place, the power of a beam to carry any load which is uniformly diffused over its length, must be inversely as the square of the length: for the power of withstanding *any* strain applied to an aliquot part of the length (which is the case here, because the load may be conceived as accumulated at its centre of gravity, the middle point of the beam) is inversely as the length; and the *actual* strain is as the length, and therefore its momentum is as the square of the length. Therefore the power of a beam to carry a weight uniformly diffused over it, is inversely as the square of the length. *N. B.* It is here understood, that the uniform load is of some determined quantity for every foot of the length, so that a beam of double length carries a double load.

561. We have hitherto supposed that the forces which tend to break a beam transversely, are acting in a direction perpendicular to the beam. This is always the case in level floors loaded in any manner; but in roofs, the action of the load tending to break the rafters is oblique, because gravity always acts in vertical lines. It may also frequently happen, that a beam is strained by a force acting obliquely. This modification of the strain is easily discussed. Suppose that the external force, which is measured by the weight W in Fig. 1. acts in the direction A *w* instead of AW. Draw C *á* perpendicular to A *w*. Then the momentum of this external force is not to be measured by W × AC, but by W × *á* C. The strain therefore by which the fibres in the section of fracture DC are torn asunder, is diminished in the proportion of CA to C *á*, that is, in the proportion of radius to the sine of the angle CA *á*, which the beam makes with the direction of the external force.

To apply this to our purpose in the most familiar manner, let AB (Fig. 3.) be an oblique rafter of a building, loaded with a weight W suspended to any point C, and thereby occasioning a strain in some part D. We have already seen, that the immediate cause of the strain on D is the reaction of the support which is given to the point B. The rafter may at present be considered as a lever, supported at A, and pulled down by the line CW. This ocsasions a pressure on B, and the support acts in the opposite direction to the action of the lever, that is, in the direction B *b*, perpendicular to BA. This tends to break the beam in every part. The pressure exerted at B is $\frac{W \times AE}{AB}$, AE being a horizontal line. Therefore the strain at D will be $\frac{W \times AE}{AB} \times BD$. Had the beam been lying horizontally, the strain at D, from the weight W suspended at C, would have been $\frac{W.AC}{AB} \times BD$. It is therefore diminished in the proportion of AC to AE, that is, in the proportion of radius to the cosine of the elevation, or in the proportion of the secant of elevation to the radius.

It is evident, that this law of diminution of the strain is the same whether the strain arises from a load on any part of the rafter, or from the weight of the rafter itself, or from any load uniformly diffused over its length, provided only that these loads act in vertical lines.

562. We can now compare the strength of roofs which have different elevations. Supposing the width of the building to be given, and that the weight of a square yard of covering is also given. Then, because the load on the rafter will increase in the same proportion with its length, the load on the slant side BA of the roof will be to the load of a similar covering on the half AF of the flat roof, of the same width, as AB to AF. But the transverse action of any load on AB, by which it tends to break it, is to that of the same load on AF as AF to

AB. The transverse strain therefore is the same on both, the increase of real load on AB being compensated by the obliquity of its action. But the strengths of beams to resist equal strains, applied to similar points, or uniformly diffused over them, are inversely as their lengths, because the momentum or energy of the strain is proportional to the length. Therefore the power of AB to withstand the strain to which it is really exposed, is to the power of AF to resist its strain as AF to AB. If, therefore, a rafter AG of a certain scantling is just able to carry the roofing laid on it, a rafter AB of the same scantling, but more elevated, will be too weak in the proportion of AG to AB. Therefore steeper roofs require stouter rafters, in order that they may be equally able to carry a roofing of equal weight per square yard. To be equally strong, they must be made broader, or placed nearer to each other, in the proportion of their greater length, or they must be made deeper in the subduplicate proportion of their length The following easy construction will enable the artist not familiar with computation to proportion the depth of the rafter to the slope of the roof.

Let the horizontal line af Fig. 4. be the proper depth of a beam whose length is half the width of the building; that is, such as would make it fit for carrying the intended tiling laid on a flat roof. Draw the vertical line fb, and the line ab having the elevation of the rafter; make ag equal to af, and describe the semicircle bdg; draw ad perpendicular to ab, ad is the required depth. The demonstration is evident.

We have now treated in sufficient detail what relates to the chief strain on the component parts of a roof, namely, what tends to break them transversely; and we have enlarged more on the subject than what the present occasion indispensably required, because the propositions which we have demonstrated are equally applicable to all framings of carpentry, and are even of greater moment in many cases, particularly in the construction of machines. These consist of levers in various forms, which are strained transversely; and similar strains frequently occur in many of the supporting and connecting parts.

563. We proceed, in the next place, to consider the other strains to which the parts of roofs are exposed, in consequence of the support which they mutually give each other, and the pressures (or *thrusts* as they are called in the language of the house carpenter) which they exert on each other, and on the walls or piers of the building.

Let a beam or piece of timber AB, Fig. 5. be suspended by two lines AC, BD; or let it be supported by two props AE, BF, which are perfectly moveable round their remote extremities E, F, or let it rest on the two polished planes KAH, LBM. Moreover, let G be the centre of gravity of the beam, and let GN be a line through the centre of gravity, perpendicular to the horizon. The beam will not be in equilibrio unless the vertical line GN either passes through P, the point in which the directions of the two lines AC, BD, or the directions of the two props EA, FD, or the perpendiculars to the two planes KAH, LBM, intersect each other, or is parallel to these directions. For the supports given by the lines or props are unquestionably exerted in the direction of their lengths; and it is as well known in mechanics that the supports given by planes are exerted in a direction perpendicular to those planes in the points of contact; and we know that the weight of the beam acts in the same manner as if it were all accumulated in its centre of gravity G, and that it acts in the direction GN perpendicular to the horizon. Moreover, when a body is in equilibrio between three forces, they are acting in one plane, and their directions are either parallel or they pass through one point.

The support given to the beam is therefore the same as

if it were suspended by two lines which are attached to the single point P. We may also infer, that the points of suspension C, D, the points of support E, F, the points of contact A, B, and the centre of gravity G, are all in one vertical plane.

When this position of the beam is disturbed by any external force, there must either be a motion of the points A and B round the centres of suspension C and D, or of the props round these points of support E and F, or a sliding of the ends of the beam along the polished planes GH and IK; and in consequence of these motions the centre of gravity G will go out of its place, and the vertical line GN will no longer pass through the point where the directions of the supports intersect each other. If the centre of gravity rises by this motion, the body will have a tendency to recover its former position, and it will require force to keep it away from it. In this case the equilibrium may be said to be *stable*, or the body to have *stability*. But if the centre of gravity descends when the body is moved from the position of equilibrium, it will tend to move still farther; and so far will it be from recovering its former position, that it will now fall. This equilibrium may be called a *tottering equilibrium*. These accidents depend on the situations of the points A, B, C, D, E, F; and they may be determined by considering the subject geometrically. It does not much interest us at present; it is rarely that the equilibrium of suspension is tottering, or that of props is stable. It is evident, that if the beam were suspended by lines from the point P, it would have stability, for it would swing like a pendulum round P, and therefore would always tend towards the position of equilibrium. The intersection of the lines of support would still be at P, and the vertical line drawn through the centre of gravity, when in any other situation, would be on that side of P towards which this centre has been moved. Therefore, by the rules of pendulous bodies, it tends to come back. This would be more remarkably the case if the points of suspension C and D be on the same side of the point P with the points of attachment A and B; for in this case the new point of intersection of the lines of support would shift to the opposite side, and be still farther from the vertical line through the new position of the centre of gravity. But if the points of suspension and of attachment are on opposite sides of P, the new point of intersection may shift to the same side with the centre of gravity, and lie beyond the vertical line; in this case the equilibrium is tottering. It is easy to perceive, too, that if the equilibrium of suspension from the points C and D be stable, the equilibrium on the props AE and BF must be tottering. It is not necessary for our present purpose to engage more particularly in this discussion.

It is plain that, with respect to the mere momentary equilibrium, there is no difference in the support by threads, or props, or planes, and we may substitute the one for the other. We shall find this substitution extremely useful, because we easily conceive distinct notions of the support of a body by strings.

Observe farther, that if the whole figure be inverted, and strings be substituted for props, and props for strings, the equilibrium will still obtain: for by comparing Fig. 5. with Fig. 6., we see that the vertical line through the centre of gravity will pass through the intersection of the two strings or props; and this is all that is necessary for the equilibrium: only it must be observed in the substitution of props for threads, and of threads for props that if it be done without inverting the whole figure, a stable equilibrium becomes a tottering one, and *vice versa*.

This is a most useful proposition, especially to the unlettered artisan, and enables him to make a practical use of problems which the greatest mechanical geniuses have found no easy task to solve. An instance will show the extent and utility of it. Suppose it were required to make a mansard or kirb roof whose width is AB (Fig. 7.), and consisting of the four equal rafters AC, CD, DE, EB. There can be no doubt but that its best form is that which will put all the parts in equilibrio, so that no ties or stays may be necessary for opposing the unbalanced thrust of any part of it. Make a chain *a c d e b* (Fig. 8.) of four equal pieces, loosely connected by pin-joints, round which the parts are perfectly moveable. Suspend this from two pins *a*, *b*, fixed in a horizontal line. This chain or festoon will arrange itself in such a form that its parts are in equilibrio. Then we know that if the figure be inverted, it will compose the frame or truss of a kirb-roof *a γ δ ε b*, which is also in equilibrio, the thrusts of the pieces balancing each other in the same manner that the mutual pulls of the hanging festoon *a c d e b* did. If the proportion of the height *d f* to the width *a b* is not such as pleases, let the pins *a*, *b*, be placed nearer or more distant, till a proportion between the width and height is obtained which pleases, and then make the figure ACDEB Fig. 7. similar to it. It is evident that this proposition will apply in the same manner to the determination of the form of an arch of a bridge; but this is not a proper place for a farther discussion.

We are now able to compute all the thrusts and other pressures which are exerted by the parts of a roof on each other and on the walls. Let AB (Fig. 9.) be a beam standing any how obliquely, and G its centre of gravity. Let us suppose that the ends of it are supported in any directions AC, BD, by strings, props, or planes. Let these directions meet in the point P of the vertical line PG passing through its centre of gravity. Through G draw lines G *a*, G *b* parallel to PB, PA. Then

$$\left.\begin{array}{l}\text{The weight of the beam}\\ \text{The pressure or thrust at A}\\ \text{The pressure at B}\end{array}\right\}\text{are proportional to}\left\{\begin{array}{l}PG\\ Pa\\ Pb.\end{array}\right.$$

For when a body is in equilibrio between three forces, these forces are proportional to the sides of a triangle which have their directions.

In like manner, if A g be drawn parallel to P b, we shall have

$$\left.\begin{array}{l}\text{Weight of the beam}\\ \text{Thrust on A}\\ \text{Thrust on B}\end{array}\right\}\text{proportional to}\left\{\begin{array}{l}\text{P}g\\ \text{PA}\\ \text{B}g\end{array}\right.$$

Or, drawing B γ parallel to P a

$$\left.\begin{array}{l}\text{Weight of beam}\\ \text{Thrust at A}\\ \text{Thrust at B}\end{array}\right\}\text{are proportional to}\left\{\begin{array}{l}\text{P}\\ \text{B}\gamma\\ \text{PB.}\end{array}\right.$$

It cannot be disputed that, if strength alone be considered, the proper form of a roof is that which puts the whole in equilibrio, so that it would remain in that shape although all the joints were perfectly loose or flexible. If it has any other shape, additional ties or braces are necessary for preserving it, and the parts are unnecessarily strained. When this equilibrium is obtained, the rafters which compose the roof are all acting on each other in the direction of their lengths; and by this action, combined with their weights, they sustain no strain but that of compression, the strain of all others that they are the most able to resist. We may consider them as so many inflexible lines having their weights accumulated in their centres of gravity. But it will allow an easier investigation of the subject, if we suppose the weights to be at the joints, equal to the real vertical pressures which are exerted on these points. These are very easily computed: for it is plain, that the weight of the beam AB (Fig. 9.) is to the part of this weight that is supported at B as AB to AG. Therefore, if W represent the weight of the beam, the vertical pressure at B will be $W \times \frac{AG}{AB}$, and the vertical pressure at A will be $W \times \frac{BG}{AB}$. In like manner, the prop BF being considered as another beam, and f as its centre of gravity, and w as it weight, a part of this weight, equal to $w \times \frac{fF}{BF}$, is supported at B, and the whole vertical pressure at B is $W \times \frac{AG}{AB} + w \times \frac{fF}{BF}$. And thus we greatly simplify the construction of the mutual thrusts of roof frames. We need hardly observe, that although these pressures by which the parts of a frame support each other in opposition to the vertical action of gravity, are always exerted in the direction of the pieces, they may be resolved into pressures acting in any other direction which may engage our attention.

All that we propose to deliver on this subject at present may be included in the following proposition.

Let ABCDE (Fig. 10.) be an assemblage of rafters in a vertical plane, resting on two fixed points A and E in a horizontal line, and perfectly moveable round all the joints A, B, C, D, E; and let it be supposed to be in equilibrio, and let us investigate what adjustment of the different circumstances of weight and inclination of its different parts is necessary for producing this equilibrium.

Let F, G, H, I, be the centres of gravity of the different rafters, and let these letters express the weights of each. Then (by what has been said above) the weight which presses B directly downwards is $F \times \frac{AF}{AB} + G \times \frac{CG}{BC}$. The weight on C is in like manner $G \times \frac{BG}{BC} + H \times \frac{DH}{CD}$ and that on D is $H \times \frac{CH}{CD} + I \times \frac{EI}{DE}$.

Let A $b\,c\,d$ E be the figure ABCDE inverted, in the manner already described. It may be conceived as a thread fastened at A and E, and loaded at b, c, and d, with the weights which are really pressing on B, C, and D. It will arrange itself into such a form that all will be in equilibrio. We may discover this form by means of this single consideration, that any part *b c* of the thread is equally stretched throughout in the direction of its length. Let us therefore investigate the proportion between the weight β which we suppose to be pulling the point *b* in the vertical direction *b* β to the weight δ, which is pulling down the point *d* in a similar manner. It is evident, that since AE is a horizontal line, and the figures A *b c d* E and ABCDE equal and similar, the lines B *b*, C *c*, D *d*, are vertical. Take *b f* to represent the weight hanging at *b*. By stretching the threads *b* A and *b c* it is set in opposition to the contractile powers of the threads, acting in the directions *b* A and *b c*, and it is in immediate equilibrio with the equivalent of these two contractile forces. Therefore make *b g* equal to *b f*, and make it the diagonal of a parallelogram *h b i g*. It is evident that *b h*, *b i*, are the forces exerted by the threads *b* A, *b c*. Then, seeing that the thread *b c* is equally stretched in both directions, make *c k* equal to *b i*; *c k* is the contractile force which is excited at *c* by the weight which is hanging there. Draw *k l* parallel to *c d*, and *l m* parallel to *b c*. The force *l c* is the equivalent of the contractile forces *c k*, *c m*, and is therefore equal and opposite to the force of gravity acting at C. In like manner, make $d n = c m$, and complete the parallelogram *n d p o*, having the vertical line *o d* for its diagonal. Then *d n* and *d p* are the contractile forces excited at *d*, and the weight hanging there must be equal to *o d*.

Therefore, the load at *b* is to the load at *d* as *b g* to *d o*. But we have seen that the compressing forces at B, C, D, may be substituted for the extending forces at *b*, *c*, *d*. Therefore the weights at B, C, D, which produce the compressions, are equal to the weights at *b*, *c*, *d*, which produce the extensions. Therefore

$$bg : do = F \times \frac{AF}{AB} + G \times \frac{CG}{BC} : H \times \frac{CH}{CD} + I \times \frac{EI}{DE}.$$

Let us inquire what relation there is between this proportion of the loads upon the joints at B and D, and the angles which the rafters make at these joints with each other, and with the horizon or the plumb lines. Produce AB till it cut the vertical C c in Q; draw BR parallel to CD, and BS parallel to DE. The similarity of the figures ABCDE and A $b c d$ E, and the similarity of their position with respect to the horizontal and plumb lines, show, without any further demonstration, that the triangles QCB and $g b i$ are similar, and that QB : BC $= gi : ib$, $= hb : ib$. Therefore QB is to BC as the contractile force exerted by the thread A b to that exerted by bc; and therefore QB is to BC as the compression of BA to the compression on BC. Then, because bi is equal to ck, and the triangles CBR and ckl are similar, CB : BR $=$ $ck : kl$, $= ck : cm$, and CB is to BR as the compression on CB to the compression on CD. And, in like manner, because $cm = dn$, we have BR to BS as the compression on DC to the compression on DE. Also BR : RS $=$ $nd : do$, that is, as the compression on DC to the load on D. Finally, combining all these ratios

$$QC : CB = gb : bi, = gb : kc$$
$$CB : BR = kc : kl, = kc : dn$$
$$BR : BS = nd : no = dn : no$$
$$BS : RS = no : do = no : do, \text{ we have finally}$$
$$QC : RS = gb : od = \text{Load at B} : \text{Load at D}.$$

Now

$$QC : BC = \text{∫}, QBC : \text{∫}, BQC, = \text{∫}, ABC : \text{∫}, ABb$$
$$BC : BR = \text{∫}, BRC : \text{∫}, BCR, = \text{∫}, CDd : \text{∫}, bBC$$
$$BR : RS = \text{∫}, BSR : \text{∫}, RBS, = \text{∫}, dDE : \text{∫}, CDE$$

Therefore

$$QC : RS = \text{∫}, ABC.\text{∫}, CDd.\text{∫}, dDE : \text{∫}, CDE.\text{∫}, ABb.\text{∫}, bBC.$$

Or

$$QC : RS = \frac{\text{ʃ, } ABC}{\text{ʃ, } ABb, \text{ʃ } CBb} : \frac{\text{ʃ, } CDE}{\text{ʃ, } d\,DC.\text{ʃ, } d\,DE}.$$

That is, the loads on the different joints are as the sines of the angles at these joints directly, and as the products of the sines of the angles which the rafters make with the plumb-lines inversely.

Or, the loads are as the sines of the angles of the joints directly, and as the products of the cosines of the elevations of the rafters jointly.

Or, the loads at the joints are as the sines of the angles at the joints, and as the products of the secants of elevation of the rafters jointly: for the secants of angles are inversely as the cosines.

Draw the horizontal line BT. It is evident, that if this be considered as the radius of a circle, the lines BQ, BC, BR, BS, are the secants of the angles which these lines make with the horizon. And they are also as the thrusts of those rafters to which they are parallel. Therefore, the thrust which any rafter makes in its own direction is as the secant of its elevation.

The horizontal thrust is the same at all the angles. For $i\iota, = k\kappa, = m\mu, = n\nu, = p\pi$. Therefore both walls are equally pressed out by the weight of the roof. We can find its quantity by comparing it with the load on one of the joints:

Thus, $QC : CB = \text{ʃ, } ABC : \text{ʃ, } AB\,b$

$BC : BT = \text{Rad.} : \text{ʃ, } BCT, = \text{Rad.} : \text{ʃ, } CB\,b$

Therefore, $QC : BT = \text{Rad.} \times \text{ʃ, } ABC : \text{ʃ, } b\,BA \times \text{ʃ, } b\,BC$.

564. It deserves remark, that the lengths of the beams do not affect either the proportion of the load at the different joints, nor the position of the rafters. This depends merely on the weights at the angles. If a change of length affects the weight, this indeed affects the form also: and this is generally the case. For it seldom happens, indeed it never should happen, that the weight on rafters of

longer bearing are not greater. The covering alone increases nearly in the proportion of the length of the rafter.

If the proportion of the weights at B, C, and D, are given, as also the position of any two of the lines, the position of all the rest is determined.

If the horizontal distances between the angles are all equal, the forces on the different angles are proportional to the verticals drawn on the lines through these angles from the adjoining angle, and the thrusts from the adjoining angles are as the lines which connect them.

If the rafters themselves are of equal lengths, the weights at the different angles are as these verticals and as the secants of the elevation of the rafters jointly.

565. This proposition is very fruitful in its practical consequences. It is easy to perceive that it contains the whole theory of the construction of arches; for each stone of an arch may be considered as one of the rafters of this piece of carpentry, since all is kept up by its mere equilibrium. We may have an opportunity of afterwards exhibiting some very elegant and simple solutions of the most difficult cases of this important problem; and we now proceed to make use of the knowledge we have acquired for the construction of roofs.

566. We mentioned by the by a problem which is not unfrequent in practice, to determine the best form of a kirb-roof. Mr Couplet, of the Royal Academy of Paris, has given a solution of it in an elaborate memoir in 1726, occupying several lemmas and theorems.

Let AE (Fig. 11.) be the width, and CF the height; it is required to construct a roof ABCDE whose rafters AB, BC, CD, DE, are all equal, and which shall be in equilibrio.

Draw CE, and bisect it perpendicularly in H by the line DHG, cutting the horizontal line AE in G. About the centre G, with the distance GE, describe the circle EDC. It must pass through C, because CH is equal to HE and the angles at H are equal. Draw HK parallel to FE, cutting the circumference in K. Draw CK, cutting GH in D. Join CD, ED; these lines are the rafters of half of the roof required.

We prove this by showing, that the loads in the angles C and D are equal. For this is the proportion which results from the equality of the rafters, and the extent of surface of the uniform roofing which they are supposed to support. Therefore produce ED till it meet the vertical FC in N; and having made the side CBA similar to CDE, complete the parallelogram BCDP, and draw DB, which will bisect CP in R, as the horizontal line KH, bisects CF in Q. Draw KF, which is evidently parallel to DP. Make CS perpendicular to CF, and equal to FG; and about S, with the radius SF, describe the circle FKW. It must pass through K, because SF is equal to CG, and CQ = QF. Draw WK, WS, and produce BC, cutting ND in O.

The angle WKF at the circumference is one half of the angle WSF at the centre, and is therefore equal to WSC, or CGF. It is therefore double of the angle CEF or ECS. But ECS is equal to ECD and DCS, and ECD is one half, of NDC, and DCS is one half of DCO, or CDP. Therefore the angle WKF is equal to NDP, and WK is parallel to ND, and CF is to CW as CP to CN; and CN is equal to CP. But it has been shown above, that CN and CP are as the loads upon D and C. These are therefore equal, and the frame ABCDE is in equilibrio.

A comparison of this solution with that of Mr Couplet will show its great advantage in respect of simplicity and perspicuity. And the intelligent reader can easily adapt the construction to any proportion between the rafters AB and BC, which other circumstances, such as garret-room, &c. may render convenient. The construction must be such that NC may be to CP as CD to $\frac{CD + DE.}{2}$. Whatever proportion of AB to BC is assumed, the point D′ will be found in the circumference of a semicircle H′ D′ h', whose centre is in the line CE, and having AB : BC = CH′ : HE′, = $c\,h'$: h' E. The rest of the construction is simple.

In buildings which are roofed with slate, tile, or shingles, the circumstance which is most likely to limit the construction is the slope of the upper rafters CB, CD. This must be sufficient to prevent the penetration of rain, and the stripping by the winds. The only circumstance left in our choice in this case is the proportion of the rafters AB and BC. Nothing is easier than making NC to CP in any desired proportion when the angle BCD is given.

567. We need not repeat that it is always a desirable thing to form a truss for a roof in such a manner that it shall be in equilibrio. When this is done, the whole force of the struts and braces which are added to it is employed in preserving this form, and no part is expended in unnecessary strains. For we must now observe, that the equilibrium of which we have been treating is always of that kind which we call the tottering, and the roof requires stays, braces, or hanging timbers, to give it stiffness, or keep it in shape. We have also said enough to enable any reader, acquainted with the most elementary geometry and mechanics, to compute the transverse strains and the thrusts to which the component parts of all roofs are exposed.

568. It only remains now to show the general maxims by which all roofs must be constructed, and the circumstances which determine their excellence. In doing this

we shall be exceedingly brief, and almost content ourselves with exhibiting the principal forms, of which the endless variety of roofs are only slight modifications. We shall not trouble the reader with any account of such roofs as receive part of their support from the interior walls, but confine ourselves to the more difficult problem of throwing a roof over a wide building, without any intermediate support; because when such roofs are constructed in the best manner, that is, deriving the greatest possible strength from the materials employed, the best construction of the others is necessarily included. For all such roofs as rest on the middle walls are roofs of smaller bearing. The only exception deserving notice is the roofs of churches, which have aisles separated from the nave by columns. The roof must rise on these. But if it is of an arched form internally, the horizontal thrusts must be nicely balanced, that they may not push the columns aside.

569. The simplest notion of a roof-frame is, that it consists of two rafters AB and BC (Fig. 12.), meeting in the ridge B.

Even this simple form is susceptible of better and worse. We have already seen, that when the weight of a square yard of covering is given, a steeper roof requires stronger rafters, and that when the scantling of the timbers is also given, the relative strength of a rafter is inversely as its length. But there is now another circumstance to be taken into the account, viz. the support which one rafter leg gives to the other. The best form of a rafter will therefore be that in which the relative strength of the legs, and their mutual support, give the greatest product. Mr Muller, in his *Military Engineer*, gives a determination of the best pitch of a roof, which has considerable ingenuity, and has been copied into many books of military education both in this island and on the Continent.

Describe on the width AC, Fig. 13. the semicircle ACF, and bisect it by the radius FD. Produce the rafter AB to the circumference in E, join EC, and draw the perpendicular EG. Now AB : AD = AC : AE, and $AE = \frac{AD \times AC}{AB}$, and AE is inversely as AB, and may therefore represent its strength in relation to the weight actually lying on it. Also the snpport which CB gives to AB is as CE, because CE is perpendicular to AB. Therefore the form which renders AE × EC a maximum seems to be that which has the greatest strength. But $AC : AE = EC : EG$, and $EG = \frac{AE \cdot EC}{AC}$, and is therefore proportional to AE.EC. Now EG is a maximum when B is in F, and a square pitch is in this respect the strongest. But it is very doubtful whether this construction is deduced from just principles. There is another strain to which the leg AB is exposed, which is not taken into the account. This arises from the curvature which it unavoidably acquires by the transverse pressure of its load. In this state it is pressed in its own direction by the abutment and load of the other leg. The relation between this strain and the resistance of the piece is not very distinctly known. Euler has given a dissertation on this subject (which is of great importance, because it affects posts and pillars of all kinds; and it is very well known that a post of ten feet long and six inches square will bear with great safety a weight, which would crush a post of the same scantling and 20 feet long in a minute); but his determination has not been acquiesced in by the first mathematicians. Now it is in relation to these two strains that the strength of the rafter should be adjusted. The firmness of the support given by the other leg is of no consequence, if its own strength is inferior to the strain. The force which tends to crush the leg AB, by compressing it in its curved state, is to its weight as AB to BD, as is easily seen by the composition of forces; and its incurvation by this force has a relation to it, which is of intricate determination. It is contained in the properties demonstrated by Bernoulli of the elastic curve. This determination also includes the relation between the curvature and the length of the piece. But the whole of this seemingly simple problem is of much more difficult investigation than Mr Muller was aware of; and his rules for the pitch of a roof, and for the sally of a dock gate, which depends on the same principles, are of no value. He is, however, the first author who attempted to solve either of these problems on mechanical principles susceptible of precise reasoning. Belidor's solutions, in his *Architecture Hydraulique*, are below notice.

Reasons of economy have made carpenters prefer a low pitch; and although this does diminish the support given by the opposite leg faster than it increases the relative strength of the other, this is not of material consequence, because the strength remaining in the opposite leg is still very great; for the supporting leg is acting against compression, in which case it is vastly stronger than the supported leg acting against a transverse strain.

570. But a roof of this simplicity will not do in most cases. There is no notice taken in its construction of the thrust which it exerts on the walls. Now this is the strain which is the most hazardous of all. Our ordinary walls, instead of being able to resist any considerable strain pressing them outwards, require, in general, some ties to keep them on foot. When a person thinks of the thinness and height of the walls of even a strong house, he will be surprised that they are not blown down by any strong blast of wind. A wall of three feet thick, and 60 feet high, could not withstand a wind blowing at the rate

of 30 feet *per* second (in which case it acts with a force considerably exceeding two pounds on every square foot), if it were not stiffened by cross walls, joists, and roof, which all help to tie the different parts of the building together.

571. A carpenter is therefore exceedingly careful to avoid every horizontal thrust, or to oppose them by other forces. And this introduces another essential part into the construction of a roof, namely the *tie* or *beam* AC, (Fig. 14.), laid from wall to wall, binding the feet A and C of the rafters together. This is the sole office of the beam; and it should be considered in no other light than as a string to prevent the roof from pushing out the walls. It is indeed used for carrying the ceiling of the apartments under it: and it is even made to support a flooring. But, considered as making part of a roof, it is merely a string; and the strain which it withstands tends to tear its parts asunder. It therefore acts with its whole absolute force, and a very small scantling would suffice if we could contrive to fasten it firmly enough to the foot of the rafter. If it is of oak, we may safely subject it to a strain of three tons for every square inch of its section. And fir will safely bear a strain of two tons for every square inch. But we are obliged to give the tie-beam much larger dimensions, that we may be able to connect it with the foot of the rafter by a mortise and tenon. Iron straps are also frequently added. By attending to this office of the tie-beam, the judicious carpenter is directed to the proper form of the mortise and tenon and of the strap. We shall consider both of these in a proper place, after we become acquainted with the various strains at the joints of a roof.

These large dimensions of the tie-beam allow us to load it with the ceilings without any risk, and even to lay floors on it with moderation and caution. But when it has a great bearing or span, it is very apt to bend downwards in the middle, or, as the workmen term it, to sway or swag; and it requires a support. The question is, where to find this support? What fixed points can we find with which to connect the middle of the tie-beam? Some ingenious carpenter thought of suspending it from the ridge by a piece of timber BD (Fig. 15.) called by our carpenters the *king-post*. It must be acknowledged that there was great ingenuity in this thought. It was also perfectly just. For the weight of the rafters BA, BC, tends to make them fly out at the foot. This is prevented by the tie-beam, and this excites a pressure, by which they tend to compress each other. Suppose them without weight, and that a great weight is laid on the ridge B. This can be supported only by the butting of the rafters in their own directions AB and CB, and the weight tends to compress them in the opposite directions, and, through their intervention, to stretch the tie-beam. If neither the rafters can be compressed, nor the tie-beam stretched, it is plain that the triangle A B C must retain its shape, and that B becomes a fixed point, very proper to be used as a point of suspension. To this point, therefore, is the tie-beam suspended by means of the king-post. A common spectator, unacquainted with carpentry, views it very differently, and the tie-beam appears to him to carry the roof. The king-post appears a pillar resting on the beam, whereas it is really a string; and an iron rod of one-sixteenth of the size would have done just as well. The king-post is sometimes mortised into the tie-beam, and pins put through the joint, which gives it more the look of a pillar with the roof resting on it. This does well enough in many cases. But the best method is to connect them by an iron strap, like a stirrup, which is bolted at its upper ends into the king-post, and passes round the tie-beam. In this way a space is commonly left between the end of the king-post and the upper side of the tie-beam. Here the beam plainly appears hanging in the stirrup; and this method allows us to restore the beam to an exact level, when it has sunk by the unavoidable compression or other yielding of the parts. The holes in the sides of the iron strap are made oblong instead of round; and the bolt which is drawn through all is made to taper on the under side; so that driving it farther draws the tie-beam upwards. A notion of this may be formed by looking at Fig. 16. which is a section of the post and beam.

It requires considerable attention, however, to make this suspension of the tie-beam sufficiently firm. The top of the king-post is cut into the form of the arch-stone of a bridge, and the heads of the rafters are firmly mortised into this projecting part. These projections are called joggles, and are formed by working the king-post out of a much larger piece of timber, and cutting off the unnecessary wood from the two sides; and, lest all this should not be sufficient, it is usual in great works to add an iron plate or strap of three branches, which are bolted into the heads of the king-post and rafters.

The rafters, though not so long as the beam, seem to stand as much in need of something to prevent their bending, for they carry the weight of the covering. This cannot be done by suspension, for we have no fixed points above them: But we have now got a very firm point of support at the foot of the king-post. *Braces*, or *struts*, E D, F D, Fig. 17. are put under the middle of the rafters, where they are slightly mortised, and their lower ends are firmly mortised into joggles formed on the foot of the king-post. As these braces are very powerful in their resistance to compression, and the king-post equally so to resist extension, the points E and F may be considered as fixed; and the rafters being thus reduced to half their former length, have now four times their former relative strength.

572. Roofs do not always consist of two sloping sides

meeting in a ridge. They have sometimes a flat on the top, with two sloping sides. They are sometimes formed with a double slope, and are called *kirb* or *mansarde roofs*. They sometimes have a valley in the middle, and are then called M roofs. Such roofs require another piece which may be called the *truss-beam*, because all such frames are called *trusses*, probably from the French word *trousse*, because such roofs are like portions of plain roofs, *troussés* or shortened.

A flat-topped roof is thus constructed. Suppose that there are three rafters AB, BC, CD (Fig. 18.) of which AB and CD are equal, and BC horizontal. It is plain that they will be in equilibrio, and the roof have no tendency to go to either side. The tie-beam AD withstands the horizontal thrusts of the whole frame, and the two rafters AB and CD are each pressed in their own directions in consequence of their butting with the middle rafter or truss-beam BC. It lies between them like the key-stone of an arch. They lean towards it, and it rests on them. The pressure which the truss-beam and its load excites on the two rafters is the very same as if the rafters were produced till they meet in G, and a weight were laid on these equal to that of BC and its load. If therefore the truss-beam is of a scantling sufficient for carrying its own load, and withstanding the compression from the two rafters, the roof will be equally strong (while it keeps its shape) as the plain roof AGD furnished with king-post and braces. We may conceive this another way. Suppose a plain roof AGD, without braces to support the middle B and C of the rafters. Then let a beam BC be put in between the rafters, butting upon little notches cut in the rafters. It is evident that this must prevent the rafters from bending downwards, because the points B and C cannot descend, moving round the centres A and D, without shortening the distance BC between them. This cannot be without compressing the beam BC. It is plain that BC may be wedged in, or wedges driven in between its ends B and C and the notches in which it is lodged. These wedges may be driven in till they even force out the rafters GA and GD. Whenever this happens, all the mutual pressure of the heads of these rafters at G is taken away, and the parts GB and GC may be cut away, and the roof ABCD will be as strong as the roof AGD furnished with the king-post and braces, because the truss-beam gives a support of the same kind at B and C as the brace would have done.

But this roof ABCD would have no firmness of shape. Any addition of weight on one side would destroy the equilibrium at the angle, would depress that angle, and cause the opposite one to rise. To give it stiffness, it must either have ties or braces, or something partaking of the nature of both. The usual method of framing is to make the heads of the rafters butt on the joggles of two side-posts BE and CF, while the truss-beam, or strut, as it is generally termed by the carpenters, is mortised square into the inside of the heads. The lower ends E and F of the side-posts are connected with the tie-beam either by mortises or straps.

This construction gives firmness to the frame; for the angle B cannot descend in consequence of any inequality of pressure, without forcing the other angle C to rise. This it cannot do, being held down by the post CF. And the same construction fortifies the tie-beam, which is now suspended at the points E and F from the points B and C, whose firmness we have just now shown.

573. But although this roof may be made abundantly strong, it is not quite so strong as the plain roof AGD of the same scantling. The compression which BC must sustain in order to give the same support to the rafters at B and C that was given by braces properly placed, is considerably greater than the compression of the braces. And this strain is an addition to the transverse strain

which BC gets from its own load. This form also necessarily exposes the tie-beam to cross strains. If BE is mortised into the tie-beam, then the strain which tends to depress the angle ABC presses on the tie-beam at E transversely, while a contrary strain acts on F, pulling it upwards. These strains however are small; and this construction is frequently used, being susceptible of sufficient strength, without much increase of the dimensions of the timbers; and it has the great advantage of giving free room in the garrets. Were it not for this, there is a much more perfect form represented in Fig. 19. Here the two posts BE, CF, are united below. All transverse action on the tie-beam is now entirely removed. We are almost disposed to say that this is the strongest roof of the same width and slope: for if the iron strap which connects the pieces BE, CF, with the tie-beam have a large bolt G through it, confining it to one point of the beam, there are five points A, B, C, D, G, which cannot change their places, and there is no transverse strain in any of the connections.

When the dimensions of the building are very great, so that the pieces AB, BC, CD, would be thought too weak for withstanding the cross strains, braces may be added as is expressed in Fig. 18. by the dotted lines. The reader will observe that it is not meant to leave the top flat externally: it must be raised a little in the middle to carry off the rain. But this must not be done by incurvating the beam BC. This would soon be crushed, and spring upwards. The slopes must be given by pieces of timber added above the strutting beam.

574. And thus we have completed a frame of a roof. It consists of these principal members: The rafters, which are immediately loaded with the covering; the tie-beam, which withstands the horizontal thrust by which the roof tends to fly out below and push out the walls; the king-posts, which hang from fixed points and serve to uphold

the tie-beam, and also to afford other fixed points on which we may rest the braces which support the middle of the rafters; and lastly, the truss or strutting-beam, which serves to give mutual abutment to the different parts which are at a distance from each other. The rafters, braces, and trusses, are exposed to compression, and must therefore have not only cohesion but stiffness. For if they bend, the prodigious compressions to which they are subjected would quickly crush them in this bended state. The tie-beams and king-posts, if performing no other office but supporting the roof, do not require stiffness, and their places might be supplied by ropes, or by rods of iron of one-tenth part of the section that even the smallest oak stretcher requires. These members require no greater dimensions than what is necessary for giving sufficient joints, and any more is a needless expence and load. All roofs, however complicated, consist of these essential parts, and if pieces of timber are to be seen which perform none of these offices, they must be pronounced useless, and they are frequently hurtful, by producing cross strains in some other piece. In a roof properly constructed there should be no such strains. All the timbers, except those which immediately carry the covering, should be either pushed or drawn in the direction of their length. And this is the rule by which a roof should always be examined.

575. These essential parts are susceptible of numberless combinations and varieties. But it is a prudent maxim to make the construction as simple, and consisting of as few parts, as possible. We are less exposed to the imperfections of workmanship, such as loose joints, &c. Another essential harm arises from many pieces, by the compression and the shrinking of the timber in the cross direction of the fibres. The effect of this is equivalent to the shortening of the piece which butts on the joint. This alters the proportions of the sides of the triangle on

which the shape of the whole depends. Now in a roof such as Fig. 18. there is twice as much of this as in the plain pent roof, because there are two posts. And when the direction of the butting pieces is very oblique to the action of the load, a small shrinking permits a great change of shape. Thus in a roof of what is called pediment pitch, where the rafters make an angle of 30 degrees with the horizon, half an inch compression of the king-post will produce a sagging of an inch, and occasion a great strain on the tie-beam if the posts are mortised into it.

We would therefore recommend Fig. 20. as a proper construction of a trussed roof, preferable to that which is generally used, and the king-post which is placed in it may be employed to support the upper part of the rafters, and also for preventing the strut-beam from bending in either direction in consequence of its great compression. It will also give a suspension for the great burdens which are sometimes necessary in a theatre. The machinery has no other firm points to which it can be attached; and the portion of the single rafters which carry this king-post are but short, and therefore may be considerably loaded with safety.

We observe in the drawings which we sometimes have of Chinese buildings, that the trussing of roofs is understood by them. Indeed they must be very experienced carpenters. We see wooden buildings run up to a great height, which can be supported only by such trussing. One of these is sketched in Fig. 21. There are some very excellent specimens to be seen in the buildings at Deptford, belonging to the victualling-office, usually called the *Red House,* which were erected about the year 1788, and we believe are the performance of Mr James Arrow of the Board of Works, one of the most intelligent artists in this kingdom.

576. Thus have we given an elementary, but a rational or scientific, account of this important part of the art of

carpentry. It is such, that any practitioner, with the trouble of a little reflection, may always proceed with confidence, and without resting any part of his practice on the vague notions which habit may have given him of the strength and supports of timbers, and of their manner of acting. That these frequently mislead, is proved by the mutual criticisms which are frequently published by the rivals in the profession. They have frequently sagacity enough (for it can seldom be called science) to point out glaring blunders; and any person who will look at some of the performances of Mr Price, Mr Wyatt, Mr Arrow, and others of acknowledged reputation, will readily see them distinguishable from the works of inferior artists by simplicity alone. A man without principles is apt to consider an intricate construction as ingenious and effectual; and such roofs sometimes fail merely by being ingeniously loaded with timber, but more frequently still by the wrong action of some useless piece, which produces strains that are transverse to other pieces, or which, by rendering some points too firm, cause them to be deserted by the rest in the general subsiding of the whole. Instances of this kind are pointed out by Price in his British Carpenter. Nothing shows the skill of a carpenter more than the distinctness with which he can foresee the changes of shape which must take place in a short time in every roof. A knowledge of this will often correct a construction which the mere mathematician thinks unexceptionable, because he does not reckon on the actual compression which must obtain, and imagines that his triangles, which sustain no cross strains, invariably retain their shape till the pieces break. The sagacity of the experienced carpenter is not, however, enough without science for perfecting the art. But when he knows how much a particular piece will yield to compression in one case, science will tell him, and nothing but science can do it, what will be the compression of the same piece in another very different case. Thus he learns how far it will now yield, and then he proportions the parts so to each other, that when all have yielded according to their strains, the whole is of the shape he wished to produce, and every joint is in a state of firmness. It is here that we observe the greatest number of improprieties. The iron straps are frequently in positions not suited to the actual strain on them, and they are in a state of violent twist, which both tends strongly to break the strap, and to cripple the pieces which they surround.

In like manner, we frequently see joints or mortises in a state of violent strain on the tenons, or on the heels and shoulders. The joints were perhaps properly shaped to the primitive form of the truss; but by its settling, the bearing on the push is changed: the brace, for example, in a very low pitched roof, comes to press with the upper part of the shoulder, and, acting as a powerful lever on the tenon, breaks it. In like manner, the lower end of the brace, which at first butted firmly and squarely on the joggle of the king-post, now presses with one corner with prodigious force, and seldom fails to splinter off on that side. We cannot help recommending a maxim of Mr Perronet the celebrated hydraulic architect of France, as a golden rule, viz. to make all the shoulders of butting pieces in the form of an arch of a circle, having the opposite end of the piece for its centre. Thus, in Fig. 18. if the joggle-joint B be of this form, having A for its centre, the sagging of the roof will make no partial bearing at the joint: for in the sagging of the roof, the piece AB turns or bends round the centre A, and the counter pressure of the joggle is still directed to A, as it ought to be. We have just now said *bends* round A. This is too frequently the case, and it is always very difficult to give the tenon and mortise in this place a true and invariable bearing. The rafter pushes in the direction BA. and the beam resists in the direction AD. The abutment should be perpendicular to neither of these but in an intermediate direction, and it ought also to be of a curved shape. But the carpenters perhaps think that this would weaken the beam too much to give it this shape in the shoulder; they do not even aim at it in the heel of the tenon. The shoulder is commonly even with the surface of the beam. When the bearing therefore is on this shoulder, it causes the foot of the rafter to slide along the beam till the heel of the tenon bears against the outer end of the mortise (See Price's *British Carpenter*, Plate C. Fig. I K). This abutment is perpendicular to the beam in Price's book, but it is more generally pointed a little outwards below, to make it more secure against starting. The consequence of this construction is, that when the roof settles, the shoulder comes to bear at the inner end of the mortises, and it rises at the outer, and the tenon taking hold of the wood beyond it, either tears it out or is itself broken. This joint therefore is seldom trusted to the strength of the mortise and tenon, and is usually secured by an iron strap, which lies obliquely to the beam, to which it is bolted by a large bolt quite through, and then embraces the outside of the rafter foot. Very frequently this strap is not made sufficiently oblique, and we have seen some made almost square with the beam. When this is the case, it not only keeps the foot of the rafter from flying out, but it binds it down. In this case, the rafter acts as a powerful lever, whose fulcrum is the inner angle of the shoulder, and then the strap never fails to cripple the rafter at the point. All this can be prevented only by making the strap very long and very oblique, and by making its outer end (the stirrup part) square with its length, and making a notch in the rafter foot to receive it. It cannot now cripple the rafter, for it will rise along with it, turning round the bolt at its inner end. We have been thus particular on this joint, because it is here that the ultimate strain of the whole roof is exerted, and its situation will not allow the excavation necessary for making it a good mortise and tenon.

Similar attention must be paid to some other straps, such as those which embrace the middle of the rafter, and connect it with the post or truss below it. We must attend to the change of shape produced by the sagging of the roof, and place the strap in such a manner as to yield to it by turning round its bolt, but so as not to become loose, and far less to make a fulcrum for any thing acting as a lever. The strains arising from such actions, in framings of carpentry which change their shape by sagging, are enormous, and nothing can resist them.

577. We shall close this part of the subject with a simple method, by which any carpenter, without mathematical science, may calculate with sufficient precision the strains or thrusts which are produced on any point of his work, whatever be the obliquity of the pieces.

Let it be required to find the horizontal thrust acting on the tie-beam AD of Fig. 18. This will be the same as if the weight of the whole roof were laid at G on the two rafters GA and GD. Draw the vertical line GH. Then, having calculated the weight of the whole roof that is supported by this single frame ABCD, including the weight of the pieces AB, BC, CD, BE, CF, themselves, take the number of pounds, tons, &c. which expresses it from any scale of equal parts, and set it from G to H. Draw HK, HL, parallel to GD, GA, and draw the line KL, which will be horizontal when the two sides of the roof have the same slope. Then ML measured on the same scale will give the horizontal thrust, by which the strength of the tie-beam is to be regulated. GL will give the thrust which tends to crush the rafters, and LM will also give the force which tends to crush the strut-beam BC.

In like manner, to find the strain on the king-post BD of Fig. 17. consider that each brace is pressed by half the weight of the roofing laid on BA or BC, and this pressure, or at least its hurtful effect, is diminished in the proportion of BA to DA, because the action of gravity is vertical, and the effect which we want to counteract by the braces is in a direction E *e* perpendicular to BA or BC. But as this is to be resisted by ne brace *f* E acting in the direction *f* E, we must draw *f e* perpendicular to E *e*, and suppose the strain augmented in the proportion of E *e* to E *f*.

Having thus obtained in tons, pounds, or other measures, the strains which must be balanced at *f* by the cohesion of the king-post, take this measure from the scale of equal parts, and set it off in the directions of the braces to G and H, and complete the parallelogram G *f* HK; and *f* K measured on the same scale will be the strain on the king-post.

578. The artist may then examine the strength of his truss upon this principle, that every square inch of oak will bear at an average 7000 pounds compressing or stretching it, and may be safely loaded with 3500 for any length of time; and that a square inch of fir will in like manner securely bear 2500. And, because straps are used to resist some of these strains, a square inch of well wrought tough iron may be safely strained by 50,000 pounds. But the artist will always recollect, that we cannot have the same confidence in iron as in timber. The faults of this last are much more easily perceived; and when timber is too weak, it gives us warning of its failure, by yielding sensibly before it breaks. This is not the case with iron; and much of its service depends on the honesty of the blacksmith.

579. In this way may any design of a roof be examined. We shall here give the reader a sketch of two or three trussed roofs, which have been executed in the chief varieties of circumstances which occur in common practice.

Fig. 2. of Plate VIII. is the roof of St Paul's Church, Covent Garden, London, the work of Inigo Jones. Its construction is singular. The roof extends to a considerable distance beyond the building, and the ends of the tie-beams support the Tuscan corniche, appearing like the mutules of the Doric order. Such a roof could not rest on the tie-beam. Inigo Jones has therefore supported it by a truss below it; and the height has allowed him to make this extremely strong with very little timber. It is accounted the highest roof of its width in London. But this was not difficult, by reason of the great height which its extreme width allowed him to employ without hurting the beauty of it by too high a pitch. The supports, however, are disposed with judgment.*

Fig. 22. is a kirb or mansarde roof by Price, and supposed to be of large dimensions, having braces to carry the middle of the rafters.

It will serve exceedingly well for a church having pillars. The middle part of the tie-beam being taken away, the strains are very well balanced, so that there is no risk of its pushing aside the pillars on which it rests.

Fig. 23. is the celebrated roof of the theatre of the university of Oxford, by Sir Christopher Wren. The span between the walls is 75 feet. This is accounted a very ingenious, and is a singular performance. The middle part of it is almost unchangeable in its form; but from this circumstance it does not distribute the horizontal thrust with the same regularity as the usual construction. The horizontal thrust on the tie-beam is about twice the weight of the roof, and is withstood by an iron strap below the beam, which stretches the whole width of the building in the form of a rope, making part of the ornament of the ceiling.

580. In all the roofs which we have considered hitherto, the thrust is discharged entirely from the walls by

---

* See Carpentry in this Vol. p. 546.

the tie-beam. But this cannot always be done. We frequently want great elevation within, and arched ceilings. In such cases, it is a much more difficult matter to keep the walls free of all pressure outwards, and there are few buildings where it is completely done. Yet this is the greatest fault of a roof. We shall just point out the methods which may be most successfully adopted.

We have said that a tie-beam just performs the office of a string. We have said the same of the king-post. Now suppose two rafters AB, BC (Fig. 24.) moveable about the joint B, and resting on the top of the walls. If the line BD be suspended from B, and the two lines DA, DC, be fastened to the feet of the rafters, and if these lines be incapable of extension, it is plain that all thrust is removed from the walls as effectually as by a common tie-beam. And by shortening BD to B *d*, we gain a greater inside height, and more room for an arched ceiling. Now if we substitute a king-post BD Fig. 25. and two stretchers or hammer beams DA, DC, for the other strings, and connect them firmly by means of iron straps, we obtain our purpose.

Let us compare this roof with a tie-beam roof in point of strain and strength. Recur to Fig. 24. and complete the parallelogram ABCF, and draw the diagonals AC, BF, crossing in E. Draw BG perpendicular to CD. We have seen that the weight of the roof, which we may call W) is to the horizontal thrust at C as BF to EC; and if we express this thrust by T, we have $T = \frac{W \times EC}{BF}$. We may at present consider BC as a lever moveable round the joint B, and pulled at C in the direction EC by the horizontal thrust, and held back by the string pulling in the direction CD. Suppose that the forces in the directions EC and CD are in equilibrio, and let us find the force S by which the string CD is strained. These forces must (by the property of the lever) be inversely as the perpendiculars drawn from the centre of motion on the lines of their direction. Therefore $BG : BE = T : S$, and $S = T \times \frac{BE}{BG}, = W \times \frac{BE.EC.}{BF.BG}$

Therefore the strain upon each of the ties DA and DC is always greater than the horizontal thrust or the strain on a simple tie-beam. This would be no great inconvenience, because the smallest dimensions that we could give to these ties, so as to procure sufficient fixtures to the adjoining pieces, are always sufficient to withstand this strain. But although the same may be said of the iron straps which make the ultimate connections, there is always some hazard of imperfect work, cracks or flaws, which are not perceived. We can judge with tolerable certainty of the soundness of a piece of timber, but cannot say so much of a piece of iron. Moreover, there is a prodigious strain excited on the king-post, when BG is very short in comparison of BE, namely, the force compounded of the two strains S and S on the ties DA and DC.

But there is another defect from which the straight tie-beam is entirely free. All roofs settle a little. When this roof settles, and the points B and D descend, the legs BA, BC, must spread further out, and thus a pressure outwards is excited on the walls. It is seldom, therefore, that this kind of roof can be executed in this simple form, and other contrivances are necessary for counteracting this supervening action on the walls. Fig. 26. is one of the best which we have seen, and is executed with great success in the circus or equestrian theatre in Edinburgh, the width being 60 feet. The pieces EF and ED help to take off some of the weight, and by their greater uprightness they exert a smaller thrust on the walls. The beam D *d* is also a sort of truss-beam, having something of the same effect. Mr Price has given another very judicious one of this kind, (*British Carpenter,* Plate IK,

Fig. C,) from which the tie-beam may be taken away, and there will remain very little thrust on the walls. Those which he has given in the following Plate K are, in our opinion, very faulty. The whole strain in these last roofs tends to break the rafters and ties transversely, and the fixtures of the ties are also not well calculated to resist the strain to which the pieces are exposed. We hardly think that these roofs could be executed.

581. It is scarcely necessary to remind the reader, that in all that we have delivered on this subject, we have attended only to the construction of the principal rafters or trusses. In small buildings all the rafters are of one kind; but in great buildings the whole weight of the covering is made to rest on a few principal rafters, which are connected by beams placed horizontally, and either mortised into them or scarfed on them. These are called *purlins*. Small rafters are laid from purlin to purlin; and on these the laths for tiles, or the skirting-boards for slates, are nailed. Thus the covering does not immediately rest on the principal frames. This allows some more liberty in their construction, because the garrets can be so divided that the principal rafters shall be in the partitions, and the rest left unincumbered. This construction is so far analogous to that of floors which are constructed with girders, binding, and bridging joists.

It may appear presuming in us to question the propriety of this practice. There are situations in which it is unavoidable, as in the roofs of churches, which can be allowed to rest on some pillars. In other situations, where partition walls intervene at a distance not too great for a stout purlin, no principal rafters are necessary, and the whole may be roofed with short rafters of very slender scantling. But in a great uniform roof, which has no intermediate supports, it requires at least some reasons for preferring this method of carcase roofing to the simpler method of making all the rafters alike.

The method of carcase-roofing requires the selection of the greatest logs of timber, which are seldom of equal strength and soundness with thinner rafters. In these the outside planks can be taken off, and the best part alone worked up. It also exposes to all the defects of workmanship in the mortising of purlins, and the weakening of the rafters by this very mortising; and it brings an additional load of purlins and short rafters. A roof thus constructed may surely be compared with a floor of similar construction. Here there is not a shadow of doubt, that if the girders were sawed into planks, and these planks laid as joists sufficiently near for carrying the flooring boards, they will have the same strength as before, except so much as is taken out of the timber by the saw. This will not amount to one-tenth part of the timber in the binding, bridging, and ceiling joists, which are an additional load; and all the mortises and other joinings are so many diminutions of the strength of the girders; and as no part of a carpenter's work requires more skill and accuracy of execution, we are exposed to many chances of imperfection. But, not to rest on these considerations, however reasonable they may appear, we shall relate an experiment made by one on whose judgment and exactness we can depend.

582. Two models of floors were made 18 inches square of the finest uniform deal, which had been long seasoned. The one consisted of simple joists, and the other was framed with girders, binding, bridging, and ceiling joists. The plain joists of the one contained the same quantity of timber with the girders alone of the other, and both were made by a most accurate workman. They were placed in wooden trunks 18 inches square within, and rested on a strong projection on the inside. Small shot was gradually poured in upon the floors, so as to spread uniformly over them. The plain joisted floor broke down with 487 pounds, and the carcase floor with 327. The

first broke without giving any warning; the other gave a violent crack when 294 pounds had been poured in.

A trial had been made before, and the loads were 341 and 482. But the models having been made by a less accurate hand, it was not thought a fair specimen of the strength which might be given to a carcase floor.

The only argument of weight which we can recollect in favour of the compound construction of roofs is, that the plain method would prodigiously increase the quantity of work, would admit nothing but long timber, which would greatly add to the expence, and would make the garrets a mere thicket of planks. We admit this in its full force; but we continue to be of the opinion that plain roofs are greatly superior in point of strength, and therefore should be adopted in cases where the great difficulty is to insure this necessary circumstance.

583. It would appear very neglectful to omit an account of the roofs put on round buildings, such as domes, cupolas, and the like. They appear to be the most difficult tasks in the art of carpentry. But the difficulty lies entirely in the mode of framing, or what the French call the *trait de charpenterie.* The view which we are taking of the subject, as a part of mechanical science, has little connection with this. It is plain, that whatever form of a truss is excellent in a square building must be equally so as one of the frames of a round one; and the only difficulty is how to manage their mutual intersections at the top. Some of them must be discontinued before they reach that length, and common sense will teach us to cut them short alternately, and always leave as many, that they may stand equally thick as at their first springing from the base of the dome. Thus the length of the purlins which reach from truss to truss will never be too great.

The truth is, that a round building which gathers in at top, like a glasshouse, a potter's kiln, or a spire steeple, instead of being the most difficult to erect with stability, is of all others the easiest. Nothing can show this more forcibly than daily practice, where they are run up without centres and without scaffoldings: and it requires gross blunders indeed in the choice of their outline to put them in much danger of falling from a want of equilibrium. In like manner, a dome of carpentry can hardly fall, give it what shape or what construction you will. It *cannot* fall unless some part of it flies out at the bottom: an iron hoop round it, or straps at the joinings of the trusses and purlins, which make an equivalent to a hoop, will effectually secure it. And as beauty requires that a dome shall spring almost perpendicularly from the wall, it is evident that there is hardly any thrust to force out the walls. The only part where this is to be guarded against is, where the tangent is inclined about 40 or 50 degrees to the horizon. Here it will be proper to make a course of firm horizontal joinings.

We doubt not but that domes of carpentry will now be raised of great extent. The old Halle au Bled at Paris, of 200 feet in diameter, was the invention of an intelligent carpenter, the Sieur Moulineau. He was not by any means a man of science, but had much more mechanical knowledge than artisans usually have, and was convinced that a very thin shell of timber might not only be so shaped as to be nearly in equilibrio, but that if hooped or firmly connected horizontally, it would have all the stiffness that was necessary; and he presented his project to the magistracy of Paris. The grandeur of it pleased them, but they doubted of its possibility. Being a great public work, they prevailed on the Academy of Sciences to consider it. The members, who were competent judges, were instantly struck with the justness of Mr Moulineau's principles, and were astonished that a thing so plain had not been long familiar to every house-carpenter. It quickly became an universal topic of conversation, dispute, and cabal, in the polite circles of Paris. But the Academy having given a very favourable report of their opinion, the project was immediately carried into execution, and soon completed, and now stands as one of the great exhibitions of Paris *.

The construction of this dome is the simplest thing that can be imagined. The circular ribs which compose it consist of planks nine feet long, 13 inches broad, and three inches thick; and each rib consists of three of these planks bolted together in such a manner that two joints meet. A rib is begun, for instance, with a plank of three feet long standing between one of six feet and another of nine, and this is continued to the head of it. No machinery was necessary for carrying up such small pieces, and the whole went up like a piece of bricklayer's work. At various distances these ribs were connected horizontally by purlins and iron straps, which made so many hoops to the whole. When the work had reached such a height, that the distance of the ribs was two-thirds of the original distance, every third rib was discontinued, and the space was left open and glazed. When carried so much higher that the distance of the ribs is one-third of the original distance, every second rib (now consisting of two ribs very near each other) is in like manner discontinued, and the void is glazed. A little above this the heads of the ribs are framed into a circular ring of timber, which forms a wide opening in the middle; over which is a glazed canopy or umbrella, with an opening between it and the dome for allowing the heated air to get out. All who have seen this dome say, that it is the most beautiful and magnificent object they have ever beheld.

* This roof has been long since destroyed, and another of smaller dimensions has been erected with ribs of iron, covered with sheet copper. Ed.

The only difficulty which occurs in the construction of wooden domes is, when they are unequally loaded, by carrying a heavy lanthern or cupola in the middle In such a case, if the dome were a mere shell, it would be crushed in at the top, or the action of the wind on the lanthern might tear it out of its place. Such a dome must therefore consist of trussed frames. Mr Price has given a very good one in his plate OP, though much stronger in the trusses than there was any occasion for. This causes a great loss of room, and throws the lights of the lanthern too far up. It is evidently copied from Sir Christopher Wren's dome of St Paul's church in London; a model of propriety in its particular situation, but by no means a general model of a wooden dome. It rests on the brick cone within it; and Sir Christopher has very ingeniously made use of it for stiffening this cone, as any intelligent person will perceive by attending to its construction. (See *Price*, Plate OP).

Fig. 27. represents a dome executed in the Register Office of Edinburgh, by James and Robert Adams, and is very agreeable to mechanical principles. The span is 50 feet clear, and the thickness is only 4½.

584. We cannot quit this subject without taking some notice of what we have already spoken of with commendation by the name of *Norman roofs*. We called them *Norman*, because they were frequently executed by that people soon after their establishment in Italy and other parts of the south of Europe, and became the prevailing taste in all the great baronial castles. Their architects were rivals to the Saracens and Moors, who about that time built many Christian churches; and the architecture which we now call Gothic seems to have arisen from their joint labours.

The principle of a Norman roof is extremely simple. The rafters all butted on joggled king-posts AF, BG, CH, &c. (Fig. 28.), and braces or ties were then disposed in the intervals. In the middle of the roof HB and HD are evidently ties in a state of extension, while the post CH is compressed by them. Towards the walls on each side, as between B and F, and between F and L, they are braces, and are compressed. The ends of the posts were generally ornamented with knots of flowers, embossed globes, and the like, and the whole texture of the truss was exhibited and dressed out.

This construction admits of employing very short timbers; and this very circumstance gives greater strength to the truss, because the angle which the brace or tie makes with the rafter is more open. We may also perceive that all thrust may be taken off the walls. If the pieces AF, BF, LF, be removed, all the remaining diagonal pieces act as ties, and the pieces directed to the centre act as struts; and it may also be observed, that the principle will apply equally to a straight or flat roof, or to a floor. A floor such as *a b c*, having the joint in two pieces *a b*, *b c*, with a strut *b d*, and two ties, will require a much greater weight to break it than if it had a continued joist *a c* of the same scantling. And, lastly, a piece of timber acting as a tie is much stronger than the same piece acting as a strut: for in the latter situation it is exposed to bending, and when bent it is much less able to withstand a very great strain. It must be acknowledged, however, that this advantage is balanced by the great inferiority of the joints in point of strength. The joint of a tie depends wholly on the pins; for this reason ties are never used in heavy works without strapping the joints with iron. In the roofs we are now describing, the diagonal pieces of the middle part only act purely as ties, while those towards the sides act as struts or braces. Indeed they are seldom of so very simple construction as we have described, and are more generally constructed like the sketch in Fig. 29. having two sets of rafters AB, *a b*, and the angles are filled up with thin planks, which give great stiffness and strength. They have also a double set of purlins, which connect the different trusses. The roof being thus divided into squares, other purlins run between the middle points E of the rafters. The rafter is supported at E by a check put between it and the under rafter. The middle point of each square of the roof is supported and stiffened by four braces, one of which springs from *e*, and its opposite from the similar part of the adjoining truss. The other two braces spring from the middle points of the lower purlins, which go horizontally from *a* and *b* to the next truss, which are supported by planks in the same manner as the rafters. By this contrivance the whole becomes very stiff and strong.

585. We hope that the reader will not be displeased with our having taken some notice of what was the pride of our ancestors, and constituted a great part of the finery of the grand hall, where the feudal lord assembled his vassals, and displayed his magnificence. The intelligent mechanic will see much to commend; and all who look at these roofs admire their apparent flimsy lightness, and wonder at their duration. We have seen a hall of 57 feet wide, the roof of which was in four divisions, like a kirb roof, and the trusses were about 16 feet asunder. They were single rafters, as in Fig. 29. and their dimensions were only eight inches by six. The roof appeared perfectly sound, and had been standing ever since the year 1425.

Much of what has been said on this subject may be applied to the construction of wooden bridges, and the centres for turning the arches of stone bridges.

ON THE

CONSTRUCTION OF ARCHES.

586. An arch is an artful disposition and adjustment of several stones or bricks, generally in a bow-like form, by which their weight produces a mutual pressure and abutment; so that they not only support each other, and perform the office of an entire lintel, but may be extended to any width, and made to carry the most enormous weights.

587. The pediment of the Greeks seems to have suggested the construction of the arch. In erecting their small houses, they could hardly fail to observe occasionally, that when two rafters were laid together from the opposite walls, they would, by leaning on each other, give mutual support, as in Fig. 1. Plate X. Nor is it unlikely that such a situation of stones as is represented in Fig. 2. would not unfrequently occur by accident to masons. This could hardly fail of exciting a little attention and reflection. It was a pretty obvious reflection, that the stones A and C, by overhanging, leaned against the intermediate stone B, and gave it some support, and that B cannot get down without thrusting aside A and C, or the piers which support them. This was an approach to the theory of an arch; and if this be combined with the observation of Fig. 1. we get the disposition represented in Fig. 3. having a perpendicular joint in the middle, and

the *principle of the arch* is completed. Observe that this is quite different from the principle of the arrangement in Fig. 2. In that figure the stones act as wedges, and one cannot get down without thrusting the rest aside; the same principle obtains in Fig. 4. consisting of five arch stones; but in Fig. 3. the stones B and C support each other by their mutual pressure (independent of their own weight), arising from the tendency of each lateral pair to fall outwards from the pier. This is the principle of the arch, and would support the keystone of Fig. 4. although each of its joints were perpendicular, by reason of the great friction arising from the horizontal thrust exerted by the adjoining stones.

This was a most important discovery in the art of building; for now a building of any width may be roofed with stone.

588. We are disposed to give the Greeks the merit of this discovery; for we observe arches in the most ancient buildings of Greece, such as the Temple of the Sun at Athens, and of Apollo at Didymos; not indeed as roofs to any apartment, nor as parts of the ornamental design, but concealed in the walls, covering drains, or other necessary openings; and we have not found any *real* arches in any monuments of ancient Persia or Egypt. Sir John Chardin speaks of numerous and extensive subterranean passages at Tchelminar, built of the most exquisite masonry, the joints so exact, and the stones so beautifully dressed, that they look like one continued piece of polished marble: but he nowhere says that they are arched; a circumstance which we think he would not have omitted—no arched door or window is to be seen. Indeed one of the tombs is said to be arch-roofed, but it is all of one solid rock. No trace of an arch is to be seen in the ruins of ancient Egypt; even a wide room is covered with a single block of stone. In the pyramids, indeed, there are two galleries, whose roofs consist of many

pieces; but their construction puts it beyond doubt that the builder did not know what an arch was: for it is covered in the manner represented in Fig. 5. where every projecting piece is more than balanced behind, so that the whole aukward mass could have stood on two pillars. The Greeks therefore seem entitled to the honour of the invention. The arched dome, however, seems to have arisen in Etruria, and originated in all probability from the employment of the augurs, whose business it was to observe the flight of birds. Their stations for this purpose were *templa*, so called *a templando*, "on the summits of hills." To shelter such a person from the weather, and at the same time allow him a full prospect of the country around him, no building was so proper as a dome set on columns; which accordingly is the figure of a temple in the most ancient monuments of that country. We do not recollect a building of this kind in Greece except that called the *Lanthern of Demosthenes*, which is of very late date, whereas they abounded in Italy. In the later monuments and coins of Italy or of Rome, we commonly find the Etruscan dome and the Grecian temple combined; and the famous Pantheon was of this form, even in its most ancient state.

589. About the middle, or rather towards the end of last century, when the Newtonian mathematics opened the road to true mechanical science, the construction of arches engrossed the attention of the first mathematicians. The first hint of a principle that we have met with is Dr Hooke's assertion, that the figure into which a chain or rope, perfectly flexible, will arrange itself when suspended from two hooks, is, when inverted, the proper form for an arch composed of stones of uniform weight. This he affirmed on the principle that the figure which a flexible festoon of heavy bodies assumes, when suspended from two points, is, when inverted, the proper form for an arch of the same bodies, touching each other

in the same points; because the forces with which they mutually press on each other in this last case, are equal and opposite to the forces with which they pull at each other in the case of suspension.

This principle is strictly just, and may be extended to every case which can be proposed. We recollect seeing it proposed, in very general terms, in the St James's Chronicle in 1759, when plans were forming for Black-friar's Bridge in London; and since it is perhaps equal in practical utility, to the most elaborate investigations of the mathematicians, our readers will not be displeased with a more particular account of it in this place.

590. Let ABC (Fig. 6.) be a parcel of magnets of any size and shape, and let us suppose that they adhere with great force by any points of contact. They will compose such a flexible festoon as we have been speaking of, if suspended from the points A and C. If this figure be inverted, preserving the same points of contact, they will remain in equilibrio. It will indeed be that kind of equilibrium which will admit of no disturbance, and which may be called a *tottering equilibrium.* If the form be altered in the smallest degree, by varying the points of contact (which indeed are points in the *figure of equilibrium*), the magnets will no more recover their former position than a needle, which we had made to stand on its point, will regain its perpendicular position after it has been disturbed.

But if we suppose planes *d e*, *f g*, *h i*, &c. drawn, that the points of mutual contact *a*, *b*, *c*, each bisecting the angle formed by the lines that unite the adjoining contacts (*f g*, for example, bisecting the angle formed by *a b*, *b c*), and if we suppose that the pieces are changed for others of the same weights, but having flat sides, which meet in the planes *d e*, *f g*, *h i*, &c. it is evident that we shall have an arch of equilibration, and that the arch will have some stability, or will bear a little change of form without tumbling down: for it is plain that the equilibrium of the original festoon obtained only in the points *a*, *b*, *c*, of contact, where the pressures were perpendicular to the touching surfaces; therefore if the curve *a*, *b*, *c*, still passes through the touching surfaces perpendicularly, the conditions that are required for equilibrium still obtain. The case is quite similar to that of the stability of a body resting on a horizontal plane. If the perpendicular through the centre of gravity falls within the base of the body, it will not only stand, but will require some force to push it over. In the original festoon, if a small weight be added in any part, it will change the form of the curve of equilibration a little, by changing the points of mutual contact. This new curve will gradually separate from the former curve as it recedes from A or C. In like manner, when the festoon is set up as an arch, if a small weight be laid on any part of it, it will bring the whole to the ground, because the shifting of the points of contact will be just the contrary to what it should be to suit the new curve of equilibration. But if the same weight be laid on the same part of the arch now constructed with flat joints, it will be sustained, if the new curve of equilibration still passes through the touching surfaces.

These conclusions, which are very obviously deducible from the principle of the festoon, shew us, without any further discussion, that the longer the joints are, the greater will be the stability of the arch, or that it will require a greater force to break it down. Therefore it is of the greatest importance to have the arch stones as long as economy will permit; and this was the great use of the ribs and other apparent ornaments in the Gothic architecture. The great projections of those ribs augmented their stiffness, and enabled them to support the unadorned compartments of the roof, composed of very small stones, seldom above six inches thick. Many old bridges are still remaining, which are strengthened in the same way by ribs.

Having thus explained, in a very familiar manner, the stability of an arch, we proceed to give the same popular account of the general application of the principle.

591. Suppose it to be required to ascertain the form of an arch which shall have the span AB (Fig. 7.), and the height F 8, and which shall have a road-way of the dimensions CDE above it. Let the figure ACDEB be inverted, so as to form a figure A *c d e* B. Let a chain of uniform thickness be suspended from the points A and B, and let it be of such a length that its lower point will hang at, or rather a little below *f*, corresponding to F. Divide AB into a number of equal parts, in the points 1, 2, 3, &c. and draw vertical lines, cutting the chain in the corresponding points 1, 2, 3, &c. Now take pieces of another chain, and hang them on at the points 1, 2, 3, &c. of the chain A*f*B. This will alter the form of the curve. Cut or trim these pieces of chain, till their lower ends all coincide with the inverted road-way *c d e*. The greater lengths that are hung on in the vicinity of A and B will pull down these points of the chain, and cause the middle point *f* (which is less loaded) to rise a little, and will bring it near to its proper height.

It is plain that this process will produce an arch of perfect equilibration; but some farther considerations are necessary for making it exactly suit our purpose. It is an arch of equilibration for a bridge, that is so loaded that the weight of the arch stones is to the weight of the matter with which the haunches and crown are loaded, as the weight of the chain A *f* B is to the sum of the weights of all the little bits of chain very nearly. But this proportion is not known beforehand; we must therefore proceed in the following manner: Adapt to the curve produced in this way a thickness of the arch stones as great as are thought sufficient to insure stability; then compute the weight of the arch stones, and the weight of the gravel or rubbish with which the haunches are to be filled up to the road-way. If the proportion of these two weights be the same with the proportion of the weights of chain, we may rest satisfied with the curve now found; but if different, we can easily calculate how much must be added equally to, or taken from, each appended bit of chain, in order to make the two proportions equal. Having altered the appended pieces accordingly, we shall get a new curve, which may perhaps require a very small trimming of the bits of chain to make them fit the road-way. This curve will be infinitely near to the curve wanted.

We have practised this method for an arch of 60 feet span and 21 feet height, the arch stones of which were only two feet nine inches long. It was to be loaded with gravel and shivers. We made a previous computation, on the supposition that the arch was to be nearly elliptical. The distance between the points 1, 2, 3, &c. were adjusted, so as to determine the proportion of the weights of chain agreeable to the supposition. The curve differed considerably from an ellipse, making a considerable angle with the verticals at the spring of the arch. The real proportion of the weights of chain, when all was trimmed so as to suit the road-way, was considerably different from what was expected. It was adjusted. The adjustment made very little change in the curve. It would not have changed it two inches in any part of the real arch. When the process was completed, we constructed the curve mathematically. It did not differ sensibly from this mechanical construction. This was very agreeable information; for it showed us that the first curve, formed by about two hours labour, on a supposition considerably different from the truth, would have been sufficiently exact for the purpose, being in no place three inches from the accurate curve, and therefore far within the joints of the intended arch stones. Therefore this process, which any intelligent mason, though ignorant of mathematical science, may go through with little trouble, will give a very proper form for an arch subject to any conditions.

592. The chief defect of the curve found in this way is a want of elegance, because it does not spring at right angles to the horizontal line; but this is the case with all curves of equilibration, as we shall see by and by. It is not material: for, in the very neighbourhood of the piers, we may give it any form we please, because the masonry is solid in that place; nay, we apprehend that a deviation from the curve of equilibration is proper. The construction of that curve supposes that the pressure on every part of the arch is vertical; but gravel, earth, and rubbish, exert somewhat of a hydrostatical pressure laterally in the act of settling, and retain it afterwards. This will require some more curvature at the haunches of an arch to balance it; but what this lateral pressure may be, cannot be deduced with confidence from any experiments that we have seen. We are inclined to think that if, instead of dividing the horizontal line AB in the points 1, 2, 3, &c. we divide the chain itself into equal parts, the curve will approach nearer to the proper form.

593. After this familiar statement of the general principle, it is now time to consider the theory founded on it more in detail. This theory aims at such an adjustment of the position of the arch stones to the load on every part of the arch, that all shall remain in equilibrio, although the joints be perfectly polished, and without any cement. The whole may be reduced to two problems. The first is to determine the vertical pressure or load on every point of a line of a given form, which will put that line in equilibrio. The second is to determine the form of a curve which shall be in equilibrio when loaded in its different points, according to any given law.

The whole theory is deducible from p. 582 of the article Roof. The fundamental proposition in that page states the proportions between the various pressures or thrusts which are exerted at the angles of an assemblage of beams or other pieces of solid heavy matter, freely moveable about those angles, as so many joints, but retaining their position by the equilibrium of those pressures. It is there demonstrated, "that the thrust at any angle, if estimated in a horizontal direction, is the same throughout, and may be represented by any horizontal line BT, Fig. 8. (Roofs, Fig. 10. Plate IX.); and that if a vertical line QTS be drawn through T, the thrust exerted at any angle D by the piece CD, in its own direction, will then be represented by BR, drawn parallel to CD; and in like manner, that the thrust in the direction ED is represented by BS, &c.; and, lastly, that the vertical thrusts or loads, at each angle B, C, D, by which all these other pressures are excited, are represented by the portions QC, CR, RS, of the vertical intercepted by those lines; that is, all these pressures are to the uniform horizontal thrust as the lines which represent them are to BT. The horizontal thrust, therefore, is a very proper unit, with which we may compare all the others. Its magnitude is easily deduced from the same proposition; for QS is the sum of all the vertical pressures of the angles, and therefore represents the weight of the whole assemblage. Therefore as QS is to BT, so is the weight of the whole to the horizontal thrust.

594. To accommodate this theory to the construction of a curvilineal arch vault, let us first suppose the vault to be polygonal, composed of the cords of the elementary arches. Let AVE (Fig. 9.) be a curvilineal arch, of which V is the vertex, and VX the vertical axis, which we shall consider as the axis or abscissa of the curve, while any horizontal line, such as HK, is an ordinate to the curve. About any point C of the curve as a centre describe a circle BLD, cutting the curve in B and D Draw the equal cords CB, CD. Draw also the horizon-

tal line CF, cutting the circle in F. Describe a circle BCDQ passing through B, C, D. Its centre O will lie in a line COQ, which bisects the angle BCD, and C *c*, which touches this circle in C, will bisect the angle *b C d*, formed by the equal chords BC, CD. Draw CLP perpendicular to *c b* and DP perpendicular to CD meeting CL in P. Through L draw the tangent GLM, meeting CD in G, and the vertical line CM in M. Draw the tangent F *a*, cutting the chords BC, CD, in *b* and *d*, and the tangent to the circle BCDQ in *c*. Lastly, draw *d* N parallel to BC.

From what is demonstrated in the article Roof, it appears, that if BC, CD, be two pieces of an equilibrated heavy polygon, and if CF represent the horizontal thrust in every angle of the polygon, C *d* and C *b* will severally represent the thrusts exerted by the pieces DC, BC, and that *b d*, or CN, will represent the weight lying on the angle BCD, by which those thrusts are balanced.

Produce *d* C to *o*, so that C *o* may be equal to C *d*. Draw *b n* to the vertical parallel to *d* B, and join *n o*. It is evident that *b n o* C is a parallelogram, and that *n* C ($= b\,d$) $=$ CN. Now the thrust or support of the piece BC is exerted in the direction C *b*, while that of DC is exerted in the direction C *o*. These two thrusts are equivalent to the thrust in the diagonal C *n*; and it is with this compound thrust that the load or vertical pressure CN is in *immediate* equilibrium.

595. Because *b* CL, NCF, are right angles, and FCL is common to both, the angles *b* CF and MCL are equal. Therefore the right angled triangles *b* CF and MCL are similar. And since CF is equal to CL, C *b* is equal to CM. It is evident that the triangles GCM and *d* CN are similar. Therefore CG : C $d =$ CM : CN, $=$ C *b* : CN. Therefore we have CN $= \frac{C\,b \times C\,d}{CG}$. But because CDP

and CLG are right angles, and therefore equal, and the angle GCP is common to the two triangles GCL, PCD, and CD is equal to CL, we have CG equal to CP. Therefore $CN = \frac{Cb \times Cd}{CP}$. Also, since CDP is a right angle, DP meets the diameter in Q, the opposite point of the circumference, and the angle DQC is equal to DC c, or c C b (because b C d is bisected by the tangent), that is, to PCQ (because the right angles b CP, c CO are equal, and c DP is common). Therefore PQ is equal to PC; and if PO be drawn perpendicular to CQ, it will bisect it, and O is the centre of the circle BCDQB.

Now let the points B and D continually approach to C (by diminishing the radius of the small circle) and ultimately coincide with it. It is evident that the circle BCDQ is ultimately the equicurve circle, and that PC ultimately coincides with OC, the radius of curvature. Also $Cb \times Cd$ becomes immediately equal to Cc^2. Therefore CN, the vertical load on any point of a curve of equilibration is $= \frac{Cc^2}{\text{Rad. Curv.}}$

It is farther evident, that CF is to C c as radius to the secant of the elevation of the tangent above the horizon. Therefore we have the load on any point of the curve always proportional to $\frac{\text{Sec.}^2\ \text{Elev.}}{\text{Rad. Curv.}}$

This load on every elementary arch of the wall is commonly a quantity of solid matter incumbent on that element of the curve, and pressing it vertically; and it may be conceived as made up of a number of heavy lines standing vertically on it. Thus, if the element E e of the curve were lying horizontally, a little parallelogram RE e r, standing perpendicularly on it, would represent its load. But as this element E e has a sloping position, it is plain that, in order to have the same quantity of heavy matter pressing it vertically, the height of the parallelogram must be increased till it meets in $\iota\rho$, the line Rι drawn parallel to the tangent EG. It is evident that the angle REρ is equal to the angle AEG. Therefore we have ER : Eρ = Rad. : Sec. Elev.

If therefore the arch is kept in equilibrio by the vertical pressure of a wall, we must have the height of the wall above any point proportional to $\frac{\text{Sec.}^3 \text{ Elev.}}{\text{Rad. of Curv.}}$

596. Cor. I. If OS be drawn perpendicular to the vertical CS, CS will be half the vertical chord of the equicurve circle. The angle OCS is equal to cCF, that is, to the angle of elevation. Therefore 1 : Sec. Elev. = CS : CO, and the secant of elevation may be expressed by $\frac{\text{CO}}{\text{CS}}$, and its cube by $\frac{\text{CO}^3}{\text{CS}^3}$. Therefore the height of wall is proportional to $\frac{\text{CO}^3}{\text{CS}^3 \times \text{CO}}$, or to $\frac{\text{CO}^2}{\text{CS}^3}$, or $\frac{\text{CO}^2}{\text{CS}^2 \times \text{CS}}$, or to $\frac{\text{Sec.}^2 \text{ of Elev.}}{\text{Vert. Chord of Curv.}}$

Cor. II. If we make the arch VC = z, the abscissa VH = x, the ordinate HC = y, the radius osculi CO = r, and the ½ vertical chord CS = s; the height of wall pressing on any point is proportional to $\frac{\dot{z}^3}{\dot{y}\, r}$; or to $\frac{\dot{z}^2}{\dot{y}^2 s}$, or $\frac{\dot{x}^2 + \dot{y}^2}{\dot{y}^2 s}$. Therefore, when the equation of the curve is given, and the height of wall on any one point of it is also given, we can determine it for any other point: for the equation of the curve will always give us the relation of $\dot{z}$, $\dot{x}$, and $\dot{y}$, and the value of r or s. This may be illustrated by an example or two. For this purpose it will generally be most convenient to assume the height above the vertex V for the unit of computation. The thickness of the arch at the crown is commonly determined by other circumstances. At the vertex the tangent to the arch is horizontal, and therefore the cube of the secant is unity or 1. Call the height of wall, at the crown, H, and let the radius of curvature in that point be R, and its half chord R (it being then coincident with the radius), and the height on any other point h. We have $\frac{1}{R} : \frac{\dot{z}^3}{\dot{y}^3 r} = H : h$, and $h = H \times \frac{\dot{z}^3}{\dot{y}^3} \times \frac{R}{r}$. The other formula gives $h = H \times \frac{\dot{z}^2}{\dot{y}^2} \times \frac{R}{s}$.

597. *Examp.* 1. Suppose the arch to be a segment of a circle, as in Fig. 10. where AE is the diameter, and O the centre. In this arch the curvature is the same throughout, or $\frac{R}{r} = 1$. Therefore $h = H \times \frac{\dot{z}^3}{\dot{y}^2}$, or $= H \times$ Cube Sec. Elev.

This gives a very simple calculus. To the logarithm of H add thrice the logarithm of the secant of elevation. The sum is the logarithm of h.

It gives also a very simple construction. Draw the vertical CS, cutting the horizontal diameter in S. Draw ST, cutting the radius OC perpendicularly in T. Draw the horizontal line T z, cutting the vertical in z. Join z O. Make C u = V v, and draw u x parallel to z O. C c must be made = C x. The demonstration is evident.

It is very easy to see that if CV is an arch of 60°, and V v is $\frac{1}{14}$th of VC, the points v and c will be on a level; for the secant of CV is twice CO, and therefore C c is 8 times V v, which is $\frac{4}{7}$th of VH.

The line $v\,g\,c\,f$ is drawn according to this calculus or construction. It falls considerably below the horizontal line in the neighbourhood of c; and then, passing very obliquely through c, it rises rapidly to an unmeasurable height, because the vertical line through A is its assymptote. This must evidently be the case with every curve which springs at right angles with a horizontal line.

It is plain that if v V be greater, all the other ordinates of the curve $v\,g\,c\,f$, resting on the circumference AVE, will be greater in the same proportion, and the curve will cut the horizontal line drawn through v in some point nearer to v than c is. Hence it appears that a circular arch cannot be put in equilibrio by building on it up to a horizontal line, whatever be its span, or whatever be the thickness at the crown. We have seen that when this thickness is only $\frac{1}{14}$ of the radius, an arch of 120 degrees will be too much loaded at the flanks. This thickness is much too small for a bridge, being only $\frac{1}{23}$ of the span CM, whereas it should have been almost double of this, to bear the inequalities of weight that may occasionally be on it. When the crown is made still thinner, the outline is still more depressed before it rises again. There is therefore a certain span, with a corresponding thickness at the crown, which will deviate least of all from a horizontal line. This is an arch of about 54 degrees, the thickness at the crown being about one-fourth of the span, which is extravagantly great. It appears in general, therefore, that the circle is not a curve suited to the purposes of a bridge or an arcade, which requires an outline nearly horizontal.

Examp. 2. Let the curve be a parabola AVE (Fig. 11.), of which V is the vertex, and DG the directrix. Draw the diameters DCF, GVN, the tangents CK, VP, and the ordinates VF and CN. It is well known that GV is to DC as VP^2 to CK^2, or as CN^2 to CK^2. Also 2 GV is the radius of the osculating circle at V, and 2 DC is one-half of the vertical chord of the osculating circle at C. Therefore $CN^2 : CK^2$ (or $y^2 : \dot{z}^2$) $= R : s$, and $s = \frac{\dot{z}^2}{\dot{y}^2}R$. But $C\,c$, or $h = H \times \frac{\dot{z}^2 R}{\dot{y}^2 S}$. Therefore

$$h = H \times \frac{\dot{z}^2\ R}{\dot{y}^2\ \frac{\dot{z}^2}{\dot{y}^2}R}, = H \times \frac{\dot{z}^2\ R}{\dot{z}^2\ R}, = H. \quad \text{Therefore}$$

$C\,c = V\,v$.

It follows from this investigation, that the back or extrados of a parabolic arch of equilibration must be parallel to the arch or soffit itself; or that the thickness of the arch, estimated in a vertical direction, must be equal throughout; or that the extrados is the same parabola with the soffit or intrados.

We have selected these two examples merely for the simplicity and perspicuity of the solutions, which have been effected by means of elementary geometry only, instead of employing the analytical value of the radius of the osculatory circle viz. $\frac{\dot{z}^3}{\dot{y}\,\ddot{x} - \dot{x}\,\ddot{y}}$, which would have involved us at least in the elements of second fluxions. We have also preferred simplicity to elegance in the investigation, because we wish to instruct the practical engineer, who may not be a proficient in the higher mathematics.

597. The converse of the problem, namely, to find the form of the arch when the figure of the back of it is given, is the most useful question of the two, at least in cases which are most important and most difficult. Of these perhaps bridges are the chief. Here the necessity of a road-way, of easy and regular ascent, confines us to an outline nearly horizontal, to which the curve of the arch must be adapted. This is the most difficult

problem of the two; and we doubt whether it can be solved without employing infinite approximating serieses instead of accurate values.

Let $a\ v\ e$ (Fig. 12.) be the intended outline or extrados of the arch AVE, and let v Q be the common axis of both curves. From c and C, the corresponding points, draw the ordinates $c\ h$, CH. Let the thickness v V at the top be a, the abscissa $v\ h$ be $= u$, and VH $= x$, and let the equal ordinates $c\ h$, CH be y, and the arch VC be z.

Then, by the general theorem, c C $= \frac{\dot{z}^3}{r\,y^3}$, r being the radius of curvature. This, by the common rules, is $= \frac{\dot{z}^3}{\dot{y}\,\ddot{x} - \dot{x}\,\ddot{y}}$. This gives us c C $\doteqdot \frac{\dot{y}\,\ddot{x} - \dot{x}\,\ddot{y}}{\dot{y}^3}$, or $= \frac{\dot{y}\,\ddot{x} - \dot{x}\,\ddot{y}}{\dot{y}^3} \times$ C; where C is a constant quantity, found by taking the real value of c C in V, the vertex of the curve. But it is evident that it is also $= a \times x - u$.

Therefore $a + x - u = \frac{\dot{y}\,\ddot{x} - \dot{x}\,\ddot{y}}{\dot{y}^3} \times \mathrm{C} = \frac{\mathrm{C}}{\dot{y}} \times$ fluxion of $\frac{\dot{x}}{\dot{y}}$.

If we now substitute the true value of u (which is given, because the extrados is supposed to be of a known form), expressed in terms of y, the resulting equation will contain nothing but x and y, with their first and second fluxions, and known quantities. From this equation the relation of x and y must be found by such methods as seem best adapted to the equation of the extrados.

Fortunately the process is more simple and easy in the most common and useful case than we should expect from this general rule. We mean the case where the extrados is a straight line, especially when this is horizontal. In this case u is equal to o.

Example. To find the form of the balanced arch, AVE (Fig. 13.), having the horizontal line $c\,v$ for its extrados.

Keeping the same notation, we have $u = 0$, and therefore $a + x = \frac{C}{\dot{y}} \times$ fluxion of $\frac{\dot{x}}{\dot{y}}$.

Assume $\dot{y} = \frac{\dot{x}}{v}$; then $\frac{\dot{x}}{\dot{y}} = v$, and $\frac{C}{\dot{y}} \times$ fluxion of $\frac{\dot{x}}{\dot{y}}$, $= \frac{C\,v\dot{v}}{\dot{x}}$, that is, $a + x = \frac{C\,v\dot{v}}{\dot{x}}$. Therefore $a\,\dot{x} + x\dot{x} = C\,v\,\dot{v}$; and by taking the fluents, we have $2\,a\,x + x^2 = C\,v^2$; and $v = \sqrt{\frac{2\,a\,x + x^2}{C}}$. Consequently, $\dot{y} = \frac{\sqrt{C}\,\dot{x}}{\sqrt{2\,a\,x \times x^2}}$ $\left(\text{being} = \frac{\dot{x}}{v}\right)$. Taking the fluent of this, we have $y = \sqrt{C} \times L\,(2\,a\,x + 2\,x^2 + 2\sqrt{2\,a\,x + x^2})$. But at the vertex, where $x = o$, we have $y = \sqrt{C} \times L\,(2\,a)$. The corrected fluent is therefore $y = \sqrt{C} \times L\,\frac{a + x + 2\sqrt{2 a x + x^2}}{a}$.

It only remains to find the constant quantity C This we readily obtain by selecting some point of the extrados where the values of x and y are given by particular circumstances of the case. Thus, when the span $2\,s$ and height h of the arch are given, we have $s = \sqrt{C} \times L\,\left(\frac{h\,a + h + \sqrt{2\,a\,h + h^2}}{a}\right)$, and consequently $\sqrt{C} = \frac{s}{L\left(\frac{a + h + \sqrt{2\,a\,h + h^2}}{a}\right)}$. Therefore the general value of $y = s \times \frac{L\left(\frac{a+x+\sqrt{2ax+x^2}}{a}\right)}{L\left(\frac{a+h+\sqrt{2ah+h^2}}{a}\right)}$;

$$= \frac{s}{L\frac{a+h+\sqrt{2ah+h^2}}{a}} \times L.\frac{a+x+\sqrt{2ax+x^2}}{a}.$$

As an example of the use of this formula, we subjoin a table calculated by Dr Hutton of Woolwich for an arch, the span of which is 100 feet and the height 40, which are nearly the dimensions of the middle arch of Blackfriars Bridge in London.

y	x	y	x	y	x
0	6,000	21	10,381	36	21,774
2	6,035	22	10,858	37	22,948
4	6,144	23	11,368	38	24,190
6	6,324	24	11,911	39	25,505
8	6,580	25	12,489	40	26,894
10	6,914	26	13,106	41	28,364
12	7,330	27	13,761	42	29,919
13	7,571	28	14,457	43	31,563
14	7,834	29	15,196	44	33,299
15	8,120	30	15,980	45	35,135
16	8,430	31	16,811	46	37,075
17	8,766	32	17,698	47	39,126
18	9,168	33	18,627	48	41,293
19	9,517	34	19,617	49	43,581
20	9,934	35	20,665	50	46,000

598. The figure for this proposition is exactly drawn according to these dimensions, that the reader may judge of it as an object of sight. It is by no means deficient in gracefulness, and is abundantly roomy for the passage of craft; so that no objection can be offered against its being adapted on account of its mechanical excellency.

The reader will perhaps be surprised that we have made no mention of the celebrated Catenarean curve, which is commonly said to be the best form for an arch; but a little reflection will convince him, that although it is the only form for an arch consisting of stones of equal weight, and touching each other only in single points, it cannot suit an arch which must be filled up in the haunches, in order to form a road-way. He will be more surprised to hear, after this, that there is a certain thickness at the crown, which will put the Catenarea in equilibrio, even with a horizontal road-way; but this thickness is so great as to make it unfit for a bridge, being such that the pressure at the vertex is equal to the horizontal thrust. This would have been about 37 feet in the middle arch of Blackfriars Bridge. The only situation therefore in which the Catenarean form would be proper, is an arcade carrying a height of dead wall; but in this situation it would be very ungraceful. Without troubling the reader with the investigation, it is sufficient to inform him that in a Catenarean arch of equilibration the abscissa VH is to the abscissa $v\,h$ in the constant ratio of the horizontal thrust to its excess above the pressure on the vertex.

599. This much will serve, we hope, to give the reader a clear notion of this celebrated theory of the equilibrium of arches, one of the most delicate and important applications of mathematical science. Volumes have been written on the subject, and it still occupies the attention of mechanicians. But we beg leave to say, with great deference to the eminent persons who have prosecuted this theory, that their speculations have been of little service, and are little attended to by the practitioner. Nay, we may add, that Sir Christopher Wren, perhaps the most accomplished architect that Europe has seen, seems to have thought it of little value: for, among the fragments which have been preserved of his studies, there are to be seen some imperfect dissertations on this very subject, in which he takes no notice of this theory, and

considers the balance of arches in quite another way. These are collected by the author of the account of Sir Christopher Wren's family. This man's great sagacity, and his great experience in building, and still more his experience in the repairs of old and crazy fabrics, had shown him many things very inconsistent with this theory, which appears so specious and safe. The general facts which occur in the failure of old arches are highly instructive, and deserve the most careful attention of the engineer; for it is in this state that their defects, and the process of nature in their destruction, are most distinctly seen. We venture to affirm, that a very great majority of these facts are irreconcileable to the theory. The way in which circular arches commonly fail, is by the sinking of the crown and the rising of the flanks. It will be found by calculation, that in most of the cases it ought to have been just the contrary. But the clearest proof is, that arches very rarely fail where their load differs most remarkably from that which this theory allows. Semicircular arches have stood the power of ages, as may be seen in the bridges of ancient Rome, and in the numerous arcades which the ancient inhabitants have erected. Now all arches which spring perpendicularly from the horizontal line, require, by this theory, a load of infinite height; and, even to a considerable distance from the springing of the arch, the load necessary for the theoretical equilibrium is many times greater than what is ever laid on those parts; yet a failure in the immediate neighbourhood of the spring of an arch is a most rare phenomenon, if it ever was observed. Here is a most remarkable deviation from the theory; for, as is already observed, the load is frequently not the fourth part of what the theory requires.

Many other facts might be adduced which shew great deviations from the legitimate results of the theory. We hope to be excused, therefore, by the mathematicians for doubting of the justness of this theory. We do not think it erroneous, but defective, leaving out circumstances which we apprehend to be of great importance; and we imagine that the defects have arisen from the very anxiety of the mechanicians to make it perfect. The arch stones are supposed to be perfectly smooth or polished, and not to be connected by any cement, and therefore to sustain each other merely by the equilibrium of their vertical pressure. The theory insures this equilibrium, and this only, leaving unnoticed any other causes of mutual action.

The authors who have written on the subject say expressly, that an arch which thus sustains itself must be stronger than another which would not; because when, in imagination, we suppose both to acquire connection by cement, the first preserves the influence of this connection unimpaired; whereas in the other, part of the cohesion is wasted in counteracting the tendency of some parts to break off from the rest by their want of equilibrium. This is a very specious argument, and would be just, if the forces which are mutually exerted between the parts of the arch in its settled state were merely vertical pressures, or, where different, were inconsiderable in comparison with those which are really attended to in the construction.

But this is by no means the case. The forms which the uses for which arches are erected oblige us to adopt, and the loads laid on the different points of the arch, frequently deviate considerably from what are necessary for the equilibrium of vertical pressures. The varying load on a bridge, when a great waggon passes along it, sometimes bears a very sensible proportion to the weight of that point of the arch on which it rests. It is even very doubtful whether the pressures which are occasioned by the weight of the stuff employed for filling up the flanks really act in a vertical direction, and in the proportion which is supposed. We are pretty certain that this is not the case with sand, gravel, fat mould, and many substances in very general use for this purpose. When this is the case, the pressures sustained by the different parts of the arch are often very inconsistent with the theory—a part of the arch is overloaded, and tends to fall in, but is prevented by the cement. This part of the arch, therefore, acts on the remoter parts, by the intervention of the parts between, employing those intermediate parts as a kind of levers to break the arch in a remote part, just as a lintel would be broken. We apprehend that a mathematician would be puzzled how to explain the stability of an arch cut out of a solid and uniform mass of rock. His theory considers the mutual thrusts of the arch stones as in the direction of the tangents to the arch. Why so? because he supposes that all his polished joints are perpendicular to those tangents. But in the present case he has no existing joints; and there seems to be nothing to direct his imagination in the assumption of joints, which, however, are absolutely necessary for employing his theory, because, without a supposition of this kind, there seems no conceiving any mutual abutment of the arch stones. Ask a common, but intelligent, mason, what notion he forms of such an arch? We apprehend that he will consider it as no arch, but as a lintel, which may be broken like a wooden lintel, and which resists entirely by its cohesion. He will not readily conceive that, by cutting the under side of a stone lintel into an arched form, and thus taking away more than half of its substance, he has changed its nature of a lintel, or given it any additional strength. Nor would there be any change made in the way in which such a mass of stone would resist being broken down, if nothing were done but forming the under side into an arch. If the lintel be so laid on the piers that it can be broken without its parts pushing the piers aside (which will be the case if it lies on the piers with horizontal joints), it will break like any other lintel; but if the joints are directed downwards, and converging to a point within the arch, the broken stone (suppose it broken at the crown by an overload in that part) cannot be pressed down without forcing the piers outwards. Now, in this mode of acting, the mind cannot trace any thing of the statical equilibrium that we have proceeded on in the foregoing theory. The two parts of the broken lintel seem to push the piers aside in the same manner that two rafters push outwards the walls of a house, when their feet are not held together by a tye-beam. If the piers cannot be pushed aside (as when the arch abuts on two solid rocks), nothing can press down the crown which does not crush the stone.

This conclusion will be strictly true if the arch is of such a form that a straight line drawn from the crown to the pier lies wholly within the solid masonry. Thus if the vault consist of two straight stones, as in Fig. 1. Pl. X. or if it consist of several stones, as in Fig. 14. disposed in two straight lines, no weight laid on the crown can destroy it in any other way but by crushing it to powder.

600. But when straight lines cannot be drawn from the overloaded part to the firm abutments through the solid masonry, and when the cohesion of the parts is not able to withstand the transverse strains, we must call the principles of equilibrium to our aid; and, in order to employ them with safety, we must consider how they are modified by the excitement of the cohering forces.

The cohesion of the stones with each other by cement or otherwise, has, in almost every situation, a bad effect. It enables an overload at the crown to break the arch near the haunches, causing those parts to rise, and then to spread outwards, just as a Mansarde or Kirb roof would do if the truss-beam which connects the heads of the lower rafters were sawn through. This can be prevented only by loading that part more than is requisite for equilibrium. It would be prudent to do this to a certain degree, because it is by this cohesion that the crown always becomes the weakest part of the arch, and suffers more by any occasional load.

We expect that it will be said in answer to all this, that the cohesion given by the strongest cement which we can employ, nay the cohesion of the stone itself, is a mere nothing in comparison with the enormous thrusts that are in a state of continual exertion in the different parts of an arch. This is very true; but there is another force which produces the same effect, and which increases nearly in the proportion that those thrusts increase, because it arises from them. This is the friction of the stones on each other. In dry freestone this friction considerably exceeds one half of the mutual pressure. The reflecting reader will see that this produces the same effect, in the case under consideration, that cohesion would do; for while the arch is in the act of failing, the mutual pressure of the arch stones is acting with full force, and thus produces a friction more than adequate to all the effects we have been speaking of.

601. When these circumstances are considered, we imagine it will appear that an arch, when exposed to a great overload on the crown (or indeed on any part), divides, of itself, into a number of parts, each of which contains as many arch stones as can be pierced (so to speak) by one straight line, and that it may then be considered as nearly in the same situation with a polygonal arch of long stones butting on each other like so many beams in a Norman roof, but without their braces and ties. It tends to break at all those angles; and it is not sufficiently resisted there, because the materials with which the flanks are filled up have so little cohesion, that the angle feels no load except what is immediate-

ly above it; whereas it should be immediately loaded with all the weight which is diffused over the adjoining side of the polygon. This will be the case, even though the curvilineal arch be perfectly equilibrated. We recollect some circumstances in the failure of a considerable arch, which may be worth mentioning. It had been built of an exceedingly soft and friable stone, and the arch stones were too short. About a fortnight before it fell, chips were observed to be dropping off from the joints of the arch stones about ten feet on each side of the middle, and also from another place on one side of the arch, about twenty feet from its middle. The masons in the neighbourhood prognosticated its speedy downfal, and said it would separate in those places where the chips were breaking off. At length it fell; but it first split in the middle, and about 15 or 16 feet on each side, and also at the very springing of the arch. Immediately before the fall a shivering or crackling noise was heard, and a great many chips dropped down from the middle between the two places from whence they had dropped a fortnight before. The joints opened above at those new places above two inches, and in the middle of the arch the joints opened below, and in about five minutes after this the whole came down. Even this movement was plainly distinguishable into two parts. The crown sunk a little, and the haunches rose very sensibly, and in this state it hung for about half a minute. The arch stones of the crown were hanging by their upper corners. When these splintered off, the whole fell down.

We apprehend that the procedure of nature was somewhat in this manner. Straight lines can be drawn within the arch stones from A (Fig. 15.) to B and D, and from those points to C and E. Each of the portions ED, DA, AB, BC, resist as if they were of one stone, composing a polygonal vault EDABC. When this is overloaded at A, A can descend in no other way than by pushing the angles

B and D outwards, causing the portions BC, DE, to turn round C and E. This motion must raise the points B and D, and cause the arch stones to press on each other at their *inner* joints *b* and *d*. This produced the copious splintering at those joints immediately preceding the total downfal. The splintering which happened a fortnight before, arose from this circumstance, that the lines AB and AD, along which the pressure of the overload was propagated, were tangents to the soffit of the arch in the points F, H, and G, and therefore the strain lay all on those corners of the arch stones, and splintered a little from off them till the whole took a firmer bed. The subsequent phenomena are evident consequences of this distribution and modification of pressure, and can hardly be explained in any other way; at least not on the theoretical principles already set forth; for in this bridge the loads at B and D were very considerably greater than what the equilibrium required; and we think that the first observed splintering at H, F, and G, was most instructive, showing that there was an extraordinary pressure at the inner joints in those places, which cannot be explained by the usual theory.

Not satisfied with this single observation, after this way of explaining it occurred to us, and not being able to find any similar fact on record, the writer of this article got some small models of arches executed in chalk, and subjected them to many trials, in hopes of collecting some general laws of the internal workings of arches which finally produce their downfal. He had the pleasure of observing the above mentioned circumstances take place very regularly and uniformly, when he overloaded the models at A. The arch always broke at some place B considerably beyond another point F, where the first chipping had been observed. This is a method of trial that deserves the attention both of the theorist and the practitioner.

If these reflections are any thing like a just account of the procedure of nature in the failure of an arch, it is evident that the ingenious mathematical theory of equilibrated arches is of little value to the engineer. We ventured to say as much already, and we rested a good deal on the authority of Sir Christopher Wren. He was a good mathematician, and delighted in the application of this science to the arts. He was a celebrated architect; and his reports on the various works committed to his charge, show that he was in the continual habit of making this application. Several specimens remain of his own methods of applying them. The roof of the theatre of Oxford, the roof of the cupola of St Paul's, and in particular the mould on which he turned the inner dome of that cathedral, are proofs of his having studied this theory most attentively. He flourished at the very time that it occupied the attention of the greatest mechanicians of Europe; but there is nothing to be found among his papers which shows that he had paid much regard to it. On the contrary, when he has occasion to deliver his opinion for the instruction of others, and to explain to the Dean and Chapter of Westminster his operations in repairing that collegiate church, this great architect considers an arch just as a sensible and sagacious mason would do, and very much in the way that we have just now been treating it. (See *Account of the Family of Wren*, p. 356, &c.) Supported therefore by such authority, we would recommend this way of considering an arch to the study of the mathematician; and we would desire the experienced mason to think of the most efficacious methods for resisting this tendency of arches to rise in the flanks. Unfortunately there seems to be no precise principle to point out the place where this tendency is most remarkable.

We are therefore highly pleased with the ingenious contrivance of Mr Mylne, the architect of Blackfriars Bridge in London, by which he determines this point with precision, by making it impossible for the overloaded arch to spring in any other place. Having thus confined the failure to a particular spot, he with equal art opposes a resistance which he believes to be sufficient; and the present condition of that noble bridge, which does not in any place show the smallest change of shape, proves that he was not mistaken. Looking on this work as the first, or at least the second, specimen of masonic ingenuity that is to be seen in the world, we imagine that our readers will be pleased with a particular account of its most remarkable circumstances.

602. The span *k a* (Fig. 16.) of the middle arch is 100 feet, and its height OV is 40, and the thickness KV of the crown is six feet seven inches. Its form is nearly elliptical; the part AVZ being an arch of a circle whose centre is C, and radius 56 feet, and the two lateral portions A *k* B and Z *a* E being arches described with a radius of 35 feet nearly. The thickness of the pier at *a b* is 19 feet. The thickness of the arch increases from the crown V to Y, where it is eight or nine feet. All the arch stones have their joints directed to the centres of their curvature. The joints are all joggled, having a cubic foot of hard stone let half way into each. By this contrivance the joints cannot slide, nor can any weight laid on the crown ever break the arch in that part, if the piers do not yield; for a straight line from the middle of KV to the middle of the joint YI is contained within the solid masonry, and does not even come near the inner joints of the arch stones. Therefore the whole resists like one stone, and can be broken only by crushing it. The joint at Z is very nearly perpendicular to a line YF drawn to the outer edge of the foundation of the pier. By this it was intended to take off all tendency of the pressure on the joint *d* Z to overset the pier; for if we suppose, according to the theory of equilibration, that this pressure is necessarily exerted perpendicularly to the joint, its di-

rection passes through the fulcrum at F, round which it is thought that the pier must turn in the act of oversetting. This precaution was adopted, in order to make the arch quite independent of the adjoining arches; so that although any of them should fall, this arch should run no risk.

Still farther to secure the independence of the arch, the following construction was practised to unite it into one mass, which should rise altogether. All below the line *a b* is built of large blocks of Portland stone, dovetailed with sound oak. Four places in each course are interrupted by equal blocks of a hard stone called *Kentish rag*, sunk half way in each course. These act as joggles, breaking the courses, and preventing them from sliding laterally.

The portion *a* Y of the arch is joggled like the upper part. The interior part is filled up with large blocks of Kentish rag, forming a kind of coursed rubble-work, the courses tending to the centres of the arch. The under corner of each arch stone projects over the one below it. By this form it takes fast hold of the rubble-work behind it. Above this rubble there is constructed the inverted arch I *e* G of Portland stone.* This arch shares the pressure of the two adjoining arches, along with the arch stones in Y *a* and in G *b*. Thus all tend together to compress and keep down the rubble-work in the heart of this part of the pier. This is a very useful precaution; for it often happens, that when the centres of the arches are struck, before the piers are built up to their intended height, the thrust of the arches squeezes the rubble-work, horizontally, after the mortar has set, but before it has dried and acquired its utmost hardness. Its bond is broken by this motion, and it is squeezed up, and never acquires its former firmness. This is effectually prevented by the pressure exerted by the back of the inverted arch.

---

* We have been informed, upon good authority, that this inverted arch does not exist in Blackfriars Bridge, and that it was inserted in the plans merely for the purpose of inspiring confidence. Ed.

Above this counter arch is another mass of coursed rubble, and all is covered by a horizontal course of large blocks of Portland stone, butting against the back of the arch stone ZI and its corresponding one in the adjoining arch. This course connects the feet of the two arches, preserves the rubble-work from too great compression, and protects it from soaking water. This last circumstance is important; for if the water which falls on the road-way is not carried off in pipes, it soaks through the gravel or other rubbish, rests on the mortar, and keeps it continually wet and soft. It cannot escape through the joints of good masonry, and therefore fills up this part like a funnel.

Supposing the adjoining arch fallen, and all tumbled off that is not withheld by its situation, there will still remain in the pier a mass of about 3500 tons. The weight of the portion VY is about 2000 tons. The directions of the thrusts RY and YF are such, that it would require a load of 4500 tons on VY to overturn the pier round F. This exceeds VY by 2500 tons; a weight incomparably greater than any that can ever be laid on it.

Such is the ingenious construction of Mr Mylne. It evidently proceeds on the principles recommended above; principles which have occurred to his experienced and sagacious mind during the course of his extensive practice. We have seen attempts by other engineers to withstand the horizontal thrusts of the arch by means of counter arches inserted in the same manner as here, but extending much farther over the main arch; but they did not appear to be well calculated for producing this effect. A counter arch springing from any point between Y and V has no tendency to hinder that point from rising by the sinking of the crown; and such a counter arch will not resist the precisely horizontal thrust so well as the straight course of Mr Mylne. *

* A Plan and Elevation, and a description of Blackfriars Bridge, will be found in the EDINBURGH ENCYCLOPÆDIA, Art. BRIDGE, see p. 485 and 526.

603. There is another species of arch which must not be overlooked, namely, the DOME or CUPOLA, with all its varieties, which include even the pyramidal steeple or spire.

604. It is evident that the erection of a dome is also a scientific art, proceeding on the principles of equilibration, and that these principles admit and require the same or similar modifications, in consequence of the cohesion and friction of the materials. At first sight, too, a dome appears a more difficult piece of work than a plain arch; but when we observe potters kilns and glass-house domes and cones of vast extent, erected by ordinary bricklayers, and with materials vastly inferior in size to what can be employed in common arches of equal extent, we must conclude that the circumstance of curvature in the horizontal direction, or the abutment of a circular base, gives some assistance to the artist. Of this we have complete demonstration in the case of the cone. We know that a vaulting in the form of a pent roof could not be executed to any considerable extent, and would be extremely hazardous, even in the smallest dimensions; while a cone of the greatest magnitude can be raised with very small stones, provided only that we prevent the bottom from flying out, by a hoop, or any similar contrivance. And when we think a little of the matter, we see plainly, that if the horizontal section be perfectly round, and the joints be all directed to the axis, they all equally endeavour to slide inwards, while no reason can be offered why any individual stone should prevail. They are all wedges, and operate only as wedges. When we consider any single course, therefore, we see that it cannot fall in, even though it may be part of a curve which could not stand as a common arch; nay, we see that a dome may be constructed, having the convexity of the curve, by the revolution of which it is formed, turned toward the axis, so that the outline is concave. We shall afterwards find that this is a stronger dome by far than if the convexity

were outwards, as in a common arch. We see also that a cone may be loaded on the top with the greatest weight, without the smallest danger of forcing it down, so long as the bottom course is firmly kept from bursting outwards. The stone lanthern on the top of St Paul's cathedral in London weighs several hundred tons, and is carried by a brick cone of eighteen inches thick, with perfect safety, as long as the bottom course is prevented from bursting outwards. The reason is evident: The pressure on the top is propagated along the cone in the direction of the slant side; and, so far from having any tendency to break it in any part, it tends rather to prevent its being broken by any irregular pressure from foreign causes.

605. For the same reasons the octagonal pyramids, which form the spires of Gothic architecture, are abundantly firm, although very thin. The sides of the spire of Salisbury cathedral are not eight inches thick after the octagon is fully formed. It is proper, however, to direct the joints to the axis of the pyramid, and to make the coursing joints perpendicular to the slunt side, because the projecting mouldings which run along the angles are the abutments on which the whole pannel depends. A considerable art is necessary for supporting those pannels or sides of the octagon which spring from the angles of the square tower. This is done by beginning a very narrow pointed arch on the square tower at a great distance below the top; so that the legs of the arch being very long, a straight line may be drawn from the top of the keystone of the arch through the whole arch stones of the legs. By this disposition the thrusts arising from the weight of these four pannels are made to meet on the massive masonry in the middle of the sides of the tower, at a great distance below the springing of the spire. This part, being loaded with the great mass of perpendicular wall, is fully able to withstand the horizontal thrust from the

legs of those arches. In many spires these thrusts are still farther resisted by iron bars which cross the tower, and are hooked into pieces of brass firmly bedded in the masonry of the sides. There is much nice balancing of this kind to be observed in the highly ornamental open spires; such as those of Brussels, Mechlin, Antwerp, &c.

606. It is now time to attend to the principle of equilibrium, as it operates in a simple circular dome, and to determine the thickness of the vaulting when the curve is given, or the curve when the thickness is given. Therefore, let B *b* A (Fig. 17.) be the curve which produces the dome, by revolving round the vertical axis AD. We shall suppose this curve to be drawn through the middle of the arch stones, and that the coursing or horizontal joints are everywhere perpendicular to the curve. We shall suppose (as is always the case) that the thickness KL, HI, &c. of the arch stones is very small in comparison with the dimensions of the arch. If we consider any portion HA *h* of the dome, it is plain that it presses on the course, of which HL is an arch stone, in a direction *b* C perpendicular to the joint HI, or in the direction of the next superior element *β b* of the curve. As we proceed downwards, course after course, we see plainly that this direction must change, because the weight of each course is superadded to that of the portion above it, to complete the pressure on the course below. Through B draw the vertical line BCG, meeting *β b*, produced in C. We may take *b c* to express the pressure of all that is above it, propagated in this direction to the joint KL. We may also suppose the weight of the course HL united in *b*, and acting in the vertical. Let it be represented by *b* F. If we form the parallelogram *b* FGC, the diagonal *b* G will represent the direction and intensity of the whole pressure on the joint KL. Thus it appears that this pressure is continually changing its direction, and that the line, which will always coincide with it, must be a curve

concave downward. If this be precisely the curve of the dome, it will be an equilibrated vaulting; but so far from being the strongest form, it is the weakest, and it is the limit to an infinity of others, which are all stronger than it. This will appear evident, if we suppose that *b* G does not coincide with the curve A *b* B, but passes without it. As we suppose the arch stones to be exceedingly thin from inside to outside, it is plain that this dome cannot stand, and that the weight of the upper part will press it down, and spring the vaulting outwards at the joint KL. But let us suppose, on the other hand, that *b* G falls within the curvilineal element *b* B. This evidently tends to push the arch stone inward, toward the axis, and would cause it to slide in, since the joints are supposed perfectly smooth and slipping. But since this takes place equally in every stone of this course, they must all abut on each other in the vertical joints, squeezing them firmly together. Therefore, resolving the thrust *b* G into two, one of which is perpendicular to the joint KL, and the other parallel to it, we see that this last thrust is withstood by the vertical joints all around, and there remains only the thrust in the direction of the curve. Such a dome must therefore be firmer than an equilibrated dome, and cannot be so easily broken by overloading the upper part. When the curve is concave upwards, as in the lower part of the figure, the line *b* C always falls below *b* B, and the point C below B. When the curve is concave downwards, as in the upper part of the figure, '*b* C' passes above, or without *b* B. The curvature may be so abrupt, that even *b*' G' shall pass without '*b* B', and the point G' be above B'. It is also evident that the force which thus binds the stones of a horizontal course together, by pushing them towards the axis, will be greater in flat domes than in those that are more convex; that it will be still greater in a cone; and greater still in a curve whose convexity is turned inwards: for in this last case the line *b* G will deviate most remarkably from the curve. Such a dome will stand (having polished joints) if the curve springs from the base with any elevation, however small; nay, since the friction of two pieces of stone is not less than half of their mutual pressure, such a dome will stand, although the tangent to the curve at the bottom should be horizontal, provided that the horizontal thrust be double the weight of the dome, which may easily be the case if it do not rise high.

607. Thus we see that the stability of a dome depends on very different principles from that of a common arch, and is in general much greater. It differs also in another very important circumstance, viz. that it may be open in the middle: for the uppermost course, by tending equally in every part to slide in toward the axis, presses all together in the vertical joints, and acts on the next course like the keystone of a common arch. Therefore an arch of equilibration, which is the weakest of all, may be open in the middle, and carry at top another building, such as a lanthern, if its weight do not exceed that of the circular segment of the dome that is omitted. A greater load than this would indeed break the dome by causing it to spring up in some of the lower courses; but this load may be increased if the curve is flatter than the curve of equilibration: and any load whatever, which will not crush the stones to powder, may be set on a truncate cone, or on a dome formed by a curve that is convex toward the axis; provided always that the foundation be effectually prevented from flying out, either by a hoop, or by a sufficient mass of solid pier on which it is set. We have mentioned the many failures which happened to the dome of St Sophia in Constantinople.* We imagine that the thrust of the great dome, bending the eastern arch outward as soon as the pier began to yield, destroyed the half dome which was leaning on it, and

* See the Edinburgh Encyclopædia, Art. Civil Architecture, Vol. VI. p. 625. and Plate CLXXIII. of that work. Ed.

thus, almost in an instant, took away the eastern abutment. We think that this might have been prevented, without any change in the injudicious plan, if the dome had been hooped with iron, as was practised by Michael Angelo in the vastly more ponderous dome of St Peter's at Rome, and by Sir Christopher Wren in the cone and the inner dome of St Paul's at London.

608. The weight of the latter considerably exceeds 3000 tons, and they occasion a horizontal thrust which is nearly half this quantity, the elevation of the cone being about 60°. This being distributed round the circumference, occasions a strain on the hoop $= \frac{7}{2 \times 22}$ of the thrust, or nearly 238 tons. A square inch of the worst iron, if well forged, will carry 25 tons with perfect safety: therefore a hoop of 7 inches broad and $1\frac{1}{2}$ inches thick will completely secure this circle from bursting outwards. It is, however, much more completely secured; for, besides a hoop at the base of very nearly these dimensions, there are hoops in different courses of the cone, which bind it into one mass, and cause it to press on the piers in a direction exactly vertical. The only thrusts which the piers sustain are those from the arches of the body of the church and the transepts. These are most judiciously directed to the entering angles of the building, and are there resisted with insuperable force by the whole lengths of the walls, and by four solid masses of masonry in the corners. Whoever considers with attention and judgment the plan of this cathedral, will see that the thrusts of these arches, and of the dome, are incomparably better balanced than in St Peter's church at Rome. But to return from this digression.

609. We have seen that if *b* G, the thrust compounded of the thrust *b* C, exerted by all the courses above HILK, and if the force *b* F, or the weight of that course, be everywhere coincident with *b* B, the element of the curve, we shall have an equilibrated dome; if it falls

within it, we have a dome which will bear a greater load; and if it falls without it, the dome will break at the joint. We must endeavour to get analytical expressions of these conditions. Therefore draw the ordinates $b \delta b''$, BDB″, C d C″. Let the tangents at b and b'' meet the axis in M, and make MO, MP, each equal to $b\,c$, and complete the parallelogram MONP, and draw OQ perpendicular to the axis, and produce b F, cutting the ordinates in E and e. It is plain that MN is to MO as the weight of the arch HA h to the thrust $b\,c$ which it exerts on the joint KL (this thrust being propagated through the course HILK); and that MQ, or its equal $b\,e$, or $\delta\,d$, may represent the weight of the half AH.

Let AD be called x, and DB be called y. Then $b\,e = \dot{x}$, and e C $= \dot{y}$ (because $b\,c$ is in the direction of the element $\beta\,b$). It is also plain, that if we make $\dot{y}$ constant, BC is the second fluxion of x, or BC $= \ddot{x}$, and $b\,e$ and BE may be considered as equal, and taken indiscriminately for x. We have also b C $= \sqrt{\dot{x}^2 + \dot{y}^2}$. Let d be the depth or thickness HI of the arch stones. Then $d\sqrt{\dot{x}^2 + \dot{y}^2}$ will represent the trapezium HL; and since the circumference of each course increases in the proportion of the radius y, $d\,y\sqrt{\dot{x}^2 + \dot{y}^2}$ will express the whole course. If $\int$ be taken to represent the sum or aggregate of the quantities annexed to it, the formula will be analagous to the fluent of a fluxion, and $\int d\,y\sqrt{\dot{x}^2 + \dot{y}^2}$ will represent the whole mass, and also the weight of the vaulting, down to the joint HI. Therefore we have this proportion $\int d\,y\sqrt{\dot{x}^2 + \dot{y}^2} : d\,y\sqrt{\dot{x}^2 + \dot{y}^2} = b\,e : b\,\text{F}, = b\,e : \dot{\text{C}}\,\text{G}, = \delta\,d : \text{CG}, = \dot{x} : \text{CG}$. Therefore CG $= \dfrac{d\,y\,\dot{x}\sqrt{\dot{x}^2 + \dot{y}^2}}{\int d\,y\sqrt{\dot{x}^2 + \dot{y}^2}}$.

If the curvature of the dome be precisely such as puts it in equilibrium, but without any mutual pressure in the vertical joints, this value of CG must be equal to CB, or to $\ddot{x}$, the point G coinciding with B. This, condition will be expressed by the equation $\frac{dy\,\dot{x}\sqrt{\dot{x}^2+\dot{y}^2}}{\int dy\sqrt{\dot{x}^2+\dot{y}^2}}$ $=\ddot{x}$, or, more conveniently, by $\frac{dy\sqrt{\dot{x}^2+\dot{y}^2}}{\int dy\sqrt{\dot{x}^2+\dot{y}^2}}=\frac{\ddot{x}}{\dot{x}}$. But this form gives only a tottering equilibrium, independent of the friction of the joints and the cohesion of the cement. An equilibrium, accompanied by some firm stability, produced by the mutual pressure of the vertical joints, may be expressed by the formula $\frac{dy\sqrt{\dot{x}^2+\dot{y}^2}}{\int dy\sqrt{\dot{x}^2+\dot{y}^2}}$ $>\frac{\ddot{x}}{\dot{x}}$, or by $\frac{dy\sqrt{\dot{x}^2+\dot{y}^2}}{\int dy\sqrt{\dot{x}^2+\dot{y}^2}}=\frac{\ddot{x}}{\dot{x}}+\frac{\dot{t}}{t}$, where t is some variable positive quantity, which increases when x increases. This last equation will also express the equilibrated dome, if t be a constant quantity, because in this case $\frac{\dot{t}}{t}$ is $=0$.

Since a firm stability requires that $\frac{dy\,\dot{x}\sqrt{\dot{x}^2+\dot{y}^2}}{\int dy\sqrt{\dot{x}^2+\dot{y}^2}}$ shall be greater than $\ddot{x}$, and CG must be greater than CB: Hence we learn, that figures of too great curvature, whose sides descend too rapidly, are improper. Also, since stability requires that we have $\frac{dy\,\dot{x}\sqrt{\dot{x}^2+\dot{y}^2}}{\ddot{x}}$ greater than $\int dy\sqrt{\dot{x}^2+\dot{y}^2}$, we learn that the upper part of the dome must not be made very heavy. This, by diminishing the proportion of b F to b C, diminishes the angle $c\,b$ G, and may set the point G above B, which will infallibly spring the dome in that place. We see here also, that the algebraic analysis expresses that peculiarity of dome-vaulting, that the weight of the upper part may even be suppressed.

The fluent of the equation $\frac{d\,y\sqrt{\dot{x}^2+\dot{y}^2}}{\int d\,y\sqrt{\dot{x}^2+\dot{y}^2}}=\frac{\ddot{x}}{\dot{x}}+\frac{\dot{t}}{t}$ is most easily found. It is $\mathrm{L}\int d\,y\sqrt{\dot{x}^2+\dot{y}^2}=\mathrm{L}\;\dot{x}+\mathrm{L}\;t$, where L is the hyperbolic logarithm of the quantity annexed to it. If we consider $\dot{y}$ as constant, and correct the fluent so as to make it nothing at the vertex, it may be expressed thus, $\mathrm{L}\int d\,y\sqrt{\dot{x}^2+\dot{y}^2}-\mathrm{L}\;a=\mathrm{L}\;\dot{x}-\mathrm{L}\;\dot{y}+\mathrm{L}\;t$. This gives us $\mathrm{L}\frac{\int d\,y\sqrt{\dot{x}^2+\dot{y}^2}}{a}=\mathrm{L}\;\frac{\dot{x}}{\dot{y}}\;t$, and therefore $\frac{\int d\,y\sqrt{\dot{x}^2+\dot{y}^2}}{a}=t\,\frac{\dot{x}}{\dot{y}}$.

This last equation will easily give us the depth of vaulting, or thickness d of the arch, when the curve is given. For its fluxion is $\frac{d\,y\sqrt{\dot{x}^2+\dot{y}^2}}{a}=\frac{\dot{t}\;\dot{x}+t\;\ddot{x}}{\dot{y}}$, and $d=\frac{a\;\dot{t}\;\dot{x}+a\;t\;\ddot{x}}{y\;\dot{y}\sqrt{\dot{x}^2+\dot{y}^2}}$, which is all expressed in known quantities; for we may put in place of t any power or function of x or of y, and thus convert the expression into another, which will still be applicable to all sorts of curves.

Instead of the second member $\frac{\ddot{x}}{\dot{x}}+\frac{\dot{t}}{t}$ we might employ $\frac{p\;\ddot{x}}{\dot{x}}$, where p is some number greater than unity.

This will evidently give a dome having stability; because the original formula $\frac{d y \dot{x} \sqrt{\dot{x}^2 + \dot{y}^2}}{\int d y \sqrt{\dot{x}^2 + \dot{y}^2}}$ will then be greater than $\ddot{x}$. This will give $d = \frac{p a \dot{x}^{p-1} \ddot{x}}{y \dot{y}^p \sqrt{\dot{x}^2 + \dot{y}^2}}$. Each of these forms has its advantages when applied to particular cases. Each of them also gives $d = \frac{a \ddot{x}}{y \dot{y} \sqrt{\dot{x}^2 + \dot{y}^2}}$ when the curvature is such as is in precise equilibrium. And, lastly, if d be constant, that is, if the vaulting be of uniform thickness, we obtain the form of the curve, because then the relation of $\ddot{x}$ to $\dot{x}$ and to $\dot{y}$ is given.

The chief use of this analysis is to discover what curves are improper for domes, or what portions of given curves may be employed with safety. Domes are generally built for ornament; and we see that there is great room for indulging our fancy in the choice. All curves which are concave outwards will give domes of great firmness: They are also beautiful. The Gothic dome, whose outline is an undulated curve, may be made abundantly firm, especially if the upper part be convex and the lower concave outwards.

The chief difficulty in the case of this analysis arises from the necessity of expressing the weight of the incumbent part, or $\int d y \sqrt{\dot{x}^2 + \dot{y}^2}$. This requires the measurement of the conoidal surface, which, in most cases, can be had only by approximation by means of infinite serieses. We cannot expect that the generality of practical builders are familiar with this branch of mathematics, and therefore will not engage in it here; but content ourselves with giving such instances as can be understood by such as have that moderate mathematical knowledge which every man should possess who takes the name of engineer.

The surface of any circular portion of a sphere is very easily had, being equal to the circle described with a radius equal to the chord of half the arch. This radius is evidently $= \sqrt{\dot{x}^2 + \dot{y}^2}$.

In order to discover what portion of the hemisphere may be employed (for it is evident that we cannot employ the whole) when the thickness of the vaulting is uniform, we may recur to the equation, or formula $\frac{\dot{d}\, y\, \dot{x} \sqrt{\dot{x}^2 + \dot{y}^2}}{\ddot{x}} = \int d\, y \sqrt{\dot{x}^2 + \dot{y}^2}$. Let a be the radius of the hemisphere. We have $\dot{x} = \frac{a\, y\, \dot{y}}{\sqrt{a^2 - y^2}}$, and $\ddot{x} = \frac{a^2\, \dot{y}^2}{\overline{a^2 - y}|^{\frac{3}{2}}}$. Substituting these values in the formula, we obtain the equation $y^2 \sqrt{a^2 - y^2} = \int \frac{a^2\, y\, \dot{y}}{\sqrt{a^2 - y^2}}$. We easily obtain the fluent of the second member $= a^3 - a^2 \sqrt{a^2 - y^2}$, and $y = a \sqrt{-\frac{1}{2} + \sqrt{\frac{5}{4}}}$. Therefore if the radius of the sphere be 1, the half breadth of the dome must not exceed $\sqrt{-\frac{1}{2} + \sqrt{\frac{5}{4}}}$, or 0,786, and the height will be 618. The arch from the vertex is about 51° 49′. Much more of the hemisphere cannot stand, even though aided by the cement, and by the friction of the coursing joints. This last circumstance, by giving connection to the upper parts, causes the whole to press more vertically on the course below, and thus diminishes the outward thrust; but it at the same time diminishes the mutual abutment of the vertical joints, which is a great cause of firmness in the vaulting. A Gothic dome, of which the upper part is a portion of a sphere not exceeding 45° from the vertex, and the lower part is concave outwards, will be very strong, and not ungraceful.

610. But the public taste has long rejected this form, and seems rather to select more elevated domes than this portion of a sphere; because a dome, when seen from a small distance, always appears flatter than it really is. The dome of St Peter's is nearly an ellipsoid externally, of which the longer axis is perpendicular to the horizon. It is very ingeniously constructed. It springs from the base perpendicularly, and is very thick in this part. After rising about 50 feet; the vaulting separates into two thin vaultings, which gradually separate from each other. These two shells are connected together by thin partitions, which are very artificially dovetailed in both, and thus form a covering which is extremely stiff, while it is very light. Its great stiffness was necessary for enabling the crown of the dome to carry the elegant stone lanthern with safety. It is a wonderful performance, and has not its equal in the world; but it is an enormous load in comparison with the dome of St Paul's, and this even independent of the difference of size. If they were of equal dimensions, it would be at least five times as heavy, and is not so firm by its gravity; but as it is connected in every part by iron bars (lodged in the solid masonry, and well secured from the weather by having lead melted all round them), it bids fair to last for ages, if the foundations do not fail.

If a circle be described round a centre placed anywhere in the transverse axis AC (Fig. 18. No. 1.) of an ellipse, so as to touch the ellipse in the extremities B, *b*, of an ordinate, it will touch it internally, and the circular arch B *a b* will be wholly within the elliptical arch B A *b*. Therefore, if an elliptical and a spherical vaulting spring from the same base, at the same angle with the horizon, the spherical vaulting will be within the elliptical, will be flatter and lighter, and therefore the weight of the next course below will bear a greater proportion to the thrust in the direction of the curve; consequently the spherical vaulting will have more stability. On the contrary, and for similar reasons, an oblate elliptical vaulting is preferable to a spherical vaulting springing with the same inclination to the horizon. (See Fig. 18. No. 2.)

611. Persuaded, that what has been said on the subject convinces the reader that a vaulting perfectly equilibrated throughout is by no means the best form, provided that the base is secured from separating, we think it unnecessary to give the investigation of that form, which has a considerable intricacy; and shall content ourselves with merely stating its dimensions. The thickness is supposed uniform. The numbers in the first column of the table express the portion of the axis counted from the vertex, and those of the second column are the lengths of the ordinates.

AD	DB	AD	DB	AD	DB
0,4	100	610,4	1080	2990	1560
3,4	200	744	1140	3442	1600
11,4	300	904	1200	3972	1640
26,6	400	1100	1260	4432	1670
52,4	500	1336	1320	4952	1700
91,4	600	1522	1360	5336	1720
146,8	700	1738	1400	5756	1740
223,4	800	1984	1440	6214	1760
326,6	900	2270	1480	6714	1780
465,4	1000	2602	1520	7260	1800

The curve delineated in Fig. 19. is formed according to these dimensions, and appears destitute of gracefulness; because its curvature changes abruptly at a little distance from the vertex, so that it has some appearance of being made up of different curves pieced together. But if the middle be occupied by a lanthern of equal, or of smaller weight, this defect will cease, and the whole will be elegant, nearly resembling the exterior dome of St Paul's in London.

612. It is not a small advantage of dome-vaulting that

it is lighter than any that can cover the same area. If, moreover, it be spherical, it will admit considerable varieties of figure, by combining different spheres. Thus, a dome may begin from its base as a portion of a large hemisphere, and may be broken off at any horizontal course, and then a similar or a greater portion of a smaller sphere may spring from this course as a base It also bears being intersected by cylindrical vaultings in every direction, and the intersections are exact circles, and always have a pleasing effect. It also springs most gracefully from the heads of small piers, or from the corners of rooms of any polygonal shape; and the arches formed by its intersections with the walls are always circular and graceful, forming very handsome spandrels in every position. For these reasons, Sir Christopher Wren employed it in all his vaultings, and he has exhibited many beautiful varieties in the transepts and the aisles of St Paul's, which are highly worthy of the observation of architects. Nothing can be more graceful than the vaultings at the ends of the north and south transepts, especially as finished off in the fine inside view published by Gwynn and Wale.

613. We conclude this article with observing, that the connection of the parts, arising from cement and from friction, has a great effect on dome-vaulting. In the same way as in common arches and cylindrical vaulting, it enables an overload on one place to break the dome in a distant place. But the resistance to this effect is much greater in dome-vaulting, because it operates all round the overloaded part. Hence it happens that domes are much less shattered by partial violence, such as the falling of a bomb or the like. Large holes may be broken in them without much affecting the rest; but, on the other hand, it greatly diminishes the strength which should be derived from the mutual pressure in the vertical joints. Friction prevents the sliding in of the arch

stones which produces this mutual pressure in the vertical joints, except in the very highest courses, and even there it greatly diminishes it. These causes make a great change in the form which gives the greatest strength; and as their laws of action are still but very imperfectly understood, it is perhaps impossible, in the present state of our knowledge, to determine this form with tolerable precision. We see plainly, however, that it allows a greater deviation from the best form than the other kind of vaulting, and domes may be made to rise perpendicular to the horizon at the base, although of no great thickness; a thing which must not be attempted in a plane arch. The immense addition of strength which may be derived from hooping, largely compensates for all defects; and there is hardly any bounds to the extent to which a very thin dome vaulting may be carried, when it is hooped or framed in the direction of the horizontal courses. *

---

* For full descriptions and drawings of the principal arches which have been made of cast iron, and for a New Theory of Arches, the reader is referred to the Edinburgh Encyclopædia, Art. Bridge. Ed.

ON THE CONSTRUCTION

OF

CENTRES FOR BRIDGES.

614. Center, or Centre, is a word borrowed from the French name *ceintre* or *cintre*, given to the frame of timber by which the brick or stone of arched vaulting is supported during its erection, and from which it receives its form and curvature.

615. It is not our intention to describe the variety of constructions which may be adopted in situations, where the arches are of small extent, and where sufficient foundation can be had in every part of it for supporting the frame. In such cases, the frequency of the props which we can set up dispenses with much care; and a frame of very slight timbers, connected together in an ordinary way, will suffice for carrying the weight, and for keeping it in exact shape. But when the arches have a wide span, and consequently a very great weight, and when we cannot set up intermediate pillars, either for want of a foundation in the soft bottom of a river, or because the arch is turned between two lofty piers, as in the dome of a stately cathedral—we are then obliged to rest every thing on the piers themselves; and the framing which is to support our arch before the keystone is set, must itself be an arch, depending on the mutual

abutment of its beams. One should think that this view of the construction of a centre, naturally derived from the erection it was to assist, would have been suggested by the slightest consideration: but it has not been so. When intermediate pillars were not employed, it was usual to frame the mould for the arch with little attention to any thing but its shape, and then to cross it and recross it in all directions with other pieces of timber, till it was thought so bound together that it could be lifted in any position, and, when loaded with any weight, could not change its shape. The frame was then raised in a lump, like any solid body of the same shape, and set in its place. This is the way still practised by many country artists, who, having no clear principles to guide them, do not stop till they have made a load of timber almost equal to the weight which it is to carry.

But this artless method, besides leading the employer into great expence, is frequently fatal to the undertaker, from the unskilfulness of the construction. The beams which connect its extremities are made also to support the middle by means of posts which rest on them. They are therefore exposed to a transverse or cross strain, which they are not able to bear. Their number must therefore be increased, and this increases the load. Some of these cross strains are derived from beams which are pressed very obliquely, and therefore exert a prodigious thrust on their supports. The beams are also greatly weakened by the mortises which are cut in them to receive the tenons of the crossing beams: and thus the whole is exceedingly weak, compared with what it might have been, by a proper disposition of the same quantity of timber.

616. The principles from which we are to derive this disposition are the general mechanical principles of carpentry, of which we have given already some account. These furnish one general rule: When we

would give the utmost strength possible to a frame of carpentry, every piece should be so disposed that it is subject to no strain but what either pushes or draws it in the direction of its length; and, if we would be indebted to timber alone for the force or strength of the centre, we must rest all on the first of these strains; for when the straining force tends to *draw* a beam out of its place, it must be held there by a mortise and tenon, which possesses but a very trifling force, or by iron straps and bolts. Cases occur where it may be very difficult to make every strain a thrust, and the best artists admit of ties; and indeed where we can admit a tie-beam connecting the two feet of our frame, we need seek no better security. But this may sometimes be very inconvenient. When it is the arch of a bridge that we are to support, such a tie-beam would totally stop the passage of small craft up and down the river. It would often be in the water, and thus exposed to the most fatal accidents by freshes, &c. Interrupted ties, therefore, must be employed, whose joint or meetings must be supported by something analogous to the king-posts of roofs. When this is judiciously done, the security is abundantly good. But great judgment is necessary, and a very scrupulous attention to the disposition of the pieces. It is by no means an easy matter to discern whether a beam, which makes a part of our centre, is in a state of compression or in a state of extension. In some works of the most eminent carpenters even of this day, we see pieces considered as struts (and considerable dependence had on them in this capacity), while they are certainly performing the office of tie-beams, and should be secured accordingly. This was the case in the boldest centre, we think, that has been executed in Europe, that of the bridge of Orleans, by Mr Hupeau. Yet it is evidently of great consequence not to be mistaken in this point; for when we are mistaken, and the piece is stretched which we imagine to be compressed, we not only are deprived of some support that we expected, but the expected support has become an additional load.

617. To ascertain this point, we may suppose the piers to yield a little to the pressure of the archstones on the centre frames. The feet, therefore, fly outwards, and the shape is altered by the sinking of the crown. We must draw our frame anew for this new state of things, and must notice what pieces must be made longer than before. All such pieces have been acting the part of tie-beams.

But a centre has still another office to sustain; it must keep the arch in its form; that is, while the load on the centre is continually increasing, as the masons lay on more courses of archstones, the frame must not yield and go out of shape, sinking under the weight on the haunches, and rising in the crown, which is not yet carrying any load. The frame must not be supple; and must derive its stiffness, not from the closeness and strength of its joints, which are quite insignificant when set in competition with such immense strains, but from struts or ties, properly disposed, which hinder any of the angles from changing its amplitude.

618. It is obvious, from all that has been said, that the strength and stiffness of the whole must be found in the triangles into which this frame of carpentry may be resolved. We have seen that the strains which one piece produces on two others with which it meets in one point, depends on the angles of their intersection; and that it is greater as an obtuse angle is more obtuse, or an acute angle more acute. And this suggests to us the general maxim, "to avoid as much as possible all very obtuse angles." Acute angles, which are not necessarily accompanied by obtuse ones, are not so hurtful; because the strain here can never exceed the straining force; whereas,

in the case of an obtuse angle, it may surpass it in any degree.

Such are the general rules on this subject. Although something of the mutual abutment of timbers, and the support derived from it, has been long perceived, and employed by the carpenters in roofing, and also, doubtless, in the forming of centres, yet it is a matter of historical fact, that no general and distinct views had been taken of it till about the beginning of this century, or a little earlier. Fontana has preserved the figure of the frames on which the arches of St Peter's at Rome were turned. The one employed for the dome is constructed with very little skill; and those for the arches of the nave and transepts, though incomparably superior, and of considerable simplicity and strength, are yet far inferior to others which have been employed in later times. It is much to be regretted, that no trace remains of the forms employed by the great architect and consummate mechanician Sir Christopher Wren. We should doubtless have seen in them every thing that science and great sagacity could suggest. We are told, indeed, that his centering for the dome of St Paul's was a wonder of its kind; begun in the air at the height of 160 feet from the ground, and without making use of even a projecting corniche whereon to rest it.

619. The earliest theory of the kind that we have met with, that is proposed on scientific principles, and with the express purpose of serving as a lesson, are two centres by M. Pitot about the beginning of this century. As they have considerable merit (greatly resembling those employed by Michael Angelo in the nave of St Peter's), and afford some good maxims, we shall give a short account of them.

What we shall describe under the name of a *centre* is, properly speaking, only one frame, truss, or rib, of a centre. They are set up in vertical planes, parallel to each other, at the distance of 5, 6, 7, or 8 feet, like the trusses or main couples of a roof. Bridging joists are laid across them.—In smaller works these are laid sparingly, but of considerable scantling, and are boarded over; but for great arches, a bridging joist is laid for every course of arch stones, with blockings between to keep them at their proper distances. The stones are not laid immediately on these joists, but beams of soft wood are laid along each joist, on which the stone is laid. These beams are afterwards cut out with the chissel, in order to separate the centre from the ring of stones, which must now support each other by their mutual abutment.

620. The centre is distinguishable into two parts, ALLB (Pl. XI. Fig. 1.) and LDL, which are pretty independent of each other, or at least act separately. The horizontal Stretcher LL cuts the semicircle ADB half way between the spring and the crown of the arch; the arches AL, LD, being 45° each. This stretcher is divided in the same proportion in the points G and H; that is, GH is one half of LL, and LG, HL, are each one-fourth of LL nearly. Each end is supported by two Struts EI, GI, which rest below on a Sole or Bed properly supported. The interval between the heads of the struts GI, HK, is filled up by the Straining Beam GH, abutting in a proper manner on the struts (see Carpentry. The extremities L, L, are united in like manner by butting joints, with the heads of the outer struts. The Arch Moulds AP, BP, are connected with the struts by cross pieces PQ, which we shall call Bridles, which come inwards on each side of the struts, being double, and are bolted to them. This may be called the lower part of the frame. The upper part consists of the king-post DR, supported on each side by the two struts or braces ML, ON, mortised into the post, and also mortised into the stretcher, at the points L, N, where it is

supported by the struts below. The arches LD, LD, are connected with the struts by the bridles P, Q, in the same manner as below.

621. There is a great propriety in many parts of this arrangement. The lower parts or haunches of the arch press very lightly on the centres. Each archstone is lying on an inclined plane, and tends to slide down only with its relative weight; that is, its weight is to its tendency to slide down the joint as radius is to the sine of elevation of the joint. Now it is only by this tendency to slide down the joint that they press on the centering, which in every part of the arch is perpendicular to the joint: But the pressure on the joint, arising from this cause, is much less than this, by reason of the friction of the joints. A block of dry freestone will not slide down at all; and therefore will not press on the centering, if the joint be not elevated 35 degrees at least. But the archstones are not laid in this manner, by sliding them down along the joint, but are laid on the centres, and slide down *their* slope, till they touch the blocks on which they are to rest; so that, in laying the archstones, we are by no means allowed to make the great deduction from their weight just now mentioned, and which Mr Couplet prescribes (Mem. Acad. Par. 1729). But there is another cause which diminishes the pressure on the centres; each block slides down the planks on which it is laid, and presses on the block below it, in the direction of the tangent to the arch. This pressure is transmitted through this block, in the same direction, to the next, and through it to the third, &c. In this manner, it is plain, that, as the arch advances, there is a tangential pressure on the lower archstones, which diminishes their pressure on the frame, and, if sufficiently great, might even push them away from it. Mr Couplet has given an analysis of this pressure, and shews, that in a semicircular arch of uniform thickness, none of the arch stones below

30° press on the frames. But he, without saying so, calculates on the supposition that the blocks descend along the circumference of this frame in the same manner as if it were perfectly smooth. As this is far from being the case, and as the obstructions are to the last degree various and irregular, it is quite useless to institute any calculation on the subject. A little reflection will convince the reader, that in this case the obstruction arising from friction *must* be taken into account, and that it *must not* be taken into account in estimating the pressure of each successive course of stones as they are laid. It is enough that we see that the pressure of the lower courses of archstones on the frame is diminished. Mr Couplet says, that the whole pressure of a semicircular arch is but $\frac{4}{9}$ths of its weight; but it is much greater, for the reason just now given.

622. We have tried, with a well made wooden model (of which the circumference was rubbed with black lead to render it more slippery), whether *any* part of the wooden blocks representing the archstones were detached from the frame by the tangential pressure of the superior blocks; but we could not say confidently that any were so detached. We perceived that all kept hold of a thin slip of Chinese paper (also rubbed with black lead) between them and the frame, so that a sensible force was required to pull it out. From a combination of circumstances, which would be tedious to relate, we believe that the centres carry more than two-thirds of the weight of the arch before the keystone is set. In elliptical and lower pitched circular arches, the proportion is still greater.

It seems reasonable enough, therefore, to dispose the framing in the manner proposed by Pitot, directing the main support to the upper mass of the arch, which presses most on the frame. We shall derive another advantage from this construction, which has not occurred to Mr Pitot.

There is an evident propriety in the manner in which he has distributed the supports of the upper part. The struts which carry the king-post spring from those points of the stretcher where it rests on the struts below: thus the stretcher, on which all depends, bears no transverse strains. It is stretched by the strut above it, and it is compressed in a small degree between the struts below it, at least by the outer ones. Mr Pitot proposes the straining beam GH as a lateral support to the stretcher, which may therefore be of two pieces: but although it *does* augment its strength, it does not seem necessary for it. The stretcher is abundantly carried by the strap, which may and should suspend it from the king-post. The great use of the straining piece is to give a firm abutment to the inner struts, without allowing any lateral strain on the stretcher. *N. B.* Great care must be taken to make the hold sufficiently firm and extensive between the stretcher and the upper struts, so that its cohesion to resist the thrusts from these struts may be much employed.

The only imperfection that we find in this frame is the lateral strains which are brought upon the upper struts by the bridles, which certainly transmit to them part of the weight of the archstones on the curves. The space between the curves and ML should also have been trussed. Mr Pitot's form is, however, extremely stiff; and the causing the middle bridle to reach down to the stretcher, seems to secure the upper struts from all risk of bending.

This centre gives a very distinct view of the offices of all the parts, and makes therefore a proper introduction to the general subject. It is the simplest that can be in its principle, because all the essential parts are subjected to one kind of strain. The stretcher LL is the only exception, and its extension is rather a collateral circumstance than a step in the general support.

623. The examination of the strength of the frame is extremely easy. Mr Pitot gives it for an arch of 60 feet span, and supposes the archstones 7 feet long, which is a monstrous thickness for so small an arch; 4 feet is an abundant allowance, but we shall abide by his construction. He gives the following scantlings of the parts:

The ring or circumference consists of pieces of oak 12 inches broad and 6 thick.

The stretcher LL is 12 inches square.

The straining piece GH is also 12 by 12.

The lower struts 10 by 8.

The king-post 12 by 12.

The upper struts 10 by 6.

The bridles 20 by 8.

These dimensions are French, which is about $\frac{1}{17}$th larger than ours, and the superficial dimensions (by which the section and the absolute strength is measured) is almost $\frac{1}{8}$th larger than ours. The cubic foot, by which the stones are measured, exceeds ours nearly $\frac{1}{5}$th. The pound is deficient about $\frac{1}{17}$th. But since very nice calculation is neither easy nor necessary on this subject, it is needless to depart from the French measures, which would occasion many fractional parts and a troublesome reduction.

The arch is supposed to be built of stone which weighed 160 pounds per foot. Mr Pitot, by a computation (in which he has committed a mistake), says, that only $\frac{11}{14}$ths of this weight is carried by the frame. We believe, however, that this is nearer the truth than Mr Couplet's assumption of $\frac{4}{5}$ths, already mentioned.

Mr Pitot farther assumes, that a square inch of sound oak will carry 8640 pounds. By his language we should imagine that it will not carry much more: but this is very far below the strength of any British oak that we have tried; so far, indeed, that we rather imagine that he means that this load may be laid on it with perfect security for any time. But to compensate for knots and other accidental imperfections, he assumes 7200 as the measure of its absolute force.

He computes the load on each frame to be 707520 pounds, which he reduces to $\frac{11}{14}$ths, or 555908 pounds.

The absolute force of each of the lower struts is 576000 (at 7200 per inch), and that of the curves 518400. Mr Pitot, considering that the curves are kept from bending outwards by the arch stones which press on them, thinks that they may be considered as acting precisely as the outer sruts EI. We have no objection to this supposition.

624. With these data we may compute the load which the lower truss can safely bear by the rule delivered in our treatise on CARPENTRY. We therefore proceed as follows:

Measure off by a scale of equal parts $a\ s$, $a\ t$, each 576000, and add $t\ v$ 518400. Complete the parallelogram $a\ v\ x\ s$, and draw the vertical $x\ c$, meeting the horizontal line a C in c. Make $c\ b$ equal to $c\ a$. Join $x\ b$, and complete the parallelogram $a\ x\ b\ y$. It is evident that the diagonal $x\ y$ will represent the load which these pieces can carry; for the line $a\ v$ is the united force of the curve AP and the strut IE, and $a\ s$ is the strength of IG. These two are equivalent to $a\ x$; $x\ b$ is, in like manner, equivalent to the support on the other side, and $x\ y$ is the load which will just balance the two supports $a\ x$ and $b\ x$.

When $x\ y$ is measured on the same scale, it will be found = 2850000 pounds. This is more than five times the load which actually lies on the frame. It is therefore vastly stronger than is necessary. Half of each of the linear dimensions would have been quite sufficient, and the struts needed only to be 5 inches by 4. Even this would have carried twice the weight, and would have borne the load really laid on it with perfect safety..

We proceed to measure the strength of the upper part. The force of each strut is 432000, and that of the curve is 518400; therefore, having drawn M v parallel to the strut ON, make M v = 432000, and M s = 432000 + 518400. Complete the parallelogram M $s\ r\ v$. Draw the horizontal line $r\ k$, cutting the vertical MC in k, and make $k\ y$ = M k. It is plain, from what was done for the lower part, that M y will measure the load which can be carried by the upper part. This will be found = 1160000. This is also greatly superior to the load; but not in so great a proportion as the other part. The chief part of the load lies on the upper part; but the chief reason of the difference is the greater obliquity of the upper struts. This shortens the diagonal M y of the parallelogram of forces. Mr Pitot should have adverted to this; and instead of making the upper struts more slender than the lower, he should have made them stouter.

The strain on the stretcher LL is not calculated. It is measured by $r\ k$, when M y is the load actually lying on the upper part. Less than the sixth part of the cohesion of the stretcher is more than sufficient for the horizontal thrust; and there is no difficulty of making the foot joints of the struts abundantly strong for the purpose.

The reader will perceive that the computation just now given does not state the proportions of the strains *actually exerted* on the different pieces, but the load on the whole, upon the supposition that each piece is subjected to a strain proportioned to its strength. The other calculation is much more complicated, but is not necessary here.

This centre has a very palpable defect. If the piers

should yield to the load, and the feet of the centre fly out, the lower part will exert a very considerable strain on the stretcher, tending to break it across between N and L, and on the other side, HKF of the lower part is firmly bound together, and cannot change its shape, and will therefore act like a lever, turning round the point F. It will draw the strut HK away from its abutment with GH, and the stretcher will be strained across at the place between H and F, where it is bolted with the bridle. This may be resisted in some degree by an iron strap uniting ON.and HK; but there will still be a want of proportional strength. Indeed, in an arch of such height (a semicircle), there is but little risk of this yielding of the piers; but it is an imperfection.

625. The centre (Pl. XI. Fig. 2.) is constructed on the same principle precisely for an elliptical arch. The calculation of its strength is nearly the same also; only the two upper struts of a side being parallel, the parallelogram M *s r v* (of Fig. 1.) is not needed, and in its stead we measure off on ON a line to represent twice its strength. This comes in place of M *r'* of Fig. 1.—*N. B.* The calculation proceeds on the supposition that the short straining piece MM.makes but one firm body with the king-post. Mr Pitot employed this piece, we presume, to separate the heads of the struts, that their obliquity might be lessened thereby: and this is a good thought; for when the angle formed by the struts on each side is very open, the strain on them becomes very great.

The stretcher of this frame is scarfed in the middle. Suppose this joint to yield a little, there is a danger of the lower strut ON losing its hold, and ceasing to join in the support; for when the crown sinks by the lengthening of the stretcher, the triangle ORN of Fig. 2. will be more distorted than the space above it, and ON will be loosened. But this will not be the case when the sinking of the crown arises from the mere compression of the struts.

Nor will it happen at all in the centre, Fig. 1. On the contrary, the strut ON will abut more firmly by the yielding of the foot of ML.

The figure of this arch of Mr Pitot's consists of three arches of circles, each of 60 degrees. As it is elegant, it will not be unacceptable to the artist to have a construction for this purpose.

626. Make BY = CD, and CZ = $\frac{1}{2}$ CY. Describe the semicircle ZÆY, and make ZS = ZÆ. S is the centre of the side arches, each of 60 degrees. The centre T of the arch, which unites these two, is at the angle of an equilateral triangle STS.

This construction of Mr Pitot's makes a handsome oval, and very near an ellipsis, but lies a little without it. We shall add another of our own, which coincides with the ellipse in eight points, and furnishes the artist, by the way, a rule for drawing an infinite variety of ovals.

Let AB, DE (Fig. 2. No. 2.) be the axes of an ellipse, C the centre, and F, *f*, the two foci. Make C *b* = CD, and describe a circle A D *b e* passing through the three given points A, D, and *b*. It may be demonstrated, that if from any point P of the arch AD be drawn a chord PD, and if a line P R *r* be drawn, making the angle DPR = PDC, and meeting the two axes in the points R and *r*, then R and *r* will be the centres of the circles, which will form a quarter APD of an oval, which has AB and DE for its two axes.

We want an oval which shall coincide as much as possible with an ellipsis? The most likely method for this is to find the very point P where the ellipsis cuts the circle AD *b e*. The easiest way for the artist is to describe an arch of a circle *a m*, having AB for its radius, and the remote focus *f* for its centre. Then set one foot of the compasses on any point P, and try whether the distance PF from the nearest focus F is exactly equal to its distance P m from that circle. Shifting the foot of the compasses from the point of the arch to another, will soon discover the point. This being found, draw PD, make the angle DP r = PD r, and R and r are the centres wanted. Then make C s = CR, and we get the centres for the other side.

The geometer will not relish this mechanical construction. He may therefore proceed as follows: Draw D d parallel to AB, cutting the circle in d. Draw e d, cutting AC in N. Draw CG parallel to e A, and make the angle CG i = AD e. Bisect CN in O, and join O i. Make OM, OM′ = O i, and draw MP, MP′ perpendicular to AB. These ordinates will cut the circle AD b e in the points P and P, where it is cut by the ellipse. We leave the demonstration as a geometrical exercise for the dilettante.

627. We said, that this centering of Mr Pitot's resembled in principle the one employed by Michael Angelo for the nave and transepts of St Peter's church at Rome. Fontana, who has preserved this, ascribes the construction of it to one of the name of San Gallo. A sketch of it is given in Fig. 3. It is, however, so much superior, and so different in principle, from that employed for the cupola, that we cannot think it the invention of the same person. It is, like Pitot's, not only divisible, but really divided into two parts, of which the upper carries by much the greatest part of the load. The pieces are judiciously disposed, and every important beam is amply secured against all transverse strains. Its only fault is a great profusion of strength. The innermost polygon a g h b is quite superfluous, because no strain can force in the struts which rest on the angles. Should the piers yield outwards, this polygon will be loose, and can do no service. Nor is the triangle g i h of any use, if the king-post above it be strapped to the tie-beam and straining sill. Perhaps the inventor considered the king-post as a

pillar, and wished to secure the tie-beam against its cross strain. This centering, however, must be allowed to be very well composed; and we expect that the well-informed reader will join us in preferring it to Mr Pitot's, both for simplicity of principle, for scientific propriety, and for strength.

There is one considerable advantage which may be derived from the actual division of the truss into two parts. If the tie-beam LL, instead of resting on the stretcher EF, had rested on a row of chocks formed like double wedges, placed above each other, head to point, the upper part of the centering might be struck independent of the lower, and this might be done gradually, beginning at the outer ends of the stretcher. By this procedure, the joints of the arch stones will close on the haunches, and will almost relieve the lower centering, so that all can be pulled out together. Thus may the arch settle and consolidate in perfect safety, without any chance of breaking the bond of the mortar in any part; an accident which frequently happens in great arches. This procedure is peculiarly advisable for low pitched or elliptical arches. But this will be more clearly seen afterwards, when we treat of the internal movements of an arch of masonry.

This may suffice for an account of the more simple construction of trussed centres; and we proceed to such as have a much greater complication of principle. We shall take for example some constructed by Mr Perronet, a very celebrated French architect.

628. Mr Perronet's general maxim of construction is to make the truss consist of several courses of separate trusses, independent, as he thinks, of each other, and thus to employ the joint support of them all. In this construction it is not intended to make use of one truss, or part of one truss, to support another, as in the former set, as is practised in the roofs of St Paul's church, Co-

vent Garden, and in Drury Lane theatre. Each truss spans over the whole distance of the piers, and would stand alone (having, however, equilibrium). It consists of a number of struts, set end to end, and forming a polygon. These trusses are so arranged, that the angles of one are in the middle of the sides of the next, as when a polygon is inscribed in a circle, and another (of the same number of sides) is circumscribed by lines which touch the circle in the angles of the inscribed polygon. By this construction the angles of the alternate trusses lie in lines pointing towards the centre of the curve. King-posts are therefore placed in this direction between the adjoining beams of the trusses. These king-posts consist of two beams, one on each side of the truss, and embrace the truss-beams between them, meeting in the middle of their thickness. The abutting beams are mortised, half into each half of the post. The other beam, which makes the base of the triangle, passes through the post, and a strong bolt is driven through the joint, and secured by a key or a nut. In this manner is the whole united; and it is expected, that when the load is laid on the uppermost truss, it will all butt together, forcing down the king-posts, and therefore pressing them on the beams of all the inferior trusses, causing them also to abut on each other, and thus bear a share of the load. Mr Perronet does not assume the invention to himself; but says, that it was invented and practised by Mr Mansard de Sagonne at the great bridge of Moulins. It is much more ancient, and is the work of the celebrated physician and architect Perrault; as may be seen in the collection of machines and inventions of that gentleman published after his death, and also in the great collection of inventions approved of by the Academy of Sciences. It is this which we propose to examine.

629. Fig 4. represents the centering employed for the bridge of Cravant. The arches are elliptical, of 60 feet

span and 20 feet rise. The arch stones are four feet thick, and weigh 176 pounds per foot. The truss-beams were from 15 to 18 feet long, and their section was 9 inches by 8. Each half of the king-posts was about 7 feet long, and its section 9 inches by 8. The whole was of oak. The five trusses were $5\frac{1}{2}$ feet asunder. The whole weight of the arch was 1350000 lbs. which we may call 600 tons (it is 558.) This is about 112 tons for each truss. We must allow near 90 tons of this really to press the truss. A great part of this pressure is borne by the four beams which make the feet of the truss, coupled in pairs on each side. The diagonal of the parallelogram of forces drawn for these beams is, to one of the sides, in the proportion of 360 to 285. Therefore say, as 360 to 285; so is 90 to $71\frac{1}{2}$ tons, the thrust on each foot. The section of each is 144 inches. We may with the utmost safety lay three tons on every inch for ever. This amounts to 432 tons, which is more than six times the strain really pressing the foot beams in the direction of their length; nay, the upper truss alone is able to carry much more than its load. The absolute strength of its foot-beam is 216 tons. It is much more advantageously placed; for the diagonal of the parallelogram of forces corresponding to its position is to the side as 438 to 285. This gives $58\frac{6}{10}$ tons for the strain on each foot; which is not much above the fourth part of what it is able to carry for ever. No doubt can therefore be entertained of the superabundant strength of this centering. We see that the upper row of struts is quite sufficient, and all that is wanted is to procure stiffness for it; for it must be carefully kept in mind, that this upper row is not like an equilibrated arch. It will be very unequally loaded as the work advances. The haunches of the frame will be pressed down, and the joints at the crown raised up. This must be resisted.

Here then we may gather, by the way, a useful lesson.

Let the outer row of struts be appropriated to the carriage of the load, and let the rest be employed for giving stiffness. For this purpose let the outer row have abundant strength. The advantages of this method are considerable. The position of the beams of the exterior row is more advantageous, when (as in this example) the whole is made to rest on a narrow foot: for this obliges us to make the last angle, at least of the lower row, more open, which increases the strain on the strut; besides, it is next to impossible to distribute the compressing thrusts among the different rows of the truss beams; and a beam which, during one period of the mason work, is acting the part of a strut, in another period is bearing no strain but its own weight, and in another it is stretched as a tie. A third advantage is, that, in a case like this, where all rests on a narrow foot, and the lower row of beams are bearing a great part of the thrust, the horizontal thrust on the pier is very great, and may push it aside. This is the most ruinous accident that can happen. An inch or two of yielding will cause the crown of the arch to sink prodigiously, and will instantly derange all the bearings of the abutting beams: but when the lower beams already act as ties, and are quite adequate to their office, we render the frame perfectly stiff or unchangeable in its form, and take away the horizontal thrust from the piers *entirely*. This advantage is the more valuable, because the very circumstance which obliges us to rest all on a narrow foot, places the foot on the very top of the pier, and makes the horizontal thrust the more dangerous.

But, to proceed in our examination of the centering of Cravant bridge, let us suppose that the king-posts are removed, and that the beams are joined by compass joints. If the pier shall yield in the smallest degree, both rows of struts must sink; and since the angles (at least the outermost) of the lower row are more open than those of the upper row, the crown of the lower row will sink more than that of the upper.

The angles of the alternate rows must therefore separate a little. Now restore the king-posts; they prevent this separation. Therefore *they are stretched;* therefore the beams of the lower row are also stretched; consequently they no longer butt on their mortises, and must be held in their places by bolts. Thus it appears that, in this kind of sagging, the original distribution of the load among the different rows of beams is changed, and the upper row becomes loaded beyond our expectation.

If the sagging of the whole truss proceed only from the compression of the timbers, the case is different, and we *may* preserve the original distribution of mutual abutment more accurately. But in this case the stiffness of the frame arises chiefly from cross strains. Suppose that the frame is loaded with arch stones on each side up to the posts HC, *b c*; Fig. 4. the angles E and *e* are pressed down, and the beams EOF, *e o* F push up the point F. This cannot rise without bending the beams EOF, *e o* F; because O and *o* are held down by the double king-posts, which grasp the beams between them. There is therefore a cross strain on the beams. Observe also, that the triangle EHF does not preserve its shape by the connection of its joints; for although the strut beams are mortised into the king-post, they are in very shallow mortises, rather for steadying them than for holding them together. Mr Perronet did not even pin them, thinking that their abutment was very great. The triangle is kept in shape by the base EF, which is firmly bolted into the middle post at O. Had these intersections not been strongly bolted, we imagine that the centres of some of Mr Perronet's bridges would have yielded much more than they did; yet some of them yielded to a degree that our artists would have thought very dangerous. Mr

Perronet was obliged to load the crown of the centering with very great weights; increasing them as the work advanced to prevent the frames from going out of shape; in one arch of 120 feet he laid on 45 tons. Notwithstanding this imperfection, which is perhaps unavoidable, this mode of framing is undoubtedly very judicious, and perhaps the best which can be employed without depending on iron work.

630. Fig. 5. represents another, constructed by Perronet for an arch of 90 feet span and 28 feet rise. The trusses were 7 feet apart, and the arch was $4\frac{1}{2}$ thick; so that the unreduced load on each frame was very nearly 225 tons. The scantling of the struts was 15 by 12 inches. The principle is the same as that of the former. The chief difference is, that in this centre the outer truss-beam of the lower row is not coupled with the middle row, but kept nearly parallel to the outer beam of the upper row. This adds greatly to the strength of the foot, and takes off much of the horizontal thrust from the pier.

Mr Perronet has shewn great judgment in causing the polygon of the inner row of truss beams gradually to approach the polygon of the outer row. By this disposition, the angles of the inner polygon are more acute than those of the outer. A little attention will shew, that the general sagging of all the polygons will keep the abutments of the lower one nearer, or exactly, to their original quantity. We must indeed except the foot-beam. It is still too oblique; and, instead of converging to the foot of the upper row, it should have diverged from it. Had this been done, this centre is almost perfect in its kind. As it is, it is at least six times stronger than was absolutely necessary. We shall have occasion to refer to this figure on another occasion.

631. This maxim is better exemplified by Mr Perronet in the centering of the bridge of St Maxence, exhi-

bited in Fig. 5. No. 2. than that of Nogent, Fig. 5. No. 1.. But we think that a horizontal truss-beam *a b* should have been inserted, in a subordinate manner, between the king-posts next the crown on each side. This would prevent the crown from rising while the haunches only are loaded, without impairing the fine abutments of *c d*, *c d*, when the arch is nearly completed. This is an excellent centering, but is not likely to be of much use in these kingdoms; because the arch itself will be considered as ungraceful and ugly, looking like a huge lintel. Perronet says, that he preferred it to the ellipse, because it was lighter on the piers, which were thin. But the failure of one arch must be immediately followed by the ruin of all. We know much better methods of lightening the piers.

632. Fig. 6. represents the centering of the bridge of Neuilly, near Paris, also by Perronet. The arch has 120 feet span, and 30 feet rise, and is 5 feet thick. The frames are 6 feet apart, and each carries an absolute (that is, not reduced to $\frac{11}{14}$ or to $\frac{4}{9}$) load of 350 tons. The strut beams are 17 by 14 inches in scantling. The king-posts are of 15 by 9 each half; and the horizontal bridles, which bind the different frames together in five places, are also 15 by 9 each half. There are eight other horizontal binders of 9 inches square.

This is one of the most remarkable arches in the world; not altogether on account of its width, for there are several much wider, but for the flatness at the crown: for about 26 feet on each side of the middle it was intended to be a portion of a circle of 150 feet radius. An arch (semicircular) of 300 feet span might therefore be easily constructed, and would be much stronger than this, because its horizontal thrust at the crown would be vastly greater, and would keep it more firmly united.

The bolts of this centre are differently placed from those of the former; and the change is judicious. Mr Perronet had doubtless found by this time, that the stiffness of his framing depended on the transverse strength of the beams; and therefore he was careful not to weaken them by the bolts. But, notwithstanding all his care, the framing sunk upwards of 13 inches before the keystones were laid; and, during the progress of the work, the crown rose and sunk, by various steps, as the loading was extended along it. When 20 courses were laid on each side, and about 16 tons laid on the crown of each frame, it sunk about an inch. When 46 courses were laid, and the crown loaded with 50 tons, it sunk about half an inch more. It continued sinking as the work advanced; and when the keystone was set it had sunk 13¼ inches. But this sinking was not general; on the contrary, the frame had risen greatly at the very haunches, so as to open the upper part of the joints, many of which gaped an inch; and this opening of the joints gradually extended from the haunches towards the crown, in the neighbourhood of which they opened on the under side. This evidently arose from a want of stiffness in the frame. But these joints closed again when the centres were struck, as will be mentioned afterwards.

We have taken particular notice of the movements and twisting of this centre, because we think that they indicate a deficiency, not only of stiffness, but of abutment, among the truss-beams. The whole has been too flexible, because the angles are too obtuse: This arises from their multiplicity. When the intercepted arches have so little curvature, the power of the load to press it inward increases very fast. When the intercepted arch is reduced to one half, this power is more than doubled; and it is also doubled when the radius of curvature is doubled. The king-posts should have been farther apart near the crown, so that the quantity of arch between them should compensate for its diminished curvature.

The power of withstanding any given inequality of

load would therefore have been greater, had the centre consisted of fewer pieces, and their angles of meeting been proportionally more acute The greatest improvement would have been, to place the foot of the lower tier of truss-beams on the very foot of the pier, and to have also separated it at the head from the rest with a longer king-post, and thus to have made the distances of the beams on the king-posts increase gradually from the crown to the spring. This would have made all the angles of abutment more acute, and would have produced a greater pressure on all the lower tiers when the frame sagged.

633. Fig. 1. of Pl. XII represents the centering of the bridge of Orleans. The arch has 100 feet span, and rises 30, and the arch-stones are 6 feet long. It is the construction of Mr Hupeau, the first architect of the bridge. It is the boldest work of the kind that we have seen, and is constructed on clear principles. The main abutments are few in number. Because the beams of the outer polygon are long, they are very well supported by straining beams in the middle; and the struts or braces which support and butt on them, are made to rest on points carried entirely by *ties*. The inventor, however, seems to have thought that the angles of the inner polygon were supported by mutual compression, as in the outer polygon. But it is plain that the whole inner polygon may be formed of iron rods. Nôt but that both polygons *may* be in a state of compression (this is very possible); but the smallest sagging of the frame will change the proportions of the pressures at the angles of the two polygons. The pressures on the exterior angles will increase, and those on the lower or interior angles will diminish most rapidly; so that the abutments on the lower polygon will be next to nothing. Such points could bear very little pressure from the braces which support the middle of the long bearings of the upper beams, and their pressures

must be borne chiefly by the joints supported by the king-posts. The king-posts would then be in a state of extension. It is difficult, however, to decide what is the precise state of the pressure at these interior angles.

634. The history of the erection of this bridge will throw much light on this point. Mr Hupeau died before any of the arches were carried farther than a very few of the first courses. Mr Perronet succeeded to the charge, and finished the bridge. As the work advanced, the crown of the frame rose very much. It was loaded; and it sunk as remarkably. This shewed that the lower polygon was giving very little aid. Mr Perronet then thought the frame too weak, and inserted the long beam DE, making the diagonal of the quadrangle, and very nearly in the direction of the lower beam *a b*, but falling rather below this line. He now found the frame abundantly strong. It is evident that the truss is now changed exceedingly, and consists of only the two long sides, and the short straining beam lying horizontally between their heads. The whole centering consists now of one great truss *a* E *e b*, and its long sides *a* E, *e b*, are trussed up at B and *f*. Had this simple idea been made the principle of the construction, it would have been excellent. The angle *a* DE might have been about 176°, and the polygon D *c g h* employed only for giving a slight support to this great angle, so as not to allow it to exceed 180°. But Mr Perronet found, that the joint *c*, at the foot of the post E *c*, was about to *draw loose*, and he was obliged to bolt long pieces of timber on each side of the joint, embracing both beams. These were evidently acting the same part as iron straps would have done; a complete proof that, whatever may have been the original pressures, there was no abutment now at the point *c*, and that the beams which met there were not in a state of compression, but were on the stretch. Mr Perronet says

that he put these cheeks to the joints *to stiffen* them. But this was not their office; because the adjoining beams were not struts, but ties, as we have now proved.

We may therefore conclude, that the outer polygon, with the assistance of the pieces, *a b*, DE, were carrying the whole load. We do not know the distance between the frames; but, supposing them seven feet apart, and the arch six feet thick, and weighing 170 pounds per foot, we learn the load. The beams were 16 inches square. If we now calculate what they would bear at the same very moderate rate allowed to the other centres, we find that the beams AB and *a b* are not loaded to one-sixth of their strength.

We have given this centre as a fine example of what carpentry is able to perform, and because, by its simplicity, it is a sort of text on which the intelligent artist may make many comments. We may see plainly that, if the lower polygon had been formed of iron rods, firmly bolted into the feet of the king-posts, it would have maintained its shape completely. The service done by the beam DE was not so much an increase of abutment as a discharge of the weight and of the *pull* at the joint *c*. Therefore, in cases where the feet of the truss are *necessarily* confined to a very narrow space, we should be careful to make the upper polygon sufficient to carry the whole load (say by doubling its beams), and we may then make the lower polygon of slender dimensions, provided we secure the joints on the king-posts by iron straps which embrace a considerable portion of the tie on each side of the joint.

635. We are far from thinking that these centres are of the best kind that could be employed in their situation; but they are excellent in their kind; and a careful study of them will teach the artist much of his profession. When we have a clear conception of the state of strain in which the parts of a frame really are, we know

what should be done in order to draw all the advantage possible from our materials. We have said in another place, that where we can give our joints sufficient connection (as by straps and bolts, or by cheeks or fishes), it is better to use ties than struts, because ties never bend.

We do not approve of Mr Perronet's practice of giving his trusses such narrow feet. By bringing the foot of the lower polygon farther down, we greatly diminish all the strains, and throw more load on the lower polygon: and we do not see any of Mr Perronet's centres where this might not have been done. He seems to affect a great span, to shew the wonders of his art; but our object is to teach how to make the best centre of a given quantity of materials; and how to make the most perfect centre, when we are not limited in this respect, nor in the extent of our fixed points.

636. We shall conclude this series of examples with one where no such affectation takes place This is the centering of the bridge at Blackfriars, London. The span of the arch is 100 feet, and its height from the spring is about 43. The drawing Plate XII. Fig. 2. is sufficiently minute to convey a distinct notion of the whole construction. We need not be very particular in our observations, after what has been said on the general principles of construction. The leading maxim, in the present example, seems to be, *that every part of the arch shall be supported by a simple truss of two legs resting, one on each pier.* H, H, &c. are called APRON PIECES to strengthen the exterior joints, and to make the RING as stiff in itself as possible. From the ends of this apron-piece proceed the two legs of each truss. These legs are 12 inches square: They are not of an entire piece, but of several, meeting in firm abutment. Some of their meetings are secured by the double king-posts, which grasp them firmly between them, and are held together by bolts. At other inter-

sections, the beams appear halved into each other ; a practice which cannot but weaken them much, and would endanger their breaking by cross strains, if it were possible for the frame to change its shape. But the great breadth of this frame is an effectual stop to any such change. The fact was, that *no sinking or twisting whatever* was observed during the progress of the mason work. Three points in a straight line were marked on purpose for this observation, and were observed every day. The arch was more than six feet thick ; and yet the sinking of the crown, before setting the keystones, did not amount to one inch.

The centre employs about one-third more timber than Perronet's great centre in proportion to the span of the arch ; but the circumference increases in a greater proportion than this, because it is more elevated. In every way of making a comparison of the dimensions, Mr Mylne's arch employs more timber : but it is *beyond all comparison* stronger. The great elevation is partly the reason of this. But the disposition of the timbers is also much more advantageous, and may be copied even in the low-pitched arches of Neuilly. The simple truss, reaching from pier to pier for the middle point of the arch, gives the strong support where it is most of all wanted ; and in the lateral points H, although one leg of the truss is very oblique, the other compensates for it by its upright position.

The chief peculiarity of this centre is to be seen in its base. This demands a more particular attention : but we must first make some observations on the condition of an arch, as it rests on the centering after the keystones are all set, and on the gradual transference of the pressure from the boards of the centering to the joints of the archstones.

637. While all the archstones lie on the centering, the lower courses are also leaning pretty strongly on each

other. But the mortar is hardly compressed in the joints; and least of all in the joints near the crown. Suppose the arch to be Catenarean, or of any other shape that is perfectly equilibrated: When the centering is gradually withdrawn, all the archstones follow it. Their wedge-like form makes this impossible, without the middle ones squeezing the lateral ones aside. This compresses the mortar between them. As the stones thus come nearer to each other, those near the crown must descend more than those near the haunches, before every stone has lessened its distance from the next by the same quantity; for example, by the hundredth part of an inch. This circumstance alone must cause a sinking in the crown, and a change of shape. But the joints near the crown are *already* more open than those near the haunches. This produces a still greater change of form before all is settled. Some masons endeavour to remedy, or at least to diminish, this, by using no mortar in the joints near the crown. They lay the stones dry, and even force them together by wedges and blocks laid between the stones on opposite sides of the crown: They afterwards pour in fine cement. This appears a good practice. Perronet rejects it, because the wedging sometimes breaks the stones. We should not think this any great harm; because the fracture will make them close where they would otherwise lie hollow. But, after all our care, there is still a sinking of the crown of the arch. By gradually withdrawing the centering, the joints close, the archstones begin to butt on each other, and to force aside the lateral courses. This abutment gradually increasing, the pressure on the haunches of the centering is gradually diminished by the mutual abutment, and ceases entirely in that course, which is the lowest that formerly pressed it: it then ceases in the course above, and then in the third, and so on. And, in this manner, not only the centering quits the arch, gradually, from the bottom to the top, *by its own retiring from it*, but the arch also quits the centering *by changing its shape*. If the centering were now pushed up again, it would touch the arch first at the crown; and it must *lift up* that part gradually before it come again in contact with the haunches. It is evident, therefore, that an arch, built on a centre of a shape perfectly suited to equilibration, will not be in equilibrio when the centering is removed. It is therefore necessary to form the centering in such a manner (by raising the crown), that it shall *leave* the arch of a proper form. This is a very delicate task, requiring a previous knowledge of the ensuing change of form. This cannot be ascertained by the help of any theory we are acquainted with.

But, suppose this attained, there is another difficulty: While the work advances, the centering is warped by the load laid on it, and continually increasing on each side. The first pressure on the centering forces down the haunches, and raises the crown. The arch is therefore less curved at the haunches than is intended: the joints, however, accommodate themselves to this form, and are close, and filled with mortar. When the masons approach the middle of the arch, the frame sinks there and rises up at the haunches. This opens all the joints in that place on the *upper* side. By the time that the keystones are set, this warping has gone farther; and the joints are open on the *under* side near the crown. It is true we are here speaking rather of an extreme case, when the centering is *very* flexible; but this occurred to Mr Perronet in the two great bridges of Neuilly and of Mantz. In this last one, the crown sunk above a foot before the key was set, and the joints at the haunches opened above an inch *above*, while some nearer the crown opened near a quarter of an inch *below*.

638. In this condition of things, it is a delicate business

to strike the centering. Were it removed in an instant, all would probably come down; for the archstones are not yet abutting on each other, and the joints in the middle are open below. Mr Perronet's method appears to us to be very judicious. He began to detach the centering at the very bottom, on each side equally, where the pressure on the centering is very slight. He cut away the blocks which were immediately under each archstone. He proceeded gradually upwards in this way with some speed, till all was detached that had been put out of shape by the bending of the centering. This being no longer supported, sunk inward, till it was stopped by the abutment which it found on the archstones near the crown, which were still resting on their blocks. During part of this process, the open joints opened still more, and looked alarming. This was owing to the removal of the load from the haunches of the centering. This allowed the crown to sink still more, by forcing out the archstones at the haunches. He now paused some days, and during this time the two haunches, now hanging in the air, gradually pressed in toward the centering, their outer joints closing in the meanwhile. The haunches were now pressing pretty hard on the archstones nearer the crown. He then proceeded more slowly, destroying the blocks and bridgings of these upper archstones. As soon as he destroyed the support of one, it immediately yielded to the pressure of the haunch; and if the joint between it and the one adjoining toward the crown happened to be open, whether on the under or the upper side, it immediately closed on it But in proceeding thus, he found every stone sink a little while it closed on its neighbour; and this was like to produce a ragged soffit, which is a deformity. He therefore did not allow them to sink so much. In the places of the blocks and bridgings which he had cut away, he set small billets, standing on their ends, between the centering and the archstones. These allowed the pendulous arch to push toward the crown without sensibly descending; for the billets were pushed out of the perpendicular, and some of them tumbled down. Proceeding in this way, he advanced to the very next course to the keystone on each side, the joints closing all the way as he advanced. The last job was very troublesome; we mean the detaching the three uppermost courses from the centering: for the whole elasticity of the centering was now trying to unbend, and pressing hard against them. He found that they were lifted up; for the joints beyond them, which had closed completely, now opened again below: but this job was finished in one day, and the centre sprung up two or three inches, and the whole arch sunk about six inches. This was an anxious time; for he dreaded the great momentum of such a vast mass of matter. It was hard to say where it would stop. He had the pleasure to see that it stopped very soon, settling slowly as the mortar was compressed, and after one or two days settling no more. This settling was very considerable both in the bridge at Neuilly and in that at Mantz. In the former, the sinking during the work amounted to 13 inches. It sunk six inches more when the blocks and bridgings were taken out, and 1½ when the little standards were destroyed, and 1¼ more next day: so that the whole sinking of the *pendulous* arch was 9½ inches, besides what it had sunk by the bending and compression of the centering *

The crown of the centering was an arch of a circle described with a radius of 150 feet; but by the sinking of the arch its shape was considerably changed, and about 60 feet of it formed an arch of a circle whose radius was 244 feet. Hence Mr Perronet infers, that a semicircle of 500 feet span may be erected. It would no doubt be stronger than this arch, because its greater horizontal thrust would keep the stones firmer together. The sinking of the arches at Mantz was not quite so great, but

* Thewhole sinking of this arch was 23½ inches. Ed.

every thing proceeded in the same way. It amounted in all to $20\frac{1}{2}$ inches, of which 12 inches were owing to the compression and bending of the centering.

639. In Fig. 5, No. 1. may be observed an indication of this procedure of the masonry. There may be noticed a horizontal line *a c*, and a diagonal *a b*. These are supposed to be drawn on the masonry as it would have stood had the frames not yielded during the building. The dotted line A *b′ c′* shews the shape which it took by the sinking of the centering. The dotted line of the other side was actually drawn on the masonry when the keystone was set: and the wavy black line on the same side shews the form which the dotted line took by the striking of the centering. The undulated part of this line cuts its former position a little below the middle, going without it below, and falling within it above This shews very distinctly the movement of the whole masonry, distinguishing the parts that were forced out, and the parts which sunk inward.

We presume that the practical reader will think this account of the internal movements of a stupendous arch very instructive and useful. As Mr Perronet observed it to be uniformly the same in several very large arches which he erected, we may conclude that it is the general process of nature. We by no means have the confidence in the durability or solidity of his arches which he prudently professes to have. We have conversed with some very experienced masons, who have also erected very great arches, and in very difficult situations, which have given universal satisfaction; and we have found them uniformly of opinion, that an arch which has settled to such a proportion of its curvature as to change the radius from 150 to 244 feet, is in a very hazardous situation. They think the hazard the greater, because the span of the arch is so great in proportion to its weight (as they express it very emphatically) or its height. The weight, say they, of the haunches is too small for forcing together the keystones, which have scarcely any wedge-like form to keep them from sliding down. This is very good reasoning, and expresses very familiar notions. The mechanician would say, that the horizontal thrust at the crown is too small. When we questioned them about the propriety of Mr Perronet s method of removing the centering, they unanimously approved of its general principle, but said that it was very ticklish indeed in the execution. The cases which he narrates were new to them. They should have almost despaired of success with arches which had gone so much out of shape by the bending of the centres; because, said they, the slope of the centering, to a great distance from the crown, was so little, that the archstones could not slide outwards along it, to close even the under side of the joints which had opened above the haunches; so that *all* the archstones were at too great a distance from each other; and a great and *general* subsiding of the whole was necessary for bringing them even to touch each other. They had *never* observed such bendings of the centerings which they had employed, having never allowed themselves to contract the feet of their trusses into such narrow spaces. They observed, that nothing but lighters with their masts down can pass under the trusses, and that the sides must be so protected by advanced works from the accidental shock of a loaded boat, that there cannot be left room for more than one. They added, that the bridges of communication, necessary for the expeditious conducting of the work, made all this supposed roominess useful: besides, the business can hardly be so urgent and crowded anywhere, as to make the passage through every arch indispensably necessary. Nor was the inconvenience of this obstruction greatly complained of during the erection of Westminster or Blackfriars bridges.

These appeared to us good reasons for preferring the

more cautious, and incomparably more secure, construction of Mr Mylne, in which the breadth given to each base of the trusses permitted a much more effective disposition of the abutting timbers, and also enabled the engineer to make it incomparably stiffer; so that no change need be apprehended in the joints which have already closed, and in which the mortar has already taken its set, and commenced an union that never can be restored if it be once broken in the smallest degree, no not even by greater compression.

640. Here we beg leave to mention our notions of the connection that is formed by mortar composed of lime or gypsum. We consider it as consisting chiefly, if not solely, in a crystallization of the line or gypsum and water. As much water is taken up as is necessary for the formation of the crystals during their gradual conversion into mild calcareous earth or alabaster, and the rest evaporates. When the free access of air is absolutely prevented, the crystallization never proceeds to that state, even although the mortar becomes extremely dry and hard. We had an opportunity of observing this accidentally, when passing through Maestricht in 1770, while they were cutting up a massy revetment of a part of the fortifications more than 300 years old. The mortar between the bricks was harder than the bricks (which were Dutch clinkers, such as are now used only for the greatest loads); but when mixed with water it made it lime water, seemingly as strong as if fresh lime had been used. We observed the same thing in one small part of a huge mass of ancient Roman work near Romney in Kent; but the rest, and all the *very old* mortar that we have seen, was in a mild state, and was generally much harder than what produced any lime water. Now when the mortar in the joints has begun its first crystallization, and is allowed to remain in perfect rest, we are confident that the subsequent crystals, whether of lime or of calcareous

earth, or of gypsum, will be much larger and stronger than can ever be produced if they are once broken; and the farther that this crystallization has been carried, that is, the harder that the mortar has become, less of it remains to take any new crystallization. Why should it be otherwise here than in every other crystallization that we are acquainted with?

641. We think therefore that it is of great consequence to keep the joints in their *first* state if possible; and that the strength (as far as it depends on the mortar) is greatly diminished by their opening; especially when the mortar has acquired considerable hardness, which it will do in a month or six weeks, if it be good. The cohesion given by mortar is indeed a mere trifle, when opposed to a force which tends to open the joints, acting, as it generally does, with the transverse force of a lever: but in situations where the overload on any particular archstones tends to push them down through between their neighbours, like wedges, the cohesion of the mortar is then of very great consequence.

We must make another observation. Mr Perronet's ingenious process tended very effectually to close the joints. In doing this, the forces which he brought into action had little to oppose them; but as soon as they were closed, the contact of the parts, formerly open, opposed an obstruction incomparably greater, and immediately balanced a force which was but just able to turn the stone gently about the two edges in which it touched the adjoining stones. This is an important remark, though seemingly very trifling; and we wish the practitioner to have a very clear conception of it; but it would take a multitude of words to explain it. It is worth an experiment. Form a little arch of wooden blocks; and form one of these so, that when they are all resting on the centering, it may be open at the outer joint—Remove the centering—Then press on the arch at some distance from the open joint.—You will find that a very small pressure will make the arch bend till that joint closes—Press a little harder, and the arch will bend more, and the next joint will open.—Thus you will find that, by pressing alternately on each side of the open joint, that stone can easily be made to flap over to either side; and that immediately after this is done the resistance increases greatly. This shews clearly, that a very moderate force, judiciously employed, will close the joints, but will not press the parts strongly together. The joints therefore are *closed*, but *no more than closed*, aud are hanging only by the edges by which they were hanging while the joints were open. The arch, therefore, though apparently close and firm, is but loose and tottering. Mr Perronet says, that his arches were firm, because hardly a stone was observed to chip or splinter off at the edges by the settlement. But he had done every thing to prevent this, by digging out the mortar from between the headers, to the depth of two inches, with saws made on purpose. But we are well informed, that before the year 1791 (twenty years after the erection) the arches at Neuilly had sunk very sensibly, and that very large splinters had flown off in several places. It could not be otherwise.

642 The original construction was too bold; we may say needlessly and ostentatiously bold. A very gentle slope of the roadway, which would not have slackened the mad gallop of a ducal carriage, nor sensibly checked the laborious pull of a loaded waggon, and a proper difference in the size of the arches, would have made this wonderful bridge incomparably stronger, and also much more elegant and pleasing to the eye. Indeed, it is far from being as handsome as it might have been. The ellipse is a most pleasing figure to every beholder; but this is concealed as much as possible, and it is attempted to give the whole the appearance of a tremendous lintel.

It has the oppressive look of danger. It will not be of long duration. The bridge at Mantz is still more exceptionable, because its piers are tall and slender. If any one of the arches fails, the rest must fall in a moment. An arch of Blackfriars bridge might be blown up without disturbing its neighbours.

643. Mr Perronet mentions another mode of striking the centering, which he says is very usual in France. Every second bridging is cut out. Some time after, every second of the remainder; after this, every second of the remainder; and so on, till all is removed. This is never practised in this country, and is certainly a very bad method. It leaves the arch hanging by a number of distant points; and it is wonderful that any arch can bear this treatment.

644. Our architects have generally proceeded with extreme caution. Wherever they could, they supported the centering by intermediate pillars, even when it was a trussed centre, having a tie-beam reaching from side to side. The centre was made to rest, not immediately on these pillars, but on pieces of timber formed like acute wedges, placed in pairs, one above the other, and having the point of the one on the thick end of the other. These wedges were well soaped and rubbed with black lead, to make them slippery. When the centres are to be struck, men are stationed at each pair of the wedges with heavy mauls. They are directed to strike together on the opposite wedges. By this operation, the whole centering descends together; or, when any part of the arch is observed to have opened its joints on the upper side, the wedges below that part are slackened. The framing may perhaps bend a little, and allow that part to subside. If any part of the arch is observed to open its joints on the under side, the wedges below that part are allowed to stand after the rest have been slackened. By this process, the whole comes down gradually, and as

slowly as we please, and the defects of every part of the arch may be attended to. Indeed the caution and moderation of our builders have commonly been such, that few defects have been allowed to shew themselves. We are but little acquainted with joints opening to the extent of two inches, and in such a case would probably lift every stone of the arch again. We have not employed trussed centerings so much perhaps as we should have done; nor do we see their advantage (speaking as mere builders) over centres supported all over, and unchangeable in their form. Such centres must bend a little, and require loading on the middle to keep them in shape. Their compression and their elasticity are very troublesome in the striking of the centres in Mr Perronet's manner. The elasticity is indeed of use when the centres are struck in the way now described.

These observations on the management of the internal movements of a great arch, will enable the reader to appreciate all the merit of Mr Mylne's very ingenious construction. We proceed therefore to complete our description.

645. The gradual enlargement of the base of the piers of Blackfriars bridge enabled the architect to place a series of five posts C, C, C, C, C, Pl. XII. Fig. 2. one on each stap of the pier; the ingenious contexture of which made it like one solid block of stone. These struts were gradually more and more oblique, till the outer one formed an obtuse angle with the lowest side of the interior polygon of the truss. On the top of these posts was laid a sloping SEAT or beam D of stout oak, the upper part of which was formed like a zig-zag scarfing. The posts were not perpendicular to the under side of the seat. The angles next the pier were somewhat obtuse. Short pieces of wood were placed between the heads of the posts (but not mortised into them), to prevent them from slipping back. Each face of the scarf was covered

with a thick and smooth plate of copper. The feet of the truss were mortised into a similar piece F, which may be called the SOLE of the truss, having its lower side notched in the same manner with the upper side of D, and like it covered with copper. Between these two lay the STRIKING WEDGE E, the faces of which correspond exactly with the slant faces of the seat and the sole. The wedge was so placed, that the corresponding faces touched each other for about half of their length. A block of wood was put in at the broad end or base of this wedge, to keep it from slipping back during the laying the archstones. Its outer end E was bound with iron, and had an iron bolt several inches long driven into it. The head of this bolt was broad enough to cover the whole wood of the wedge within the iron ferrule.

We presume that the reader, by this time, foresees the use of this wedge. It is to be driven in between the sole and the seat, having first taken out the block at the base of the wedge. As it advances into the wider spaces, the whole truss must descend, and be freed from the arch; but it will require prodigious blows to drive it back. Mr Mylne did not think so, founding his expectation on what he saw in the launching of great ships, which slide very easily on a slope of 10 or 12 degrees. He rather feared, that taking out the block behind would allow the wedge to be pushed back at once, so that the descent of the truss would be too rapid. However, to be certain of the operation, he had prepared an abundant force in a very ingenious manner. A heavy beam of oak, armed at the end with iron, was suspended from two points of the centre like a battering ram, to be used in the same manner. Nothing could be more simple in its structure, more powerful in its operation, or more easy in its management. Accordingly the success was to his wish. The wedge did not slip back of itself; and very moderate blows of the ram drove it back with the greatest ease.

The whole operation was over in a very few minutes. The spectators had suspected, that the space allowed for the recess of the wedge was not sufficient for the settlement of the arch; but the architect trusted to the precautions he had taken in its construction. The reader, by turning to the article ARCH, will see, that there was only the arch LY which could be expected to settle: accordingly, the recess of the wedge was found to be much more than was necessary. However, had this not been the case, it was only necessary to take out the pieces between the posts below the seat, and then to drive back the heads of the struts; but this was not needed, we believe, in any of the arches. We are well assured that none of the arches sunk an inch and a half. The great arch of 100 feet span did not sink one inch at the crown. It could hardly be perceived whether the arch quitted the centering gradually or not, so small had been the changes of shape.

646. We have no hesitation in saying, that (if we except some waste of great timber by uncommon joggling) the whole of this performance is the most perfect of any that has come to our knowledge.

647. The subject which we have been considering is very closely connected with the construction of wooden bridges. These are not always constructed on the sole principles of equilibrium, by means of mutual abutment. They are stiff frames of carpentry, where, by a proper disposition, beams are put into a state of extension, as well as of compression, so as to stand in place of solid bodies as big as the spaces which the beams inclose; and thus we are enabled to couple two, three, or four of these together, and set them in abutment with each other like mighty archstones. We shall close this article, therefore, with two or three specimens of wooden bridges, disposed in a series of progressive composition, so as to serve as a sort of introduction to the art in gene-

ral, and furnish a principle which will enable the intelligent and cautious artist to push it with confidence as far as it can go.

The general problem is this. Suppose that a bridge is to be thrown over the space AB (Pl. XII. Fig. 3.), and that this is too wide for the strength of the size of timber which is at our command; how may this beam AB be supported with sufficient effect? There are but two ways in which the middle point C (where the greatest strain is) can be supported: 1. It may be suspended by two ropes, iron rods, or wooden ties, DC, EC, made fast to two firm points D, E, above it; or it may rest on the ridge of two rafters *d* C, *e* C, which rest on two firm points *d*, *e*, below it. 2. It may be supported by connecting it with a point so supported; and this connection may be formed, either by suspending it from this point, or by a post resting on it. Thus it may hang, by means of a rod or a king-post FC, from the ridge F of two rafters AF, BF; or it may rest on the strut C *f*, whose lower extremity *f* is carried by the ropes, rods, or wooden ties A *f*, B *f*.

Whichsoever of these methods we employ, it follows, from the principles of carpentry, that the support given to the point C is so much the more powerful, as we make the angle DCE, or *d* C *e*, or the equivalent angles AFB, or A *f* B, more acute.

Each of these methods may be supposed equally strong. Our choice will depend chiefly on the facility of finding the proper points of support D, E, *d*, *e*; except in the second case, where we require no fixed points but A and B. The simple forms of the first case require a great extent of figure. Very rarely can we suspend it from points situated as D and E. It is even seldom that we have depth enough of bank to allow the support of the rafters *d* C, *e* C; but we can always find room for the simple truss AFB. This therefore is the most usually practised.

648. In the construction, we must follow the maxims and directions prescribed in our articles Carpentry and Roof. The beams FA, FB, must be mortised into AB, in the firmest manner, and there secured with straps and bolts; and the middle must hang by a strap attached to the king-post FC, or to the iron rod that is used for a king-post. No mortising in the point C must be employed; it is unnecessary, and it is hurtful, because it weakens the beam, and because it lodges water, and soon decays by rot. The best practice is not to suspend the beam immediately by this strap, but to let it rest, as in Fig. 4. on a beam C, which crosses the bridge below, and has its other end supported in the same manner by the other truss.

It is evident that the length of the king-post has no effect on the support of C. We may therefore contract every thing, and preserve the same strength of support, by finding two points *a* and *b* (Fig. 5.) in the banks, at a moderate distance below A and B, and setting up the rafters *a* F, *b* F, and suspending C from the shortened king-post. In this construction, when the beam AB rests on a cross bearer, as is drawn here, the struts *a* F, *b* F, are kept clear of it. No connection between them is necessary, and it may be hurtful, by inducing cross strains on both. It will, however, greatly increase the stiffness of the whole. This construction may safely be loaded with ten times the weight that AB can carry alone.

649. Suppose this done, and that the scantling of AB is too weak for carrying the weight which may be brought on the parts AC, CB. We may now truss up each half, as in Fig. 6. and then the whole will form a handsome bridge, of the simplest construction possible. The intersections of the secondary braces with

those of the main truss will form a hand-rail of agreeable figure.

We are not confined to the employment of an entire piece AB, nor to a rectilineal form. We may frame the bridge as in Fig. 7. and in this form we dissuade from allowing any connection with the middle points of the main braces. This construction also may be followed till each beam AC and CB is loaded to ten times what it can safely bear without the secondary trussing.

650. There is another way by which a bridge of one beam may be supported beyond the power of the first and simplest construction. This is represented in Fig. 8. and Fig 9. The truss beam FG should occupy one-third of AB. The advantage of this construction is very considerable. The great elevation of the braces (which is a principal element of the strength) is preserved, and the braces are greatly shortened.

This method may be pushed still farther, as in Fig. 10.

651. And all these methods may be combined, by joining the constructions of Fig. 8. and Fig. 9. with that of Fig. 10.

In all of them there is much room for the display of skill, in the proper adjustment of the scantling of the timber, and the obliquity of the braces to the lengths of the different bearings. A very oblique strut, or a slender one, will suffice for a small load, and may often give an opportunity to increase the general strength; while the great timbers and upright supports are reserved for the main pressures. Nothing will improve the composition so much as reflecting progressively, and in the order of these examples, on the whole. This alone can preserve the great principle in its simplicity and full energy.

652. These constructions are the elements of all that

can be done in the art of building wooden bridges, and are to be found more or less obviously and distinctly in all attempts of this kind. We may assert, that the more obviously they appear, the more perfect the bridge will be. It is astonishing to what extent the principle may be carried We have seen a bridge of 42 feet span formed of two oak trusses, the biggest timber of which did not exceed six inches square, bearing with perfect steadiness and safety a waggon loaded with more than two tons, drawn by four stout horses. It was framed as Fig. 16. nearly, with the addition of the dotted lines, and was near thirty years old; protected, however, from the weather by a wooden roof, as many bridges in Germany are.

We recollect another in the neighbourhood of Stettin, which seemed constructed with great judgment and spirit. It had a carriage-road in the middle, about 20 feet (we think) wide, and on each side a foot-way about five feet wide. The span was not less than 60 feet, and the greatest scantling did not appear to exceed 10 inches by 6.

This bridge consisted of four trusses, two of which formed the outside of the bridge, and the other two made the separation between the carriage-road and the two foot-ways. We noticed the construction of the trusses very particularly, and found it similar to the last, except in the middle division of the upper truss, which, being very long, was double-trussed, as in Fig. 17.

The reader will find in that volume of Leupold's *Theatrum Machinarum*, which he calls *Theatrum Pontificum*, many specimens of wooden bridges, which are very frequent in the champain parts of Germany. They are not, in general, models of mechanic art; but the reflecting reader, who considers them *carefully*, will pick up here and there subordinate hints, which are ingenious, and may sometimes be useful.

What we have now exhibited are not to be considered as models of construction, but as elementary examples and lessons, for leading the reader systematically into a thorough conception of the subject

653. We cannot quit the subject without taking notice of a very wonderful bridge at Wittengen in Switzerland, slightly described by Mr Coxe (*Travels*, vol. I. 132.) It is of a construction more simple still than the bridges we have been describing. The span is 230 feet, and it rises only 25. The sketch (Fig. 18.) will make it sufficiently intelligible. ABC is one of the two great arches, approaching to a Catenarian shape, built up of seven courses of solid logs of oak, in lengths of 12 or 14 feet, and 16 inches or more in thickness. These are all picked of a natural shape, suited to the intended curve; so that the wood is nowhere cut across the grain to trim it into shape. These logs are laid above each other, so that their abutting joints are alternate, like those of a brick-wall; and it is indeed a wooden-wall simply built up, by laying the pieces upon each other, taking care to make the abutting joints as close as possible. They are not fastened together by pins or bolts, or by scarfings of any kind. They are, however, held together by iron straps, which surround them, at the distance of five feet from each other, where they are fastened by bolts and keys.

These two arches having been erected (by the help, we presume, of pillars, or a centering of some kind), and well butted against the rock on each side, were freed from their supports, and allowed to settle. They are so placed, that the intended road *a b c* intersects them about the middle of their height. The road-way is supported by cross joists, which rest on a long horizontal summer beam. This is connected with the arches on each side by uprights bolted into them. The whole is covered with a roof, which projects over the arches on each side, to defend them from the weather. Three of the spaces between these uprights have struts or braces, which give the upper work a sort of trussing in that part.

This construction is simple and artless; and appears, by the attempt to truss the ends, to be the performance of a person ignorant of principle, who has taken the whole notion from a stone arch. It is, however, of a strength much more than adequate to any load that can be laid on it. Mr Coxe says, but does not explain how, that it is so contrived that any part of it can be repaired independent of the rest. It was the last work of one Ulrich Grubenmann of Tuffen, in the canton of Appenzel, a carpenter without education, but celebrated for several works of the same kind; particularly the bridge over the Rhine at Schaffhausen, consisting of two arches, one of 172 and the other of 193 feet span, both resting on a small rock near the middle of the river.*

While writing this article, we got an account of a wooden bridge erected in North America, in which this simple notion of Grubenmann's is mightily improved. The span of the arch was said to exceed 250 feet, and its rise exceedingly small. The description we got is very general, but sufficient, we think, to make it perfectly intelligible.

654. In Fig. 19. DD, EE, FF, are supposed to be three beams of the arch. They consist of logs of timber of small lengths, suppose of 10 or 12 feet, such as can be found of a curvature suited to its place in the arch without trimming it across the grain. Each beam is double, consisting of two logs applied to each other, side to side, and *breaking joint*, as the workmen term it. They are kept together by wedges and keys driven through them at short intervals, as at K, L, &c.

The manner of joining and strongly binding the two side pieces of each beam is shown in Fig. 20. The mortise

---

* Drawings of this remarkable bridge, which is now destroyed, will be found in the EDINBURGH ENCYCLOPÆDIA, vol. IV. p. 358, Plates LXXXIX and XC.—ED.

a i c b and *d c i o*, which is cut in each half beam, is considerably longer on the outside than on the inside, where the two mortises meet. Two keys, BB and CC, are formed, each with a notch *b c d*, or *a i o*, on its side; which notch fits one end of the mortise. The inner side of the key is straight, but so formed, that when both keys are in their places, they leave a space between them wider at one end than at the other. A wedge AA, having the same taper as the space just mentioned, is put into it and driven hard. It is evident that this must hold the two logs firmly together.

This is a way of uniting timber not mentioned in the article CARPENTRY; and it has some peculiarities worthy of notice. In the first place, it may be employed so as to produce a very strong lateral connexion, and would then co-operate finely with the other artificial methods of scarfing and tabling that we described in the article referred to. But it requires nice attention to some circumstances of construction to secure this effect. If the joints are accurately formed to each other, as if the whole had been one piece divided by an infinitely thin saw, this manner of joining will keep them all in their places. But no driving of the wedge AA will make them firmer, or cause one piece to press hard on the other. If the abutment of two parts of the half beam is already close, it will remain so; but if open in the smallest degree, the driving of the wedge will not make it tighter. In this respect, therefore, it is not so proper as the forms described in CARPENTRY.

In order that the method now described may have the effect of *drawing* the halves of the beams together, and of keeping them hard squeezed on each other, the joints must be made so as not to correspond exactly. The prominent angle *a i o* (Fig. 21.), formed by the ends of the two half mortises, must be made a little more obtuse than the angle *a f o* of the notch of the key which this prominence is intended to fill up. Moreover, the opposite side *e t* of this key should not be quite straight, but a very little convex.

With these precautions, it is easy to see that, by driving the wedge AA, we cause the notch *a f o* to take hold, first at the two points *a* and *o*, and then, by continuing to drive the wedge, the sides *a f*, *o f*, of the notch gradually compress the wood of the half beams, and press them on each other. By continuing to drive the wedge, the mutual compression of the key and the beam squeezes all together, and the space *a f o i* is completely filled up. We may see, from this process, that the mutual compression and drawing together of the timber will be greater in proportion as we make the angle *a i o* more prominent, and its corresponding angle *a f o* more deep; always taking care that the key shall be thick enough not to break in the narrow part.

This adjustment of the keys to the mortise is necessary on another account. Supposing the joints to fit each other exactly before driving the wedge, and that the whole shrinks a little by drying—by this the angle *a i o* will become more prominent, and the angle *a f o* will become more shallow; the joint will open at *a* and *o*, and the mutual compression will be at an end.

We may also observe, that this method will not give any additional firmness to the abutments of the different lengths employed to piece out the arch-beam; in which respect it differs materially from the other modes of joining timber.

Having shown how each beam is pieced together, we must now show how a number of them are united, so as to compose an arch of any thickness. This is done in the very same way. The beams have other mortises worked out of their inner sides, half out of each half of the beam. The ends of the mortises are formed in the same way with those already described. Long keys BB, CC, (Fig. 19.) are made to fit them properly, the notches being placed so as to keep the beams at a proper distance from each other. It is now plain that driving in a long wedge AA will bind all together.

In this manner may an arch be extended to any span, and made of any thickness of arching. The bridge over Portsmouth river, in North America, was more than 250 feet in length, and consisted of several parallel arches of beams. The inventor (we think that his name is Bludget) said that he found the strength so great, that he could with perfect confidence make one of four times the span.

We admire the ingenuity of this construction, and think it very effectual for bringing the timbers into firm and uniform abutment; but we imagine that it requires equilibration, because it is extremely flexible. There is nothing to keep it from bending, by an inequality of load, but the transverse strength of the beams. The keys and wedges can have very little power to prevent this bending. The distance between the beams will also contribute little or nothing to the stiffness; nay, we imagine that a great distance between them will make the frame more flexible. Could the beams be placed so near each other that they could be somehow joggled on each other, the whole would be stiffer; but at present they will bend like the plates of a coach-spring. But nothing hinders us from adding diagonal pieces to this construction, which will give it any degree of stiffness, and will enable it to bear any inequality of loading. When completed in this manner, we imagine that it will be at least equal to any construction that has been yet thought of. One advantage it possesses that is very precious: any piece that fails may be taken out, and replaced by another, without disturbing the rest, and without the smallest risk. On the whole, we think it a very valuable addition to British carpentry. The method here practised, both for joining the parts of one beam and for framing the different beams together, suggests the most firm and light constructions for dome-roofs that can be conceived; incomparably superior to any that have yet been erected. The whole may be framed, without a nail or a spike, into one net-like shell that cannot even be pulled in pieces. We may perhaps consider this in another article; at present we return to the consideration of trussed bridges.

When the width of the river exceeds what is thought practicable by a single truss, we must then combine, either by simple addition or by composition, different trusses together. We compose a bridge by simple addition when we make a frame of carpentry of an unchangeable and proper shape, to serve as one of the arch-stones of a bridge of masonry. This may be easily comprehended by looking at Fig. 22. Each of the frames A, B, C, D, must be considered as a separate body, and all are supported by their mutual abutment. The nature of the thing is not changed, although we suppose that the rails of the frame B, instead of being mortised into an upright *b′ b′* unconnected with the frame C, is mortised into the upright *c c* of that frame, the direction and intensity of the mutual pressures of the two frames are the same in both cases; accordingly this is a very common form of small wooden bridges. It is usual, indeed, to put diagonal battens into each: but we believe that this is more frequently done to please the eye than to produce an unalterable shape of each frame.

To an unskilful carpenter this bridge does not seem essentially different from the centering of Mr Hupeau for the bridge of Orleans; and indeed, in many cases, it requires reflection, and sometimes very minute reflection, to distinguish between a construction which is only an addition of frame to frame till the width be covered, from a construction where one frame works on the adjoining one transversely, pushing it in one part and drawing it in another. The ready way for an unlettered artist to form a just notion of this point, is to examine whether he may saw through the connecting piece *b′ b′* from one end to the other, and make them two separate frames. Whenever this cannot be done without that part opening, it is a construction by composition. Some of the beams are on the stretch; and iron straps, extending along both pieces, are necessary for securing the joint The bridge is no longer a piece of masonry, but a performance of pure carpentry, depending on principles peculiar to that art. Equilibration is necessary in the first construction; but, in the second, any inequality of loading is made ineffectual for hurting the edifice, by means of the stretch that is made to operate on some other piece. We are of opinion that this most simple employment of the distinguishing principle of carpentry, by which the beams are made to act as ties, will give the most perfect construction of a wide bridge. One polygon alone should contain the whole of the abutments; and one other polygon should consist entirely of ties; and the beams which form the radii, connecting the angles of the two polygons, complete the whole. By confining the attention to these two simple objects, the abutments of the outer polygon, and the joints of the inner one, may be formed in the most simple and efficient manner, without any collateral connexions and dependencies, which divide the attention, increase the complication, and commonly produce unexpected and hurtful strains. It was for this reason that we have so frequently recommended the centering of the bridge of Orleans. Its office will be completely performed by a truss of the form of Fig. 23; where the polygon ABCDEF, consisting of two layers of beams (if one is not sufficient), contains the whole abutments, and the other A *b c d e* F is nothing but an iron rod. In this construction, the obtuseness of the angles of the lower polygon is rather an advantage. The braces G *c*, G *d*, which are wanted for trussing the middle of the outer beams, will effectually secure the angles of the exterior polygon against all risk of change. The reader must perceive that we have now terminated in the construction of the Norman roof. We indeed think it the best general form, when some moderate declivity is not an insuperable objection. When this is the case, we recommend the general plan of the centering of the bridge of Orleans. We would make the bridge (we speak of a great bridge) consist of four trusses; two to serve as the outsides of the bridge, and two inner trusses, separating the carriage-way from the foot-paths. The road should follow the course of the lower polygon, and the main truss should form the rails. It might look strange; but we are here speaking of strength; and evident, but not unwieldy, strength, once it becomes familiar, is the surest source of beauty in all works of this kind.

END OF VOLUME FIRST.

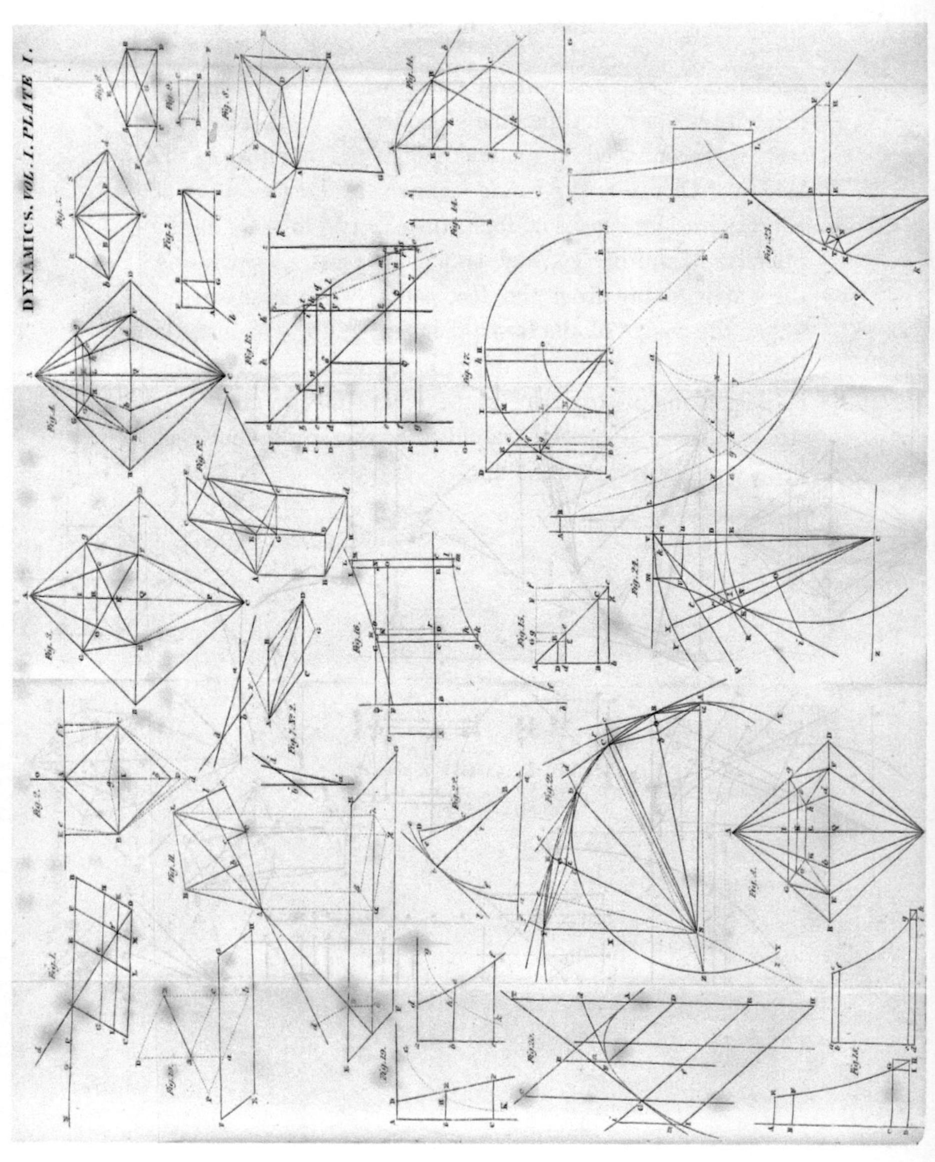

The material originally positioned here is too large for reproduction in this reissue. A PDF can be downloaded from the web address given on page iv of this book, by clicking on 'Resources Available'.

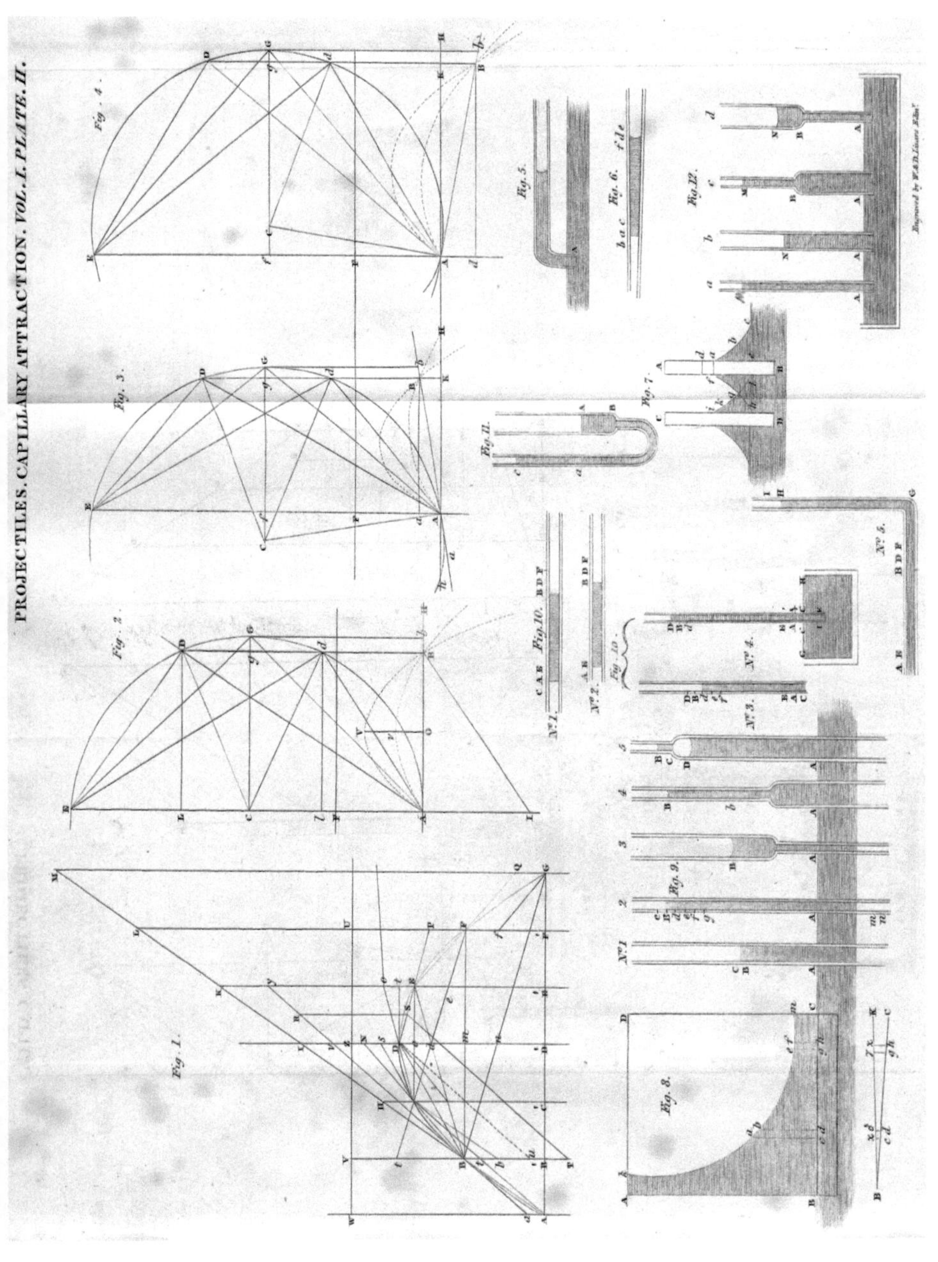

The material originally positioned here is too large for reproduction in this reissue. A PDF can be downloaded from the web address given on page iv of this book, by clicking on 'Resources Available'.

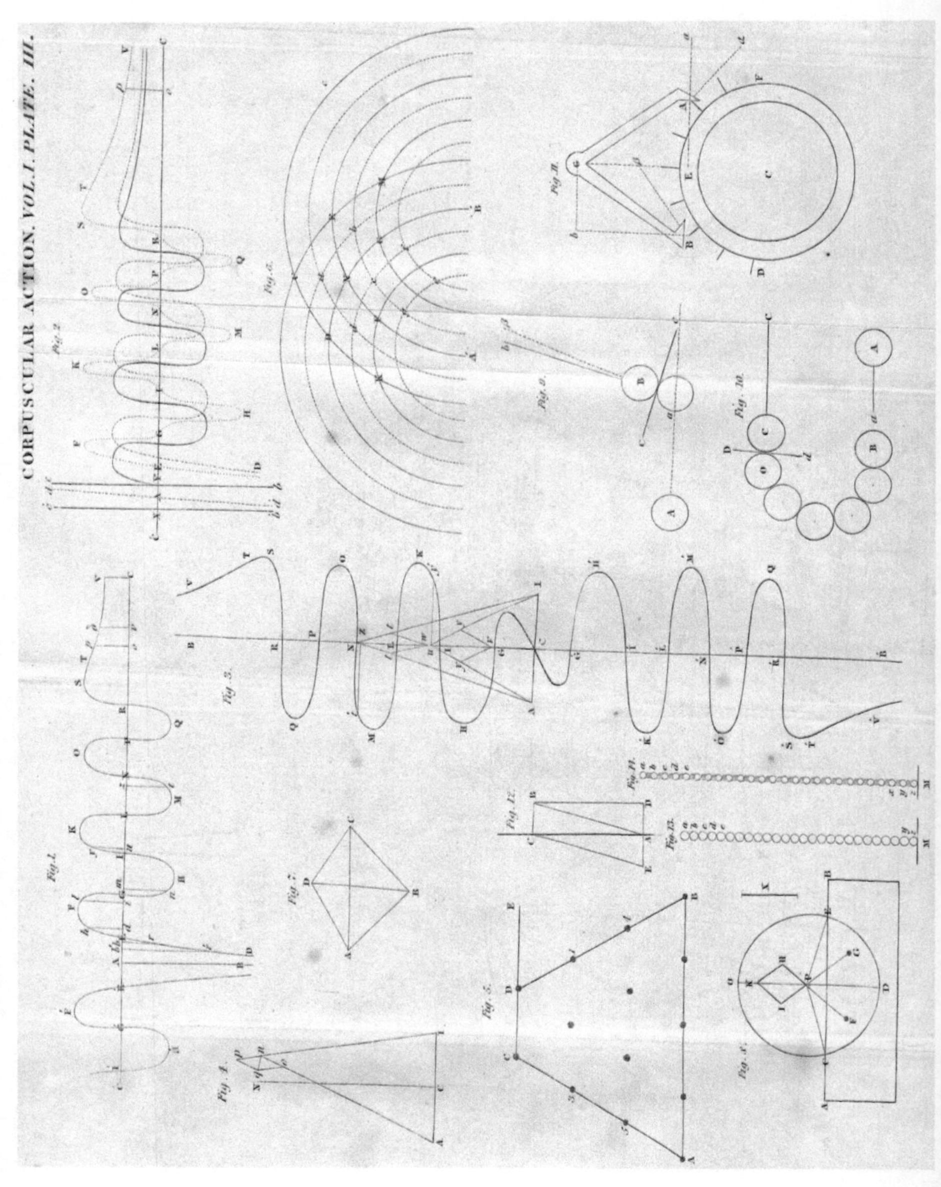

The material originally positioned here is too large for reproduction in this reissue. A PDF can be downloaded from the web address given on page iv of this book, by clicking on 'Resources Available'.

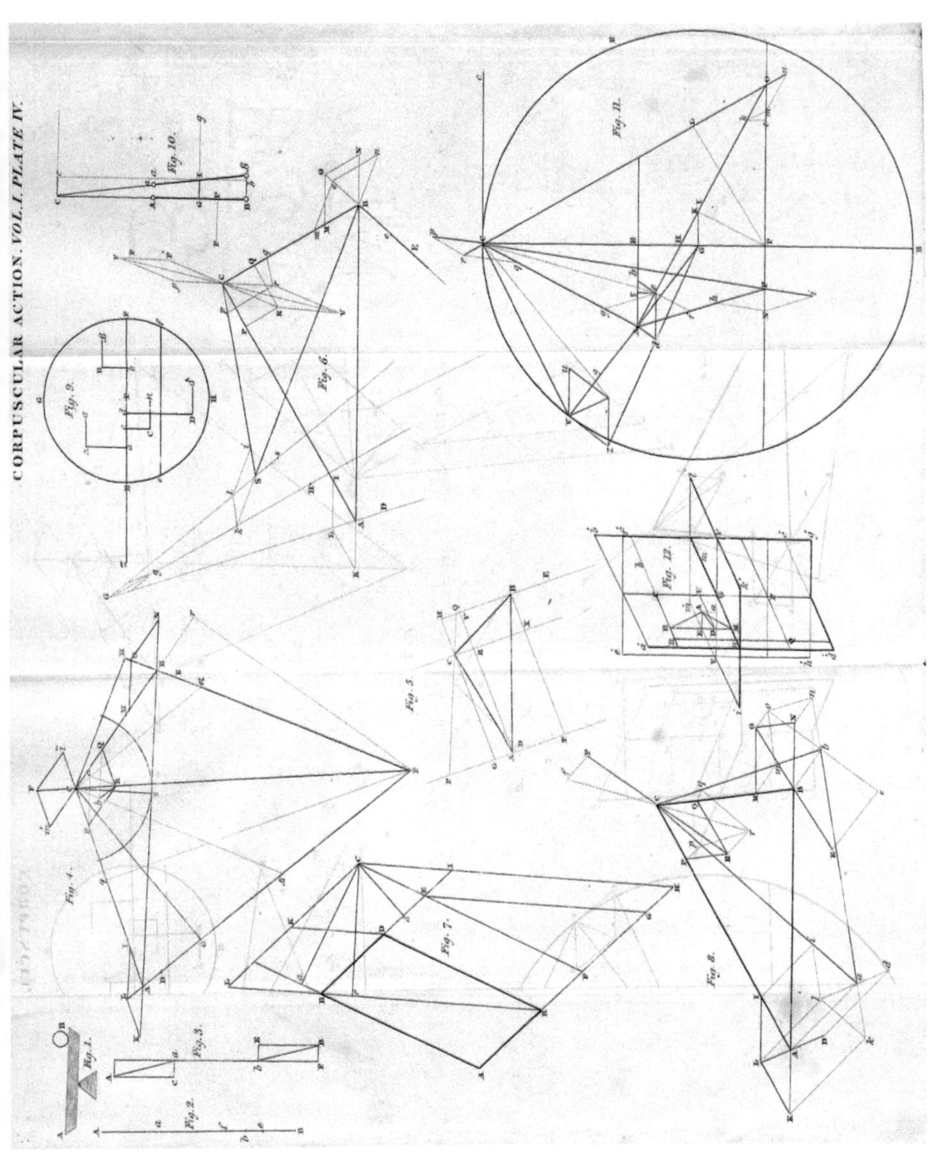

The material originally positioned here is too large for reproduction in this reissue. A PDF can be downloaded from the web address given on page iv of this book, by clicking on 'Resources Available'.

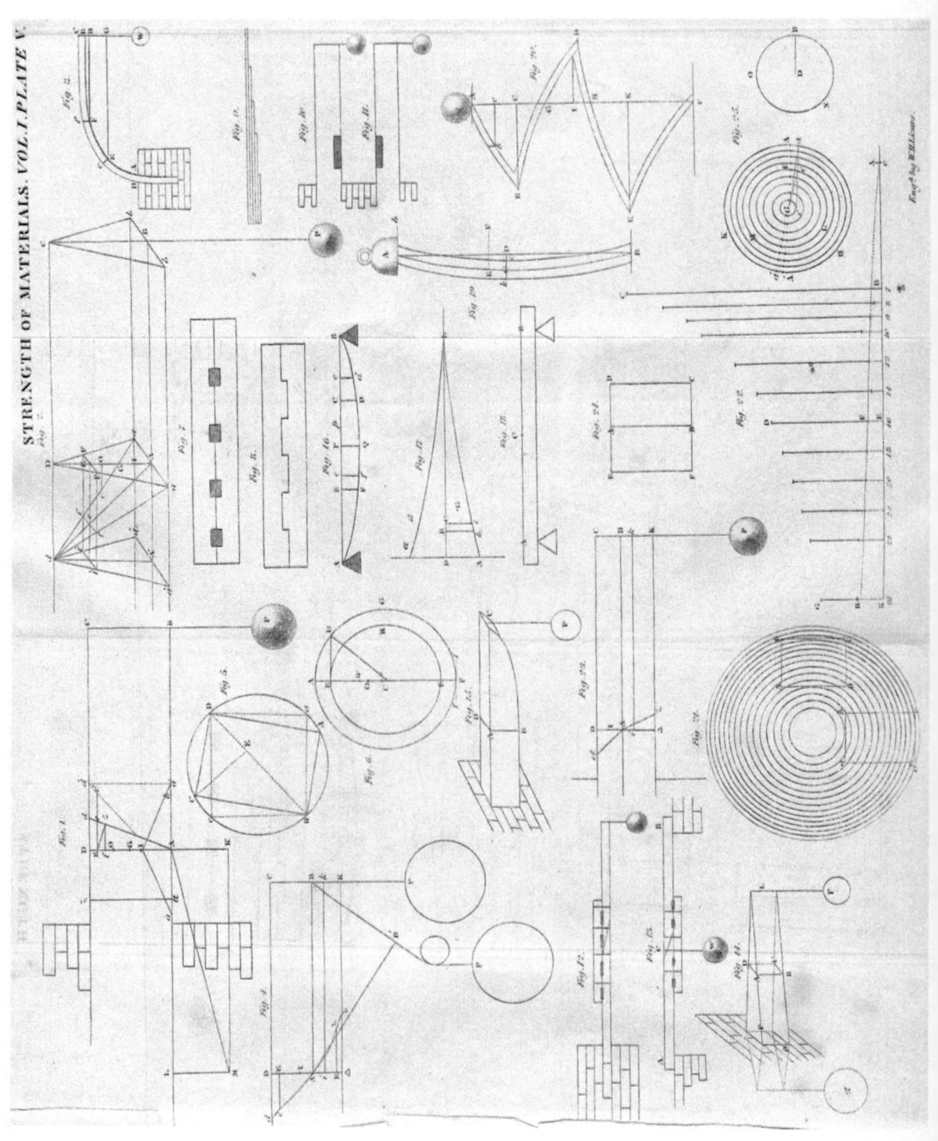

The material originally positioned here is too large for reproduction in this reissue. A PDF can be downloaded from the web address given on page iv of this book, by clicking on 'Resources Available'.

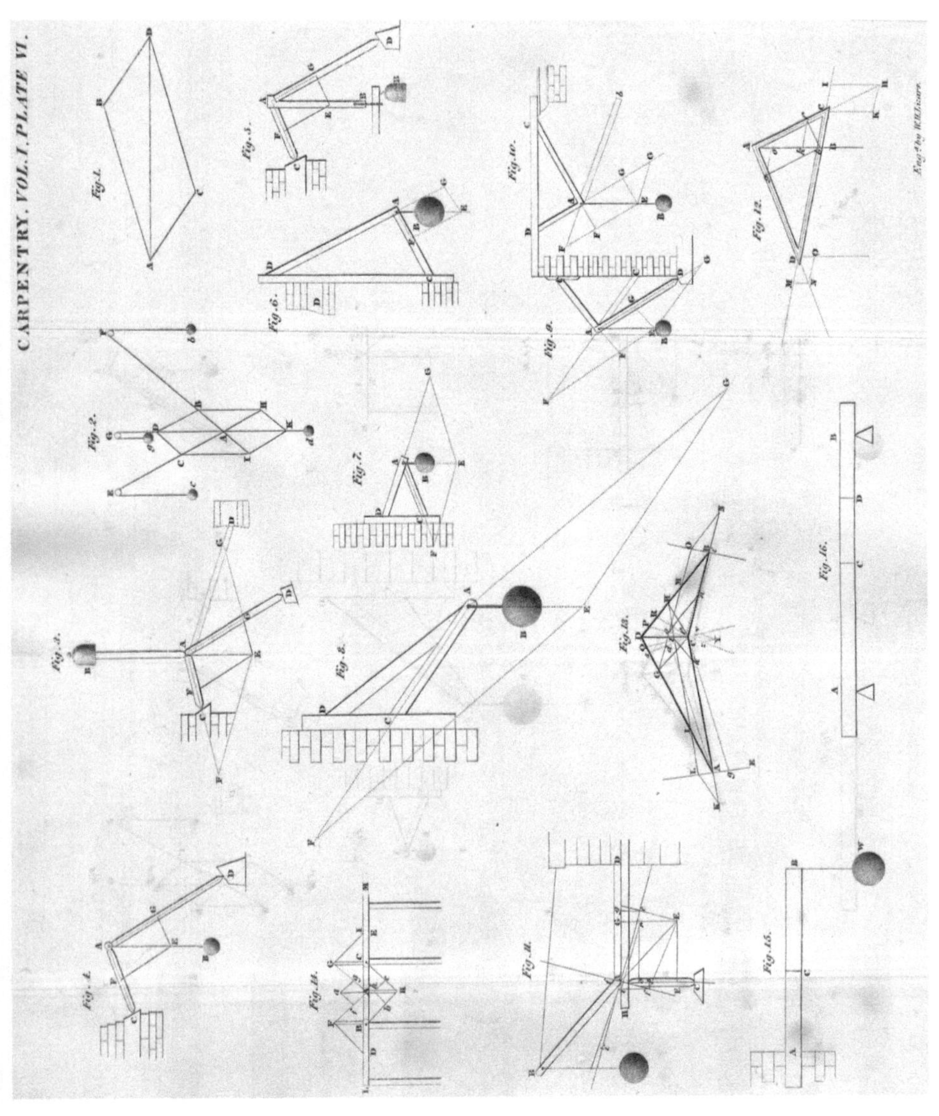

The material originally positioned here is too large for reproduction in this reissue. A PDF can be downloaded from the web address given on page iv of this book, by clicking on 'Resources Available'.

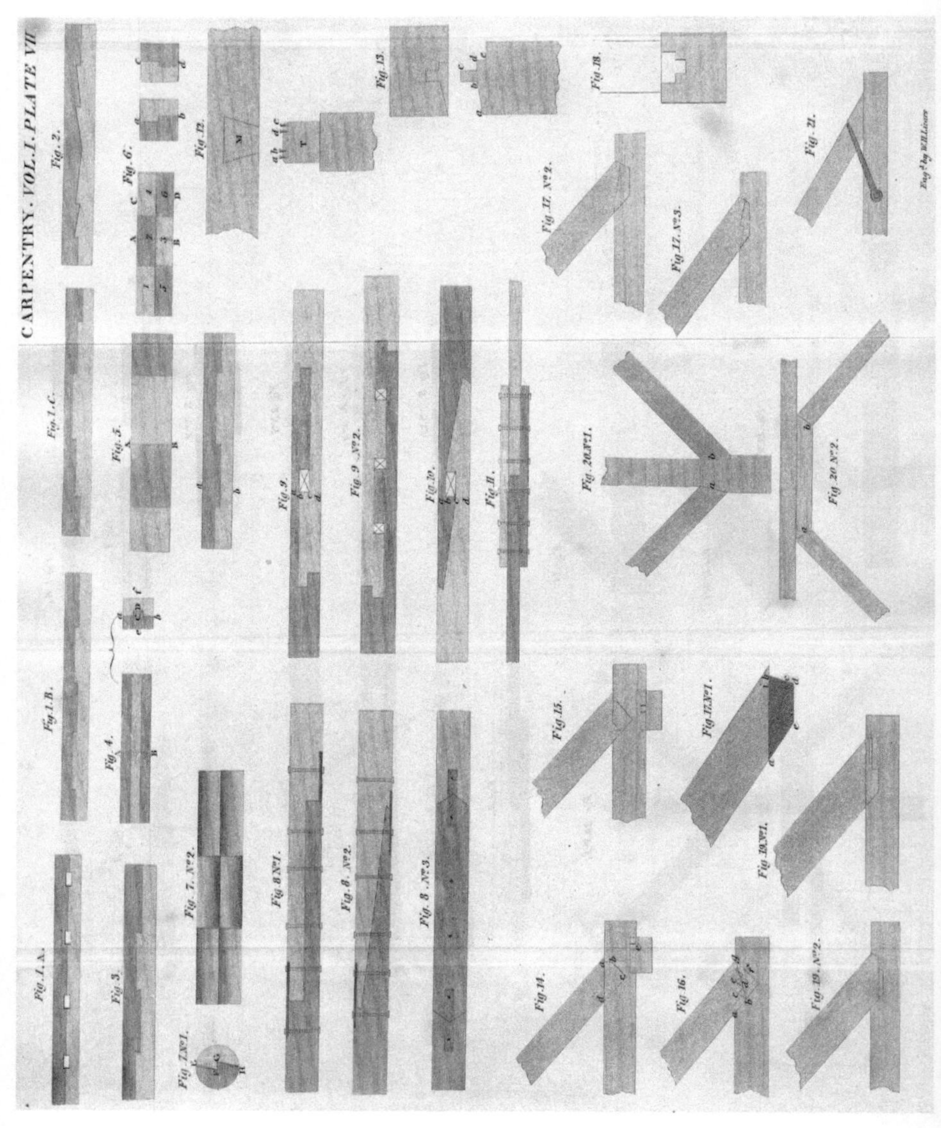

The material originally positioned here is too large for reproduction in this reissue. A PDF can be downloaded from the web address given on page iv of this book, by clicking on 'Resources Available'.

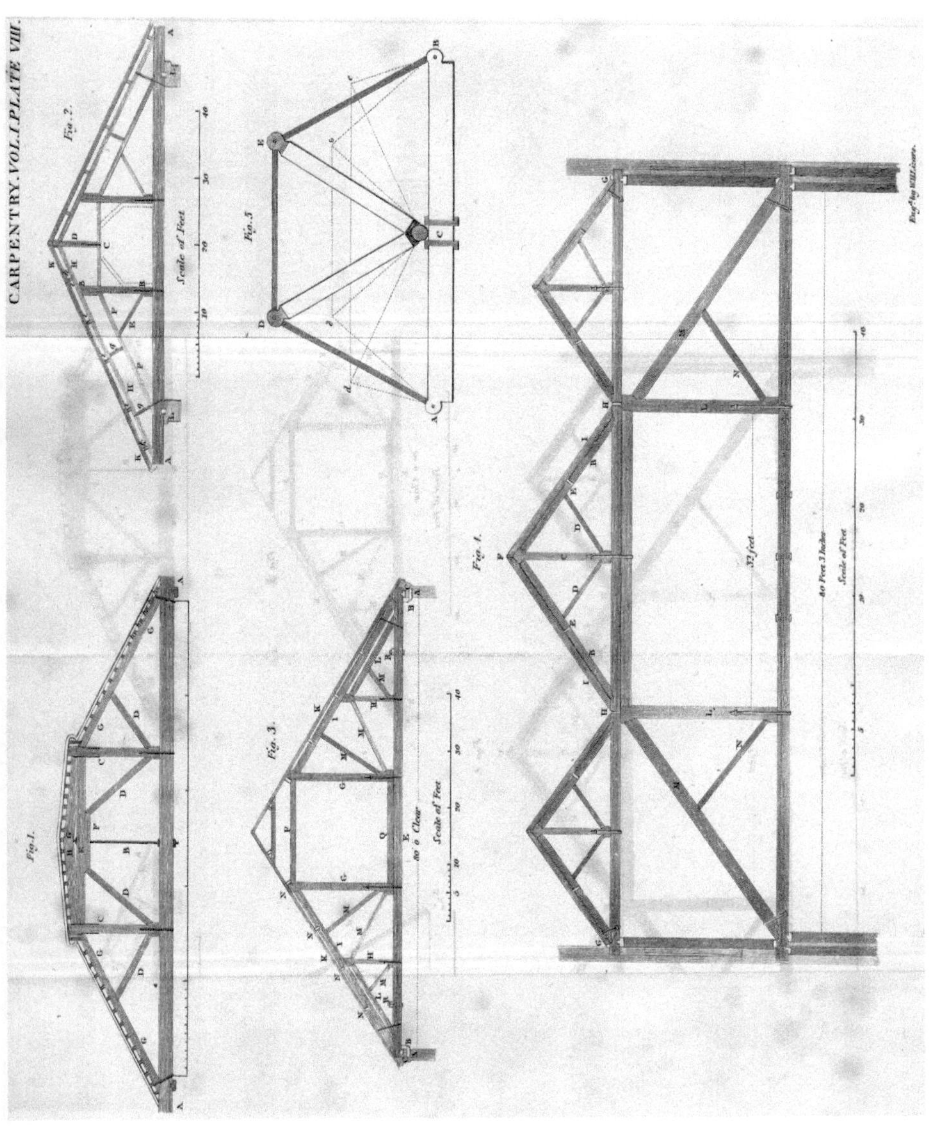

The material originally positioned here is too large for reproduction in this reissue. A PDF can be downloaded from the web address given on page iv of this book, by clicking on 'Resources Available'.

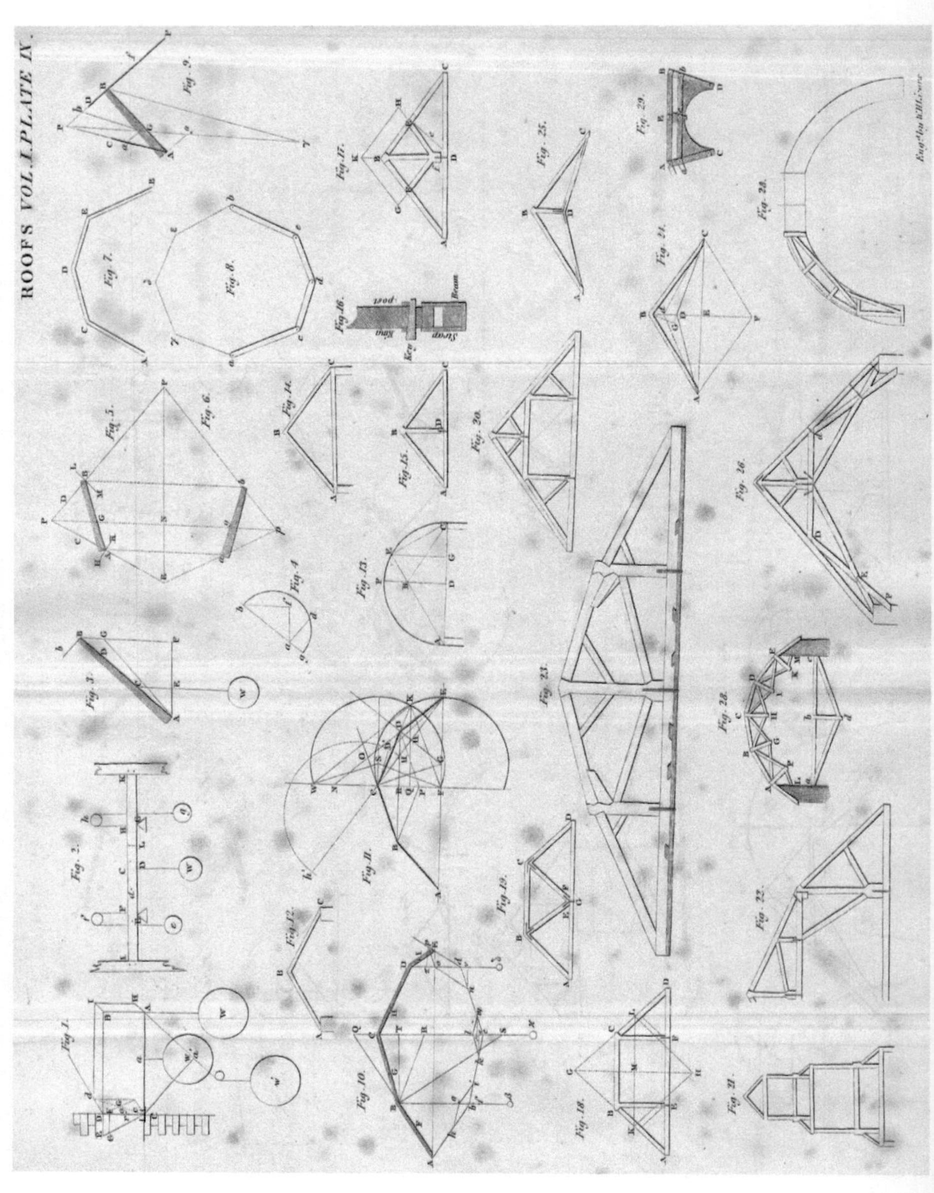

The material originally positioned here is too large for reproduction in this reissue. A PDF can be downloaded from the web address given on page iv of this book, by clicking on 'Resources Available'.

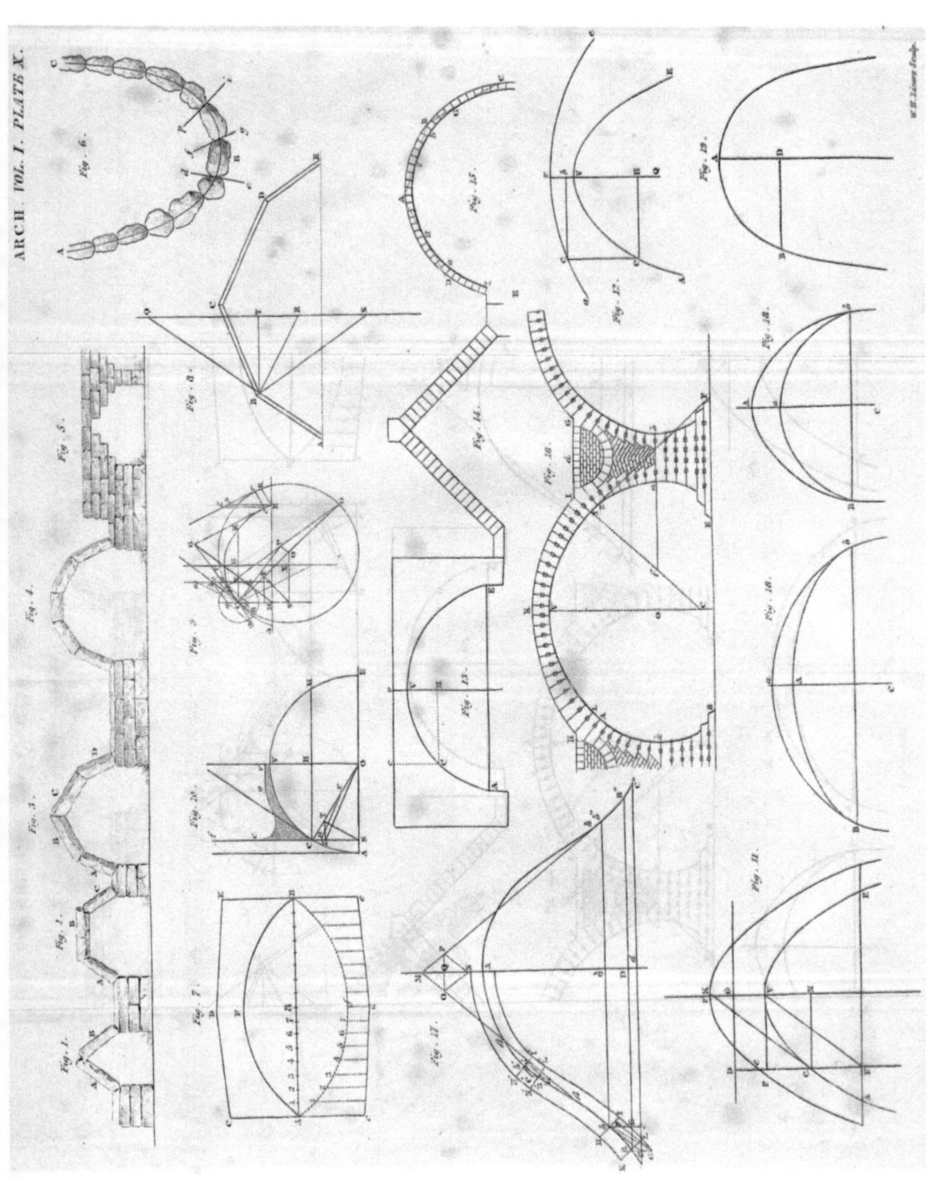

The material originally positioned here is too large for reproduction in this reissue. A PDF can be downloaded from the web address given on page iv of this book, by clicking on 'Resources Available'.

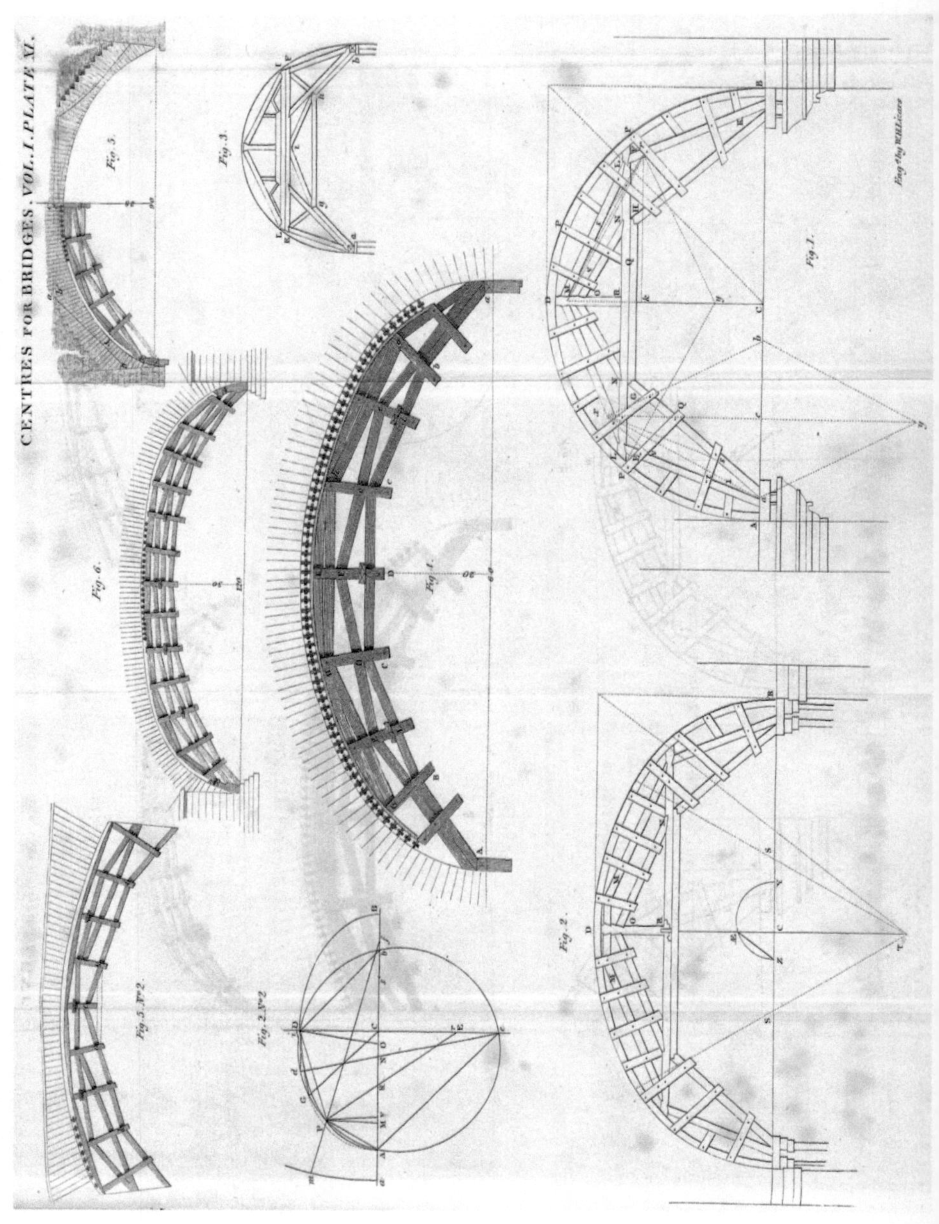

The material originally positioned here is too large for reproduction in this reissue. A PDF can be downloaded from the web address given on page iv of this book, by clicking on 'Resources Available'.

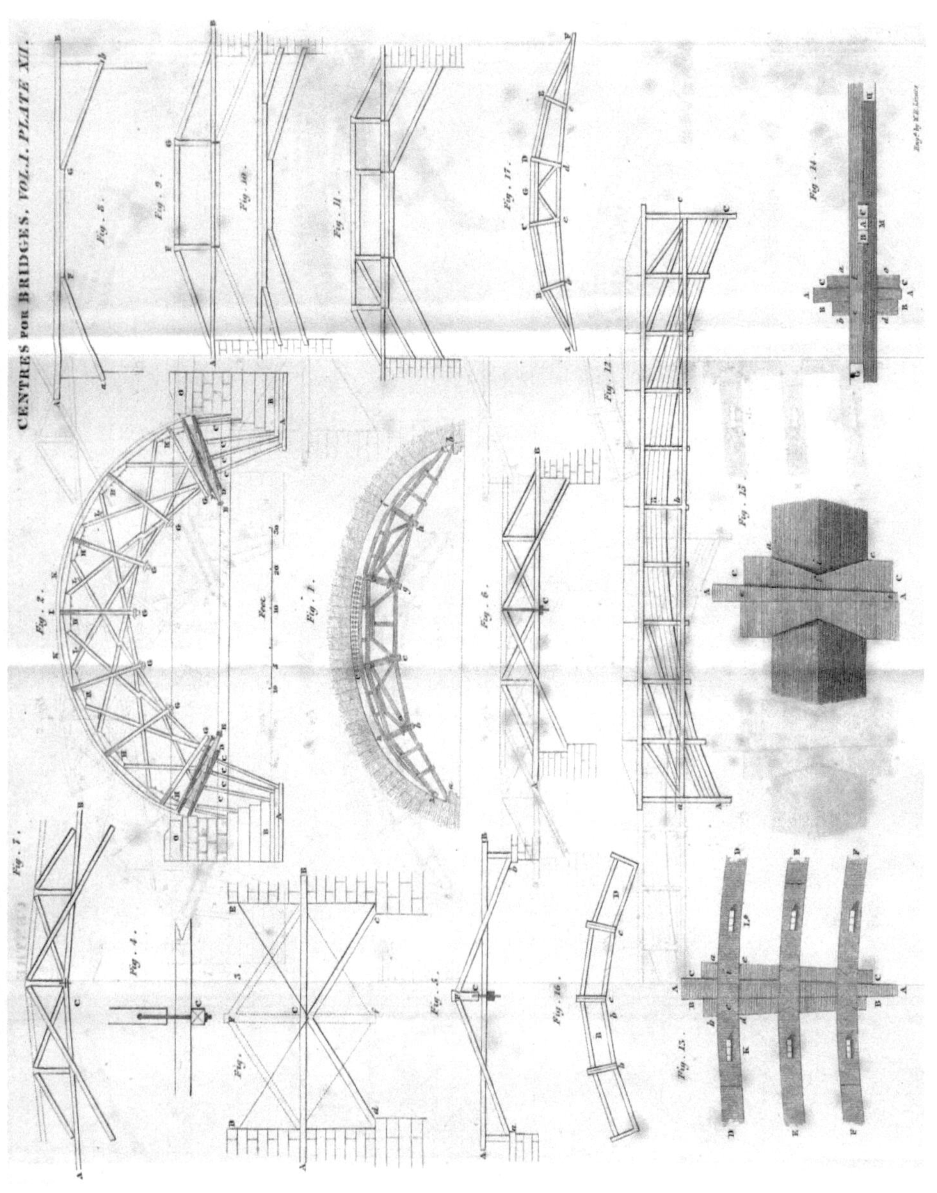

The material originally positioned here is too large for reproduction in this reissue. A PDF can be downloaded from the web address given on page iv of this book, by clicking on 'Resources Available'.

For EU product safety concerns, contact us at Calle de José Abascal, 56–1°, 28003 Madrid, Spain or eugpsr@cambridge.org.

www.ingramcontent.com/pod-product-compliance
Ingram Content Group UK Ltd.
Pitfield, Milton Keynes, MK11 3LW, UK
UKHW042207080726
473066UK00007B/282

9781108070379